硅 藻
基础与应用
Diatoms
Fundamentals and Applications

【以】约瑟夫·塞克巴奇 (Joseph Seckbach)
【美】理查德·戈登 (Richard Gordon) ◎主编

张育新 高立洪 谢更新 ◎译

上海交通大学出版社
SHANGHAI JIAO TONG UNIVERSITY PRESS

内容提要

本书汇聚了来自中国、加拿大、美国等多个国家的学者对硅藻研究的介绍,包括对一些业余的硅藻爱好者的描述,硅藻的形态和视觉之美,中国学术领域对硅藻的研究现状,从细胞形态介绍到光子生物学在硅藻领域的应用,以及硅藻在法医学、生物医学、能源及农业等方面的应用。本书可作为硅藻知识的科学普及读物,也可作为硅藻领域科研工作者和对硅藻感兴趣人员的参考书。

图书在版编目(CIP)数据

硅藻:基础与应用/(以)约瑟夫·塞克巴奇(Joseph Seckbach),(美)理查德·戈登(Richard Gordon)主编;张育新,高立洪,谢更新译. — 上海:上海交通大学出版社,2025.1

书名原文:Diatoms: Fundamentals and Applications

ISBN 978-7-313-30380-6

Ⅰ.①硅… Ⅱ.①约…②理…③张…④高…⑤谢… Ⅲ.①硅藻土材料 Ⅳ.①TU55

中国国家版本馆 CIP 数据核字(2024)第 050879 号

硅藻:基础与应用
GUIZAO: JICHU YU YINGYONG

主　　编:[以] 约瑟夫·塞克巴奇(Joseph Seckbach) [美] 理查德·戈登(Richard Gordon)	译　　者:张育新　高立洪　谢更新
出版发行:上海交通大学出版社	地　　址:上海市番禺路951号
邮政编码:200030	电　　话:021-64071208
印　　制:浙江天地海印刷有限公司	经　　销:全国新华书店
开　　本:710mm×1000mm　1/16	印　　张:24.5
字　　数:502千字	
版　　次:2025年1月第1版	印　　次:2025年1月第1次印刷
书　　号:ISBN 978-7-313-30380-6	
定　　价:168.00元	

版权所有　侵权必究

告读者:如发现本书有印装质量问题请与印刷厂质量科联系

联系电话:0573-85509555

推荐序

英国诗人 William Blake 曾写道:"To see a world in a grain of sand(一沙一世界)",意喻微小中蕴含宏大,微观处可现繁复。这一表述也精确适用于硅藻。

硅藻是自然界中常见的微生物。在池塘里随手掬水,取其中一滴放在光学显微镜下观察,一个富含微米数量级硅藻的微观世界就呈现在眼前。若将硅藻颗粒挑出置于电子显微镜下,更会发现,一粒硅藻便是一座精巧的微型"宫殿",其结构中孔洞丰富,表面"纹饰"纷呈。实际上,硅藻的种属和形态丰富多彩,并且在水体、沉积物、土壤等诸多环境中均存在。因此,尽管硅藻貌似微不足道,却以其独特的微结构和出色的环境适应能力,诠释了地球生态系统的多样性和复杂性,堪称奇妙。

硅藻早在侏罗纪时期便亮相于地球舞台。自此至今,硅藻种群经历了地球的板块飘移、山川沧桑,见证了恐龙的消逝和人类的崛起,却始终位列水体微生物的主流。如今,硅藻的细分种类已达约 20 万种,数量和种属都十分庞大。作为地球上最重要的光合微生物,硅藻通过光合作用产生的氧气,贡献了地球氧气总量的五分之一以上,对支撑地球生态体系起着关键作用。尽管硅藻如此重要,在 Antonie van Leeuwenhoek 发明出能够直接观察微生物的光学显微镜之前,人类从未察觉到地球上生存着硅藻。在侏罗纪到人类纪长达亿年的时间长河里,硅藻默默进化、默默塑造地球生态。直到 1703 年,在伦敦皇家学会收到的一篇稿件中首次出现显微镜观察硅藻的记录,硅藻的显微形态精美,观察硅藻成为当时的一种雅趣甚至时尚。这驱使了人们对硅藻的日益关注,并留下了丰富的观察记录。1904 年,Ernst Haeckel 的《自然界的艺术形态》收录了硅藻显微图,已称得上多姿多彩,令读者领略硅藻之美。这些古典时期的光学薄片标本的一部分保存至今,对当今的硅藻研究也颇具宝贵价值。

与硅藻显微观察的兴盛几乎同期,随着工业革命的技术突破和产业繁荣,硅藻沉积成岩形成的矿产——硅藻土也开始得到广泛应用。其中,最突出的例子是 Alfred Bernhard Nobel 利用硅藻土作为稳定硝化甘油炸药的载体,研发出取得巨大商业成功的新型炸药。就此角度而言,硅藻土对诺贝尔奖的贡献可谓巨大。20 世纪开启了从简单观察硅藻到基于硅藻开展多门类科研的新篇章,特别是 20 世纪 30 年代扫描电镜和透射电镜的发明,帮助人们首次实现在微米尺度上观察硅藻的精细结构。水生硅藻、土壤硅藻及其生态环境效应研究,乃至硅藻或硅藻土的应用研究,相继兴起。其中极富创造性的一个应用,当属法医学上通过研究逝者身体中的硅藻来判断其是否属于溺亡。随后,基于 20 世纪下半叶以来多种测试分析新技术(精细谱学、基因测序等)的运用,硅藻和硅藻土的生物、化学、物理和材料学研究在近几十年来得到了快速发展。研究者如今已能够利用聚焦离子束(FIB)减薄技术和飞行时间-二次离子质谱(TOF-SIMS)等技术来研究硅藻壳体内部的超精细结构。研究者在硅藻光合作用的关键机制方面也取得了突破,解析出了硅藻主要捕光天线蛋白的高分辨结构。此外,基于硅藻的先进能源和材料研究(如生物燃料和太阳能电池材料)正在兴起。近年来,研究者还提出了基于硅藻-矿物相互作用的水体二氧化碳固碳策略假说。总体上,硅藻和硅藻土研究呈现出蓬勃发展的喜人局面。

  本书原著作者 Joseph Seckbach 博士和 Richard Gordon 博士,他们从事藻类微生物的研究和教学已超过半个世纪,在耄耋之年仍笔耕不辍,推出这本内容丰富的硅藻著作。本书的译者重庆大学张育新教授团队在硅藻土材料领域的研究深耕多年,本书的出版无疑将推动我国硅藻相关研究的持续深入,对提高我国硅藻和硅藻土研究水平起到重要作用。

袁鹏

2024 年 8 月 14 日

译者序

硅藻是一种独特的微观生物，因其在生物地球化学循环、生态系统平衡以及现代科技中的广泛应用而逐渐引起了科学界的高度重视。本人之前翻译出版的《硅藻纳米技术的进展和新兴应用》，该书详细探讨了硅藻的物种起源以及它们在极端环境中的生存机制，受到了学术界的广泛好评，不仅促进了硅藻研究的普及，也推动了许多年轻学者投身于这一领域的研究工作。

《硅藻：基础与应用》原著由 Joseph Seckbach 和 Richard Gordon 主编，全书详尽地介绍了硅藻的生物学特征、生态作用以及在现代科技中的广泛应用，展示了硅藻在各种极端环境中的适应性，并揭示了它们在生物燃料等新兴领域的应用潜力。书中还介绍了 Karsten、Holzinger、Witkowski 等多位科学家的研究成果，生动展现了硅藻双原子细胞壁的美丽结构。硅藻不仅在科学研究中具有重要的学术价值，其应用前景也在不断扩大。早在 2017 年，广州市刑事科学技术研究所的硅藻检验技术为诊断溺死原因提供了重要理论支撑，其"法医硅藻检验关键技术及设备"项目获得了国家科技进步二等奖。翻译出版本书的初衷，是希望将这一领域的前沿研究成果引入国内学术界，使更多的研究者、学生和科技工作者能够从中受益，并推动相关研究的发展。作为一个致力于硅藻研究的团队，我们提出了"硅藻-硅藻土-硅酸盐"的硅三角循环体系，在这一领域取得了显著的"顶天-立地-下海"成果："边修复边增产"的硅藻有机肥（立地）、硅藻基超级电容器电极材料、腐蚀防护（下海）和隐身防护（顶天）的硅藻微胶囊等产品都相继问世，先后四次获得重庆市自然科学奖一等奖、科技进步奖一等奖等省部级奖励。与此同时，团队培养出了许多优秀的本硕博毕业生，被《人民日报》《重庆日报》等宣传报道。此外，我们的研究团队在中国国际"互联网+"等创新创业大赛中屡获殊荣，其中"硅根结缔"项目获得了第八届中国国际"互联网+"大学生创新创业大赛"青年红色筑梦之旅"金奖和唯一的"最佳乡村振兴奖"。

在此，特别感谢重庆市农业科学院农业工程研究所所长高立洪研究员和重庆大学环境与生态学院谢更新教授对本书的建议和指导，并参与了相关章节的翻译和校稿。高立洪研究员鼓励我们将硅藻用于生态农业和智慧农业领域，谢更新教授鼓励我们将硅藻应用于地外生态圈的改造，这都是非常好的建议，我们倍感振奋。还要感谢广东工业大学袁鹏教授为本书作序，他在硅藻研究领域取得了卓越的成绩，其硅藻的课题研究荣获广东省自然科学一等奖。正是有了袁鹏教授及众多学术前辈的支持与鼓励，才让我们能够在硅藻研究领域不断前行。

最后感谢硅藻（硅藻土）研究的同仁们：中科院广州地化所刘冬研究员、同济大学包志豪教授、东南大学张友法教授和王松副研究员、上海交通大学顾佳俊教授、北京工业大学王金淑教授和吴俊书副研究员等，以及中国科学技术馆、中国科协、英国皇家化学会、重庆市科协、重庆市教科院、沙坪坝区科协、重庆大学校科协、重庆市南渝中学校和重庆大学附中（重庆七中），他们对我们硅藻科普工作的大力支持，使更多的中小学生可以关注到硅藻，喜欢上硅藻。

硅藻的研究正在迅速发展，希望本书能够为研究者们提供宝贵的参考资源，激发更多的研究灵感，并促进国内外学术交流。

张育新

2024 年 10 月

序 言

Joseph Seckbach 和 John P. Kociolek[1]主编的硅藻领域研究进展综述《硅藻世界》出版至今，已经过去许多年，毋庸置疑，这一时间阶段硅藻研究也取得了重大进展，由此获得的大量数据出版《硅藻：基础与应用》。本书除了讲述硅藻生物学的基本问题外，还包括瓣膜形态发生、有性生殖和细胞周期、生态学和生物多样性，以及硅藻研究应用方面的众多成果。

关于应用方面的部分本书首先回顾了硅藻土的应用，包括其商业用途和未来趋势[2]。本书还讨论了生物二氧化硅的光子特性，这些特性被带入硅藻瓣膜错综复杂的图案中，启发了无数的硅藻研究学者和硅藻业余爱好者，以及化学家和物理学家[3]。本书介绍了硅藻在生物医学领域的应用，硅藻土作为一种潜在的可生物降解的药物载体，一方面用于组织工程和出血控制的药物输送[4]，另一方面用于在法医学中分子法的应用[5]。法医学中的硅藻这一章介绍是一个飞跃，因为它可以使硅藻学家在进行法医检查时不必再处理溺水者的内脏器官，转基因编码的应用不会解决硅藻物理存在的问题，例如，在肺部它将支持硅藻的识别，从而促进受害者溺水栖息地的识别，这可能是转基因编码胜过经典光学显微镜（LM）检查和载玻片上处理过的硅藻薄膜计数的另一个标志。值得读者注意的是，英国在 2017 年放弃在水质评估中使用基于 LM 的硅藻瓣膜计数[6]，所有对江河湖泊监测中经典硅藻指数持怀疑态度的从业人员，都希望转基因编码能够成为首选的替代方案。

硅藻脂尽管被认为是非常好的生物柴油来源，有些物种甚至被命名为含油形式（如 *Fistulifera solaris*），但其他微藻在大规模生物质生长和石油生产中仍然是首选。本书介绍了"Diafuel"（硅藻生物燃料）的新术语和相应的商标，其中 *Fistulifera saprophila* 的瓣膜图形作为核心部分[7]。与其他微藻不同，硅藻细胞器和脂质液滴被包裹在硅质盒状硅藻壳中。与"软体"微藻不同，硅藻不会受损，除非挤压油的压力超过临界强度。这是否允许对硅藻进行压榨呢？让我们拭目以待。然而，迄今纳米压痕的一个研究领域正在寻求获得一种不会杀死硅藻的榨取压力，并允许它们恢复。预测硅藻未来作为人类能源使用，这还需要几代人的努力。"Diafuel"项目提供的一个独特技术过程是其二次产品——生物二氧化硅，由于其光子特性可以在能源生产中"回收"，例如太阳能面板或作为具有独特光子特性的新型材料[8]。培养大量硅藻用于生物燃料的一个缺点是其必须在可能被空气传播的矿物颗粒和微生物废物污染的开放式竞赛池塘和维护成本更高的封闭式光生物反应器之间进行选择。Gordon

等[9]提出在泡沫包装中培养硅藻,可能是这两种解决方案的替代方案。如果在某些条件下成功实施,"气泡农业"生物燃料的生产成本可能低于矿物燃料。

中国正在进行大规模的硅藻应用研究[10]。讨论了硅藻在新材料、生物、能源生产和储存、废水处理、复合材料、硅藻基陶瓷等方面的应用研究实例。中国的许多实验室致力于硅藻的基础研究,包括生态学、生物学、分类学和系统发育学,总体研究规模巨大。

美丽的镶框硅藻版画有极大的商业价值。然而,似乎没有被任何商业涉及的最原始的硅藻之美本身就是一种价值。硅藻学者与硅藻业余爱好者保持友好关系是很常见的,硅藻爱好者是光学显微镜专家,经常使用非常复杂的显微镜系统。令人惊讶的是,随着技术的变化,拥有扫描电镜(SEM)的硅藻业余爱好者的数量也在增加。硅藻之美的章节介绍中包含美术作品[11]。硅藻壳的形态介绍是另一章的主题[12]。尽管我们对瓣膜形成的细胞机制的研究取得了进展,但其遗传控制在很大程度上仍然未知。

硅藻研究的基本方面包括有性生殖和生命周期,以及对这些问题的最新看法[13],这些研究需要科学家大量的、耐心的实验,对理解硅藻分类学和系统学的许多方面至关重要。尽管硅藻研究享有盛誉,但很少有年轻学者愿意学习这项技术,而且需要在倒置显微镜下花费数天、数月甚至数年的时间来分离相似菌株的克隆培养物,以发现它们的相容性并进行成功的杂交实验。当然,分子工具的使用使寻找潜在的相容的克隆体变得更加容易,但是不能保证一定会有后代。关于硅藻共生的研究不多见,幸运的是,本书中有一章是关于硅藻细胞(蓝藻)中的内共生生物,以及硅藻作为内共生生物(甲藻)的介绍[14]。在给硅藻下定义时,通常会使用这样的语句:硅藻是单细胞光合生物,存在于所有提供有足够的光照和最低限度的水分的地方。硅藻的生命和瓣膜形态发生与光合作用息息相关,尽管硅藻在水生和陆生生态系统中扮演着如此重要的角色,但硅藻光合作用这一章的笔者强调,硅藻在细胞器层面上的光合作用知之甚少[15]。

Raven[16]综述了铁在硅藻生理学中的作用,尽管铁很重要,但现有的知识很少,而且大多局限于海洋浮游生物,本书是关于硅藻作为对人类和其他生物体有危害的毒素(如软骨藻酸等)的潜在生产者的最新综述,除了相当多的拟菱形藻的代表外,只有两种海洋菱形藻被检测到会产生毒素。然而,我们应该期待毒素及其生产者的名单会不断增加。形成水华的毒素生产者并不总是出现在人类严重影响的环境中(如虾养殖),有些是栖息在北极和南极的海水或寒冷的洋流中。Bates等[17]介绍了有毒硅藻的复杂生物学、它们的分布和检测方法。

本书有几个部分充满纪念意义,纪念那些去世的硅藻学家。有一篇是Joseph Seckbach纪念他的朋友Lawrence Bogorad(芝加哥大学的教授),整本书都是关于他展开[18],这是对Joseph Seckbach在芝加哥大学的硕士和博士导师的致敬。Joseph毕业后,他们保持着非常友好的联系。Lawrence Bogorad对光合色素的研究对我们理解叶绿体起源和光合作用产生了相当大的影响。另一篇是 Wladyslaw Altermann 总结了著名古生物学家和原生生物学家 Alex Volker Altenbach 的生活和科学生涯[19]。还有一篇是 Janice Pappas 写的感人的文章,献给 Frithjof A. S. Sterrenburg,他是一位硅藻爱好者,我们许多人过去都曾与他合作过[20]。就笔者个人而言,笔者在法兰克福会见了 Frithjof,当时他正在访问 Horst Lange-Bertalot。我们和 Frithjof 度过了难忘的几天,我很珍惜他给我打的无数个电话,每当一个紧急的分类学问题需要答复的时候,他都会接听。我们两次为 Kinker 收藏项目争取资金的共同努力都以失败告终。我有一个特别的回忆,那就是 Frithjof 和他的父亲发表了一篇联合论文[21]。大概很少有硅藻学家知道 *Nitzschia nienhuisii* 是 1990 年在毛里塔尼亚海岸被一个儿子和父亲描述为 Frithjof,这种独特而美丽的硅藻常见于大西洋和印度洋的非洲海岸,似乎需要建立一个新的(尚未命名的)属,见证 Frithjof 的骄傲多年后,笔者发现自己在和儿子发表联合论文时,也被类似的情感所打动[22]。

本书一定会成为硅藻基础研究和应用研究之间的纽带,笔者认为材料科学研究以及材料学中研究仪器的使用[如聚焦离子束(FIB)][23]在硅藻的基础研究中非常有启发,而且硕果累累。

参考文献

[1] Seckbach, J. and Kociolek, J. P. (eds.) (2011). The Diatom World. Springer London, Limited.

[2] Ghobara, M. M. and Mohamed, A. (2019). Diatomite in use: Occurrence, characterization, modification, and prospective trends. In: Diatoms: Fundamentals & Applications [DIFA, Volume 1 in the series: Diatoms: Biology & Applications, series editors: Richard Gordon & Joseph Seckbach]. J. Seckbach and R. Gordon, (eds.) Wiley-Scrivener, Beverly, MA, USA: pp. 471-510.

[3] Ghobara, M. M., Mazumder, N., Vinayak, V., Reissig, L., Gebeshuber, I. C., Tiffany, M. A. and Gordon, R. (2019). On light and diatoms: A photonics and photobiology review. In: Diatoms: Fundamentals & Applications [DIFA, Volume 1 in the series: Diatoms: Biology & Applications, series editors: Richard Gordon & Joseph Seckbach]. J. Seckbach and R. Gordon, (eds.) Wiley-Scrivener, Beverly, MA, USA: pp. 127-188.

[4] Maher, S., Aw, M. S. and Losic, D. (2019). Diatom silica for biomedical applications. In: Diatoms: Fundamentals & Applications [DIFA, Volume 1 in the series: Diatoms: Biology & Applications, series editors: Richard Gordon & Joseph Seckbach]. J. Seckbach and R. Gordon, (eds.) Wiley-Scrivener, Beverly, MA, USA: pp. 511-536.

[5] Vinayak, V. and Gautama, S. (2019). Diatoms in forensics: A molecular approach to diatom testing in forensic science. In: Diatoms: Fundamentals & Applications [DIFA, Volume 1 in the series: Diatoms: Biology & Applications, series editors: Richard Gordon & Joseph Seckbach]. J. Seckbach and R. Gordon, (eds.) Wiley-Scrivener, Beverly, MA, USA: pp. 435-470.

[6] Mann, D. G., Kelly, M. G., Walsh, K., Glover, R., Juggins, S., Sato, S., Boonham, N. and Jones, T. (2017). Development and adoption of a next-generation-sequencing approach to diatom-based ecological assessments in the UK [Abstract]. Phycologia 56(4, Supplement), 125-126.

[7] Vinayak, V., Joshi, K. B. and Sarma, P. M. (2019). Diafuel© (diatom biofuel) vs electric vehicles, a basic comparison: A high potential renewable energy source to make India energy independent. In: Diatoms: Fundamentals & Applications [DIFA, Volume 1 in the series: Diatoms: Biology & Applications, series editors: Richard Gordon & Joseph Seckbach]. J. Seckbach and R. Gordon, (eds.) Wiley-Scrivener, Beverly, MA, USA: pp. 537-582.

[8] Ghobara, M. M., Mazumder, N., Vinayak, V., Reissig, L., Gebeshuber, I. C., Tiffany, M. A. and Gordon, R. (2019). On light and diatoms: A photonics and photobiology review. In: Diatoms:

Fundamentals & Applications [DIFA, Volume 1 in the series: Diatoms: Biology & Applications, series editors: Richard Gordon & Joseph Seckbach]. J. Seckbach and R. Gordon, (eds.) Wiley-Scrivener, Beverly, MA, USA: pp. 127-188.

[9] Gordon, R., Merz, C. R., Gurke, S. and Schoefs, B. (2019). Bubble farming: Scalable microcosms for diatom biofuel and the next Green Revolution. In: Diatoms: Fundamentals & Applications [DIFA, Volume 1 in the series: Diatoms: Biology & Applications, series editors: Richard Gordon & Joseph Seckbach]. J. Seckbach and R. Gordon, (eds.) Wiley-Scrivener, Beverly, MA, USA: pp. 583.

[10] Zhang, Y. X. (2019). Current diatom research in China. In: Diatoms: Fundamentals & Applications [DIFA, Volume 1 in the series: Diatoms: Biology & Applications, series editors: Richard Gordon & Joseph Seckbach]. J. Seckbach and R. Gordon, (eds.) Wiley-Scrivener, Beverly, MA, USA: pp. 41-96.

[11] Tiffany, M. A. and Nagy, S. S. (2019). The beauty of diatom cells in light and scanning electron microscopy. In: Diatoms: Fundamentals & Applications [DIFA, Volume 1 in the series: Diatoms: Biology & Applications, series editors: Richard Gordon & Joseph Seckbach]. J. Seckbach and R. Gordon, (eds.) Wiley-Scrivener, Beverly, MA, USA: pp. 31-40.

[12] Bedoshvili, Y. D. and Likhoshway, Y. V. (2019). Cellular mechanisms of diatom valve morphogenesis. In: Diatoms: Fundamentals & Applications [DIFA, Volume 1 in the series: Diatoms: Biology & Applications, series editors: Richard Gordon & Joseph Seckbach]. J. Seckbach and R. Gordon, (eds.) Wiley-Scrivener, Beverly, MA, USA: pp. 97-112.

[13] Poulíčková, A. and Mann, D. G. (2019). Diatom sexual reproduction and life cycles. In: Diatoms: Fundamentals & Applications [DIFA, Volume 1 in the series: Diatoms: Biology & Applications, series editors: Richard Gordon & Joseph Seckbach]. J. Seckbach and R. Gordon, (eds.) Wiley-Scrivener, Beverly, MA, USA: pp. 243-270.

[14] Stancheva, R. and Lowe, R. (2019). Diatom symbioses with other photoauthotrophs. In: Diatoms: Fundamentals & Applications [DIFA, Volume 1 in the series: Diatoms: Biology & Applications, series editors: Richard Gordon & Joseph Seckbach]. J. Seckbach and R. Gordon, (eds.) Wiley-Scrivener, Beverly, MA, USA: pp. 223-242.

[15] Scarsini, M., Marchand, J., Manoylov, K. M. and Schoefs, B. (2019). Photosynthesis in diatoms. In: Diatoms: Fundamentals & Applications [DIFA, Volume 1 in the series: Diatoms: Biology & Applications, series editors: Richard Gordon & Joseph Seckbach]. J. Seckbach and R. Gordon, (eds.) Wiley-Scrivener, Beverly, MA, USA: pp. 189-210.

[16] Raven, J. A. (2019). Iron and ferritin in diatoms. In: Diatoms: Fundamentals & Applications [DIFA, Volume 1 in the series: Diatoms: Biology & Applications, series editors: Richard Gordon & Joseph Seckbach]. J. Seckbach and R. Gordon, (eds.) Wiley-Scrivener, Beverly, MA, USA: pp. 211-222.

[17] Bates, S. S., Lundholm, N., Hubbard, K. A., Montresor, M. and Leaw, C. P. (2019). Toxic and harmful marine diatoms. In: Diatoms: Fundamentals & Applications [DIFA, Volume 1 in the series: Diatoms: Biology & Applications, series editors: Richard Gordon & Joseph Seckbach]. J. Seckbach and R. Gordon, (eds.) Wiley-Scrivener, Beverly, MA, USA: pp. 388-434.

[18] Seckbach, J. (2019). Dedication to Lawrence Bogorad. In: Diatoms: Fundamentals & Applications [DIFA, Volume 1 in the series: Diatoms: Biology & Applications, series editors: Richard Gordon & Joseph Seckbach]. J. Seckbach and R. Gordon, (eds.) Wiley-Scrivener, Beverly, MA, USA: pp. v-vi.

[19] Altermann, W. (2019). Alex Altenbach—in memoriam of a friend. In: Diatoms: Fundamentals & Applications [DIFA, Volume 1 in the series: Diatoms: Biology & Applications, series editors: Richard Gordon & Joseph Seckbach]. J. Seckbach and R. Gordon, (eds.) Wiley-Scrivener, Beverly, MA, USA: pp. 27-30.

[20] Pappas, J. L. (2019). A memorial to Frithjof Sterrenburg: The importance of the amateur diatomist. In: Diatoms: Fundamentals & Applications [DIFA, Volume 1 in the series: Diatoms: Biology & Applications, series editors: Richard Gordon & Joseph Seckbach]. J. Seckbach and R. Gordon, (eds.) Wiley-Scrivener, Beverly, MA, USA: pp. 1-26.

[21] Sterrenburg, F. A. S. and Sterrenburg, F. J. G. (1990) An outline of the marine littoral diatom biocoenosis of the Banc-d'Arguin, Mauritania, West Africa. Botanica Marina 33(5), 459-465.

[22] Dabek, P., Witkowski, J., Witkowski, A. and Riaux-Gobin, C. (2015). Morphology of Biddulphia seychellensis (Grunow in Van Heurck) FW Mills and the generic limits of Biddulphia Gray. Nova Hedwigia 144(Supplement), 97-105.

[23] Witkowski, A. (2019). Application of focused ion beam in studies of ultrastructure of diatoms. In: Diatoms: Fundamentals & Applications [DIFA, Volume 1 in the series: Diatoms: Biology & Applications, series editors: Richard Gordon & Joseph Seckbach]. J. Seckbach and R. Gordon, (eds.) Wiley-Scrivener, Beverly, MA, USA: pp. 113-126.

前　言

硅藻简介：基础与应用

Joseph Seckbach

《硅藻：基础与应用》是"硅藻：生物学与应用系列"的第一部，原著由 Wiley-Scrivener 出版，由 Joseph Seckbach 和 Richard Gordon 主编。

本书进一步补充了上一部（《硅藻世界》，由 Joseph Seckbach 和 J. Patrick Kociolekeds 主编，于 2011 年由 Springer 出版）中介绍的知识，以及一些新的主题。硅藻是什么？它们是迷人的微观单细胞或群落、微观真核藻类。它们普遍分布在水生栖息地，被认为是浮游植物的主要部分。它们存在于淡水、咸水环境、咸水和海洋区域，是生物燃料的来源。它们生活在高温和低温以及不同的 pH 值中。它们的细胞分为两半，它们的细胞壁是硅化的。本书介绍了硅藻的许多方面，可能你从来没有遇到过，也不知道它们的存在。硅藻利用大气中 20% 的 CO_2 并通过光合作用过程释放出大气中对所有生命都至关重要的 O_2。它们的叶绿体组成独特，不同于其他绿藻和高等植物，它们不具备其他绿色植物所具有的叶绿素 b。

据推测，硅藻在进化过程中发生过一次共生事件，根据这一理论，宿主真核原始细胞吸收了蓝藻类型的细胞，并使用这种客体（或穿透物）作为其真核质体实体的一部分，与宿主细胞核进行一些遗传物质交换。为了充分的光合作用，铁是必需的。否则，如果其营养中缺乏铁，绿藻和高等植物会变得苍白和褪绿。藻类铁蛋白是光合作用反应和其他细胞铁需求的铁储存库。

Karsten、Holzinger、Witkowski 和其他人展示了硅的双原子细胞壁的细胞学研究，展示了最美丽的硅藻外观。此外，Tiffany 和 Nagy 在这一章中特别推崇双原子壁的"理想之美"。在生态环境中，硅藻无处不在，生活在淡水、极地冷水、温泉水和高山湖泊中。

本书献给我的三位亲密的朋友：Lawrence Bogorad，他是我在芝加哥大学攻读硕士和博士学位的导师；我的同事 Alex Altenbach，我在慕尼黑路德维希马克西米利安大学地质系攻读硕士学位时相识；Frithjof，一位来自荷兰的电子显微镜学家和硅藻业余爱好者。上述三人都参与了藻类和硅藻的研究。

目 录

第1章　纪念 Frithjof A. S. Sterrenburg：硅藻业余爱好者的重要性 ·········· 001
   1.1　简介 ················ 001
   1.2　背景和兴趣 ············ 002
   1.3　硅藻业余爱好者的性格 ······· 004
   1.4　硅藻业余爱好者和收藏的重要性 ··· 007
   1.5　硅藻业余爱好者是贸易的专家 ···· 008
   1.6　硅藻业余爱好者作为同行评审 ···· 010
   1.7　结束语 ··············· 014

第2章　纪念一位朋友——Alex Volker Altenbach ······················ 020

第3章　硅藻之美 ·············· 023
   3.1　硅藻观察的早期历史 ········ 023
   3.2　活硅藻 ·············· 024
   3.3　形状和结构 ············ 025
   3.4　不同尺度下的硅藻之美 ······· 026
   3.5　形态发生过程中的瓣膜 ······· 027
   3.6　Jamin-Lebedeff 干涉对比显微镜 ··· 028
   3.7　结论 ················ 029

第4章　中国硅藻研究现状 ········· 030
   4.1　用于能量转换和储存的硅藻 ····· 030
      4.1.1　简介 ············ 030
      4.1.2　硅藻二氧化硅：结构、性能及其优化 ··· 032
      4.1.3　用于锂离子电池材料的硅藻 ···· 033
      4.1.4　用于储能的硅藻：超级电容器 ·· 035
      4.1.5　用于太阳能电池的硅藻 ······ 040
      4.1.6　用于储氢的硅藻 ········· 041

　　　　4.1.7 用于热能储存的硅藻 ·· 042
　　4.2 水处理用硅藻 ·· 043
　　　　4.2.1 制备硅藻土基吸附复合材料的载体 ·· 044
　　　　4.2.2 制备多孔碳材料的催化剂和模板 ·· 045
　　　　4.2.3 表面和多孔结构的改性 ·· 047
　　　　4.2.4 制备硅藻土基金属氧化物复合材料的载体 ································ 054
　　4.3 基于硅藻壳结构的复合凹坑摩擦学性能研究 ·· 063

第 5 章　硅藻瓣膜形态发生的细胞机制 ·· 072
　　5.1 简介 ·· 072
　　5.2 瓣膜对称性 ·· 072
　　5.3 瓣膜硅化顺序 ·· 074
　　5.4 SDV 中的二氧化硅 ·· 075
　　5.5 宏观形态发生控制 ·· 075
　　5.6 形态发生的细胞骨架控制 ·· 076
　　5.7 囊泡在形态发生中的作用 ·· 077
　　5.8 瓣膜胞吐作用和 SDV 起源 ·· 078
　　5.9 结论 ·· 079

第 6 章　聚焦离子束技术在硅藻超微结构分类学研究中的应用 ························ 084
　　6.1 简介 ·· 084
　　6.2 材料与方法 ·· 085
　　6.3 结果 ·· 085
　　　　6.3.1 复杂条纹超微结构 ·· 085
　　6.4 讨论 ·· 089
　　　　6.4.1 培养标本与野生标本 ·· 090
　　6.5 结论 ·· 090

第 7 章　光与硅藻：光子学和光生物学综述 ·· 093
　　7.1 简介 ·· 093
　　7.2 硅藻壳独特的多尺度结构 ·· 093
　　7.3 硅藻壳的光学性质 ·· 099
　　　　7.3.1 作为光子晶体壁盒的壳 ·· 101
　　　　7.3.2 光聚焦现象 ·· 104
　　　　7.3.3 光致发光特性 ·· 106
　　　　7.3.4 硅藻壳在硅藻光生物学中的可能作用 ·· 107

7.4 硅藻光生物学 ·········· 108
7.4.1 水下光场 ·········· 108
7.4.2 细胞周期光调节 ·········· 108
7.4.3 羽纹纲中的趋光现象 ·········· 108
7.4.4 叶绿体迁移(核营养不良) ·········· 110
7.4.5 蓝光及其对细胞微管的影响 ·········· 110
7.4.6 高光强度下的光调节策略 ·········· 112
7.4.7 紫外线辐射照射下的光调节策略 ·········· 112
7.4.8 硅藻和弱光 ·········· 113
7.4.9 硅藻和无光 ·········· 113
7.4.10 光导管与细胞视觉 ·········· 114
7.5 硅藻和光的应用 ·········· 115
7.5.1 在光催化中的应用 ·········· 115
7.5.2 基于生物的紫外线过滤器 ·········· 116
7.5.3 在太阳能电池中的应用 ·········· 116
7.5.4 基于发光特性的应用 ·········· 117
7.5.5 隐形硅藻 ·········· 117
7.6 结论 ·········· 118

第8章 硅藻的光合作用 ·········· 126
8.1 简介 ·········· 126
8.2 叶绿体结构反映了内共生的两个步骤 ·········· 128
8.3 光合色素 ·········· 129
8.3.1 叶绿素 ·········· 129
8.3.2 类胡萝卜素 ·········· 130
8.4 光合器官的组织 ·········· 130
8.5 非光化学猝灭 ·········· 132
8.6 碳吸收和固定 ·········· 134
8.7 结论 ·········· 135

第9章 硅藻中的铁 ·········· 142
9.1 简介 ·········· 142
9.2 硅藻对铁的获取 ·········· 142
9.3 硅藻含铁蛋白质与铁利用的经济性 ·········· 143
9.4 铁储存 ·········· 146
9.5 结论 ·········· 147

第 10 章　硅藻与其他光合自养生物共生 ································· 151
  10.1　简介 ··· 151
  10.2　具有固氮球状蓝藻内共生体的硅藻 ······························· 152
  10.3　具有固氮丝状异细胞蓝藻内共生体的硅藻 ························· 156
  10.4　附生、内源性和内生附生硅藻 ··································· 158
  10.5　鞭毛藻的藻内共生体 ··· 159

第 11 章　硅藻有性繁殖和生命周期 ····································· 165
  11.1　简介 ··· 165
  11.2　辐射型硅藻 ··· 166
    11.2.1　生命周期和繁殖 ······································ 166
    11.2.2　配子发生和配子结构 ·································· 168
    11.2.3　产卵 ·· 169
  11.3　羽状硅藻的生命周期和繁殖 ····································· 170
  11.4　生长孢子的发育与结构 ··· 174
    11.4.1　初期阶段 ·· 174
    11.4.2　外壁 ·· 175
  11.5　性生殖诱导 ··· 176

第 12 章　北极底栖生物的生态学、细胞生物学和超微结构 ················ 185
  12.1　简介 ··· 185
  12.2　北极的环境背景 ··· 185
  12.3　生长随温度的函数 ··· 186
  12.4　长期暗培养后的生长情况 ······································· 187
  12.5　长期暗培养后的细胞生物学性状 ································· 188
  12.6　超微结构特征 ··· 190
  12.7　结论 ··· 191

第 13 章　淡水硅藻的生态学——当前的发展趋势和应用 ··················· 194
  13.1　简介 ··· 194
  13.2　硅藻分布 ··· 196
  13.3　硅藻扩散能力 ··· 196
  13.4　硅藻生态学中的功能分类 ······································· 197
  13.5　空间生态学和集合群落 ··· 200
  13.6　水生生态系统的生物监测 ······································· 200
  13.7　结论 ··· 202

第 14 章 法医学中的硅藻：法医学中硅藻检验的分子方法 ····· 210
- 14.1 简介 ····· 210
- 14.2 尸检取证措施 ····· 212
- 14.3 溺水受害者与死于其他原因的差异 ····· 212
- 14.4 生物样品中硅藻的鉴定技术 ····· 213
  - 14.4.1 水样的形态学分析 ····· 214
  - 14.4.2 位点特异性硅藻的作用 ····· 214
- 14.5 案例研究 ····· 215
  - 14.5.1 案例 1 ····· 215
  - 14.5.2 案例 2 ····· 215
  - 14.5.3 案例 3 ····· 216
- 14.6 利用分子工具鉴定组织和水样中的硅藻 ····· 217
- 14.7 溺水受害者组织中硅藻 DNA 的分化 ····· 218
- 14.8 聚合酶链反应(PCR) ····· 219
- 14.9 从一名溺水受害者的生物样本中提取硅藻 DNA ····· 221
  - 14.9.1 生物样本 ····· 221
  - 14.9.2 利用硅胶梯度和酚氯仿法提取 DNA 从组织中分离浮游生物/硅藻 ····· 222
- 14.10 硅藻诊断溺水的最佳条形码标记 ····· 222
  - 14.10.1 细胞色素 C 氧化酶亚基 1(COI) ····· 223
  - 14.10.2 核 rDNA ITS 区 ····· 223
  - 14.10.3 核小亚基 rRNA 基因 ····· 224
- 14.11 DNA 测序 ····· 224
- 14.12 测序技术的进步导致了数据解释技术的进步 ····· 225
- 14.13 结论 ····· 225

第 15 章 使用中的硅藻土：性质、修改、商业应用和未来趋势 ····· 234
- 15.1 硅藻土的性质 ····· 234
  - 15.1.1 硅藻土的形成 ····· 235
  - 15.1.2 硅藻对溶解的抵抗力(它们能保存数百万年的原因) ····· 235
- 15.2 发现和古代应用的历史 ····· 237
- 15.3 硅藻土的发生和分布 ····· 238
- 15.4 硅藻土的开采和加工 ····· 239
- 15.5 硅藻土的特征 ····· 240
- 15.6 硅藻壳改性 ····· 240
- 15.7 硅藻土改性 ····· 241

	15.7.1	基于硅藻土的过滤技术	242
	15.7.2	保温用硅藻土	244
	15.7.3	硅藻土基建筑材料	245
	15.7.4	硅藻土作为一种杀虫剂	246
	15.7.5	硅藻土作为土壤改良剂	246
	15.7.6	硅藻土作为填料	247
	15.7.7	硅藻质土作为硅砂材料	247
	15.7.8	硅藻土作为动物和人类的食物添加剂	248
	15.7.9	硅藻土和纳米技术	248
	15.7.10	非工业应用	250
15.8	硅藻土的制造及未来的发展前景		251
15.9	结论		251

第 16 章　生物医学用硅 ································ 263

16.1	简介	263
16.2	硅藻:用于治疗药物输送的天然二氧化硅微胶囊	264
	16.2.1　结构	264
	16.2.2　硅藻的表面修饰	265
	16.2.3　硅藻原子作为药物载体的应用	266
	16.2.4　硅藻作为药物输送应用的可生物降解载体的来源	272
	16.2.5　用于其他生物医学应用的硅藻	274
16.3	结论	277

第 17 章　Diafuel™(硅藻生物燃料)与电动汽车的基本比较:使印度能源独立的高潜力可再生能源 ································ 282

17.1	简介	282
17.2	关于温室气体排放(GHG)与 CO_2 和温度关系的探讨	283
17.3	2015 年巴黎协议的结果	284
17.4	印度的能源需求	285
17.5	批评家谈论电动汽车进入市场	288
17.6	电动汽车与 Diafuel™ 大型汽车的比较	289
	17.6.1　电动汽车	289
	17.6.2　Diafuel™	291
17.7	电动汽车的发电来源	297
	17.7.1　具有零碳排放的资源	297
17.8	电动汽车与汽油驱动汽车的 CO_2 排放量	301

17.9 耗尽地球金属来运行电动汽车与 Diafuel™ 丰富资源的比较 ·········· 302
　　17.9.1 Diafuel™ 能成为答案吗 ·········· 303
　　17.9.2 从硅藻中获取 Diafuel™ ·········· 304
17.10 目前状况 ·········· 305
　　17.10.1 EV 与 Diafuel™ 的数据分析与比较 ·········· 306
17.11 结论 ·········· 309

第 18 章 泡沫农业：硅藻生物燃料和下一次绿色革命的可扩展缩影 ·········· 316
18.1 简介 ·········· 316
　　18.1.1 泡沫农业概念 ·········· 319
　　18.1.2 可能通过无人机进行气泡注射、取样、收获和密封 ·········· 321
　　18.1.3 方法 ·········· 322
18.2 机械性能 ·········· 323
　　18.2.1 最佳气泡尺寸 ·········· 323
18.3 光学性质 ·········· 325
18.4 表面性质 ·········· 325
　　18.4.1 气体交换特性 ·········· 326
18.5 毒性限制 ·········· 330
　　18.5.1 藻类油液滴特性 ·········· 331
18.6 生物膜 ·········· 332
18.7 细菌共生体 ·········· 332
　　18.7.1 土壤作为 CO_2 的来源 ·········· 333
18.8 需求 ·········· 334
　　18.8.1 硅藻和其他藻类的选择 ·········· 334
18.9 指数增长与平稳阶段 ·········· 336
18.10 碳回收 ·········· 337
18.11 包装 ·········· 337
　　18.11.1 农民的作物选择 ·········· 338
　　18.11.2 泡沫养殖与光生物反应器和跑道之间 ·········· 338
18.12 结论 ·········· 339

附录 ·········· 364

第 1 章
纪念 Frithjof A. S. Sterrenburg：硅藻业余爱好者的重要性

Janice L. Pappas

1.1 简介

纵观科学，业余爱好者为许多学科的知识体系做出了贡献。业余爱好者致力于追求与自己特定兴趣相关的知识，对于硅藻领域来说，也不例外。在 19 世纪，从事硅藻研究的是不同领域的爱好者[1]，例如，Friedrich Traugott Kützing 是一名药剂师和学校教师，后来成为一名硅藻学家，他发现硅藻由二氧化硅组成，其外壳有两个部分，一部分是"原生的"，另一部分是"次生的"，并在当时杰出的动物学家和硅藻学家 C. G. Ehrenburg 的帮助下发表了他的发现[2]。一些业余爱好者成立了显微镜协会或俱乐部，作为追求他们对微观世界的共同兴趣的爱好者（如 Quekett 显微镜俱乐部）。在维多利亚时代后期，硅藻研究已经专业化，目前，个人可以找到专业导师进行正式培训，成为硅藻研究员。

话虽如此，现代公民科学家仍在为硅藻研究做出贡献。只有很少的人接受过正规的培训、获得认证和报酬并专门从事硅藻研究，但硅藻研究正处于一个阶段，大量必要的工作可能涉及数十万个物种，硅藻业余爱好者可以发挥至关重要的作用。从事这一角色的业余爱好者很快意识到，需要与专业人士一起合作，尤其是在分类学、命名法、显微镜和显微摄影的技术问题上。

现代硅藻业余爱好者的典型例子是 Frithjof （见图 1.1），他于 2016 年 3 月 11 日去世，并在硅藻研究领域中留下了自己的印记，他对硅藻研究充满热情和奉献精神，这为他赢得了国际专业人士的认可和尊重。Frithjof 在硅藻分类和命名以及显微镜和显微摄影方面的工作让专业和硅藻业余爱好者受益匪浅。他的贡献将继续对硅藻研究

图 1.1　Frithjof 在显微镜下观察

人员从事和开展研究的方式产生影响。无论人们如何理解成为一名业余爱好者意味着什么，Frithjof 以其自己的方式超越了这一概念。

从历史上看，硅藻业余爱好者没有报酬，他们通常从事各种职业[1-2]。一个典型的例子是 Astrid Cleve-Euler，她是 19 世纪末瑞典乌普萨拉大学第一位获得科学博士学位的女性。然而，在她的一生中，尽管她在化学、植物学、地质学和硅藻研究方面做出了贡献，但她从未被聘为瑞典自然历史博物馆的科学家。即使是商人和政治家 van Leeuwenhoek（17～18 世纪）也没有因为研究微观世界而获得报酬[3-4]。

其他专门从事硅藻研究的业余爱好者。例如，宽角斜纹藻，最初由 John Thomas Quekett[5]命名为宽角舟形藻，他是一名硅藻业余爱好者，同时也是一名显微镜学家和组织学家。还有一位硅藻研究业余爱好者的典型代表是 William Smith 牧师，他在维多利亚时代是一位杰出的硅藻研究员[2]。作为一项常见的休闲活动，硅藻被装裱成各种装饰，载玻片或由个人购买或自己准备，装裱后的硅藻可以在维多利亚时代的客厅里用显微镜观察[6]。硅藻装裱师如 Johann Diedrich Möller（硅藻装裱艺术的鼻祖）[7]和牙医 William Gatrell[8]等都是非常受欢迎的装裱师。显微镜的出现和普遍使用为硅藻装饰奠定了基础，并引发了硅藻业余爱好者对硅藻兴趣的激增。

从某种意义上来说，Frithjof 是硅藻研究者中的后起之秀。然而，他发展了自己的风格，并超越外科医生 John Redmayne[9]和教师 John Albert Long[10]等业余爱好者。与 Quekett 一样，Redmayne 和 Long 在 19 世纪末和 20 世纪初作为硅藻业余爱好者，他们买卖硅藻样本以及制作装裱物，并从 *Hardwicke's Science Gossip* 等流行出版物获取硅藻的最新信息[11]，在其 1885 年的第二十一卷中，介绍了一篇关于 Jacques-Joseph Brun 的文章《阿尔卑斯山和侏罗山的硅藻》，其中药理学家和硅藻学家 Brun[12]谈到了硅藻随时间的沉积和硅藻土的形成[11]。他与同时代的硅藻化石研究人员一样，在分类学方面的研究成果推动了 Adolph Schmidt[13]于 1874 年开始编制《硅藻图集》，该图集一直由其他人提供资料，直到 1959 年才得以完成，这些人中包括 Friedrich Hustedt，他从一名教师转变为专业硅藻学家，对硅藻研究产生了巨大影响[14]。Frithjof 的硅藻研究轨迹与 Smith、Brun 和 Hustedt 相似，与这些前辈一样，Frithjof 从业余爱好者逐渐成为专家，他的专业知识赢得了人们的尊重。Frithjof 将自己的贡献提升到了科学同行评审的水平，而不是只能在大众出版物上发表。

1.2 背景和兴趣

Frithjof 出生于 1934 年，他最初在阿姆斯特丹大学学习医学，他自学成才的天性与正规化的教育方法相冲突，于是他开始自学，通过博览群书，学习各种难度级别的科目，并以此作为终生的努力方向。对于 Frithjof 来说，生活和学习都是一次冒险。

　　Frithjof 沉浸于广泛的学习，他的才能表现在音乐[15]和天文学等领域，年轻时他在管弦乐队中演奏小号（见图 1.2）、萨克斯、单簧管和钢琴，他还为这样的管弦乐队编曲，他曾多年担任爵士乐手。Frithjof 拥有许多望远镜（见图 1.3），他在 20 世纪 80 年代初出版的一些著作介绍了如何更清晰地观察星星[16-17]，他喜欢与许多人分享关于星星的对话，其中包括老朋友 Michael Stringer 和 Wulf Herwig，因为他乐于看到我们眼前世界以外的东西。

图 1.2　1968 年，Frithjof 与 Paul Whiteman 管弦乐团的长号手 Bill Rank 一起演奏小号

图 1.3　1980 年，Frithjof 在自家后院向一位朋友夸赞夜空

　　Frithjof 是一个工匠，他对所有电子、电气和机械方面的东西都很在行[18]。Frithjof 与 Wulf 和老朋友 Hein de Wolf 一样，对电气设备有着浓厚的兴趣。他喜欢飞机，喜欢修复老式汽车和摩托车以及军用无线电接收器，他收集的军用无线电接收器让他对军事更加着迷。他撰写了有关接收器的手册[19]，并担任专业电子期刊的编辑委员会成员[20]。Michel Poulin 在参观 Frithjof 家时注意到，他还是一个狂热的摩托车骑手。当 Frithjof 出现在 Hein 的办公室时，Hein 总是感到惊讶他会穿着他的摩托车装备还是西装。他对历史知识感兴趣[21]并对历史藏品记录和保存，其中包括军事历史[22-23]，特别是关于第二次世界大战的军事史。关于《奥斯陆报告》中包含的德语 Peenemünde 文件的解码，他撰写一份具有洞察力的资料[24]。

　　Frithjof 的人生观在很大程度上受到他青年时期经历的影响，他成长于 20 世纪 30 年代和 40 年代，成年时世界正处于经济萧条时期，随后又爆发了第二次世界大战。他的个人回忆汇编于 2009 年刊登在一个关于第二次世界大战飞机的网络论坛上[25]，他非常了解人类历史上的那个时期，尽管他对那个时代对自己的影响很敏感，但他也对其他人如何受到影响感兴趣[26]。

　　多年来，Frithjof 和他亲爱的妻子 José（见图 1.4）一起建立了美好的生活，他们享受着他们最大的修复工程，在北荷兰的 Sijbekarspel 小农庄，他们对 1890 年的维多

图 1.4　1978 年，Frithjof 和 José 在树木和鲜花中放松

利亚式房屋进行了最大规模的修复。这是一项爱的事业，他们在那里幸福生活了 35 年。

正如 Michel Poulin 回忆的那样，他们的家"维护得很好，有一个漂亮的花园"。40 多年来，Frithjof 一直从事通信、广告和公共关系方面的个体经营[27]，为通信、医疗仪器、航空、国防、制药、能源等领域的研究中心和高科技产业担任顾问。他喜欢与客户互动，以热情洋溢的态度完成每个项目，并密切关注应有的细节。Frithjof 和 José 喜欢在整个欧洲旅行，尤其是在巴尔干半岛、希腊、葡萄牙、太平洋岛屿（拉罗汤加和西萨摩亚）、印度尼西亚、澳大利亚、新西兰、美国西部，其中包括加利福尼亚（尤其是旧金山）、内华达、犹他州、科罗拉多州、亚利桑那州和佛罗里达州。他假期是在德国和英国度过的，有一次是骑着老式摩托车，有一次是在澳大利亚东海岸骑着现代摩托车。出于工作他前往阿拉伯联合酋长国。Frithjof 和 José 到过世界上许多地方，体验了许多文化和生活方式。

作为一个有讲故事天赋的多才多艺的人（见图 1.5），Frithjof 直率、意志坚强，不惧怕有争议的话题（如科学与宗教），他的观点、阐释和论述雄辩有力，在任何特定主题上的立场都不会出错[28-29]，他根据自己的坚定信念行事。尽管他和 José 在 2003 年北美硅藻研讨会期间搬到了荷兰，但由于他反对 1990 年的海湾战争，Frithjof 也对前往美国持保留态度；相反，他发送了一个带有"最后祝酒词"的视频，供在会议上播放。Frithjof 是一个见多识广的人，他见证了各种生活方式，并沉浸在他所遇到的大量想法中，他头脑灵活，能将这些想法转化为自学的知识，并将这些知识组合成精妙的成果，这种态度在他从事硅藻科学研究的过程中发挥了很好的作用。

图 1.5　1985 年，Frithjof 在演讲

1.3　硅藻业余爱好者的性格

无论你是客户、朋友还是硅藻爱好者，Frithjof 总是有时间回答你的询问，他以他一

贯的亲和力、敏锐的洞察力、专注的精神和干练、犀利的幽默感来回答你的问题。他对语言的掌握和熟练程度[30]总是以娴熟的演讲展示出来,改变你对正在讨论话题的看法。

在与Frithjof进行对话时,有时你会被带入一场不知道的即将发生的"旅程"。在编写《硅藻纳米技术》特刊期间与Frithjof通信以及在即将出版遇到的困难时,我发现自己被卷入了一场关于古希腊历史和神话中隐喻讨论中,和神话中的隐喻来应对这一局面,我们就特洛伊的Helen、Medea和Aegeus交换了沉思。编辑部的一位老编辑告诉我,我是重新开始出版过程的动力,所以我最终成为主角,由Frithjof决定由特洛伊的Helen变成Medea。参与特刊的人可能记得在这场"悲喜剧"期间发生的电子邮件往来,并可能猜测谁可能是对手,Frithjof和我确定他不是Aegeus,因此不是Medea的对手。

当Paul Hargraves给Frithjof发电子邮件并告诉他自己不记得在会议上见过他时,Frithjof发送了一张"他自己"的照片,Paul很快确定这张照片是奥地利海军上将Hermann Freiherr von Spau。令他失望的是,这个诡计很快就被识破,Frithjof收到了一张身份不明的皇室贵族照片,他被要求从中认出Paul。Frithjof知道这张照片是在温莎为爱德华七世国王举行葬礼的九位君主,他将比利时一世Albert认成了Paul。然而,正如Paul所说,他不知道与Albert有任何关系。Frithjof的幽默感总能给与他交流的人带来乐趣。

Frithjof拥有令人钦佩的《旧约全书》知识,他用英语或德语与Joseph Sechback交换了引文,他称Joseph Sechback为Methuselah。Sechback和他的妻子在参观一家奶酪工厂的途中乘坐了Frithjof的跑车。Richard Gordon从Frithjof那里发现,作为一名流动生物学家,可能会导致蜕变成一个类似Erdös的生物[31]!

2016年春季,在荷兰-佛兰德硅藻协会的一次有说服力的演讲中Herman van Dam纪念了Frithjof。在Vlaamse Kring van Diatomisten(NVKD)成立期间,Herman遇到了Frithjof,他们建立了长久的友谊。他谈到了Frithjof对硅藻的热爱,并重申了Frithjof所特有的兴趣、努力和知识的兼收并济。他谈到了Frithjof与Hein de Wolf对《硅藻学报》的贡献,Frithjof为此撰写了六篇文章。Herman强调了Frithjof贡献了许多关于布纹藻属和斜纹藻属以及其他分类群的出版物,并指出Frithjof是公认的硅藻研究领域的国际专家。Herman指出Frithjof是一个有强烈意见和原则的人。Herman还指出,Frithjof具有宽宏大量的品质,例如,他愿意在硅藻研究方面帮助和指导其他人。

Paul Hargraves指出,他贡献了他的时间和专业知识(如在识别布纹藻属方面),并且被收录在关于该属的许多出版物中,他慷慨地将我列为合著者,就像我们后来在Haslea项目上的合作一样。Wulf Herwig发现Frithjof非常聪明、睿智和幽默,而且总是乐于助人。Jackie White-Reimer记得Frithjof是一个热情、亲切的人,当他们一家人去Frithjof家里做客时,Frithjof和José竭尽全力让他们感到舒适并带他们四处参观,甚至租了一辆足够容纳他们所有人的汽车。Reimer和Frithjof夫妇在彼此的家中度过了一段时间,Frithjof是"体贴的房客"。Frithjof甚至给当时Reimer的12

岁和10岁的女儿Laura和Emilie留下了不可磨灭的印象,她们记得他有英国口音、性格随和、友好、友善和快乐,Emilie记得Frithjof偷偷给了她们一罐姜饼饼干,这让她们高兴不已。

一些人有幸收到了Frithjof的节日祝福和精彩的摄影作品(见图1.6~图1.9),有令人眼花缭乱的硅藻和超凡脱俗的实体。他慷慨大方,直接或通过网络为专业人士和业余爱好者鉴定硅藻和提供显微镜方面的指导性意见。他的名字经常出现在致谢中,以表彰对他正确识别和正确命名所考虑的硅藻分类群的贡献。当人们询问他关于布纹藻属和斜纹藻属物种的专业知识时,他总是不吝赐教。

图1.6　2006年Frithjof发给朋友的数字圣诞卡

图1.7　2008年Frithjof发给朋友的数字圣诞卡

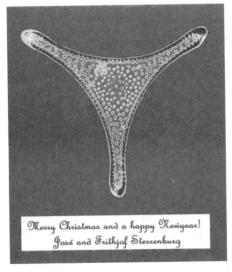

图1.8　2010年Frithjof发给朋友的数字圣诞卡

图1.9　2011年Frithjof发给朋友的数字圣诞卡

Michael Stringer 对 Frithjof 的评价尤其高,他和 Frithjof 有许多共同的兴趣。正如 Michael 赞赏地说:"他总是在那里帮助我,他改变了我对今天存在的硅藻的研究。"Michael 还说 Frithjof 最喜欢的甜食是英式甘草糖、圣诞布丁和掺有威士忌和白兰地的圣诞蛋糕。

Frithjof 体现了多方面的个性,他有很多朋友,他们都记得那些使他成为一个独特的人的品质,Klaus Kemp 和 Phil Basson 在网站上向 Frithjof 表示敬意,称他"为硅藻研究做出了巨大贡献,并通过来自苏拉威西岛和其他地方的捐赠对硅藻业余爱好者给予了极大的支持,"并称 Frithjof"是一位真正的绅士"[32]。

1.4 硅藻业余爱好者和收藏的重要性

Frithjof 是荷兰莱顿国家自然历史博物馆"Naturalis"的助理研究员。他参观和使用了各个博物馆的藏品,并了解它们的价值。作为博物馆的倡导者,Frithjof 无论是在报刊还是网络上都在积极发声[33],赞扬了藏品的优点、藏品的获取、藏品的记录以及藏品的管理。1990 年,他阐明了集合作为分类学、形态学和生态学研究的基础的意义,特别是在硅藻研究中[33-34]。1991 年,Hein de Wolf 和 Frithjof 承担了更新 Fryxell 1975 年论文的任务,通过对硅藻收集进行国际调查[35]。正如 Frithjof 所说:

"分类学不是集邮,而是重新定义所描述的分类群的生物。只有当你检查原始材料来定义明确的参考标准("类型")时,才能可靠地做到这一点,这显示了博物馆藏品和维护这些藏品的中坚力量的巨大价值[36]。为了使此类研究有意义,良好的分类学是不可或缺的要求,尽管财政拮据,但硅藻分类学在过去 25 年中蓬勃发展(就像 19 世纪下半叶一样)。人们现在普遍认为,分类学必须基于对最初描述物种的原始材料的调查,即所谓的分型过程,因为随着时间的推移,错误的鉴定不断积累,导致物种范式的错误转变。因此,收集原始材料至关重要[37]。"

Herman van Dam 表示 Frithjof 致力于收藏。Frithjof 与 Hein de Wolf 一起发表了关于 Johannes Kinker 收藏的论文[37-39]。Kinker 是一名职业股票经纪人,也是维多利亚时代享有盛誉的硅藻业余爱好者,他收藏了大量硅藻藏品[37]。他没有发表过文章,但实业家、化学家、硅藻业余爱好者、显微镜学家和植物学家 Henri-Ferdinand van Heurck[40]曾表示 Kinker 的收藏包含"杰作"[8]。该藏品的发现证明了 Frithjof 和 Hein 的奉献精神,使人们能够认识到 Kinker 作为业余爱好者对硅藻研究的贡献以及博物馆藏品的重要性。他们更普遍地向硅藻业余爱好者表示敬意,指出,

人们时常把这些维多利亚时代的研究者描绘成古灵精怪的"长腿叔叔",不着边际地追逐着神秘的生物已经成为一种时尚,但这种形象是错误的。奠定科学基础从来都不是无关紧要的,这些早期研究人员的工作为目前的硅

藻分类和硅藻研究的实际科学应用奠定了良好的基础。Kinker 根本不用谦虚,因为他"是一位业余爱好者"。他同时代的大多数人也是如此,"业余爱好者"的地位在他那个时代是可敬的,甚至是杰出的,事实上,如果科学资助的现代趋势继续下去,我们很可能会在有生之年看到这种情况的重现[37]。

1.5 硅藻业余爱好者是贸易的专家

Frithjof 很小的时候就对自然世界感兴趣,尤其是野花、昆虫和硅藻。他好奇的天性以及对电子和物理的迷恋只是他兴趣发展的一个前兆,正如一个简短的网页[41]中记载的那样,当他 12 岁时,他就有了一台维多利亚时代的鼓式显微镜。当时,他家住在一所房子里,地窖里有一英尺深的污水,作为一个对细菌充满好奇的小伙子,他采样了一些有气味的混合物,将其放在盖玻片上,然后将其固定在加拿大香脂中,他把载玻片带给他的家庭医生,他的家庭医生有一台蔡司显微镜,他在查看时发现存在螺旋藻。他的父亲向他介绍了显微镜,并为 Frithjof 提供了一个全新世界的窗口。

Frithjof 善于表达、机智、博学、敏锐、自信,他将自己沉浸在任何适合他兴趣的知识中。由于对微观世界的迷恋,他对硅藻产生了浓厚的兴趣,他在光学显微镜、样品和载玻片制备以及显微摄影方面具有的能力是他多次反复试验的结果,在研究硅藻的过程中,他的决心和奉献精神使他不断取得成果。

图 1.10　1982 年 Frithjof 和一个学生在佛罗里达州取样

Frithjof 看到硅藻研究涉及很多方面,野外工作和收集水样只是开始(见图 1.10),在显微镜下观察这些样本是一个相当大的考虑因素,Frithjof 很快了解到处理样品和材料预处理之间的联系,以最大限度地减少硅藻的损坏或损失。无论是使用水样、云母碎片还是干燥材料,Frithjof 都了解各种最合适的处理方案[24]。

Frithjof 制订了详细的装裱工艺规程,以达到最佳效果。他对制作和观看具有形态学意义的硅藻载玻片的指导非常明确。正如 Frithjof 会提醒我们的那样,"千万不要在载玻片上制备,一定要在盖玻片上制备,在载玻片上制备会导致图像质量严重下降。"[42-43]。

Frithjof 在显微镜方面的专业知识是首屈一指的[42-43],并且他对基本、正确使用光学显微镜的概念具有深刻的指导意义[46]。在与 Paul Hamilton 的电子邮件中,Frithjof 甚至担心显微镜臂的位置(即面向或远离用户),因为这会影响如何读取载玻片位置刻度。他对不同物镜的成像质量进行了比较[47-48]，并且他一直对光学感兴趣[47-49]，包括衡量硅藻的硅化厚度以及镶嵌介质的折射率对图像分辨率的影响。Frithjof 总结了显微镜光学的一些"重点"，包括相位对比[50]和差分干涉对比或诺马斯基棱镜[50]：

（1）通过观察物镜，检查你的显微镜是否有"无限校正"物镜，或者校正了 160 mm 镜筒长的物镜，它要么提到"160"，要么给出无穷大符号。你不能将在这方面不同的物镜组合在一起。

（2）通常相衬物镜提供更好的对比度，但分辨率较低。

（3）平面目标与硅藻无关，它们给出边缘的平面图像，但硅藻通常只覆盖图像的中心。还有硅藻本身不是平的。

（4）超过数值孔径（NA）为 1.0 的冷凝器必须始终用浸渍油涂在载玻片上。这是凌乱和不方便的，所以 90% 的人都使用"干"冷凝器。在这种情况下，它的有效 NA 永远不会大于 1.0，即使规范说它的 NA 为 1.4。

（5）此外，对比度将随有效聚光器 NA 的增加而降低，低对比度硅藻的最佳选择是 DIC，但这需要特殊的光学器件（物镜和聚光器），而且价格昂贵。

（6）如果你也能做扫描电镜，则不必追求尽可能高的分辨率。如果没有扫描电镜，许多硅藻属甚至无法识别。

（7）总而言之，除非你的工作旨在对困难且结构非常精细的分类群进行关键的分类分离，否则我会推荐一个好的现代 NA 为 1.3 目标。如果你追求极致，扫描电镜支持的数字图像相关技术（DIC）是最好的选择[51]。

Frithjof 在理解显微镜光学和显微镜学方面非常有条理。照明是显微技术的一个重要因素，Frithjof 认为环形倾斜照明是获得相位对比效果的一种方式[44,50,52]。Frithjof 处于照明知识的最前沿，他测试了具有各种显微镜光学装置的发光二极管（LED）的实用性，并报道了自 2004 年以来的案例：

> 从那时起，出现了更合适的 LED。我已经测试了一些 1 W（大约 3.5 V，0.3 A）或更小的白色 LED，它的发光面积约为 3 平方毫米。对于 Köhler 照明，如果显微镜聚光镜有一个大底透镜，这可能太小了；使用较小直径的 Leitz 聚光镜镜头，包括 Heine，就足够了。但是，不必使用 Köhler 照明，临界照明就相当令人满意。在这种情况下，只需拧下顶部镜头，即可使用非常低的功率而不会出现任何问题。与文章中提到的相反，最高 NA 的物镜也可以毫无问题地使用。即使是小型 LED 与良好的收集器相结合，也太亮，无法在简单的明场中以全功率视觉使用，至于视场的不均匀照明，现在可以通过使用允许背景减法的图像处理软件来简单地补救[53]。

正确的光学元件和适当的照明使光学显微镜学家能够发展使用硅藻支架在分辨率极限下进行观察的技能。Frithjof 精通这门手艺，正如他所说，显微镜用户会熟悉宽角斜纹藻或菱形肋缝藻之类的好听的名字,是因为它们已经成为检查镜片质量的标准测试对象,尤其是检查显微镜专业知识的质量[34],以及硅藻通常用于测试物镜的质量,甚至更多用于测试显微镜学家的专业知识[48]。

作为他对显微镜的兴趣和能力的自然结果,Frithjof 和 Peter Höbel[54] 阐明了当材料不可用于扫描电镜时,使用紫外线或蓝光显微照相技术获得清晰的 3D 效果的技术[54]。Frithjof 与 Hamilton 和 Williams[55] 共同开发了另一种技术,可以在不考虑显微镜品牌的情况下在显微镜载玻片上定位标本,并且该技术避免了在标本周围进行标记或墨迹圈或购买 England Finder 的需要[55-56]。他利用自己的知识和创造性思维设计了清晰观察硅藻所需的设备、显微镜镜头[46,57] 或照明器[58]。他热衷于技术改进,并将他积累的知识的精细细节应用于显微镜和载玻片制备技术,以进行示例性硅藻分类学研究。Frithjof 认为,分类学谜题是他迎接科学侦探挑战的机会,因为在寻找谜题的过程中会遇到各种曲折[36]。

1.6 硅藻业余爱好者作为同行评审

Frithjof 善于利用他的众多才能解决问题,他注意到过去的业余爱好者在硅藻研究中起到的作用,并且了解显微镜的出现和人们对硅藻的兴趣交织在一起的历史。硅藻成为他的特别兴趣,是因为他看到 1853 年 William Smith 牧师对 *Gyrosigma tenuissimum* 的描述后,硅藻成了他的特别兴趣所在。有待解决的难题——最初的描述说瓣膜的侧面很深,其他描述说它有直边。几十年来,Smith 的材料都无法获得,但当 Frithjof 凭借敏锐的眼光、敏捷的头脑和技术能力,找到了与 Smith 最初描述相反的直边瓣膜,解决了这个难题。发现之路充满曲折,但与 Myriam de Haan 和 Wulf Herwig 一起,出现了一个"幸运突破"[59],以便解决这个难题。因此,Frithjof 对硅藻研究着迷。

Frithjof 于 20 世纪 70 年代首次发表有关硅藻的文章,他在推进硅藻分类学和许多分类群的命名研究方面发挥了重要作用[60]。在来自 AlgaeBase[61] 的一个表格中,Frithjof 单独或与同事命名了 39 个物种(和/或变种):16 种布纹藻、8 种斜纹藻、8 种海氏藻、2 种肋缝藻(与 Horst Lange-Bertalot)[62]、2 种偏缝藻、1 种菱形藻、1 种双菱藻、1 种岩网藻等[63]。许多其他分类群都经过 Frithjof 的仔细评估。作为 Frithjof 的证明,*Gyrosigma sterrenburgii* Stidolph、*Pleurosigma sterrenburgii* Stidolph、*Neidium sterrenburgii* Metzeltin 和 Lange-Bertalot 以及 *Rhopalodia sterrenburgii* Krammer 都是以他的名字命名[61]。

Frithjof 在硅藻分类学和命名法方面最突出的贡献是他在布纹藻和斜纹藻方面的工作,其独自[20,64-76] 或者合作[77-87] 发表了许多文章。正如 Paul Hamilton 所观察到的那样:Frithjof 的工作彻底而精确,他对细节的关注至关重要,值得在文章发表之前花时间来进行研究。Frithjof 命名和纠正分类错误以及设计新的形态特征以帮助识别布纹藻属和斜纹藻属[67,88]。

在Frithjof发表的关于2012年关岛中型斜纹藻的文章中,可以看出他在硅藻分类学方面的洞察力,他指出:

在LM中,*P. intermedium*具有更宽的顶点和(箭头)"距",在关岛标本中不存在。该距对应于扫描电镜中的附属裂缝,在扫描电镜中,*P. intermedium*有一排内部边缘的复极晕,围绕着顶点延续。在关岛标本中,顶端边缘没有穿孔。在*P. intermedium*中,内部中央中缝结节被一个相当宽的非相关区域包围,在关岛标本中,复极晕一直延伸到中央结节,还要注意中央内缝裂末端的形状差异。这些差异在形态发生上是完全独立的,因此分化是基于多形性的特征。关岛标本似乎与*P. acus* Mann 1925的类型不匹配,因为这显示了不同的条纹密度比[89]。

Frithjof在两篇文章中[90-91]对*Haslea brittanica*发表评论,表达了他对正确命名法与分类学问题不同的惊愕:

谢谢你对我关于*Haslea brittannica*几个问题的回答,所有答复者都提到了《硅藻图像学》,第7卷(2000年)。我有两个问题,一是有一个光学显微镜(LM)图,显示垂直排列的条纹;二是这不是来自材料类型。在海氏藻中可以看到垂直排列的条纹,但在非海氏藻中也可以看到,例如Hustedt的Naviculae orthostichae部分。是否不需要扫描电镜来验证在LM中不可见的通用字符的存在?通用转移不需要在类型材料中验证吗?可能有相似之处[90]。虽然我同意海氏藻是可能的,以及这种情况下的扫描电镜,转移是否在形式上有效。

总结我收到的许多回复:根据守则,转让在形式上有效(只需要引用基础名和参考文献),但它在分类学上是否正确是完全不同的事情(类型调查,扫描电镜)[91]。

海氏藻与布纹藻和斜纹藻的比较研究包括确定形态特征和分离这些硅藻属的明确描述和论据[92]。

有时,命名法的修订并没有进入想要鉴定硅藻标本的专业人士和业余爱好者的通用术语。Frithjof总是随时查看分类群的正确命名,并在出版物和网络上提供此类修订[48]。将*Surirella patrimonii* Sterrenburg转移到*Petrodictyon*就是一个很好的例子,它说明了Frithjof能够将相衬显微调查与类型材料检查相结合,以命名适当的属,并随后在印度尼西亚苏拉威西海岸的海洋沿岸生境中鉴定该分类单元[93]。

Frithjof非常清楚在硅藻类群的命名[94]和分类群名称中未解决的问题,命名分类群取决于一个人的使用显微镜的能力、玻片制备技术以及在检查硅藻时能够看到精细结构细节的鉴别力。Frithjof 会煞费苦心地通过这些步骤来产生硅藻分类单元的有效命名[74]。

他与同事 Stuart Stidolph、Kathryn Smith 和 Alexandra Kraberg 在一个项目上的工作最终出现在了美国地质调查局出版的 Stidolph 地图集上，该地图集包含 400 多种海洋沿岸硅藻物种的 1 000 张显微照片[95-96]。与往常一样，Frithjof 确保分类和命名是最新和正确的。Frithjof 认为向硅藻界提供地图集和其他资源很有价值，因为他知道关于硅藻的大型汇编很少[97]，而且文献往往是分散的，有时很难获得。

Frithjof 认为，良好的分类的结果是"消除已建立的错误的分类"，但关于引入新属的"长期更新状态"已经引起了大量的命名变化。分类学和命名法变化的文档散布在整个文献中，他提倡使用基于网络的方法来收集此类信息并使所有人都可以轻松访问[98]。

Frithjof 曾是荷兰生态研究所河口和海岸生态中心的研究员，他认识到生态研究的好坏取决于分类群名称的合理性，他了解分类学研究的多用途价值，并参与了来自不同地区的海草附生硅藻群的研究[99]。他与父亲 F. J. G. Sterrenburg 一起发表了关于海洋沿岸分类群的论文，包括直丝舟形藻[100]和新物种尼氏菱形藻的命名[101]。硅藻群是世界性的。然而，通过确定此前在毛里塔尼亚海岸附近的潮间带海草栖息地未发现的硅藻群的特定名称，使 Frithjof 和他的父亲将生态特征描述影响的生态[101]。

Frithjof 和 Michael Stringer 对研究英国埃塞克斯两树岛上一系列海洋潮汐泻湖中的硅藻种群动态很感兴趣，他们发现当时硅藻的数量急剧减少，然而到 2010 年冬天，硅藻又繁盛起来，Frithjof 和 Michael 使用样本的载玻片记录了这些变化，Frithjof 确定了存在的分类群。Frithjof 说，在如此寒冷的条件下发现硅藻并非闻所未闻，因为许多硅藻物种属于"嗜冷极端生物"[102]，Frithjof 和他的同事将重新审视硅藻的环境耐受性。

关于物种命名和生物多样性研究的更大的问题总是在 Frithjof 的脑海中，即环境条件会影响硅藻形态。他观察到瓣膜中的畸形，包括以前未检测到的不对称、异常弯曲、边缘扭曲和不规则条纹图案，他想知道这对物种命名有什么影响[103]。环境对硅藻多样性以及生物多样性的影响是学术圈和政界的一个普遍话题，他非常清楚硅藻分类学对确定硅藻的现状和确定这些分类群可能受到的威胁产生持久影响[104]。

在许多方面，Frithjof 对分类学决定的细节和复杂性的关注是他作为硅藻学家的专长，他指出，

> 人们普遍存在一种误解，认为分类学包括对新物种或新变种的描述，认识到分类学可能必须从消除已建立但无效的"伪物种"开始，这可能会让不专攻该学科的工作者感到惊讶。同理，只有明确地确定了有机体的特性，才有可能确定其生物学特征，如生理学、营养需求、栖息地和生物地理学。只有知道了这些，有机体才能成为了解地球历史信息的来源。硅藻尤其如此，其硅

质外骨骼在分类学上具有丰富的信息和持久性，可以得出数百万年的结论[104]。

对 Frithjof 来说，硅藻细胞学[106]和形态发生与分类学、命名法和生态学一样有趣。他对作为布纹藻中的形态发生特征的缝裂偏转和手性进行了研究[105]。与 Mary Ann Tiffany 和 María Esther Meave del Castillo 一起，Frithjof 参与了斜纹藻的瓣膜形态发生研究[83,106]，目的是确定瓣膜在内层和外层中具有纳米结构的椭圆形柱，这些柱夹在一起并且在结构上没有腔室。Frithjof 看到了二氧化硅沉积在硅藻壳构造及其与形态学的联系方面的重要性，他认为这一点应在分类学中得到反映。

Frithjof 进一步探讨了瓣膜的形态发生及其与结构形态的关系，由于对硅藻二氧化硅壳及其形态的"工程"感兴趣，Frithjof 产生了"晶体宫殿"及其在纳米技术中的作用的想法[107]。由于硅藻研究与使用技术进步，特别是关于 SEM，其可以详细研究硅藻在二氧化硅纳米结构形成中的结构、功能和形态发生。Frithjof 与其同事 Tiffany 和 Meave del Castillo 继续研究斜纹藻瓣膜形态发生以记录瓣膜形成的步骤[108]，他们还发现布纹藻属瓣膜形成具有没有位点的三明治结构，尽管与斜纹藻属略有不同，并且发现这些结构比其他硅藻的位点纳米结构弱[108]。Frithjof 和其同事发现斜纹藻属和潜在的布纹藻属，在相邻的柱子之间构建了"减速带"，以弥补没有位点设计的缺点[108]。对硅藻形态发生和模式形成的研究表明，Frithjof 有能力超越他已经了解的研究领域以及他如何继续挑战自己。Frithjof、Richard Gordon、Kenneth Sandhage 是 2005 年出版的《纳米科学与纳米技术》硅藻纳米技术专刊的主编，他的纳米结构成果也出现在那里[107,109-110]。

为了进一步采用"构造范式"，Frithjof、Richard Gordon、Mary Ann Tiffany 和 Stephen Nagy 为《天体生物学》撰写了一章（由 Joseph Seckbach 编辑），探讨硅藻的构造形态及其对硅藻是不是极端微生物的影响[111]。Frithjof 及其同事描述了各种栖息地，包括温泉、泥炭沼泽、高盐度或污染水域，以及海洋沿岸地区的泥滩，并想知道在不太极端的栖息地中也发现的此类物种是否可以被称为极端微生物。从其几何设计的纳米结构水平来看，硅藻可以承受影响其硅藻壳的各种物理和化学环境，硅藻表现出严格的形态发生控制和物种内的恒定性。Frithjof 知道这种"构造范式"仍然不足以解释硅藻的形态多样性，但这种多样性证明了硅藻在各种栖息地中的生存，尽管如此，此时 Frithjof 及其同事仍不愿将硅藻称为极端微生物[111]。

从所涵盖的各种学科及其相互关系中，Frithjof 展示了他作为硅藻业余爱好者的发展和成长，他为显微镜、样品和载玻片制备方法、显微照相技术做出了贡献，并阐述了各种生物领域的复杂性和相互关系以及它们如何影响硅藻研究。毫无疑问，通过对他最喜欢的硅藻属（布纹藻属和斜纹藻属）的研究，他产生了最直接和持久的影响，在他的硅藻研究方法中，他有能力看到问题的本质、从事此类研究的基本原理以及如何寻找解决方案。Frithjof 为什么是业余爱好者在硅藻研究中很重要的缩影，来源于他雄辩的口才、迷人的个性和和蔼可亲的品格，使 Frithjof 不仅是一位独特的贡献者,而且还超越了"硅藻业余爱好者"的称号,成为一名教师、导师、作家和硅藻研究专家。正如 Frithjof 所说,硅藻分类学是一门专业[48],特别是对于像他那样成为认真的研究硅藻的人。

1.7 结束语

Frithjof 从一个完美的业余爱好者变成了一个专业人士,无论他是把你当作硅藻研究的同事、导师、朋友还是熟人,他留下来的财富是多方面的,他既了解显微镜应用以及载玻片制备和显微摄影的复杂性,又理解并支持分类学和解决命名问题的重要性,以及这些研究如何为硅藻生态学、生物地层学和细胞学研究提供合法性,他一直致力于硅藻研究,来源于他的手稿和参与的尚未发表的成果[112-116]。具体而言,Frithjof 被称为布纹藻属和斜纹藻属的权威来源。

为了所有研究硅藻的人的利益,Frithjof 的收藏品已捐赠给德国不来梅港的弗里德里希赫斯特硅藻研究中心的阿尔弗雷德韦格纳研究所。正如 José 所说,60 年来,他是一位非常亲爱、充满爱心、始终开朗和乐观的丈夫。按照 Frithjof 的意愿,他的骨灰现在安息在海底的硅藻中。Frithjof 总是说他的墓志铭应该写成"永不沉闷的时刻",他的愿望得到了满足。正如 Hein de Wolf 在文章中所说的那样[117],来自 José 的消息:

永不沉闷,这就是 Frithjof 希望人们记住的。
恶性肿瘤袭击了他聪明的的大脑。
3 月 11 日,结束了他幸福而有价值的生活。
但最重要的是:他带着满足的心情离开了人世(他自己的话)。

José Sterrenburg

我们可以毫不含糊地说,因为认识 Frithjof A. S. Sterrenburg,我们的精神世界都更加富足。我们想念你并感谢你的工作。

参考文献

[1] Bahls, L. L. (2015). The role of amateurs in modern diatom research, Diatom Research, 30(2), 209-210.
[2] Werner, D. (ed.) (1977). The Biology of Diatoms, Botanical Monographs, Vol. 13. University of California Press, Berkeley and New York.
[3] Pedrotti, P. W. (2016). Thonis Philipszoon, "Antonj van Leeuwenhoek" 1632-1723 A.D, http://www.vanleeuwenhoek.com/, accessed on 2 December 2016.
[4] Sterrenburg, F. A. S. (1982). Anton van Leeuwenhoek: Pioneer or Loner (in Dutch, Organorama, 19(2).
[5] Sterrenburg, F. A. S. (1990a). The quest for Quekett—in search of Navicula angulata Quekett 1848, Microscopy, 468.
[6] Lynk, H. (2016). A Cabinet of Curiosities: A Selection of Antique Microscope Slides from the Victorian Era c. 1830s - 1900, http://www.victorianmicroscopeslides.com/history.htm, accessed on 28 November 2016.

[7] Walker, D. (2009). Enjoying a Möller 80 form diatom type-slide with a microphotograph setting, http://www.microscopy-uk.org.uk/mag/indexmag.html? http://www.microscopy-uk.org.uk/mag/artjan09/dw-moller.html.
[8] Stevenson, B. (2009). William Gatrell (1864-1902), Victorian Era Microscope Specimen Mounter, http://www.microscopy-uk.org.uk/mag/indexmag.html? http://www.microscopyuk.org.uk/mag/artjan09/bs-gatrell.html.
[9] Stevenson, B. (2013). John thomas redmayne, 1846-1880, http://microscopist.net/RedmayneJT.html.
[10] Walker, D. (2012). The diatomist John Albert Long (1863-1945*): Notes on aspects of his life and work with examples of his prepared slides, http://www.microscopy-uk.org.uk/mag/artdec11/dw-long.html.
[11] Taylor, J.E. (ed.) (1885). Hardwicke's science-gossip; an illustrated medium of interchange and gossip for students and lovers of nature, Vol. XXI. Chatto and Windus, Piccadilly, London.
[12] JStorGlobal Plants, Jacques-Joseph Brun. (1826-1908). http://plants.jstor.org/stable/10.5555/al.ap.person.bm000392757, accessed on 2 December 2016.
[13] Schmidt, A., Schmidt, M., Fricke, F., Heiden, H., Müller, O., Hustedt, F. (1874-1959). Atlas der Diatomaceenkunde. Aschersleben, Leipzig.
[14] Alfred Wegener Institute. (2015). FriedrichHustedt, http://www.awi.de/en/science/biosciences/polar-biological-oceanography/main-research-focus/hustedt-diatom-study-centre/fried rich-hustedt.html.
[15] Sterrenburg, F.A.S. (1967). Fifty Years of Jazz Records (in Dutch). Stichting IVIO, Amsterdam.
[16] Sterrenburg, F.A.S. (1983a). IRAS-Mission Invisible, Astronomy, 11, 66.
[17] Sterrenburg, F.A.S. (1983b). A Phased Approach to Astronomy, Astronomy, 11(6), 24.
[18] Sterrenburg, F.A.S. (1979). Report on medical electronics (in Dutch). Kluwer, Deventer.
[19] Sterrenburg, F.A.S. (1970a). Receivers. Instructions for the Advanced Amateur on the Art of Receivers, Amplifiers, Antennas, Measuring and the Like, 1st edition (in Dutch), De Muiderkring, Bussum, (2nd through 5th editions (in Dutch) published after 1970 until 1980).
[20] Sterrenburg, F.A.S. (2007). Basionym of Gyrosigma scalprum, Diatom Research, 22(2), 495.
[21] Hargraves, P.E. (2016). Frithjof Sterrenburg, https://list.indiana.edu/sympa/arc/diatom-l.
[22] Sterrenburg, F.A.S. (1970b). Electronica en de Battle of Britain, Radio Bulletin, Aug, 1970, 321.
[23] Weaver, W., Sterrenburg, F.A.S. (1968). Science and imagination: a selection from the work of Warren Weaver (in Dutch). Wetenschappelijke Uitgeverij, Amsterdam.
[24] Sterrenburg, F.A.S. (Year unknown). How to prepare diatom samples, https://www.scribd.com/document/155323338/Clean-Diatoms, accessed on 16 December 2016.
[25] Sterrenburg, F.A.S. (2009a). Personal recollections of WW2, https://ww2aircraft.net/forum/search/133066/.
[26] Sterrenburg, F.A.S., Toonder, M. (1972). Letters. F.A.S. Sterrenburg to Marten Toonder (1912-2005).
[27] Sterrenburg, F.A.S. (1983c). Computers in medicine, Organorama, 20.
[28] Sterrenburg, F.A.S. (1976). Silence around a Nobel investigation (in Dutch), Organorama, 13(1).
[29] Sterrenburg, F.A.S. (1977). Harvest the Past (in German), Organorama, 14(3), 11.
[30] Hargraves, P.E. (2016). Frithjof Sterrenburg, https://list.indiana.edu/sympa/arc/diatom-l.
[31] Gordon, R. (2011). Cosmic Embryo #1: My Erdös Number Is 2i, http://www.science20.com/cosmic_embryo/cosmic_embryo_1_my_erd%C3March6s_number_2i.
[32] Kemp, K., Basson, P. (2016). Never a dull moment, https://list.indiana.edu/sympa/arc/diatom-l.
[33] Sterrenburg, F.A.S. (1990b). Diatom collections—legacy or legend? Diatom Research, 5(2).
[34] Sterrenburg, F.A.S. (2002a). A second look at some well-known test diatoms, http://www.microscopy uk.org.uk/mag/indcxmag.html? http://www.microscopy-uk.org.uk/mag/artjul02/fsdiatom.html.
[35] de Wolf, H., Sterrenburg, F.A.S. (2003). International Survey of Diatom Collections, http://home.planet.nl/~wolf0334/.
[36] Sterrenburg, F.A.S. (2011a). Pandora's box. The diatoms of Sullivant & Wormley 1859, http://www.microscopy-uk.org.uk/mag/indexmag.html? http://www.microscopy-uk.org.uk/mag/artsep11/fs-pandora.html.
[37] Sterrenburg, F.A.S., de Wolf, H. (1993). The Kinker collection: preliminary investigation, Quekett Journal of Microscopy, 37, 35.

[38] Sterrenburg, F. A. S., de Wolf, H. (2004). The Kinker diatom collection: discovery-exploration-exploitation. In: VII International Symposium "Cultural Heritage in Geosciences, Mining and Metallurgy: Libraries-Archives-Museums": "Museums and their collections", C. F. Winkler Prins & S. K. Donovan (eds.), pp. 253-260. 19-23 May 2003, Scripta Geologica Special Issue, Leiden, The Netherlands.
[39] de Wolf, H., Sterrenburg, F. A. S. (1993). The legacy of the Dutch diatomist J. Kinker (1823-1900), Quekett Journal of Microscopy 37.
[40] Robbrecht, E. (2007). Botanic Garden Meise History, Henri Van Heurck (Antwerp 1839-1909), http://www.plantentuinmeise.be/PUBLIC/GENERAL/HISTORY/vanheurck.php.
[41] Sterrenburg, F. A. S. (2014a). My favourite slide, http://www.microscopy-uk.org.uk/mag/artfeb14/fs-favourite.html.
[42] Sterrenburg, F. A. S. (2002b). Microscopy primer, http://www.microscopy-uk.org.uk/index.html? http://www.microscopy-uk.org.uk/primer/.
[43] Sterrenburg, F. A. S. (2006a). Cleaning diatom samples, http://www.microscopy-uk.org.uk/mag/indexmag.html?http://www.microscopy-uk.org.uk/mag/artaug06/fs-diatoms.html.
[44] Sterrenburg, F. A. S. (1975). Guidance on Microscopy (in Dutch). Kluwer, Deventer.
[45] Sterrenburg, F. A. S. (2013a). An imaging conundrum in diatoms, http://www.microscopy-uk.org.uk/mag/artfeb13/fs-diatom-conundrum.html.
[46] Sterrenburg, F. A. S. (2011b). Advanced Techniques for visualization of diatom structures? Micros. Today 19.
[47] Sterrenburg, F. A. S. (2012a). What price optics? Micscape Magazine, 198, http://www.microscopy-uk.org.uk/mag/artapr12/fs-optics.html
[48] Sterrenburg, F. A. S. (2012b). Diatoms and microscope optics—some thoughts, http://www.microscopy-uk.org.uk/mag/artjan13/fs-diatom-micro.html.
[49] Sterrenburg, F. A. S. (2009b). Zernike's colour phase-contrast, http://www.microscopy-uk.org.uk/mag/indexmag.html? http://www.microscopy-uk.org.uk/mag/artjul09/fs-phase.html.
[50] Sterrenburg, F. A. S. (1978). Enhancing the visibility of diatoms, Microscopy, 33(6), 384.
[51] Sterrenburg, F. A. S. (2012c). Objectives, https://list.indiana.edu/sympa/arc/diatom-l.
[52] Sterrenburg, F. A. S. (2010a). Extreme annular illumination, http://www.microscopy-uk.org.uk/mag/indexmag.html?http://www.microscopy-uk.org.uk/mag/artapr10/fs-pseudo-phase.html.
[53] Sterrenburg, F. A. S. (2009c). LED lighting, https://list.indiana.edu/sympa/arc/diatom-l.
[54] Höbel, P., Sterrenburg, F. A. S. (2011). UV photomicrography of diatoms, Diatom Research, 26(1-2).
[55] Sterrenburg, F. A. S., Hamilton, P., Williams, D. (2012). Universal coordinate method for locating light-microscope specimens, Diatom Research, 27(2).
[56] Hamilton, P. B., Williams, D. M., Sterrenburg, F. A. S. (2013). Some notes on locating specimens in the microscope, Diatom Research, 28(4).
[57] Sterrenburg, F. A. S. (2006b). A compilation of LOMO microscope resources on Micscape. Supplemental page to Micscape article (4: Optics, correction collar maintenance), http://www.microscopy-uk.org.uk/mag/artoct06/iw-JenaObjsup.html.
[58] Sterrenburg, F. A. S. (2002c). A heavy caliber microscope lamp, http://www.microscopy-uk.org.uk/mag/indexmag.html? http://www.microscopy-uk.org.uk/mag/artoct02/fslamp.html.
[59] Sterrenburg, F. A. S., de Haan, M., Herwig, W. E. (2014). A lucky break. How past vandalism favoured modern diatom research, www.microscopy-uk.org.uk/mag/artjul14/fs-lucky-break.docx.
[60] Sterrenburg, F. A. S. (1988). Observations on the genus Anorthoneis Grunow, Nova Hedwigia, 47(3-4), 363.
[61] Guiry, M. D., Guiry, G. M. (2017). AlgaeBase. World-wide electronic publication, National University of Ireland, Galway, http://www.algaebase.org.
[62] Lange-Bertalot, H., Sterrenburg, F. A. S. (2004). New Frustulia species (Bacillariophyceae) from fossil freshwater deposits in Florida, U.S.A, Nova Hedw., 78(3-4).
[63] Kooistra, W. H. C. F., Forlani, G., Sterrenburg, F. A. S., Stefano, M. D. (2004). Molecular phylogeny and morphology of the marine diatom Talaroneis posidoniae gen. et sp. nov. (Bacillariophyta) advocate the return of the Plagiogrammaceae to the pennate diatoms, Phycologia, 43(1).
[64] Sterrenburg, F. A. S. (1989). Studies on tube-dwelling Gyrosigma populations, Diatom Research, 4(1).
[65] Sterrenburg, F. A. S. (1990c). Studies on the genera Gyrosigma and Pleurosigma (Bacillariophyceae). A

new phenomenon: co-existence of dissimilar raphe structures in populations of several species. In: Ouvrage dedié la Mémoire du Professeur Henry Germain, M. Ricard (ed.). Koeltz Scientific Books, Königstein, Germany.

[66] Sterrenburg, F. A. S. (1991a). Studies on the genera Gyrosigma and Pleurosigma (Bacillariophyceae). The typus generis of Pleurosigma, some presumed varieties and imitative species, Botanica Marina, 34(6), 561.

[67] Sterrenburg, F. A. S. (1991b). Studies on the genera Gyrosigma and Pleurosigma (Bacillariophyceae). Light-microscopical criteria for taxonomy, Diatom Research 6(2), 367.

[68] Sterrenburg, F. A. S. (1992). Studies on the genera Gyrosigma and Pleurosigma (Bacillariophyceae). The type of the genus Gyrosigma and other Attenuati sensu Peragallo, Diatom Research, 7(1), 137.

[69] Sterrenburg, F. A. S. (1995a). Studies on the genera Gyrosigma and Pleurosigma (Bacillariophyceae). Gyrosigma acuminatum (Kützing) Rabenhorst, G. spenceri (Quekett) Griffith et Henfrey and G. rautenbachiae Cholnoky, Proceedings of the Academy of Natural Sciences of Philadelphia, 146, 467.

[70] Sterrenburg, F. A. S. (1995b). Studies on the genera Gyrosigma and Pleurosigma (Bacillariophyceae). Gyrosigma balticum (Ehrenberg) Rabenhorst, G. pensacolae sp. n. and simulacrum species, Botanica Marina, 38(1-6), 401.

[71] Sterrenburg, F. A. S. (1997). Studies on the genera Gyrosigma and Pleurosigma (Bacillariophyceae). Gyrosigma kutzingii (Grunow) Cleve and G. peisonis (Grunow) Hustedt, Proceedings of the Academy of Natural Sciences of Philadelphia, 148, 157.

[72] Sterrenburg, F. A. S. (2000). Studies on the genera Gyrosigma and Pleurosigma (Bacillariophyceae). Gyrosigma reversum (Gregory) Hendey and G. naja (Meister) Sterrenburg, nov. comb, Proceedings of the Academy of Natural Sciences of Philadelphia, 150, 301.

[73] Sterrenburg, F. A. S. (2001a). Studies on the genera Gyrosigma and Pleurosigma (Bacillariophyceae). The types of Shadbolt and related taxa, Proceedings Academy of Natural Sciences of Philadelphia, 151, 121.

[74] Sterrenburg, F. A. S. (2002d). Nulla vestigia retrorsum. The case of Pleurosigma aequatoriale Cleve, Constancea, 83(1), 14.

[75] Sterrenburg, F. A. S. (2003a). Studies on the diatom genera Gyrosigma and Pleurosigma (Bacillariophyceae). Pleurosigma strigosum W. Smith and some presumptive relatives. Celebrating Norman I. Hendey's Centennial, Micropaleontology, 49(2).

[76] Sterrenburg, F. A. S. (2003b). Studies on the genera Gyrosigma and Pleurosigma (Bacillariophyceae). Pleurosigma obscurum W. Smith revisited, Diatom Research, 18(2), 323.

[77] Jahn, R., Sterrenburg, F. A. S. (2003). Gyrosigma sinense (Ehrenberg) desikachary: typification and emended species description, Diatom Research, 18(1).

[78] Jahn, R., Sterrenburg, F. A. S., Kusber, W.-H. (2005). Typification and taxonomy of Gyrosigma fasciola (Ehrenberg) J W. Griffith et Henfrey, Diatom Research, 20(2).

[79] Sterrenburg, F. A. S., de Wolf, H. (2004). The Kinker diatom collection: discovery-exploration-exploitation. In: VII International Symposium "Cultural Heritage in Geosciences, Mining and Metallurgy: Libraries-Archives-Museums": "Museums and their collections", C. F. Winkler Prins &. S. K. Donovan (eds.), pp.253-260. 19-23 May 2003, Scripta Geologica Special Issue, Leiden, The Netherlands.

[80] Sterrenburg, F. A. S., Underwood, G. J. C. (1997). Studies on the genera Gyrosigma and Pleurosigma (Bacillariophyceae). The marine "Gyrosigma spenceri" records: Gyrosigma limosum Sterrenburg et Underwood nov. sp, Proceedings Academy of Natural Sciences of Philadelphia, 148, 165.

[81] Sterrenburg, F. A. S., Tiffany, M. A., Lange, C. B. (2000). Studies on the genera Gyrosigma and Pleurosigma (Bacillariophyceae). Species from the Salton Sea, California, USA, including Pleurosigma ambrosianum, nov sp, Proceedings of the Academy of Natural Sciences of Philadelphia, 150, 305.

[82] Sterrenburg, F. A. S., Meave del Castillo, M. E., Tiffany, M. A. (2003a). Studies on the genera Gyrosigma and Pleurosigma (Bacillariophyceae). Pleurosigma species in the plankton from the Pacific coast of Mexico, with the description of P. gracilitatis sp nov, Cryptogam. Algol., 24(4), 291.

[83] Sterrenburg, F. A. S., Meave del Castillo, M. E., Tiffany, M. A. (2003b). Valve morphogenesis in the diatom genus Pleurosigma W. Smith (Bacillariophyceae): an engineering paradigm. In NADS2003. E Gaiser (ed). North American Diatom Society (abstract).

[84] Stevenson, B. (2013). John thomas redmayne, 1846-1880, http://microscopist.net/Redmayne JT.

html.
- [85] Sar, E. A., Hinz, F., Sterrenburg, F. A. S., Lavigne, A. S., Lofeudo, S., Sunesen, I. (2012). Species of Pleurosigma (Pleurosigmataceae) with lanceolate or slightly sigmoid valve outlines: analysis of type material, Diatom Research, 27(4).
- [86] Sar, E. A., Sterrenburg, F. A. S., Lavigne, A. S., Sunesen, I. (2013). Diatoms from marine coastal environments of Argentina. Species of the genus Pleurosigma (Pleurosigmataceae, Boletin De La Sociedad Argentina De Botanica 48(1).
- [87] Liu, B., Sterrenburg, F. A. S., Huang, B. (2015). Gyrosigma xiamenense sp. nov. (Bacillariophyta) from the middle intertidal zone, Xiamen Bay, southern China, Phytotaxa, 222(4).
- [88] Sterrenburg, F. A. S. (1993b). Studies on the genera Gyrosigma and Pleurosigma (Bacillariophyceae). Rules controlling raphe fissure morphogenesis in Gyrosigma, Diatom Research, 8(2).
- [89] Sterrenburg, F. A. S. (2012d). Guam diatoms, https://list.indiana.edu/sympa/arc/diatom-l.
- [90] Sterrenburg, F. A. S. (2014b). Haslea brittanica, https://list.indiana.edu/sympa/arc/diatom-l.
- [91] Sterrenburg, F. A. S. (2014c). Haslea brittanica (follow-up message), https://list.indiana.edu/sympa/arc/diatom-l.
- [92] Sterrenburg, F. A. S., Tiffany, M. A., Hinz, F., Herwig, W. E., Hargraves, P. E. (2015). Seven new species expand the morphological spectrum of Haslea. A comparison with Gyrosigma and Pleurosigma (Bacillariophyta), Phytotaxa, 207(2), 143.
- [93] Sterrenburg, F. A. S. (2001b). Transfer of Surirella patrimonii Sterrenburg to the genus Petrodictyon, Diatom Research, 16(1).
- [94] Sterrenburg, F. A. S. (2010b). Nitzschia singalensis vs N. firthii. What's in a diatom name? http://www.microscopy-uk.org.uk/mag/indexmag.html? http://www.microscopy-uk.org.uk/mag/artapr10/fs-nitzschia.html.
- [95] Sterrenburg, F. A. S. (2012e). Stidolph Diatom Atlas, https://list.indiana.edu/sympa/arc/diatom-l.
- [96] Stidolph, S. R., Sterrenburg, F. A. S., Smith, K. E. L., Kraberg, A., Stuart, R. (2012). Stidolph diatom atlas, U.S. Geological Survey Open-File Report 2012-1163, http://pubs.usgs.gov/of/2012/1163/.
- [97] Sterrenburg, F. A. S. (2013b). Book review: An introduction to the microscopical study of diatoms. Robert B. McLaughlin. Edited by John Gustav Delly & Steve Gill, http://www.microscopy-uk.org.uk/mag/artfeb13/fs-review.html.
- [98] Sterrenburg, F. A. S. (2002e). Taxonomy and the web? Diatom Research, 17(2).
- [99] Sterrenburg, F. A. S., Erftemeijer, P. L. A., Nienhuis, P. H. (1995). Diatoms as epiphytes on seagrasses in South Sulawesi (Indonesia). Comparison with growth on inert substrata, Botanica Marina, 38(1).
- [100] Sterrenburg, F. A. S., Sterrenburg, F. J. G. (1991). Studies on a widely distributed marine littoral diatom-Navicula orthoneoides Hust (Bacillariophyceae, Nova Hedwigia, 52(3-4), 411.
- [101] Sterrenburg, F. A. S., Sterrenburg, F. J. G. (1990). An outline of the marine littoral diatom biocoenosis of the Banc-d'Arguin, Mauritania, West Africa, Botanica Marina, 33(5), 459.
- [102] Stringer, M. J., Sterrenburg, F. A. S. (2011). Phoenix rising: eclipse and resurrection of a diatom sanctuary, http://www.microscopy-uk.org.uk/mag/indexmag.html? http://www.microscopyuk.org.uk/mag/artjan11/ms-fs-phoenix.html.
- [103] Sterrenburg, F. A. S. (1973). Extreme malformation and the notion of species, Microscopy, 32.
- [104] Sterrenburg, F. A. S. (2005a). Taxonomy and ecology—an inseparable pair, Proceedings of the California Academy of Sciences, 56, 156.
- [105] Sterrenburg, F. A. S. (1993b). Studies on the genera Gyrosigma and Pleurosigma (Bacillariophyceae). Rules controlling raphe fissure morphogenesis in Gyrosigma, Diatom Research, 8(2).
- [106] Sterrenburg, F. A., Tiffany, M. A., del Castillo, M. E. (2005). Valve morphogenesis in the diatom genus Pleurosigma W. Smith (Bacillariophyceae): Nature's alternative sandwich, J. Nanosci. Nanotechnol., 5(1).
- [107] Sterrenburg, F. A. S. (2005b). Crystal palaces-diatoms for engineers, J. Nanosci. Nanotechnol., 5(1).
- [108] Sterrenburg, F. A., Tiffany, M. A., del Castillo, M. E. (2005). Valve morphogenesis in the diatom genus Pleurosigma W. Smith (Bacillariophyceae): Nature's alternative sandwich, J. Nanosci. Nanotechnol., 5(1).
- [109] Gordon, R., Kling, H. J., Sterrenburg, F. A. S. (2005a). A guide to the diatom literature for diatom nanotechnologists, J. Nanosci. Nanotechnol., 5(1).

[110] Gordon, R., Sterrenburg, F. A. S., Sandhage, K. H. (2005b). A Special Issue on Diatom Nanotechnology, J. Nanosci. Nanotech., 5(1).

[111] Sterrenburg, F. A. S., Gordon, R., Tiffany, M. A., Nagy, S. S. (2007). Diatoms: living in a constructal environment. In Algae and Cyanobacteria in Extreme Environments. Series: Cellular Origin, Life in Extreme Habitats and Astrobiology. Vol. 11, J. Seckbach (ed). Dordrecht, The Netherlands: Springer.

[112] Sterrenburg, F. A. S. (1996). Cytoplasmic inheritance in diatoms, unpublished manuscript.

[113] Sterrenburg, F. A. S. (2002f). Taxonomy-ecology-microscopy. A true triad, unpublished manuscript.

[114] Sterrenburg, F. A. S. (2004). Habitat specificity of diatoms, unpublished manuscript.

[115] Sterrenburg, F. A. S. (2013c). The search of fine detail: the history of diatom imaging. In: Diatoms are Forever: Growing your Nanotechnology, R Gordon, F. A. S Sterrenburg, M. A Tiffany, S. S. Nagy & I. C Gebeshuber (eds.), unpublished manuscript.

[116] Sterrenburg, F. A. S. (2013d). The naming of the beasts: making an inventory of the realm of diatoms. In Diatoms are Forever: Growing your Nanotechnology R. Gordon, F. A. S. Sterrenburg, M. A. Tiffany, S. S. Nagy and I. C. Gebeshuber (eds.), unpublished manuscript.

[117] de Wolf, H. (2016). Frithjof Sterrenburg, https://list.indiana.edu/sympa/arc/diatom-l.

第 2 章
纪念一位朋友
——Alex Volker Altenbach

Wladyslaw Altermann

Alexander Volker Altenbach 在长期患病后于 2015 年去世，在 Alexander 英年早逝近两年后，应另一位朋友 Joseph Seckbac 的邀请，我才勉强写下这篇讣告。我觉得，在两年后，在发表了纪念 Alex 的文章之后[1]，没有必要再写这篇讣告了，甚至可能会有失偏颇。但 Joseph 坚持要这样做，毕竟在 2006 年 Joseph 获得 DAAD 学术奖学金访问当时我所在的路德维希-马克西米利安-慕尼黑大学的系时(LMU)，是我介绍 Joseph 和 Alex 认识的，这种关系产生了一本有趣而受称赞的书，这本书由 Alex、Joan M Bernhard 和 Joseph 主编[2]。

Alex 生于 1953 年，离法兰克福很近，只比我大几个月，但在事业和科学方面要领先我很多。当我们于 1994 年在伦敦大学初次见面时，他刚刚成为古生物学的教授，并且已经因其在有孔虫方面的工作而享誉国际。我正忙于在太古宙叠层石的适应训练(DSc)，而 Alex 的适应训练程序即将完成，我们很容易找到了一种共同语言，很快，与伦敦大学的另一位年轻教授 Wolfgang Heckl 一起，发现了对前寒武纪生命和寻找太古宙生物圈调查新方法的共同兴趣，他是纳米技术专家，如今是慕尼黑著名的世界最大科技博物馆德国博物馆的馆长。

这是一个非常富有成效的时间，虽然 Alex 后来对太古代古生物学不太感兴趣，但他在我们的工作组中始终是一个令人振奋的角色和伟大思想的传递者。Alex 是一位影响深远的学者，他在计算、编程、统计方法以及海洋生物学和沉积学方面具有丰富的经验，他对所有学科都感兴趣，并深入参与了同事和学生的工作。他在法兰克福歌德大学学习地质学和古生物学，此后在丹麦的一个石油钻井平台和北海的油气勘探中获得了经验，但他也从事环境地质学，甚至参与了一家私人咨询公司和一家总部位于汉堡的地理软件公司。他后来主要在西班牙工作，在那里一直奋斗到

图 2.1　Alexander Volker Altenbach (1953—2015)

晚年。他于20世纪80年代加入基尔大学,在那里他对海洋和海底相互作用产生了兴趣,这是基尔大学新成立的海洋地球科学和地球物理研究中心的一个项目。他在1985年的博士论文是关于底栖有孔虫的生物量[3]和他在测量有孔虫有机碳和生物量方面的研究,这项工作是Alex[4]在国际上的一项突破,他花了数月时间研究热带和极地海域许多海岸的有孔虫与沉积物之间的关系,以及巴塔哥尼亚、纳米比亚和澳大利亚之间进行研究航行和活动,并就碳通量、稳定同位素分馏等各个地层时期的有孔虫发表了大量文章和生物量计算,其中一些研究集中在纳米比亚硅藻土带上[5-8]。

Alex于1995年获得了伦敦大学的教授职位,从1994年起,他就代替同样提前离职的微古生物学者Konrad Weidich教授任教,作为一名新员工,他立即开始主动联络,寻找新的合作伙伴。我们就是这样认识的,出乎意料的是,Alex没有等待邀请,而是在地球科学学院的各个研究所里,挨家挨户地敲门,一天深夜,他发现我正在写适应训练论文,他坐下来,自我介绍,开始提问和回答问题,后来,他向我介绍了他的研究以及他非常重视和喜爱的年轻同事和学生,如Ulli Struck、Christoph Mayr、Carola Leiter等。在LMU的研究所和地球科学学院的支持下,他始终关注他人的进步、学习和教学。

Alex的研究是由好奇心和发现传统思维之外事物的愿望驱动的,他的文章有时会引起争议,正如他自己写的那样,有些评论否定了他的新发现,因为它们与公认的模型相矛盾,促使他开展新的活动,并激发了更多关于这一主题的工作。因此,国际期刊拒绝了他关于在硫化条件下茁壮成长的有孔虫的手稿("……缺氧有孔虫似乎不合理……"),导致他写了一本关于真核生物缺氧策略的书[2]。在这本书中,我们共同发表了一篇关于缺氧和凝集有孔虫与太古代真核生物进化相关性的推测性文章,这更像是一个哲学概念,而不是科学工作和数据,但它写起来很有趣,当然也很有教育意义[9]。在同一本书中,Alex与他的妻子Maren发表了关于元古代有孔虫和氧化还原条件的文章[9],以及关于有孔虫中的碳和氮分馏的文章[5]。这本书收录了各学科领导者的优秀科学成果,如Tom Fenchel、Aharon Oren、Jürgen Schieber等,以及大量年轻研究人员,包括Alex及其同事的学生。

他建立了一个稳定同位素实验室,后来担任了学院院长,他认为这是一项必要且有趣的工作,但他讨厌官僚作风。当学院分裂并成立新部门时,他感到非常失望,所有这些都以不健康的竞争结束,取代了多学科合作,并导致同事之间的利己主义斗争。他退出学院,并得到了巴伐利亚州立收藏馆和路德维希·马克西米利安大学古生物学和历史地质学研究所的支持,在那儿他是一名热情的教授。他和他的许多学生以及他的妻子Maren一起工作。他与Maren一起出版了一本关于澳大利亚异国情调以及印度尼西亚和菲律宾蜥蜴和其他爬行动物的生态和行为的旅行指南[10-11]。不久,他病倒了,开始了与癌症的长期斗争。当时我已经在南非比勒陀利亚大学,但我们继续见面和合作,每次我们在慕尼黑见面时,我都能看到他的身体和健康发生了令人震惊的变化。Alex对自己要求很高,也是一名坚强的战士,他仍然尽可能多地在办公室里与他的研究人员和学生一起工作。他仍然梦想着和他的妻子 Maren 一起来看我,并调查我在野外为他拍摄的卡鲁和奥兰治河的大型蜥蜴,可惜这个梦想一直没有实现。Alex 在 2015 年 8 月 24 日去世,我们深深怀念他。

Wlady Altermann,比勒陀利亚,2017 年 7 月

参考文献

[1] Schiebel, R., Schönfeld, J., Struck, U. (2016). Memorial to Alexander Volker Altenbach (1953-2015), Journal of Foraminiferal Research, 46(1), 4-6.

[2] Altenbach, A. V., Bernhard, J. M., Seckbach, J. (eds.). (2012a). Anoxia: Evidence for Eukaryote Survival and Paleontological Strategies. Cellular Origin, Life in Extreme Habitats and Astrobiology (COLE). vol. 21, Dordrecht: Springer.

[3] Altenbach, A. V. (1985). Die Biomasse der benthischen Foraminiferen: Ph. D. Thesis Kiel University, 167.

[4] Altenbach, A. V. (1987). The measurement of organic carbon in foraminifera, The Journal of Foraminiferal Research, 17(2), 106-109.

[5] Altenbach, A. V., Leiter, C., Mayr, C., Struck, U., Hiss, M., Radic, A. (2012b). Carbon and nitrogen isotopic fractionation in foraminifera: Possible signatures from anoxia. In: Anoxia: Evidence for Eukaryote Survival and Paleontological Strategies. Cellular Origin, Life in Extreme Habitats and Astrobiology (COLE), Vol. 21, A. V. Altenbach, J. M. Bernhard & J Seckbach (eds.), pp. 518-535. Springer, Dordrecht.

[6] Leiter, C., Altenbach, A. V. (2010). Benthic foraminifera from the diatomaceous mud belt off namibia: Characteristic species for severe anoxia, Palaeontol. Electronica, 13(2).

[7] Schönfeld, J., Altenbach, A. V. (2005). Late Glacial to Recent distribution pattern of deep-water Uvigerina species in the north-eastern Atlantic, Mar. Micropaleontol., 57(1-2), 1-24.

[8] Struck, U., Altenbach, A. V., Emeis, K.-C., Alheit, J., Eichner, C., Schneider, R. (2002). Changes of the upwelling rates of nitrate preserved in the $\delta 15N$-signature of sediments and fish scales from the diatomaceous mud belt of Namibia, Geobios, 35(1), 3-11.

[9] Altenbach, A. V., Gaulke, M. (2012). Did redox conditions trigger test templates in Proterozoic Foraminifera? In: Anoxia: Evidence for Eukaryote Survival and Paleontological Strategies. Cellular Origin, Life in Extreme Habitats and Astrobiology (COLE), Vol. 21, A. V. Altenbach, J. M. Bernhard & J. Seckbach (eds.), pp.592-614. Springer, Dordrecht.

[10] Gaulke, M., Altenbach, A. V. (1994). Contribution to the knowledge of the snake fauna of Masbate (Philippines) (Squamata: Serpentes), Herpetozoa, 7, 63-66.

[11] Gaulke, M., Altenbach, A. V. (2007) Australia: Natur und TierVerlag Gmbh. p. 391. Münster, Germany.

第 3 章
硅藻之美

Mary Ann Tiffany，Stephen S. Nagy

3.1 硅藻观察的早期历史

自从首次使用显微镜以来，微观硅藻的美就已经被人们注意到。1703 年，一位伦敦皇家学会会员观察可能是平板藻属的东西时描述到，它们是"许多漂亮的树枝，由长方形和正方形组成"[1]。

自此，特别是在维多利亚时代，研究和绘制世界各地的硅藻成为欧洲国家一些绅士的流行时尚。这些早期的硅藻业余爱好者在许多聚会中制作了永久载玻片，现在被认为是珍贵的古董。示例参见 https://www.antiquemicroscopesandslides.com/diatoms。

用暗场照明并在低功率下观察布满载玻片的清洁硅藻壳可以显示出意想不到的颜色[2]：

> 在中低功率下用深色地面照明检查某些硅藻时，会注意到它们显示出最可爱的色调，像精致的蓝色或发光的彩虹色，例如，多种形式的辐环藻属、辐裥藻属、具粉被眼纹藻、梯楔藻属、鳞盘藻属、斜纹藻属和柄链藻属。这些颜色不存在于壳中，而是衍射的结果，或与穿过硅藻微小标记的光发生干涉，就像穿过衍射光栅一样，当标记在更高的能量下分解成点和球体时，它们就消失了。

显微镜制造商曾经为较低功率的物镜提供暗场聚光器，其光锥比现在的更窄。现代暗场聚光器通常具有 1.2～1.4 NA 的照明锥，设计用于中高功率消色差物镜。虽然它们很难定位，但 NA 为 0.7～0.85、0.8～0.95 的暗场聚光镜在搜索稀有形式的散布时特别有用，然后可以通过寻找它们的折射颜色将其定位在散布场中。最好的暗场/倾斜照明工具可能是 Leitz Heine 聚光镜，它有一个无限可变的照明锥，可以提供暗场到圆形的倾斜照明，并且可以针对每个单独的物镜进行调整。虽然这些较旧的暗场聚光镜可能是为与使用的框架或支架不同而制造的，但在大多数情况下，设计一个安装在不同于制造商支架上的适配器并不难。

Carpenter 在 1856 年观察到：

> 在另一组硅藻（被 Ehrenberg 和许多其他博物学家认为是微生物）中，不仅

植物的形态通常非常明显,而且它们的表面表现出非凡的美感和对称性,这些纹理是最好的测试对象,可以用来提高显微镜的能力。

此外,硅藻瓣膜以奇妙的几何设计进行了复杂的安排,这些需要极大的耐心和技巧。由于载玻片是商业销售的,硅藻贴片者对其方法保密,只有 Meakin 和 Swatman 于 20 世纪 30 年代发表的几篇详细介绍方法和技术的文献。Klaus Kemp 10 岁移民英国,16 岁开始在 Flatters and Garnett 工作,有幸被 Wilfred Garnett 聘用,负责文物保护部门,还收留他学习生物分类,他与 Gordon McKechne 一起检查了公司制作和拥有的载玻片,发现自己"惊讶于单个载玻片的精美程度"。当 Moeller 向他展示一份 400 张展览用的载玻片时,他决定"我也要做这个",因为他看到了手里拿着的载玻片上被光线照亮的小点。他在文物保护部门工作了多年,直到公司倒闭,公司解散后,Wilfred Garnett 授权他保存 Calloway 牧师归还的硅藻收藏品,Calloway 曾为公司制作硅藻幻灯片,但视力不佳导致 Calloway 放弃了这个收藏品,该收藏品实际上最初来自 J. A. Long。Klaus 负责保存收藏以供后代和个人使用,他一生都在重新发明收集、识别和排列硅藻的工艺,他继续重新发明制作展览载玻片的材料和方法,并复制了 Moeller 华丽的布置,在布置中替换了一种形式,以说明这是复制品,而不是 Moeller 制作的。此后,他循环排列了大约 1 500 份表格(见图 3.1),他还发现了新物种。

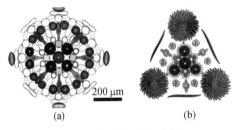

图 3.1　排列的硅藻幻灯片
(a) Brightfield 光学显微镜(LM);(b) 极化暗场图像

在现代,一些作者强调硅藻研究在生态和进化研究中的作用[3],但如果这些作者被质疑,他们总是会说硅藻迷人的美学吸引了他们。人眼欣赏几何设计的对称性和纯粹性,这是硅藻的显著特征。

3.2　活硅藻

活的硅藻具有内在的美感,完整的硅藻壳可以在普通的明场显微镜下成像。它们通常呈可爱的金黄色,质体中含有岩藻黄素和类胡萝卜素以及常见的黄绿素[4]。硅藻细胞通常比较厚。使用聚焦叠加技术和 Photoshop 在不同聚焦水平上拍摄 2～6 幅图像,可以克服光学显微镜固有的景深限制,在同一幅图像中同时显示瓣膜和叶绿体[见图 3.2(a)～(d)]。这种技术常用于昆虫和花卉的宏观摄影。差分干涉显微镜[DIC,见图 3.2(e)和(f)]清楚地显示了细胞器,尤其适用于轻微硅化物种的成像。将图 3.2(c)和(d)中的相同物种与图 3.2(e)和(f)进行比较,由环带和瓣膜组成的硅藻壳显示出比明场更多的对比度和细节。

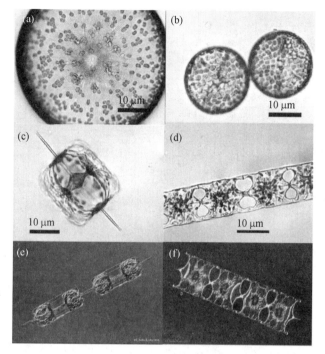

图 3.2 活硅藻

(a) 威氏圆筛藻;(b) 辐桐藻属;(c) 布氏双尾藻;(d) 活动齿状藻;(e) 布氏双尾藻,微分干涉对比度;(f) 活动齿状藻

3.3 形状和结构

硅藻有各种各样的形状和大小,从微小的小盘藻(直径约为 3 μm)到一些肉眼几乎可见的硅藻。扫描电镜可以对硅藻壳、瓣膜和环带进行三维观察。许多硅藻瓣膜简单但有针状的结构[见图 3.3(a)]。其他的则在不同的平面上起伏[见图 3.3(b)和(c)],例如,斜纹藻属和布纹藻属等是轻微的沙漏形[见图 3.3(d)]。一种特别不寻常的硅藻化石,来自新西兰奥马鲁的 Kittonia elaborata,其管状刺类似于所谓的火星天线[见图 3.3(e)]。另一种带刺的硅藻是布氏双尾藻,它也有一条玻璃条纹状的裙子[见图 3.3(f)]。华壮双菱藻的形状类似于正在建造的带有肋骨和龙骨的船[见图 3.3(g)]。双生双楔藻具有不寻常的石棺形状或经典的可口可乐瓶状[见图 3.3(h)]。膨突角状藻的重瓣具有扭曲的外观,带有大的单眼和刺[见图 3.3(i)]。硅藻瓣的几何形状通常是圆形或足球形,但有些物种是具有三边或更多边的多边形[见图 3.3(j)~(l)]。Adolph Schmidt 认为来自莫尔斯的直链藻属是最美丽的硅藻,但 Klaus Kemp 更喜欢胸隔藻属,因为它复杂的内部结构是该属独有的。

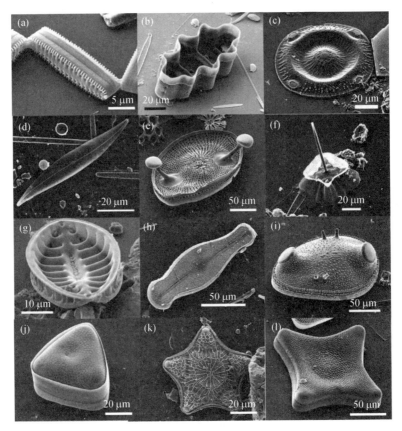

图 3.3　各种形状的硅藻 SEM 图

(a) 菱形海线藻；(b) 黏质哑铃藻；(c) *Glyphodiscus stellatus*；(d) 斜纹藻属；(e) *Kittonia elaborate*；(f) 布氏双尾藻；(g) 华壮双菱藻；(h) 双生双楔藻；(i) 膨突角状藻；(j) *Trigonium arcticum*；(k) 五角星型三角藻；(l) 四角变型三角藻

3.4　不同尺度下的硅藻之美

硅藻瓣膜有两个侧面，即内表面和外表面，有些甚至具有"典型"的培养皿结构，其中培养皿的侧面称为地幔，我们将使用蛛网藻作为例子，说明硅藻的两侧在不同尺度下的美丽，整个瓣膜的内部视图[见图 3.4(a)]显示的结构很像带有辐射肋的彩色玻璃"玫瑰"窗。两个异瓣膜壳之一具有中央径向狭缝[见图 3.4(b)和(d)]，另一个更简单但仍然有吸引力（未显示）。肋连接肋，形成蛛网状外观（因此得名）。在扫描电镜中，外部瓣膜没有显示肋，但可以有中央狭缝[见图 3.4(c)和(d)]。与瓣膜表面相比，覆盖层较陡，掌部较小（Round 等 1990 年使用的术语，指孔隙闭塞）[见图 3.4(e)]。在高倍率下，单个纹路非常精美，边缘有卷曲的狭缝和微小的球体。此外，瓣膜表面呈现出精细的网状装饰[见图 3.4(f)]。

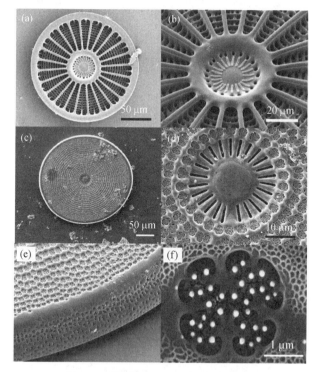

图 3.4　蛛网藻在不同尺度上的 SEM 图

(a) 整个瓣膜的内部视图；(b) 内部瓣膜的中心部分，注意径向肋的三维性；(c) 整个瓣膜的外部视图；(d) 带有径向狭缝的外部瓣膜的中心部分；(e) 外部瓣膜边缘视图；(f) 外部筛孔的高度放大视图，带有微小的硅球

3.5　形态发生过程中的瓣膜

　　硅藻有两个瓣膜和许多构成硅藻壳的环带。在无性分裂过程中，每个新的子细胞保留一个原始瓣膜并形成一个新瓣膜。作为此过程的一部分，二氧化硅以独特的图案添加到基底层[5-6]。成型瓣膜还没有完全补充二氧化硅，因此看起来与成熟的瓣膜不同，这些未完成的结构有自己的美。

　　星纹藻属是一个具有射线和精致星状图案的属[7]。通常在现代南方星纹藻中出现 7 条射线[见图 3.5(a)和(b)]，偶尔会看到 6 条射线[1]，当在扫描电镜中以高倍率观察不同阶段未成熟瓣膜的外表面时，可以发现复杂的结构[见图 3.5(c)～(f)]。从基底层产生的最早阶段是六边形壁图案，每个六边形的直径略小于 1 μm[5]。六边形可以在奇特的三角形结构下方被辨别出来[见图 3.5(c)和(d)]。随着硅化的进展和三角形的合并，设计变成花卉，这些可能会形成迷人的壁纸图案[见图 3.5(e)和(f)]。成熟瓣膜的表面就不那么有趣了，因为这些图案几乎完全被二氧化硅覆盖[7]。

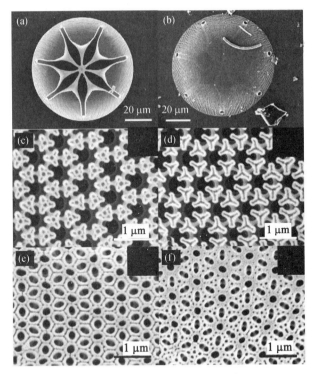

图 3.5　南方星纹藻未成熟瓣膜在增大阶段和高倍镜下的 SEM 图
(a) 整个成熟瓣膜的内部视图；(b) 未成熟瓣膜的外部视图；(c)～(f) 瓣膜成熟时的表面

3.6　Jamin-Lebedeff 干涉对比显微镜

使用专业的显微照明技术，可以制作出非常漂亮的硅藻图片，将暗场与偏振光技术相结合，可以在中等放大倍数下显示出美丽而微妙的干涉色。硅藻的相位对比图像可以使用数字编辑程序进行处理，以显示出折射和干涉的潜在色彩，但没有任何技术可以与使用 Jamin-Lebedeff 干涉对比设备和白光产生的图像相提并论。早在 20 世纪初，英国和德国的一些显微镜制造商就制造了这种设备，但这种设备经常处理不当导致无法使用或者很难使用。来自聚光器的偏振照明光束被聚光器中的方解石板分成两束平行光束，一束靠近硅藻，另一束穿过硅藻壳，两束光束在装有相同但相对的方解石板的物镜中重新组合，可以操纵所得图像以产生干涉色，干涉色根据被照射的硅藻部分的光路长度而变化。Nagy 利用该技术对硅藻进行了专门的成像。图 3.6 将三个物种的彩色 Jamin-Lebedeff 图像与使用扫描电镜获得的图像进行了对比，图 3.7 比较了另外三个物种。在两者中可以观察到相同的结构，扫描电镜图像显示了更多的 3D 效果，而 Jamin-Lebedeff 图像可以说更漂亮。

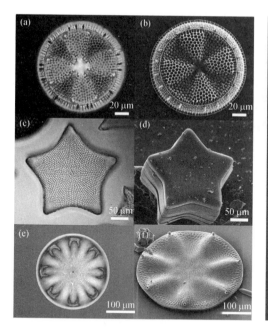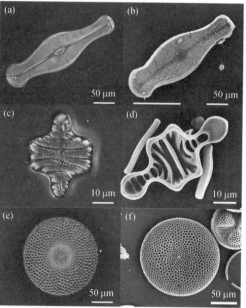

图 3.6 三种硅藻物种的 Jamin-Lebedeff 图像和 SEM 对比

(a)和(b) *Actinoptychus heliopelta*；(c) 和(d) *Trigonium graeffeanum*；(e) 和(f) *Aulacodiscus janischii*

图 3.7 另外三种硅藻物种的 Jamin-Lebedeff 图像和 SEM 对比

(a)和(b) *Didymosphenia geminata*；(c) 和(d) *Tetracyclus emarginatus*；(e) 和(f) *Coscinodiscus* sp

3.7 结论

硅藻的美丽肯定会继续吸引科学家和更多的人。尽管许多现代研究人员正在使用基因组技术来研究这些生物，但它们硅质外壳的美丽将继续引起人们的兴趣，并且很可能是它们的第一吸引力，也是最持久的吸引力。

参考文献

[1] Round, F. E., Crawford, R. M., Mann, D. G. (1990). The diatoms, biology & morphology of the genera. Cambridge University Press, New York.
[2] Taylor, F. B. (1929) Notes on Diatoms: An Introduction to the Study of the Diatomaceae. Bournemouth, UK.
[3] Cox, E. J. (2012). Ontogeny, homology, and terminology—wall morphogenesis as an aid to character recognition and character state definition for pennate diatom systematics, J. Phycol., 48(1), 1-31.
[4] Kuczynska, P., Jemiola-Rzeminska, M., Strzalka, K. (2015). Photosynthetic pigments in diatoms, Mar. Drugs, 13(9), 5847-5881.
[5] Tiffany, M. A. (2008). Valve development in Aulacodiscus, Diatom Research, 23(1), 185-212.
[6] Tiffany, M. A. (2015). Valve and girdle band morphogenesis in the pseudocellate diatom species Biddulphia biddulphiana J. E. Smith (Boyer) and Isthmia nervosa Kütz, Nova Hedwigia, Beiheft, 144, 61-95.
[7] Tiffany, M. A., Hernandez-Becerril, D. U. (2005). Valve morphogenesis in the diatom family Asterolampraceae H. L. Smith, Micropaleontology, 15, 217-258.

第 4 章
中国硅藻研究现状

YuXin Zhang

4.1 用于能量转换和储存的硅藻

4.1.1 简介

在过去的 20 年里,人们广泛探索了具有纳米级尺寸和独特性能的新型合成材料,以开发用于能源生产和储存的新解决方案[1-2]。其中,对由碳、硅、无机氧化物及聚合物等材料制成的具有不同形貌的多孔纳米结构的研究越来越多。这种结构具有一些独特的性能,例如用于离子传输的高可用表面积结构,以及具有优异的机械、电学、电化学、磁性和光学性质用于能量转换或储存的合成材料[3-6]。这些合成材料的主要缺点是生产成本高、批量生产时间长、生产过程中需使用有毒化学药品并产生危险废弃物,会对环境造成污染,不符合当下提倡可持续发展的主旋律[7-8]。环境对它们的生产至关重要,例如,涂有氧化还原循环有机化学品(如醌类)的超细过渡金属颗粒(如铁和钒)或含有金属杂质的碳纳米管可能会对健康造成负面影响[9]。为了解决这些问题,人们非常重视使用加工成本低、对环境影响较小的天然或生物材料替代这些危险材料,作为能量转换和储存的替代解决方案。

在自然界中,有许多自然生成的具有多种功能的复杂生物材料,这些生物材料的性能有时比人工材料好得多[10-11]。从数百万年的进化中发展起来的自然分子自组装过程能够在温和的环境条件和更低的能耗下,以高精度和可重复性创建这些独特的生物结构。这些生物材料在具有复杂结构和独特性能的新型纳米结构材料的仿生工程方面引起了极大的兴趣,同时也作为低成本天然材料的来源,只需最少的加工即可使用。大多数生物体,如细菌、藻类、鱼类、昆虫、植物、动物和人类(骨骼),都能够将这种类型的无机结构或其有机复合物合成为具有有序的微米到纳米级特征的复杂结构,这些特征通过现有的工程或化学合成工艺是不可能复制的[12-14]。其中,称为硅藻的单细胞藻类的无定形二氧化硅外骨骼(硅藻壳)是生物衍生纳米结构材料的最引人注目的例子之一[15]。据估计,100 000 种硅藻中的每一种都有一个称为硅藻壳的特定三维二氧化硅壳,其特有的形状装饰着独特的纳米尺寸特征图案,如孔、脊、穗和刺[15-18]。

每个硅藻壳结构具有多层多孔膜或结构,其形状、大小和图案各不相同。从图 4.1 中可以获得硅藻形状和孔隙结构的显著多样性,其中包括几种最典型的硅藻形状。这些多样的形状和有序的多孔结构无可辩驳地展示了自然设计在微米和纳米尺度上的精确性,为将这种材料的广泛应用提供了巨大的机会。新术语"硅藻纳米技术"最近被创造出来,用来描述探索这些独特材料及其应用的新兴领域,涉及分子生物学、材料科学、生物技术、纳米技术和光子学等不同学科[18]。硅藻二氧化硅的各种潜在应用,包括光学、光子学、催化、生物传感器、药物输送、微流体、分子分离、过滤、吸附、生物封装和免疫隔离以及纳米材料的模板合成已经被提出和进行了探索[19-21]。值得注意的是,硅藻二氧化硅可通过大量培养硅藻获得,但它们巨大的数量来自一种叫作硅藻土的低成本硅藻化石,这是一种由纯硅藻壳组成的白色矿物粉末,可从采矿业获得数千吨。

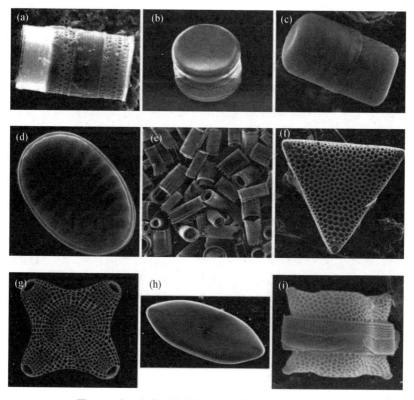

图 4.1　由二氧化硅构成的不同形状和三维结构的硅藻

因此,这些基于其独特结构的有前景的天然材料被考虑用于能量转换和储存也就不足为奇了。本章介绍了硅藻基纳米结构(天然硅藻、硅藻壳、硅藻复制品及其复合材料)在锂离子电池材料、超级电容器、太阳能电池、储氢性能和热能储存性能等能源相关领域的应用进展,尽管该领域仍处于早期阶段,但它正在快速发展,并期待硅藻纳米技术对该领域做出巨大贡献。

4.1.2 硅藻二氧化硅:结构、性能及其优化

由纯二氧化硅组成的硅藻或硅藻壳结构用于保护细胞,它由两个瓣膜组成,这两个瓣膜通过环绕并固定在一起。瓣膜由堆叠的六角形腔室组成,这些腔室由二氧化硅板隔开。瓣膜的整体结构可概括如下:称为肋的二氧化硅线分叉,偶尔从成核点分支,呈三角形硅藻的线性中脉或中心的圆形中环[22-26]。这些硅藻的复杂分级二氧化硅结构是通过进化开发和优化的,以提供更多的功能,包括具有强大机械结构的细胞保护和对抗捕食者的力量,在水环境中容易移动,用于营养物质吸收的分子筛孔结构,具有光学和光子特性的特定纳米到微米级孔模式,用于光和能量收集以及可能的传感和通信。硅藻硅壁(硅藻壳),以各种纹理和形状出现,大多是对称的。硅藻壳的微米级尺寸和孔结构的纳米级尺寸与光的波长相似,使它们增强了光散射特性并被用作光电器件[27-31]。更重要的是,硅藻的光合受体位于靠近硅藻壳的叶绿体中,因此其二氧化硅结构的光通道和聚焦特性可以帮助更多的光传输和收集到光感受器中,形成硅藻壳,有助于提高光合作用[29,32-35]。

由于硅藻二氧化硅的特性有许多限制,例如电阻率高,不利于能量转换和储存等应用,大量的研究致力于将二氧化硅改性或转化为其他材料,且保留硅藻结构。这些修改涉及许多材料,包括金属、半导体、碳和聚合物[18]。已经提出的几种基于金属(Au、Ag、Pt)和纳米粒子涂层的方法,使用水热转化、溶胶-凝胶化学气相沉积和原子层沉积将二氧化硅表面转化为具有新的和更高效的光学、电学和磁学特性的复合材料[18]。采用水热处理和热退火相结合的方法,用 $ZnFe_2O_4/SiO_2$ 共形包覆硅藻。这些涂层显示出由 Mn^{2+} 离子中的 4G-6S 跃迁产生的绿色激发光[35-37]。此外,溶胶-凝胶表面涂层工艺,结合结构导向剂,已经通过各种氧化物提供了硅藻的保形涂层[37]。而且,通过二氧化钛的原子层沉积,在保持孔形状的同时减小硅藻膜孔的尺寸以产生具有光催化活性的硅藻二氧化硅[38]。具有过渡金属氧化物改性结构的分级多孔硅藻被认为在循环过程中表现出优异的比电容和电容保持力,并且通常具有精细的细节和图案化的纳米级特征[18,20,39-42]。

另一种策略是将硅藻二氧化硅完全转化为另一种材料,其中几种方法已证明可以转化为非天然金属(Au,Ag)、聚合物和 Si,而不改变生物组装的 3D 形貌[18]。这种被称为 BaSIC(生物碎屑和保持形状的无机转化)的策略是由 Snandhage 和他的团队首创的,包括气/硅置换反应、保形涂层或这些的组合。使用单质气体反应物进行氧化/还原反应的气-硅置换,或使用卤化物-气体反应物进行复分解反应,分别将硅基颗粒转化为 MgO 和 TiO_2 副本[43-46]。将置换反应和溶液涂层方法相结合,已证明可生成一系列具有多种功能化学性质的复合副本,包括 $MgO/BaTiO_3$、$MgO/BaTiO_3$(掺杂 Eu^{3+})、$BaTiO_3$ 和 $SrTiO_3$。氮化硼共形包覆在硅藻胞片上,随后,底层的硅藻颗粒产生了独立的氮化硼结构,这为大规模制备纳米结构非氧化物陶瓷提供

了希望,包括尚未探索的能源。

最著名的硅藻改性工艺之一是使用气态镁作为还原剂,通过镁热还原法将硅藻二氧化硅转化为硅藻硅,并精确保留其三维多孔结构。首先经过650℃的热处理,硅藻二氧化硅转化为连续的硅和氧化镁纳米晶体混合物,然后选择性溶解氧化镁,产生具有初始硅藻结构形式的硅纳米晶体互连网络[47]。与电化学工艺制备的多孔硅相比,这种工艺用于制备多孔硅具有成本低、时间短、性能先进和可扩展性强等诸多优点。随后,为获得更好的电化学性能而将硅藻硅与碳涂层相结合的做法引起了人们的广泛关注,人们设计出了用于多种能源应用的新型电极材料,其中少数材料已经得到了证实[48]。

从这些介绍的方法中,我们了解到硅藻硅材料可以设计和制造出许多新材料和新性能。然而,在能源生产和转换应用方面,对它们的探索还很有限。表4.1总结了目前开发的硅藻生物二氧化硅结构装置的概念和与能源相关的应用,考虑到这些材料的巨大潜力和特性,我们预计未来将有巨大的发展空间。

表4.1 基于硅藻生物硅结构设备的能源相关应用研究

设备应用	改性工艺	改进
锂电池	镁热还原、浸渍和碳化	卓越的循环性能和高容量保留
超级电容器	碱性金属氧化物	高比电容和高循环稳定性
染料敏化太阳能电池	TiO_2嵌入或沉积	提高光伏转换效率
氢化物储存	酸性热过滤处理	氢吸附容量
热能储存	材料改良	高拉伸延展性和蓄热能力

4.1.3 用于锂离子电池材料的硅藻

低成本、高能量密度和长寿命的可充电锂离子电池的性能在很大程度上取决于锂储存材料。过去,石墨因其优异的循环性能而被认为是锂离子电池(LIBs)负极材料的潜在候选材料。然而,石墨的理论容量仅为 $372\ mA\cdot h\cdot g^{-1}$。因此开发具有更高功率和能量密度的电极材料至关重要。硅由于具有 $4\,200\ mA\cdot h\cdot g^{-1}$ 的高容量,是锂离子电池极具吸引力的负极材料之一[49-52]。

在过去的几年里,人们探索了许多不同形式的硅用于锂电池应用,包括硅膜、纳米颗粒、电化学生产的多孔硅等[50-51]。其中,具有独特分层结构和高比表面积的多孔硅被认为是非常有吸引力的电极材料,由于其对液态电解质的高接触表面积,能够促进锂离子的快速传输,从而可提供优异的倍率特性,并在充电/放电循环期间保持良好的电子导电性。为了提高多孔硅电极的性能,开发具有新型多孔纳米结构的电极至关重要,该结构能够通过降低晶体应变和增加可用于离子传输的比表面积来大大提高性能。通过锂离子的嵌入和脱嵌过程,二氧化硅作为负极材料受到严重的体积膨胀和快速容量衰减的严重阻碍,导致电极结构粉化和循环性能差[53]。

因此,为了提高导电性、电荷存储能力、电子传输能力以及抑制颗粒粉化,有必要通过具有高比表面积和孔隙率的新纳米级形态(如纳米线、纳米管[22],纳米片和纳米球[23])来改善硅阳极的性能。使用不同的硅结构获得制备的锂离子电池以下放电容量性能,如三维多孔硅颗粒(约 2 600 mA·h·g^{-1})、硅纳米管阵列(约 1 800 mA·h·g^{-1})、硅纳米管(约 1 000 mA·h·g^{-1})[54]。另一种提高硅基阳极性能的方法是使用含硅的聚合物[19]进行涂覆[25],但是它们的反应条件非常费力、耗时且需要复杂的处理,例如需要高压或高温的反应条件以及昂贵的原材料。碳涂层被认为是最优化的解决方案,因为与裸多孔硅相比,碳涂层的分级多孔硅提供了足够的空间来适应硅的体积变化,提高循环稳定性,并提高电子电导率,从而促进形成稳定的固体电解质中间相(SEI)层[55]。分级多孔 Si/C 复合材料在第一次循环时表现出最高的可逆容量,约为 1 628 mA·h·g^{-1},在接下来的循环中具有出色的容量保持率循环[21]。用于锂离子电池应用的硅和多孔硅负极的最大限制之一是这些材料的高生产成本,这使得它们在为新兴电动汽车行业制造电池方面的吸引力较小。

为了改善锂离子电池合成硅的这些问题,硅藻土被认为是制备多孔硅负极的有前途的原材料。随后,将硅与碳涂层结合,可以减轻硅的体积变化并保持多孔硅颗粒之间的电接触[48]。多孔硅颗粒是通过镁热还原商业硅藻土获得。由于硅颗粒和孔隙之间的空隙空间,每个硅颗粒都被设计为提供足够的空间来适应充电和放电过程中硅的体积变化,从而大大提高了循环稳定性。Campbell 等[56]在最近的工作中展示了碳涂层和转化硅藻硅作为高倍率锂离子电池负极的首次演示之一。硅藻硅是通过使用镁热工艺从硅藻二氧化硅还原工艺制备的,然后用聚丙烯酸(PAA)进行碳化,总结在图 4.2 中。所得硅藻转化的纳米硅表现出 162.6 cm^2·g^{-1} 的高 BET 比表面积,而原始 DE 的值则为 7.3 cm^2·g^{-1}。DE 包含 SiO$_2$ 结构,可为纳米级硅制作理想的生物衍生模板。基于 DE 的纳米硅负极表现出良好的循环性能,在 C/5(0.7 A·g^{-1}Si)和高面积负载(2 mg·cm^{-2})(见图 4.3)。这项工作还展示了基于 DE 的 Si 负极的首次

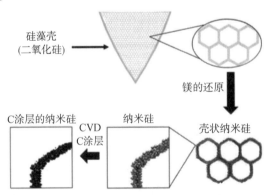

图 4.2 获得用作锂离子阳极活性材料的 C 涂层、DE 衍生、圆台状纳米 Si 结构的过程示意图[56]

倍率能力测试,测试了 C/30～4C 的 C 速率。负极保持了 654.3 mA·h·g^{-1} 的比容量,比石墨的理论值(372 mA·h·g^{-1})高出近 2 倍。

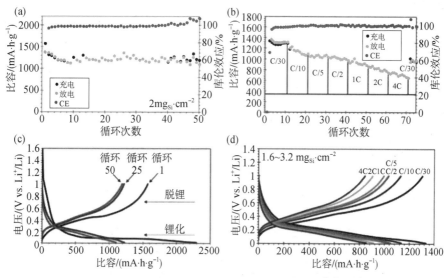

图 4.3　DE 衍生纳米 Si 基电极的电化学表征[56]

(a) 基于 Si 的 C/5 下 50 次循环的充放电循环性能;(b) C/30～4C 下 75 次循环的 C 速率测试;(c) 循环 1、25 和 50 C/5 下充放电数据的电压剖面,各种 C 速率的电压剖面;(d) 循环 1～10 的 CV

在 Wang 等的另一工作中,在 Si 转化硅藻获得的多孔 Si/C 复合材料负极上,在 0.1 A·g^{-1} 的电流密度下观察到比放电和充电容量约为 1 700 mA·h·g^{-1}。负极还表现出比裸多孔硅(205 mA·h·g^{-1})更好的循环能力。结果表明,颗粒粉碎的限制成为提高硅负极锂离子电池性能的关键[12,48,57-63]。这些例子确实,证明硅转化硅藻是锂离子电池的有前途的负极材料,我们可以期待更多的工作来进一步推进这些电极以提高容量和循环性。

4.1.4　用于储能的硅藻:超级电容器

超级电容器因其高功率密度、快速充电/放电速率、可持续循环寿命(数百万次循环)和优异的循环稳定性而成为下一代电源设备最有希望的候选者。特别是,基于过渡金属氧化物的准电容器表现出比基于碳质材料和导电聚合物的准电容器高得多的比电容,因为它们可以提供多种氧化态以进行有效的氧化还原电荷转移[21,63-72]。特别是过渡金属氧化物准电容器比碳质材料和导电聚合物准电容器具有更高的比电容,因为它们可以提供多种氧化态以实现有效的氧化还原电荷转移。过渡金属氧化物作为超级电容器的电极材料,包括 CuO[73]、MnO$_2$[74-75]、NiO[76]、Fe$_2$O$_3$[77]、MoO$_3$[78]、V$_2$O$_5$[79]、Co$_3$O$_4$[70]等已被证明可以提高超级电容器的能量和功率密度。然而,大多数金属氧化物也具有很大的体积,并且具有低电子电导率、低离子扩散常数和低结构磁化率,这些限制了它们的应用[80-81]。通过提供具有高孔隙率的可靠模板来最大化利用金属氧化物的准电容是设计金属氧化物基电化学超级电容器的高性能电极的基本标准之一。

基于 MnO_2 的电化学超级电容器由于制造成本低、比电容高(理论容量为 $1\,370\,F·g^{-1}$)、丰富的可用性、环境相容性以及在碱性/中性介质中的高循环稳定性而引起了广泛的关注[82-83]。具有各种结构和形态的锰氧化物超级电容器,例如纳米线[84]、纳米片[85]、纳米管[86]、纳米花[87]和中空纳米球[88]已通过电化学和化学途径制造,它们的电化学性质也已得以研究[82]。

Zhang 等率先提出了使用硅藻土 3D 结构与锰和镍氧化物相结合并制造用于电化学电容器应用的复合电极的概念,表明这些发展的宝贵潜力[89-92]。图 4.4 显示了用于制造超级电容器电极的 MnO_2 改性硅藻结构的典型形态,证实了 MnO_2 涂层在硅藻二氧化硅结构上具有纳米纤维结构,并保留了它们的孔和整个形状[93]。

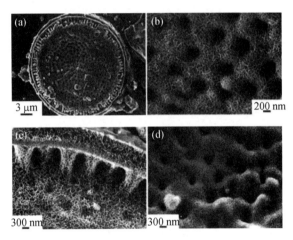

图 4.4　用作超级电容器电极的二氧化锰改性硅藻土复合材料的 SEM 图[93]

纯化后的硅藻表现出 $8\,F·g^{-1}$ 的比电容能力,而通过一锅水热法获得的分级多孔 MnO_2 改性硅藻表现出更高的功率能力,为 $202.6\,F·g^{-1}$。由于硅藻表面的独特结构,观察到 MnO_2 改性的纳米片在纯化的硅藻土上垂直生长,增加了电极的比表面积,从而构建了分层结构。此外,蚀刻硅藻土后的 MnO_2 纳米结构具有更高的功率容量($297.8\,F·g^{-1}$)和良好的循环稳定性(5 000 次循环后保留 95.92%)。随后,他们提出,直链藻属型的 MnO_2 图案在 $0.5\,A·g^{-1}$ 的扫描速率下表现出 $371.2\,F·g^{-1}$ 的比电容和良好的循环稳定性[在 5 A 的扫描速率下 2 000 次循环后电容保持率为 93.1% g^{-1}(见图 4.5)][90]。基于这些结果,分级多孔 MnO_2 改性硅藻土复合材料被清楚地证明是一种有前途的低成本、环保和电化学稳定的活性材料超级电容器。

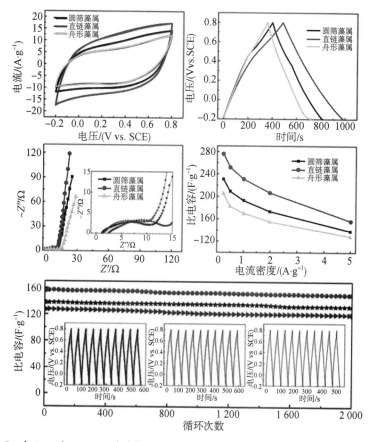

图 4.5 在 1 mol/L Na_2SO_4 溶液中测量的硅藻二氧化硅核壳结构电极的电化学性能[90]

此外,证明将 MnO_2 改性硅藻与其他材料结合可以提高超级电容器的性能。例如,合成并探索了中空硅藻二氧化硅结构、TiO_2 纳米球和 MnO_2 介孔纳米片在高性能超级电容器中的应用,该概念如图 4.6 所示[91]。该混合物在 0.2 A·g^{-1} 的扫描速率下表现出 425 F·g^{-1} 的高功率电容和长循环稳定性(在 2 000 次循环后保持

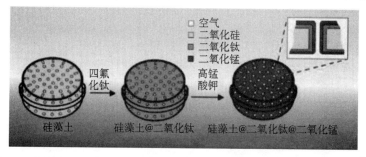

图 4.6 硅藻@TiO_2、硅藻@TiO_2@MnO_2 三维复合超级电容器合成过程示意图[91]

孔隙结构的横截面显示了 TiO_2 和 MnO_2 纳米复合材料涂层内外硅藻表面

94.1%)。由于 TiO_2 纳米球和硅藻结构层提供丰富的界面和开孔通道,MnO_2 纳米片的电子传输正在增加(见图 4.7)。此外,MnO_2 纳米结构、氧化石墨烯纳米片(GO)和多孔硅藻土(DE)微粒的独特组合在 160 ℃下表现出更大的比电容(152.5 F·g^{-1})和相对更好的循环稳定性(83.3%的电容保持率,在以 2 A·g^{-1} 的扫描速率进行 2 000 次循环后)[92]。这些研究表明,独特的硅藻结构混合物作为超级电容器的活性材料具有广阔的前景,总结在表 4.2 中。

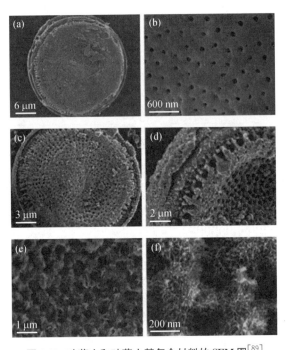

图 4.7 硅藻土和硅藻土基复合材料的 SEM 图[89]
(a)和(b) 纯化硅藻土;(c)和(d) NiO 改性硅藻土复合材料;(e)和(f) 硅藻土壳中心的孔隙

表 4.2 二氧化锰硅藻基超级电容器的电化学性能比较

材料	%/循环次数	比电容/(F·g^{-1})	条件/(A·g^{-1})
纯净硅藻土[89]		8	
二氧化锰/硅藻土[90]	93/1 000	371.2	0.5
氧化镍/硅藻土[89]	90.61/1 000	218.7	0.25
二氧化锰/硅藻土[89]	95.92/5 000	202.6	0.25
二氧化锰(刻蚀硅藻土)[89]	90.48/5 000	297.8	0.25
二氧化锰@二氧化钛@二氧化锰[91]	94.1/2 000	425	0.2
硅藻土@二氧化锰[92]	83.3/2 000	152.5	2

氧化镍(NiO)因其高比电容、高化学/热稳定性、易得性、环境友好性和低成本而在超级电容器中得到了很好的研究[94-95]。然而,大部分 NiO 的离子扩散常数和结构敏感性较低,这限制了它们的应用[81,96]。为了提高具有 NiO 纳米线的超级电容器的比电容,制备了分层多孔 NiO 苔藓装饰的硅藻土。高倍图像显示,硅藻壳一侧有数百个大孔隙规则排列,孔隙中几乎没有任何离散的杂质。发现独特的 NiO 改性硅藻土结构表现出 218.7 F·g^{-1} 的比电容和出色的循环稳定性(1 000 次循环后保留率达 90.61%)[89]。基于这些电化学结果得出的结论是,分层结构对电解质扩散有影响,从而改善电化学性能,使 NiO 改性硅藻土成为高性能超级电容器的有吸引力的电极。

先前使用 MnO$_2$ 和 NiO 改性硅藻进行电极应用的研究表明,控制这些对性能有显著影响的硅藻表面的氧化物的形态非常重要。结果表明,在硅藻表面修饰的金属氧化物/氢氧化物通过精确控制形貌和成分具有高功率密度、更快的充放电速率、可持续的循环寿命和优异的循环稳定性。理论上,这种独特的纳米结构可以解决电极材料在长期循环过程中的聚集和体积膨胀问题,有利于纳米结构的结构稳定性。

Zhang 和 Losic 的团队证明了硅藻基超级电容器开发的进一步改进,基于硅藻硅形态,通过使用镁热还原将硅藻二氧化硅转化为硅,然后使用水热工艺生长 MnO$_2$ 纳米结构层。应用于高性能超级电容器的新型 3D 硅-硅藻土@二氧化锰电极的制造示意图如图 4.8 所示。

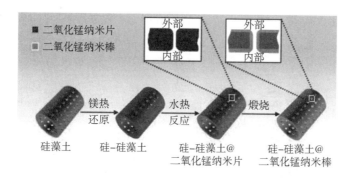

图 4.8　硅-硅藻土@二氧化锰超级电容器材料制造示意图

硅-硅藻土@二氧化锰纳米片的性能表现出的电化学性能最好,在 0.5 A·g^{-1} 的电流密度下提供 341.5 F·g^{-1} 的高比电容,良好的倍率性能(47.7% 保持率随电流增加约 20 倍),以及稳定循环特性(2 000 次循环后仍保持 84.8%)。基于硅-硅藻土@二氧化锰纳米片作为正极、AGO 作为负极的非对称超级电容器的最大功率密度为 2.22 kW·kg^{-1},能量密度为 23.2 W·h·kg^{-1}。这些优异的电化学性能可归因于硅藻良好的导电性及其独特的纳米结构,这扩大了表面积并增加了暴露在电解质中的活性位点。考虑到硅藻可以通过可扩展的转换过程从廉价且可用的自然资源中获得,这些结果表明硅-硅藻土@二氧化锰电极显示出用作超级电容器低成本和高性能电极材料的巨大潜力。与其他复合材料的结合有望进一步提高这些性能,并有可能将这些电极转化为实际的实际应用。

4.1.5 用于太阳能电池的硅藻

太阳能电池可分为硅太阳能电池、染料敏化纳米晶太阳能电池、塑料太阳能电池、有机太阳能电池等[97-103],其中硅太阳能电池具有独特的可调谐光学和电子特性,成为太阳能应用中的主导。硅半导体不是良好的导电体,导致电阻大,同时质量损失大。由于成本高,技术不成熟,没有大规模应用。

1991年,Grätzel发表了一份引人注目的报告,提出了一种新型太阳能电池,称为染料敏化太阳能电池(DSSC),基于纳米晶TiO_2的能量转化效率为7%,随后将效率提高到10%[104-105]。从那时起,DSSCs引起了研究人员的极大兴趣,因为它们提供了提高高效太阳能转换和低成本的可能性。TiO_2[101]、ZnO[106]、SnO_2[107]和Nb_2O_5[108]等金属氧化物材料被应用于DSSC。其中,TiO_2是至关重要的,因为它具有吸引人的光学、电学和生物特性,可以产生最佳性能并被紫外线激活。然而,TiO_2薄膜缺乏纳米颗粒提供的高表面积[38]。

然而,一些合成加工涉及复杂的沉积系统或化学有毒元素。开发环境友好、低毒的生物材料至关重要。此外,复合功能纳米结构材料有助于提高效率和降低生产成本[8,105,109],如图4.9所示。由于大量染料分子吸附在纳米结构表面上,硅藻的比表面积可以提高太阳能电池的效率[1,40,110]。硅藻的折射率为1.43[111]而多孔TiO_2薄膜的折射率为1.7~2.5[32],这使TiO_2硅藻层在孔阵列中具有高介电层对比度和光散射。也就是说,TiO_2-硅藻复合DSSCs提高了DSSC的效率[112]。

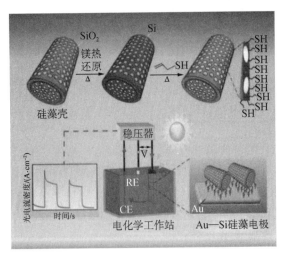

图4.9 硅藻壳体到纳米结构硅的镁热转换及其在光电化学能量转换中的应用示意图[112]

在第二步中,将巯基改性硅藻壳体连接到金电极表面,并用于设置用于进行光电流测量的三电极电池中硅藻在太阳能电池中的应用历史并不长。Chandrasekaran等[113]在2014年首次表明,由硅藻制备的半导体和高表面积3D硅复制品可维持光电流并实现太阳能转换,如图4.10所示。之后,许多科学家开始了这方面的研究,包括中国的研究人员。

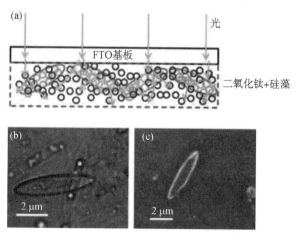

图 4.10　硅藻电极示意图和光学显微镜下观察到的硅藻图[112]

(a) TiO_2 硅藻工作电极;(b) 硅藻;(c) 硅藻壳体

Chen等还进行了理论和实验研究,以了解用于增强太阳能的硅藻壳的纳米光子捕光材料。他发现增强的吸收效率发生在400～500 nm和650～700 nm区域内,硅藻可以通过耦合到局域波导模式有效地操纵入射光,这将提高硅藻中的太阳能转换效率。

在以往的研究中,硅藻不仅应用于DSSC,还应用于其他类型的太阳能电池。Chen等介绍了具有更好热稳定性的硅藻叶绿素提取物,作为表面纹理硅太阳能电池上的旋涂抗反射层的效果。他们发现,在350～1 100 nm的光谱区域内,稍微沉积一层薄薄的硅藻提取物层可以将反射降低多达13%。然而,硅藻在太阳能电池中的应用由于其低成本、天然丰富和环境友好的性质,是一个很有前途的方向,我们仍然需要提高硅藻太阳能电池的能量转换效率。

4.1.6　用于储氢的硅藻

氢能已被提议作为清洁高效能源生产的替代燃料。然而,储氢特别具有挑战性,因为在室温和大气压下很难建立安全有效的系统。天然矿物作为储氢材料,例如组成为 $Zn_4O(BDC)_3$ 的金属有机骨架5(MOF-5)[114]、微孔金属配位材料(MMOMs)、单壁碳纳米管(SWNT)[110]、硅酸盐纳米管、氢化镁(MgH_2)已被广泛探索[1,115]。高离解温度、缓慢的氢化-脱氢动力学和氧化反应性限制了其中一些材料在氢相关应用中的使用。令人惊讶的是,高孔隙率、大表面积、小粒径、强吸附性和优异的热稳定性使硅藻土成为储氢的理想选择[116]。

在另一项研究中，Jin 等[117]发现原始硅藻土在 2.63 MPa 和 298 K 下的氢吸附容量为 0.463%，这是已知吸附剂中最高的。由于活化具有适当的孔隙特性，酸热活化硅藻土的氢吸附容量可达 0.833%。结果证实了氢吸附能力强烈依赖于硅藻土天然孔隙特性的重要证据。为了进行进一步的研究，他们创造了一种有效的金属改性方法，该方法已开发用于将 Pd 和 Pt 纳米粒子分散在硅藻土上。通过掺入 0.5% 的 Pt 和 Pd，氢吸附能力分别提高到 0.696% 和 0.980%。

因此，具有大表面积、适当孔容和小孔径的硅藻土矿物是一种很有前景的室温储氢物理吸附材料。硅藻土特有的多孔微结构对氢解吸行为的影响为进一步提高这些装置的性能打开了窗口。

4.1.7 用于热能储存的硅藻

热能储存（TES）作为能源的多功能、清洁和高效利用的中间步骤，受到越来越多的全球关注和越来越多的研究兴趣[118-121]。因此，为了获得舒适的居住环境，需要使用空调系统来控制室内环境温度的变化，这也导致了大量的能源消耗。

在各种储能方式中，热储能被公认为关键技术之一，是未来调整电力供需时差的有效方法。TES 方法有显热储能、潜热储能和可逆化学反应储能三种[122-123]。在 TES 方法中，使用相变材料（PCM）实现的潜热储能是最有效的技术，因为其在热能充/放电过程中具有高能量储存密度和窄温度变化的明显优势[122-124]。PCM 主要有两种：无机 PCM 和有机 PCM。无机相变材料是指基于脱水和水合过程中的潜热储存的无机盐水合物，具有较高的储能密度和较高的热导率[118-126]。然而，它们也有一些限制其应用的缺点，如 PCM 在固液变化过程中的泄漏问题[125,127]。为了克服这个问题，引入了形状稳定支撑来制造形状稳定的复合 PCM[128]。PCM 的稳定支持通常包括微囊容器、聚合物微囊壳和多孔材料[129]，例如膨胀石墨[127,130]、脂肪酸酯[131]、石蜡膨润土[132]、颗粒[133]、珍珠岩[134]、石膏[135]、硅藻土[136-137]、蛭石[138-139]、凹凸棒土[140]和黏土矿物[141]。

表 4.3　PCM 和 PCM/硅藻土复合材料的熔点和潜热

PCM 样本	熔点/℃	潜热/(J·g^{-1})
煤油/煅烧硅藻土[156]	33.04	89.54
煤油	27.47	201.50
PCM-DP600-碳纳米管[158]	27.12	89.40
癸酸	31.5	155.5
月桂酸	44	175.8
棕榈酸[160]	63	212.1
硬脂酸	69.6	222.2

(续表)

PCM 样本	熔点/℃	潜热/(J·g^{-1})
石蜡/硅藻土[161]	41.11	70.51
GHM	45.98	172.80
GHL	40.21	157.61
GHM/硅藻土[143]	45.86	96.21
GHL/硅藻土	39.03	63.08
聚乙二醇(50%)/硅藻土[149]	27.7	87.09
十六烷/硅藻土	23.68	120.1
十六烷/硅藻土/剥离型	—	
类石墨烯材料(xGnp)[162]	22.09	120.8
十八烷/硅藻土	31.29	116.8
十八烷/硅藻土/类石墨烯	30.20	126.1

值得注意的是,当生活环境中的相对湿度发生变化时,多孔材料可以吸收或释放水蒸气。所以多孔材料可以调节室内环境的相对湿度,让人感觉舒适,降低能耗[142]。考虑到这一点,硅藻土是掺入 PCM 用于热能储存的一种经济且重量轻的材料的可行候选物[143]。在过去的 20 年中,已经研究了将 PCMs 与硅藻结合作为最小化能量消耗的潜在技术。石蜡/硅藻土复合 PCMs(2 000×)的 SEM 形貌[144]。

Xu 等[145]报道了一种具有高拉伸延展性和储热能力的石蜡/硅藻土复合相变材料,可用作生产新型热能储存工程水泥基复合材料的细骨料(见表 4.3)。他们的团队还制造了新的石蜡/硅藻土/多壁碳纳米管复合 PCM,其熔化温度为 27.12℃,潜热为 89.40 J·g^{-1}。此外,这种复合 PCM 显示出良好的化学相容性和热稳定性[146]。此外,Li 等[147]展示了硅藻土与石蜡的可组装相变材料复合物。Karaman 等[137]制备了一种聚乙二醇(PEG)/硅藻土复合材料,作为一种用于热能储存的新型形状稳定相变材料。结果表明,复合相变材料的熔化温度和潜热分别为 27.70℃和 87.09 J·g^{-1}。添加膨胀石墨提高了复合材料的热导率。Li 等[140]使用融合吸附法制备了多种二元脂肪酸/硅藻土形状稳定相变材料。结果表明,癸酸-月桂酸/硅藻土的潜热降低至癸酸-月桂酸 PCMs 的 57%,相变温度从 16.36℃小幅上升至 16.74℃。综上所述,改性硅藻土复合相变材料具有以下显著特点:相变温度区表观比热大,导热系数合适,相变过程中保持形状稳定,无须容器。

4.2 水处理用硅藻

硅藻土是一种天然的生物成因矿物,主要由无定形二氧化硅(矿物学上属于蛋白石-A)组成。硅藻土具有发达的孔隙,尤其是大孔,以及丰富的表面硅烷醇(Si—OH),在吸附和载体的利用方面显示出前景。

了解硅藻土的表面特性对于改善其作为功能材料的应用至关重要。结合1氢魔角旋转核磁共振(1H MAS NMR)和傅里叶变换红外光谱(FTIR)方法研究了硅藻土表面羟基物质及其在热处理条件下的转变。结果表明,硅藻土表面羟基包括孤立羟基和氢键羟基,其中羟基来源于吸附在硅藻土表面的水。吸附的水在热处理后解吸,NMR和FTIR曲线中孤立羟基和氢键羟基的质子信号增加并在1 000 ℃时达到最大值。1 100 ℃煅烧后,归类为强氢键的羟基与孤立的羟基基本发生缩合。但是一些弱氢键的羟基可能仍然存在于微孔中。

此外,在确认硅藻土表面和结构特性的基础上,重点关注直接应用和改性以提高客体分子/离子的吸附和负载能力。

4.2.1 制备硅藻土基吸附复合材料的载体

通过共沉淀法(样品表示为MagDt-P)和水溶胶法(MagDt-H)制备了硅藻土负载/未负载的磁铁矿纳米颗粒。无负载和有负载的磁铁矿纳米粒子的平均尺寸分别约为25 nm和15 nm。负载的磁铁矿纳米颗粒存在于硅藻壳的表面或孔隙内部,比未负载的具有更好的分散性和更少的共聚。硅藻土二氧化硅的羟基结构和脱水过程示意图如图4.11所示。

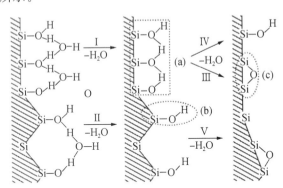

图4.11 硅藻土二氧化硅的羟基结构和脱水过程示意图

合成的磁铁矿纳米颗粒对六价铬Cr(Ⅵ)的吸附主要受物理化学过程的控制,其中包括静电吸引,然后是氧化还原过程,其中Cr(Ⅵ)被还原为三价铬Cr(Ⅲ),如图4.12所示。Cr(Ⅵ)的吸附高度依赖于pH值,吸附动力学遵循准二级动力学模型。硅藻土负载/非负载磁铁矿的吸附数据与朗缪尔等温线方程拟合良好(见图4.13)。

根据伪二级动力学方程计算,通过水溶胶法合成的硅藻土负载的磁铁矿表现出比未负载的纳米级磁铁矿($10.6 \text{ mg} \cdot \text{g}^{-1}$)和微米级磁铁矿($5.3 \text{ mg} \cdot \text{g}^{-1}$)更高的单位质量磁铁矿吸附容量($11.4 \text{ mg} \cdot \text{g}^{-1}$)。这些材料每单位质量(g)的磁铁矿含量的单层吸附容量(从使用朗缪尔方程模拟获得)遵循类似的顺序:硅藻土负载的纳米级磁铁

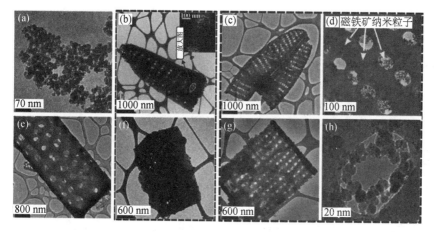

图 4.12　硅藻土负载材料的 TEM 图

(a)磁铁矿纳米粒子；(b)硅藻土；(c)和(d) MagDt-P 中的两个硅藻壳；(e)和(f) MagDt-H 中的两个硅藻壳；(g)和(h)高分辨率的样品 MagDt-H

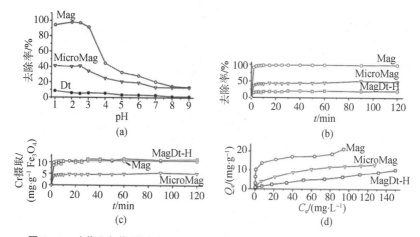

图 4.13　硅藻土负载/非负载磁铁矿的吸附数据与朗缪尔等温线方程拟合曲线

(a)初始溶液 pH 值对 Mag、MicroMag 和 Dt 上 Cr(Ⅵ)去除的影响；(b)搅拌时间对 Mag、MicroMag 和 MagDt-H 去除 Cr(Ⅵ)效率的影响；(c)搅拌时间对归一铬摄取到实际 Fe_3O_4 含量的影响；(d) Mag、MicroMag 和 MagDt-H 的 Cr(Ⅵ)吸附等温线

矿(69.2 mg·g^{-1})＞未负载的纳米级-鳞片磁铁矿(21.7 mg·g^{-1})＞微米级磁铁矿(14.6 mg·g^{-1})。这些结果表明硅藻土载体可用于提高 Cr(Ⅵ)吸附能力。此外，硅藻土的负载有助于磁铁矿纳米颗粒的储存和造粒。这些基本结果表明，硅藻负载/未负载的磁铁矿纳米颗粒很容易制备，使得从水溶液中去除 Cr(Ⅵ)具有广阔的应用前景。

4.2.2　制备多孔碳材料的催化剂和模板

硅藻土被用作合成大孔碳的模板。然而，在硅藻土基多孔碳的制备中，总是使用硫酸作为催化剂来催化碳前驱体。硫酸的加入对环境有潜在危害，也增加了制备成本。在我们之前的研究中,我们发现硅藻土本身含有固体酸位:Brønsted 酸位来自一些氢键键合的硅烷醇,Lewis 酸位来自非常小的粘土矿物颗粒,牢固地粘附在硅藻壳上。因此,利用硅藻土(来自圆筛藻属)的这些固有酸性位点而不是添加酸来催化碳前驱体,并研究了固体酸度对多孔碳最终结构的关键影响。

结果表明,固体酸度使初始/活化硅藻土成为生成多孔碳(所得碳表示为 C/Dt$_{CE}$)的催化剂,并且碳产物的多孔参数强烈依赖于硅藻土模板的固体酸度。硅藻土的形态也显著影响多孔碳的纹理结构。碳产品中的两种大孔结构,部分实心柱和有序空心管,分别来源于硅藻壳中心孔和边缘孔的复制(见图 4.14)。中孔和尺寸均匀的虫洞状微孔分别是由于模板去除过程中石墨微晶的堆积和碳膜的结构重构的结果。硅藻土的酸度显著影响碳的孔隙率。硅藻土模板的酸性位点越多,碳产物的表面积和孔体积越大,表明硅藻土模板的酸度与硅藻土模板碳的孔隙参数呈正相关,如图 4.15 所示。

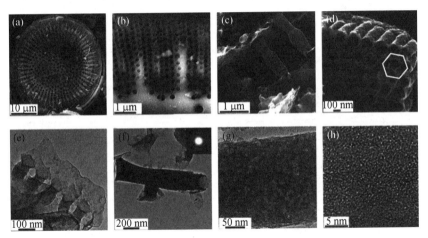

图 4.14　硅藻土负载材料的 SEM 和 TEM 图

(a) 硅藻壳;(b) 硅藻壳边缘大孔;(c) 碳柱的横截面图;(d) C/Dt$_{CE}$ 的有序大孔;(e) 有序大孔的电子显微 TEM 图;(f) 碳柱(插图:选区电子衍射 SAED 图);(g) 碳柱中孔;(h) 碳壁的微孔

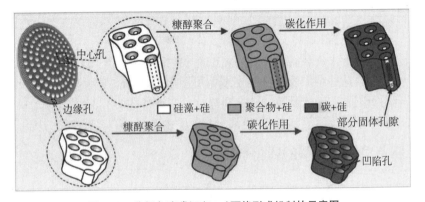

图 4.15　分级多孔碳(C/Dt$_{CE}$)可能形成机制的示意图

分级多孔碳显示出高比表面积(高达 426 $m^2 \cdot g^{-1}$)和孔体积(高达 0.195 $cm^3 \cdot g^{-1}$)以及对溶剂石脑油和 H_2 具有良好的吸附能力,在吸附和储气方面具有潜在的应用前景,如图 4.16 所示。

来自不同硅藻属的硅藻土表现出不同的形态和孔隙率,导致获得的硅藻土模板炭具有不同的结构和对亚甲基蓝(MB)的吸附能力。末端封闭的碳管来源于针杆藻属硅藻壳的复制(获得的碳表示为 C/Dt_{SE},图 4.15)。

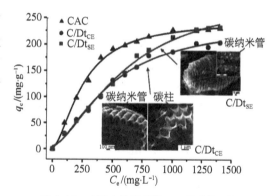

图 4.16　硅藻土模板碳的 MB 吸附等温线

三种吸附剂的比表面积单层吸附容量遵循 C/Dt_{SE} (333 $mg \cdot g^{-1}$)>C/Dt_{CE}=商业活性炭(CAC, 250 $mg \cdot g^{-1}$)的顺序。硅藻土模板碳比商业活性炭具有更高的 MB 吸附能力,在吸附和净化领域显示出良好的应用前景。

然而,必须指出的是,使用氟化氢(HF)作为蚀刻剂对硅藻土模板的蚀刻对环境具有潜在威胁。因此,在我们的研究中选择了氢氧化钾(KOH)来蚀刻硅藻土模板,我们发现 KOH 还可以作为活化剂来提高硅藻土模板化碳的孔隙率。分离出的碳产品在其表面负载 KOH,在管式炉中于 N_2 环境下再次碳化后,碳的孔隙率得到改善。KOH 活化的硅藻土模板碳比原始碳和 CO_2 活性炭具有更高的比表面积(988 $m^2 \cdot g^{-1}$)、总孔体积(0.675 $cm^3 \cdot g^{-1}$)和 MB 吸附容量(645.2 $mg \cdot g^{-1}$)。此外,以 KOH 蚀刻硅藻土模板得到的含硅和含钾溶液为硅源和钾源,制备了分子筛副产物 K-H 沸石(Z/Dt),如图 4.17 所示。沸石产品具有棒状形态和纳米尺寸的颗粒,具有以中孔为主的多孔结构。因此,硅藻土不仅作为催化剂和模板剂,还作为硅源制备碳材料和沸石,为废液回收利用提供了一种经济可行的简便方法。

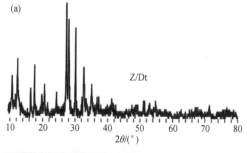

图 4.17　K-H 沸石的 XRD 图、SEM 图和 TEM 图
(插图:多孔形态)

4.2.3　表面和多孔结构的改性

具有中孔/大孔结构的天然硅藻土已被用于吸附水溶液中的金属离子或有机分子。然而,硅藻土对这些分子或离子的吸附能力一般,接近甚至低于天然沸石或黏土

矿物,如高岭石和蒙脱石。这是由于硅藻壳表面硅烷醇(即Si—OH)对上述吸附质不具有强的吸附亲和力,并且这些吸附质太小而不能被硅藻土的微米级或纳米级通道过滤;此外,硅藻土的离子交换能力远小于沸石或蒙脱石。

1. 表面硅烷化

为了提高对离子的吸附能力,采用硅烷化改变硅藻土的表面性质。

1) 用于吸附 Cu(Ⅱ) 离子的表面硅烷化

具有金属亲和基团的有机硅烷已应用于多种类型的合成二氧化硅以改善其重金属吸附,这已被证明是改性硅藻土表面的有效方法。有机组分和硅藻壳之间的共价键使有机部分能够持久地固定在硅烷化产物上,并防止它们浸入周围的溶液中。

通过在所使用的有机硅烷试剂中引入特殊的官能团,可以显著提高合成材料对污染物的吸附选择性。γ-氨基丙基三乙氧基硅烷(APTES)被选择用于硅藻土的表面硅烷化,其衍生物对 Cu(Ⅱ) 显示出高吸附能力。

水解后的 APTES 低聚化,然后通过氢键或范德华相互作用附着在硅藻土表面,形成含有高含量 APTES 物种的交联结构。在高温(800℃)下加热会去除物理吸附水和氢键封端水,从而暴露出硅藻土表面分离的硅烷醇(见图 4.18)。这些硅烷醇充当 APTES 分子接枝的底物。在接枝主导结构中,形成了相对薄的 APTES 层,接枝的 APTES 表现出非常高的热稳定性。接枝 APTES 的分解温度约为 540℃。

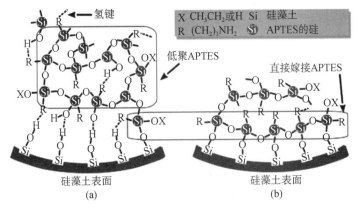

图 4.18 APTES 修饰的相关机制示意图

(a) 寡聚主导硅烷化;(b) 嫁接主导硅烷化

由于氨基对铜离子的高亲和力,APTES 改性硅藻土上的 Cu(Ⅱ) 吸附量是未改性硅藻土上的 13 倍。接枝主导的改性硅藻土比低聚主导的硅藻土具有显著更高的 Cu(Ⅱ) 吸附效率。这种更高效率的主要原因是由低聚 APTES 形成的交联网络中的氮原子受到氢键的强烈影响,这削弱了它们与铜的配位,如图 4.19 所示。

2) 用于某些有机分子的表面硅烷化

为了改善有机分子的吸附,使用带有苯基官能团的苯基三乙氧基硅烷[PTES, $C_6H_5Si(OCH_2CH_3)_3$]对硅藻土表面进行改性。硅烷化后,一个官能团(—C_6H_5,苯

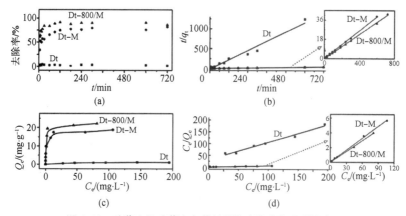

图 4.19 硅藻土及硅藻土负载材料的去除率和吸附拟合图

(a) 搅拌时间对 Dt(硅藻土)、Dt/M(APTES 接枝硅藻土)和 Dt-800/M(硅烷化前在 800℃加热硅藻土的 APTES 嫁接衍生物)的 Cu(Ⅱ)去除效率的影响；(b) 基于拟二级动力学模型的 Cu(Ⅱ)吸附线性拟合图；(c) Dt、Dt/M 和 Dt-800/M 的 Cu(Ⅱ)吸附等温线；(d) 基于 Langmuir 等温线模型的 Cu(Ⅱ)吸附线性拟合图

基)被成功地引入硅藻土(Dt 和硅烷化产物表示为 PTES-Dt)的表面上。PTES-Dt 表现出疏水性，水接触角(WCA)高达 120°±1°，而 Dt 具有超亲水性，WCA 为 0°。Dt 和 PTES-Dt 的苯吸附数据与 Langmuir 等温线方程拟合良好。苯在 PTES-Dt 上的 Langmuir 吸附容量为 28.1 mg·g^{-1}，是 Dt 上的 4 倍多。

此外，吸附动力学结果表明，在 0.118～0.157 的相对压力范围内 PTES-Dt 比 Dt 更快地达到平衡。PTES-Dt 优异的苯吸附性能归因于苯基与苯分子之间强烈的 p 系统相互作用以及 PTES-Dt 的大孔隙度。

2. 苯吸附多孔结构的改性

为了增加多孔结构，微孔/大孔被引入硅藻土上，用于制备具有有机分子吸附能力的分级多孔结构的复合材料，PTES 在硅藻土表面的构象示意图以及衍生物对苯的吸附能力如图 4.20 所示。

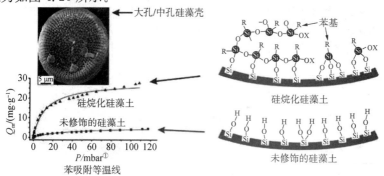

图 4.20 PTES 在硅藻土表面的构象示意图以及衍生物对苯的吸附能力

---

① 1 bar = 10^5 Pa。

1) 表面蚀刻和微孔改性

采用简便的 NaOH 蚀刻方法处理硅藻壳载体,然后在预先接种纳米晶硅沸石-1 的硅藻壳表面水热生长 MFI 型沸石,制备分层多孔硅藻土/MFI 型(Dt-Z)复合材料。NaOH 蚀刻对优化 Dt/Z 复合材料的孔隙率参数产生双重影响:①保留了更多的载体大孔隙率;②通过均匀分布的沸石涂层引入大量微孔,而不会由于蚀刻导致硅藻壳上的孔扩大而阻塞硅藻壳中的大孔。如图 4.21 所示,在 NaOH 蚀刻后,硅藻壳的盘状形态得到很好的保留。中心大孔的直径增加到 400~1 000 nm。

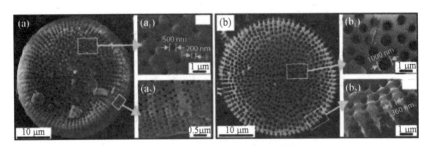

图 4.21 Dt 和 Dt-E 的 SEM 图
(a) Dt;(b) Dt-E

在有或没有 NaOH 蚀刻(Dt/Z 或 Dt-E/Z)的复合材料的制备步骤中发生的形态变化如图 4.22 所示。晶种处理后,纳米硅沸石-1 颗粒均匀地包覆在硅藻壳表面以及中央大孔的内壁上[见图 4.22(a)和(d)]。由于纳米晶体覆盖了内孔壁的表面[见图 4.22(a)和(d)],晶种硅藻土(Dt_{seeded})和 NaOH 蚀刻晶种($Dt-E_{seeded}$)的中心大孔尺

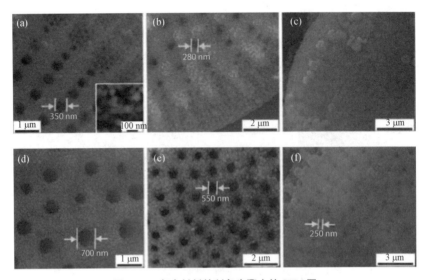

图 4.22 复合材料的制备步骤中的 SEM 图
(a) Dt_{seeded};(b) Dt/Z_2;(c) Dt/Z_4;(d) $Dt-E_{seeded}$;(e) $Dt-E/Z_2$;(f) $Dt-E/Z_4$

寸减小了大约 160 nm。2 天的水热处理明显增大了 MFI 型颗粒,并且它们在复合材料中的尺寸(分别为 Dt/Z_2 和 Dt-E/Z_2)增长到大约 150 nm。4 天的水热处理进一步增大了沸石颗粒。通过沸石晶体的相互生长,在硅藻壳表面形成了连续的沸石薄膜。Dt/Z_4 完全丧失了硅藻的大孔隙度[见图 4.22(c)],而 Dt-E/Z_4 保留了圆盘形态和大孔隙度[见图 4.22(f)]。

Dt-E/Z_2 的 TEM 图[见图 4.23(a)]显示硅藻壳表面完全且均匀地被纳米颗粒覆盖,并且硅藻土的大孔结构仍然清晰可见。SAED 图案中的圆形条纹[见图 4.23(a)的插图]表明随机取向的结晶沸石纳米晶体涂覆在硅藻壳表面。Dt-E/Z_4 的 TEM 图显示了共生沸石颗粒的致密膜(暗区)[见图 4.23(b)]。晶体的 SAED 图案[见图 4.23(b)的插图]与先前报道的 MFI 沸石晶体的 SAED 图案相匹配,进一步证实了 MFI 型沸石涂覆在硅藻壳表面。

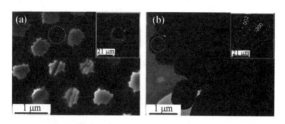

图 4.23　Dt-E/Z_2 和 Dt-E/Z_4 的 TEM 图

(a) Dt-E/Z_2(插图为虚线圆圈中区域的 SAED 模式);(b) Dt-E/Z_4(插图为虚线圆圈中区域的 SAED 图案)

突破测量是一种直接的方法,旨在阐明低浓度 VOCs 的动态吸附性能。根据图 4.24 中的穿透曲线评估,Dt/Z 复合材料比纯硅沸石纳米颗粒表现出更高的单位质量沸石苯吸附能力和更小的传质阻力。

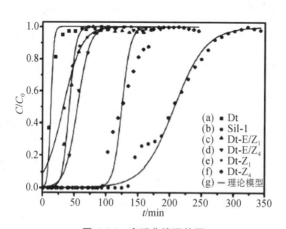

图 4.24　穿透曲线评估图

(a) Dt 的突破曲线和数学模型;(b) Sil-1;(c) Dt-E/Z_1;(d) Dt-E/Z_2;(e) Dt-Z_1;(f) 25℃下的 Dt-E/Z_4(标准化为 0.2 g 固体);吸附期间,以 1.00 mL/min 的速度将含苯蒸气(C_0=1.51 mmol/L,P/P_0=0.27)的干氮通过柱

2) 表面带电和微孔改性

如图 4.25 所示,通过简单的涂层工艺制备了一种新型的分级多孔硅藻土/硅藻土-1(M-Dt/Sil-1$_{nano}$)复合材料,具有高效的苯去除效率。首先对硅藻土的表面电荷进行改性,然后在温和的低温回流反应条件下,在预改性的硅藻土载体表面原位合成硅石-1 纳米颗粒(Sil-1$_{nano}$)。

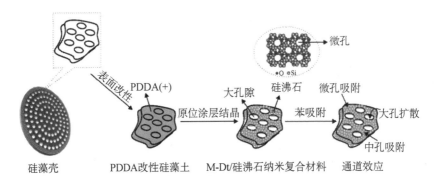

图 4.25 硅藻土/硅藻土的合成和苯吸附评估过程示意图

如图 4.26 所示,原位涂覆工艺后获得的 M-Dt/Sil-1$_{nano}$ 复合材料的表面涂覆有硅沸石-1 纳米颗粒(80 nm)。M-Dt/Sil-1$_{nano}$ 复合材料的大孔清晰可见,孔径为 100～200 nm。透射电子显微镜结果表明,在 M-Dt/Sil-1$_{nano}$ 复合材料中,硅藻土载体的大孔明显,在孔的内边缘涂有一层硅沸石-1 纳米颗粒。晶体的 SEAD 图案与先

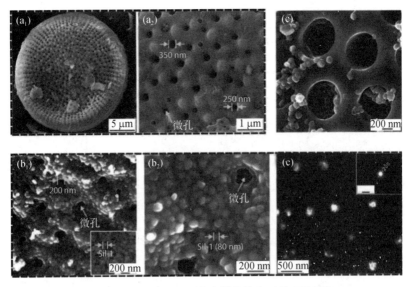

图 4.26 M-Dt/Sil-1$_{nano}$ 复合材料的 SEM 和 TEM 图

(a) Dt 的 SEM 图;(b) M-Dt/Sil-1$_{nano}$;(c) Dt/Sil-1$_{nano}$;(d) M-Dt/Sil-1$_{nano}$ 的 TEM 图

前报道的 MFI 沸石的晶体结构相匹配,进一步证实了硅藻土上的 Sil-1$_{nano}$ 涂层。未经硅藻土预改性的 Dt/Sil-1$_{nano}$ 复合材料的 SEM 显示只有少量 Sil-1$_{nano}$ 不均匀地涂覆在硅藻土上,表明表面充电预改性方法的优越性。

与合成的 Sil-1$_{nano}$ 和商业 ZSM-5 相比,硅藻土/硅质岩-1 复合材料每单位质量沸石表现出显著更高的静态和动态苯吸附容量[分别为 94.9 mg·g^{-1}(Sil-1$_{nano}$)和 246.0 mg·g^{-1}(Sil-1$_{nano}$)](见图 4.27)。图 4.28 中的穿透曲线结果表明复合材料比 Sil-1$_{nano}$ 和商业 ZSM-5 更低的传质阻力和明显更快的渗透率,这是由于复合材料中的大孔隙降低了气体渗透阻力。

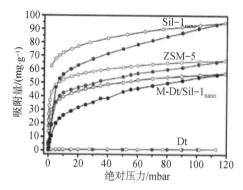

图 4.27 各种样品的静态苯吸附-解吸等温线

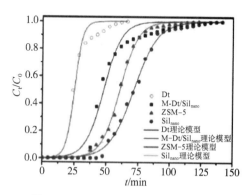

图 4.28 Dt、M-Dt/Sil-1$_{nano}$、ZSM-5 和 Sil-1$_{nano}$ 对苯吸附的突破曲线

3) 硅藻土基陶瓷/硅质岩-1 复合材料

上述方法为提高硅藻土对苯的吸附性能提供了有效途径。然而,在以前的研究中,合成的复合材料是粉末形式。并且它们的气体吸附性能容易受到流体阻力的干扰,导致传质效率低。因此,一种新型复合材料(Sil-PCS)结合了具有独特三维网状结构的分级大孔陶瓷载体和微孔硅沸石-1 纳米颗粒(Sil-1)涂层的优点,使用简便的方法合成。首先,采用以硅藻土为陶瓷骨架,聚氨酯泡沫为牺牲模板的聚合物海绵法制备了具有三维网状结构的多孔陶瓷载体。随后,在温和的条件下,在陶瓷表面上原位均匀涂覆 Sil-1。纳米复合材料的分级孔隙率是由于 Sil-1 固有的微孔、由 Sil-1 堆积产生的中孔以及陶瓷载体的分级大孔(见图 4.29)。

由于原位硅沸石-1 涂层工艺有利于 Sil-1 在陶瓷载体改性表面的稳定性和分散性,以及复合材料的分级多孔整体结构有利于苯吸附的传质效率,因此与商业微米级 ZSM-5 产品(66.5 mg·g^{-1})和合成 Sil-1(Sil$_{SYN}$,94.7 mg·g^{-1})相比,复合材料(Sil-PCS)显示出高得多的苯吸附能力[133.3 mg·g^{-1}(Sil-1)](见图 4.30)。此外,复合材料吸附苯的动态过程与线性驱动力(LDF)模型非常吻合,并且通过 IG 仪器评估,复合材料的吸附-解吸速率常数分别比 ZSM-5 和 Sil$_{SYN}$ 高 3 倍和 5 倍(见图 4.31)。

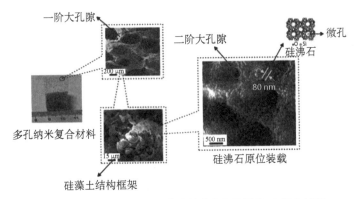

图 4.29 硅藻土基陶瓷/硅藻土的分级多孔结构-1 复合材料

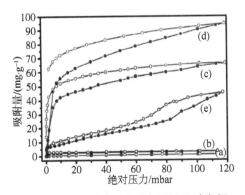

图 4.30 (a) Dt,(b) PCS,(c) ZSM-5,(d) Sil_{SYN} 和 (e) Sil-PCS 的苯吸附(填充符号)和解吸(开放符号)等温线

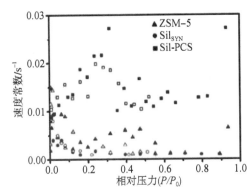

图 4.31 苯吸附(填充符号)和解吸(开放符号)的速率常数随 ZSM-5、Sil_{SYN} 和 Sil-PCS 上相对压力的变化

4.2.4 制备硅藻土基金属氧化物复合材料的载体

通过一种新方法合成的硅藻土-Fe_2O_3 杂化材料,是一种圆孔蜂窝状的多相催化剂用于光芬顿降解有机污染物(见图 4.32)。硅藻土-Fe_2O_3 催化剂的催化活性通过在可见光照射下(波长大于 420 nm)在过氧化氢存在下降解有机染料来评估(见图 4.33)。结果表明,该催化剂对罗丹明 B 的脱色率为 99.14%,总有机碳(TOC)去除率为 73.41%,表现出优异的催化性能,这可能是由于硅藻土的吸附能力和非均相光芬顿反应产生的羟基自由基的协同作用。此外,催化剂使用 5 次循环后脱色效率仍高于 90%。因此,杂化材料的简单性和成本效益对于合成降解有机污染物的高效可见光光催化剂提供了广阔的前景。

固定在天然硅藻土上的钯(Pd)纳米颗粒是通过简单的程序实现的。催化剂对 Heck 和 Suzuki 反应具有高活性,可以多次回收和重复使用。并对催化过程进行了

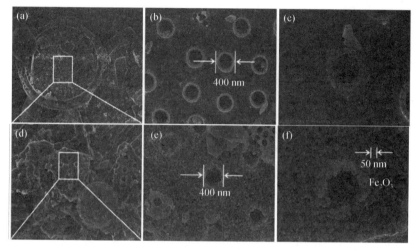

图 4.32 硅藻土[(a),(b),(c)]和硅藻土-Fe_2O_3[(d),(e),(f)]的 SEM 图

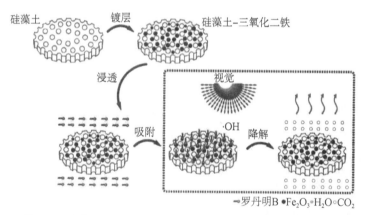

图 4.33 硅藻土-Fe_2O_3/H_2O_2/vis 系统中 RhB 光 Fenton 降解的机理

研究。硅藻土负载的 Pd 纳米粒子的合成是对先前报道的在二氧化硅球上合成银纳米粒子的程序进行了轻微修改(见图 4.34)。形成的 Pd 纳米颗粒的尺寸在 20~100 nm 范围内(见图 4.35)。硅藻土负载 Pd 纳米粒子催化剂中钯含量为 3.66%。上面制备的硅藻土负载的 Pd 首先用于 Heck 偶联反应,这是有机合成中碳-碳键形成的通用方法。

$$\text{硅藻土} \xrightarrow[CF_3COOH]{SnCl_2 \cdot 2H_2O} \xrightarrow[PVP]{H_2PdCl_4} \text{硅藻土支撑的聚对苯二甲酸丁二酯}$$

图 4.34 硅藻土负载钯纳米粒子的制备

纳米银/硅藻土纳米复合材料是用一种简便、有效的原位还原方法开发的(见

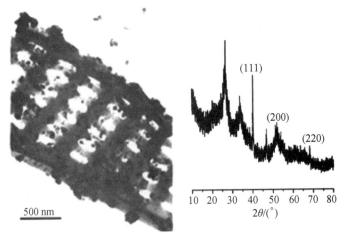

图 4.35 硅藻土支撑钯纳米粒子的 TEM 图和硅藻土负载钯的 XRD 图

图 4.36)。所制备的纳米银/硅藻土纳米复合材料对革兰氏阳性菌和革兰氏阴性菌显示出惊人的抗菌性能。发现原位生成的银纳米颗粒均匀地分散在硅藻土中。它可以在 5 min 内杀死≥99.999% 的细菌(0.5 g 纳米硅藻土处理 100 mL 水),并且最少的银离子浸出到处理过的水中,在水净化行业具有广阔的应用前景。硅醇离子交换过程的示意图如图 4.37 所示。

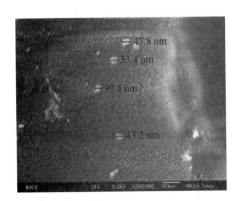

图 4.36 纳米银硅藻土的 SEM 图

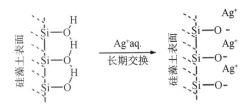

图 4.37 硅醇离子交换过程的示意图

为了大大提高吸附能力,硅藻土由多羟基铝支撑。柱状硅藻土对 Pb^{2+} 和 Cd^{2+} 的吸附量分别比天然硅藻土提高了 23.79% 和 27.36%。柱状硅藻土的表面性质比天然硅藻土更利于离子吸附。结果表明,硅藻土可以通过多羟基铝柱状改性,大大提高其吸附性能。天然硅藻土和不同溶液浓度制备的柱状硅藻土的 SEM 图如图 4.38 所示,柱状原理和柱状硅藻土制备的主要步骤如图 4.39 所示。

以 $Mn(Ac)_2$、$KMnO_4$ 和/或 $MnSO_4$ 为锰源,硅藻土为载体,采用水热法制备了花状、线状和片状 MnO_2/硅藻土。由于使用不同的制备条件(锰来源、反应物、反应

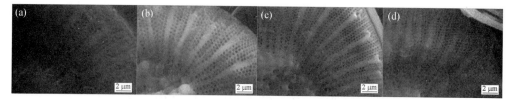

图 4.38　天然硅藻土和不同溶液浓度制备的柱状硅藻土的 SEM 图
(a) 天然硅藻土;(b) 0.1 mol/L;(c) 0.2 mol/L;(d) 0.5 mol/L

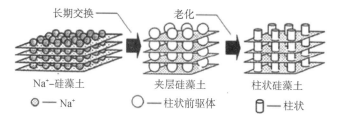

图 4.39　柱状原理和柱状硅藻土制备的主要步骤

温度和反应时间),产生了不同形状的 MnO_2。硅藻土样品呈圆盘状,具有均匀有序的多孔结构,每个圆盘的直径约为 25 μm。大量的花状、线状和片状 MnO_2 纳米粒子随机沉积在硅藻土表面,如图 4.40 所示。结果表明,不同形态的 MnO_2 沉积硅藻土样品具有高表面积和丰富的表面羟基(尤其是线状 MnO_2/硅藻土样品)。线状 MnO_2/硅藻土样品在去除 Cr(Ⅵ) 方面表现出最佳性能,最大 Cr(Ⅵ) 吸附容量为 101 mg·g^{-1}。

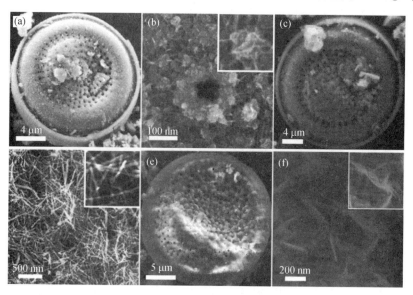

图 4.40　花状、线状和片状 MnO_2/硅藻土 SEM 图
(a)和(b) 花状 MnO_2/硅藻土;(c)和(d) 线状 MnO_2/硅藻土;(e)和(f) 片状 MnO_2/硅藻土

通过在室温下将天然硅藻土与硫酸亚铁溶液混合，在一锅中合成了一种新的氧化铁-硅藻土体系，铁(29%)结晶成针铁矿相(α-FeOOH)。图 4.41 进行的分析进一步强调了针铁矿在硅藻土表面的优先存在，这涉及硅藻土表面硅烷醇基团和离子物种之间的氢键。针铁矿的形成机制被认为是通过羟基离子沉淀亚铁离子，然后铁物质氧化，首先产生纤铁矿(γ-FeOOH)，然后是热力学更稳定的针铁矿。硅藻铁对亚砷酸盐物种阳离子的修复特性在自动平衡 pH 值下进行 24 h 的批量罐试验研究。与天然硅藻土(0.5 mg·g^{-1})相比，这些材料显示出明显更高的 As(Ⅲ) 吸附容量(16 mg·g^{-1})。这些结果可以通过亚砷酸盐与针铁矿的强亲和力来解释，而氧化铁晶体在硅藻土表面的分散进一步强调了这一点。

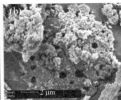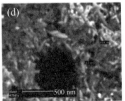

图 4.41　不同放大倍数下硅藻-铁的 SEM 图

粒状掺氮 TiO$_2$/硅藻土(GNTD)复合材料将氮掺杂与颗粒固定相结合，用于水处理和细菌灭活，发现预处理的硅藻土是一种很好的 TiO$_2$ 载体，因为它具有高吸附能力，可以富集 TiO$_2$ 纳米颗粒周围的目标分子，从而提高其光氧化效率，并且造粒增加了其可回收性。N 掺杂剂显示出其将 TiO$_2$ 的光响应区域扩展到可见光区域的效果。这种结构可以有效地缩小 TiO$_2$ 的带隙并扩大其吸附边缘，从而使其对可见光具有响应性。污染物(包括 RhB 和细菌)首先被吸附到硅藻土表面，然后被 N-TiO$_2$ 产生的活性物质降解，如图 4.42 所示。由于预处理硅藻土的吸附能力和 N-TiO$_2$ 的光活性，整个过程在可见光照射下表现出良好的效率。

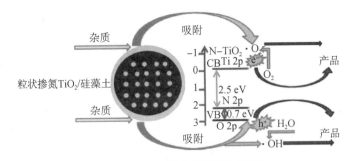

图 4.42　GNTD 的光催化活性机理图

通过沉淀/水热法结合碱处理合成了新型磁性 MnFe$_2$O$_4$/硅藻土复合材料。在温和的条件下，通过水热法成功地将磁性 MnFe$_2$O$_4$ 纳米粒子固定在纯化的硅藻土上。连续碱处理是为了进一步提高其吸附性能。吸附等温线符合朗缪尔等温线模型,在 318 K 时对亚甲基蓝和碱性品红的最大吸附容量分别达到 104.06 mg·g^{-1} 和 284.09 mg·g^{-1}。$MnFe_2O_4$/硅藻土复合材料的吸附行为是一个吸热和自发过程,磁性 $MnFe_2O_4$/硅藻土复合材料可作为吸附剂去除废水中的阳离子染料。搅拌时间对初始浓度为 50 mg/L 的 $MnFe_2O_4$/DE-NaOH,$MnFe_2O_4$,DE-NaOH,DE 和 $MnFe_2O_4$/DE 复合物吸附 MB(a)和 BF(b)的影响如图 4.43 所示。

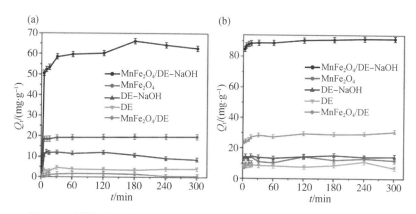

图 4.43 搅拌时间对初始浓度为 50 mg/L 的 $MnFe_2O_4$/DE-NaOH,$MnFe_2O_4$,DE-NaOH,DE 和 $MnFe_2O_4$/DE 复合物吸附 MB(a)和 BF(b)的影响

使用 $KMnO_4$ 和 $(NH_4)_2S_2O_8$ 的水热法制备了 MnO_2 纳米线沉积的硅藻土样品,在 90 ℃ 水热处理不同时间后获得的 MnO_2/硅藻土样品的 SEM 图如图 4.44 所示。这些 MnO_2 纳米线/硅藻土样品显示出高表面积和极好的 Cr(Ⅵ) 和 As(Ⅴ) 吸附行为,最大 Cr(Ⅵ) 和 As(Ⅴ) 吸附容量分别为 197.6 mg·g^{-1} 和 108.2 mg·g^{-1}。Cr(Ⅵ) 或 As(Ⅴ) 吸附机制被证实为 MnO_2 纳米线/硅藻土表面羟基与 Cr(Ⅵ) 或 As(Ⅴ) 物种之间的离子交换。

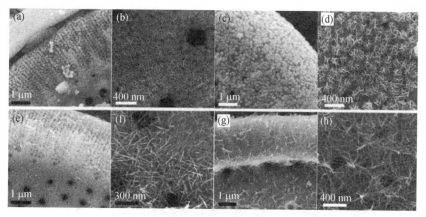

图 4.44 在 90 ℃ 水热处理不同时间后获得的 MnO_2/硅藻土样品的 SEM 图
(a)和(b) 6 h;(c)和(d) 8 h;(e)和(f) 10 h;(g)和(h) 12 h

TiO_2-硅藻土光催化剂是通过溶胶-凝胶法用各种预改性硅藻土制备的。通过紫外光和可见光照射下甲基橙(MO)染料的降解率来评价不同样品的光催化活性。磷酸预处理法制备的样品表现出最高的光催化活性。不同光照下存在 Dt、TiO_2 和 TiO_2-Dt 的 MO 的 C/C_0 随时间变化曲线如图 4.45 所示,紫外线照射 90 min 后,约 90% 的 MO 被最有效的光催化剂分解。在可见光照射 8 h 后,近 60% 的 MO 被同一样品分解。进一步的机理研究表明,H_3PO_4 预处理过程可以明显改变硅藻土载体的表面特征,导致 Si—O—Ti 键的形成,增加 TiO_2 与硅藻土的结合强度,抑制负载 TiO_2 的晶体生长,从而在颗粒空间形成热稳定的介孔结构。它有助于在 TiO_2-硅藻土中构建微孔、中孔和大孔分级多孔结构,提高催化过程中的电荷和传质效率,从而显著提高磷酸预处理的 TiO_2-硅藻土的光催化活性。

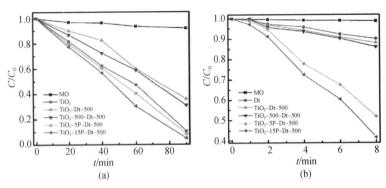

图 4.45 不同光照下存在 Dt、TiO_2 和 TiO_2-Dt 的 MO 的 C/C_0 随时间变化曲线

(a) 可见光;(b) 紫外光

以 $TiCl_4$ 为前驱体,采用典型的水解沉淀法将 TiO_2 纳米粒子固定在硅藻土上。平均粒径为 7~14 nm 的 TiO_2 纳米颗粒很好地沉积在硅藻土表面,硅藻土和 TiO_2 硅藻土的 TEM 图如图 4.46 所示。在紫外光下证明了对还原水性 Cr(Ⅵ) 的光催化活性。与商业二氧化钛(P25,Degussa)相比,TiO_2/硅藻土复合材料由于其相对较高的吸附能力而具有更好的反应活性。此外,根据可重复使用性测试,制备的光催化剂表现出相对良好的光催化稳定性。

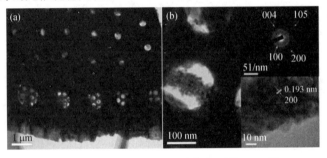

图 4.46 硅藻土(a)和 TiO_2 硅藻土(b)的 TEM 图

采用改进的溶胶-凝胶法合成了不同钒浓度的掺钒(V)TiO₂/硅藻土复合光催化剂,纯 TiO₂ 和 TiO₂/硅藻土的 TEM 图如图 4.47 所示。硅藻土使 TiO₂ 纳米粒子在基体上的分散良好,从而抑制团聚。V-TiO₂/硅藻土杂化物在 TiO₂ 吸收边出现红移,吸收强度增强,如图 4.48 所示。最重要的是,由于 V^{4+} 离子取代了 Ti^{4+} 位点,在 TiO₂ 带隙中形成了掺杂能级。与未掺杂样品(3.13 eV)和其他掺杂样品(3.05 eV)相比,0.5% V-TiO₂/硅藻土光催化剂显示出更窄的带隙(2.95 eV)。与未掺杂的样品相比,V-TiO₂/硅藻土样品在太阳光照射下降解罗丹明 B 的光催化活性显著提高,如图 4.49 所示。在我们的实验中,掺入 TiO₂ 晶格中的 V^{4+} 离子负责增加可见光吸收和电子转移到吸附在 TiO₂ 表面的超氧自由基,而以 V₂O₅ 形式存在于 TiO₂ 颗粒表面的 V^{5+} 离子有助于 e^-—h^+ 分离。此外,由于硅藻土作为载体的组合,这种混合光催化剂可以通过自然沉降快速从溶液中分离,并表现出良好的可重复使用性。

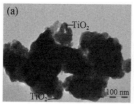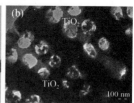

图 4.47 纯 TiO₂(a)和 TiO₂/硅藻土(b)的 TEM 图

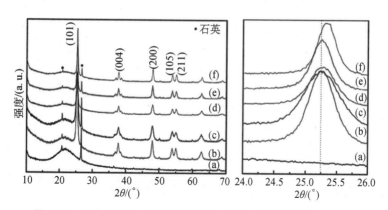

图 4.48 硅藻土、TiO₂/硅藻土和 V-TiO₂/硅藻土复合材料的 XRD 图
(a) DE;(b) TD;(c) 0.25% V/TD;(d) 0.5% V/TD;(e) 1.0% V/TD;(f) 1.5% V/TD

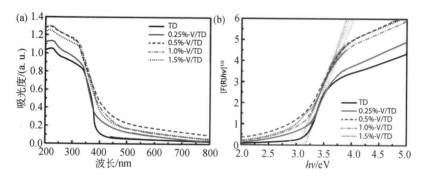

图 4.49 未掺杂和 V 掺杂 TiO₂/硅藻土复合材料的紫外-可见漫反射光谱(DRS)(a)和转换的 Kubelka Munk 函数与光能的关系图(b)

通过简便的浸渍-煅烧方法成功合成了新型可见光响应硅藻土/g-C_3N_4 复合材料。图 4.50 显示了 g-C_3N_4、原始硅藻土、碱处理硅藻土和硅藻土/g-C_3N_4 复合材料的尺寸和形态。可以看出,硅藻土样品具有盘状和高度发达的大孔结构。NaOH 洗涤后,原始硅藻土的形态得以保留,而中央大孔的尺寸增加[见图 4.50(a)和(b)]。将硅藻土加载到层状 g-C_3N_4 的表面后,可以观察到硅藻土/g-C_3N_4 复合材料呈大量的蓬松薄片状[见图 4.50(c)]。图 4.50(d)是含 2.32% 硅藻土的 g-C_3N_4 的 TEM 图像,我们可以清楚地观察到一些硅藻土颗粒负载在层状结构的 g-C_3N_4 的表面。通过 HR-TEM 图可以清楚地观察到硅藻土和 g-C_3N_4 的边界[见图 4.50(f)],表明硅藻土层与 g-C_3N_4 结合得很好。硅藻土/g-C_3N_4 复合材料对 RhB 的降解表现出很高的效率。在可见光照射下,光反应动力学常数值约为 g-C_3N_4 的 1.9 倍。增强的光活性主要归因于 g-C_3N_4 与带负电荷的硅藻土之间的静电相互作用,协同效应导致 g-C_3N_4 光生电子和空穴的有效迁移。

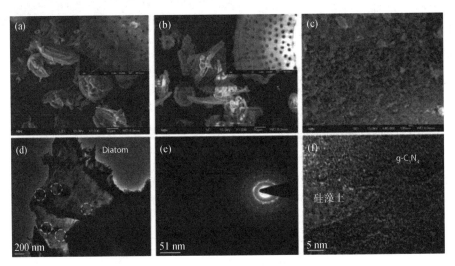

图 4.50 不同硅藻土样品的 FE-SEM 和 TEM 图

(a) 硅藻土;(b) 碱洗硅藻土;(c) 硅藻土/g-C_3N_4 复合材料;(d) 硅藻土/g-C_3N_4 复合材料;(e) g-C_3N_4 的 SEAD 模式;(f) 鳞片状 g-C_3N_4 和硅藻土颗粒

以植酸为分子黏合剂,通过逐层(LBL)组装方法在大孔硅藻土表面沉积了粒径为 10 nm 的 TiO_2 胶体,如图 4.51 所示。LBL 组装的基础主要是从溶液中沉积到胶体球上的带相反电荷的物质之间的静电吸引力,这非常适合生产胶体稳定、均匀包覆的颗粒。这种灵活而简便的程序允许用不同成分的均匀层涂覆各种形状和尺寸的胶体。为了继续沉积过程并形成 TiO_2 纳米颗粒的三维涂层,可以使用"黏合剂"分子,如图 4.52 所示。在这里,植酸被选为 TiO_2 的理想"黏合剂"分子。植酸是众所周知的天然存在的酸,六个磷酸官能团对称地连接到环己烷氧化环上。

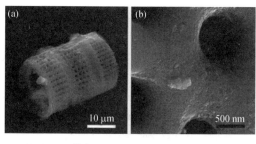 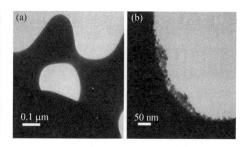

图 4.51　带有 TiO$_2$ 涂层的硅藻土的 SEM 图

图 4.52　初始硅藻土(a)和带有 TiO$_2$ 涂层的硅藻土(b)的 TEM 图

研究者制备了一种高铁酸钾缓释复合体系(缓释 K$_2$FeO$_4$)并应用于水处理,如图 4.53 所示。本研究的目的是通过使用硅藻土增强 K$_2$FeO$_4$ 的稳定性来最大限度地提高 K$_2$FeO$_4$ 对水处理的有效性,他们发现 K$_2$FeO$_4$ 的分解速率明显降低,大大提高了释放的 K$_2$FeO$_4$ 与污染物的接触速率。通过降解作为模型污染物的甲基橙,研究者研究了完全缓释 K$_2$FeO$_4$ 体系中 K$_2$FeO$_4$ 含量的影响因素。缓释 K$_2$FeO$_4$ 系统中的最佳 K$_2$FeO$_4$ 含量约为 70%。在天然水样中,缓释 K$_2$FeO$_4$ 用量为 0.06 g/L,反应时间为 20 min,去除了 36.84% 的可溶性微生物产物和 17.03% 的简单芳香蛋白,这些去除率优于传统氯消毒后观察到的去除率。

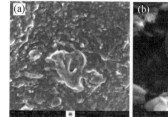   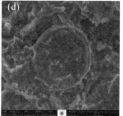

图 4.53　缓释 K$_2$FeO$_4$ 体系的 SEM 图

(a) K$_2$FeO$_4$;(b) 硅藻土;(c)和(d) 硅藻土/K$_2$FeO$_4$

4.3　基于硅藻壳结构的复合凹坑摩擦学性能研究

随着对减少摩擦和能源消耗的需求不断增加,表面纹理由于其作为润滑剂的流体动力学作用而引起了广泛关注。随着仿生技术的发展,可以通过模拟生物系统的高级结构来获得先进的表面纹理机制。现有研究表明,自然界中超过 25% 的生产力是由硅藻提供的[148],其硅藻壳(即壳)由多层孔隙结构组成[34,97]。即使在同一孔的底部,有时也会出现一个或多个孔。这可以通过代表性圆筛藻属未定种的两级圆柱孔来证明。如图 4.54 所示,其中最外面的圆柱形孔的直径约为 1.212 μm,而在其底部

2.23 μm 约为第二级圆柱形孔的直径。

图 4.54　圆筛藻属的 SEM 图

(a) 圆筛藻属的外壳;(b) 圆筛藻属的双层多孔结构:从内部观察,显示孔隙的第二层和外层

硅藻孔形状主要是立方体、圆柱形或六边形[149-150]。正是精细的多层次孔隙结构,使硅藻具有较高的回弹性和拉伸性能等良好的力学性能,从而在进化中得以生存。例如,代表性的圆筛藻属未定种的每一层孔隙的弹性模量分别高达 3.4 GPa、1.7 GPa 和 15.61 GPa[18,24]。测试表明圆筛藻属未定种的硬度达到 0.12 GPa[151]。此外,已证实硅藻壳可以承受 150~680 N·mm^{-1} 范围内的更高应力[152]。迄今为止,通过实验对硅藻的摩擦学特性进行的研究还很少。用原子力显微镜观察发现,硅藻成分之间的摩擦磨损可以通过其自润滑来克服,Aulacoseira granulate 和 Ellerbeckia arenaria(即两种硅藻)的条带可以充当滚珠轴承或固体润滑剂[153-154]。

尽管上面对简单的凹坑效应和硅藻壳进行了研究,但对硅藻多层次孔结构的摩擦学性能,特别是其在工程表面的仿生应用的模拟研究还很少。

作为对上述表面织构研究中使用的简单凹坑的扩展探索,利用硅藻结构的二级孔及其延伸形状对平行滑动表面进行织构,在本研究中称为复合凹坑。每个复合凹坑由两级孔组成,其对表面摩擦学性能的影响通过流固相互作用(FSI)方法进行研究。这种方法的应用没有考虑到硅藻壳在水中移动时由于环境高水压而引起的变形。该方法在求解水膜压力时将硅藻壳变形纳入水膜厚度中,得到的水膜压力在进一步分析硅藻壳变形时考虑,这是一个双向耦合过程。

Meng 等研究了具有不同复合和简单凹坑形状的物理模型。基于这些模型,复合凹坑的流体动力润滑作用用 Navier-Strokes 方程求解,因为它可以克服 Reynolds 方程在润滑膜惯性力明显或膜厚比(即润滑膜厚度与匹配表面之间的间隙之比)较小的条件下预测摩擦副润滑性能的局限性,其中润滑剂容易发生湍流[155-157]。同时,利用固体本构方程求解织构化表面的变形和应力等力学性能。接下来,在典型复合凹坑和简单凹坑形状之间以及在复合凹坑形状之间对平行滑动表面的摩擦学性能进行比较,以找到最佳复合凹坑。

参考文献

[1] Schlapbach, L., Züttel, A. (2001). Hydrogen-storage materials for mobile applications, Nature, 414 (6861), 353-358.
[2] Aricò, A. S., Bruce, P., Scrosati, B., Tarascon, J.-M., van Schalkwijk, W. (2005). Nanostructured materials for advanced energy conversion and storage devices, Nat. Mater., 4(5), 366-377.
[3] Nakajima, T., Volcani, B. E. (1969). 3, 4-Dihydroxyproline: A new amino acid in diatom cell walls, Science, 164(3886), 1400-1401.
[4] Sumper, M., Lorenz, S., Brunner, E. (2003). Biomimetic control of size in the polyamine-directed formation of silica nanospheres, Angew. Chem. Int. Ed., 42(42), 5192-5195.
[5] Hirscher, M. (2004). Nanoscale materials for energy storage, Materials Science and Engineering: B, 108 (1-2), 1-1.
[6] Klaine, S. J., Alvarez, P. J. J., Batley, G. E., Fernandes, T. F., Handy, R. D., Lyon, D. Y. et al. (2008). Nanomaterials in the environment: behavior, fate, bioavailability, and effects, Environ. Toxicol. Chem., 27(9), 1825-1851.
[7] Biswas, P., Wu, C.-Y. (2005). Nanoparticles and the Environment, J. Air Waste Manage. Assoc., 55 (6), 708-746.
[8] Cerneaux, S., Zakeeruddin, S. M., Pringle, J. M., Cheng, Y.-B., Grätzel, M., Spiccia, L. (2007). Novel nano-structured silica-based electrolytes containing quaternary ammonium iodide moieties, Adv. Funct. Mater., 17(16), 3200-3206.
[9] Nel, A., Xia, T., Madler, L., Li, N. (2006). Toxic potential of materials at the nanolevel, Science, 311 (5761), 622-627.
[10] Tibbitt, M. W., Dahlman, J. E., Langer, R. (2016). Emerging frontiers in drug delivery, J. Am. n Chem. Soc., 138(3), 704-717.
[11] Wang, Y., Santos, A., Evdokiou, A., Losic, D. (2015). An overview of nanotoxicity and nanomedicine research: principles, progress and implications for cancer therapy, J. Mater. Chem. B, 3(36), 7153-7172.
[12] Wang, C., Li, Y., Ostrikov, Kostya (Ken)., Yang, Y., Zhang, W. (2015). Synthesis of SiC decorated carbonaceous nanorods and its hierarchical composites Si@SiC@C for high-performance lithium ion batteries, J. Alloys Compd., 646, 966-972.
[13] Maher, S., Kumeria, T., Wang, Y., Kaur, G., Fathalla, D., Fetih, G. et al. (2016). From the mine to cancer therapy: Natural and biodegradable theranostic silicon nanocarriers from diatoms for sustained delivery of chemotherapeutics, Adv. Healthc. Mater., 5(20), 2667-2678.
[14] Aw, M. S., Simovic, S., Addai-Mensah, J., Losic, D. (2011). Silica microcapsules from diatoms as new carrier for delivery of therapeutics, Nanomedicine, 6(7), 1159-1173.
[15] Maher, S., Alsawat, M., Kumeria, T., Fathalla, D., Fetih, G., Santos, A. et al. (2015). Luminescent silicon diatom replicas: self-reporting and degradable drug carriers with biologically derived shape for sustained delivery of therapeutics, Adv. Funct. Mater., 25(32), 5107-5116.
[16] Mann, D. G. (1999). The species concept in diatoms, Phycologia, 38(6), 437-495.
[17] De Stefano, M., De Stefano, L. (2005). Nanostructures in diatom frustules: functional morphology of valvocopulae in cocconeidacean monoraphid taxa, J. Nanosci. Nanotechnol., 5(1), 15-24.
[18] Losic, D., Mitchell, J. G., Voelcker, N. H. (2009). Diatomaceous lessons in nanotechnology and advanced materials, Adv. Mater., 21(29), 2947-2958.
[19] Yu, Y., Addai-Mensah, J., Losic, D. (2011). Chemical functionalization of diatom silica microparticles for adsorption of gold (iii) ions, J. Nanosci. Nanotechnol., 11(12), 10349-10356.
[20] Yu, Y., Addai-Mensah, J., Losic, D. (2010). Synthesis of self-supporting gold microstructures with three-dimensional morphologies by direct replication of diatom templates, Langmuir, 26(17), 14068-14072.
[21] Bao, L., Zang, J., Li, X. (2011). Flexible Zn_2SnO_4/MnO_2 core/shell nanocable-carbon microfiber hybrid composites for high-performance supercapacitor electrodes, Nano Lett., 11(3), 1215-1220.
[22] Gordon, R., Drum, R. W. (1994). Int. Rev. Cytol., 150, 243-372.
[23] Lewin, R. A. (1990). Skeletons on the coffee table, Nature, 346(6285), 619-620.
[24] Losic, D., Short, K., Mitchell, J. G., Lal, R., Voelcker, N. H. (2007). AFM nanoindentations of diatom biosilica surfaces, Langmuir, 23(9), 5014-5021.
[25] Nassif, N., Livage, J. (2011). From diatoms to silica-based biohybrids, Chem. Soc. Rev., 40(2), 849-859.
[26] Sumper, M. (2002). A phase separation model for the nanopatterning of diatom biosilica, Science, 295

(5564), 2430-2433.
[27] Li, C., Wang, F., Yu, J. C. (2011). Semiconductor/biomolecular composites for solar energy applications, Energy Environ. Sci., 4(1), 100-113.
[28] Zhang, Q., Chou, T. P., Russo, B., Jenekhe, S. A., Cao, G. (2008). Aggregation of ZnO nanocrystallites for high conversion efficiency in dye-sensitized solar cells, Angew. Chem. Int. Ed., 47(13), 2402-2406.
[29] Jeffryes, C., Gutu, T., Jiao, J., Rorrer, G. L. (2008). Metabolic insertion of nanostructured TiO_2 into the patterned biosilica of the diatom pinnularia sp. by a two-stage bioreactor cultivation process, ACS Nano, 2(10), 2103-2112.
[30] Noll, F., Sumper, M., Hampp, N. (2002). Nanostructure of diatom silica surfaces and of biomimetic analogues, Nano Lett., 2(2), 91-95.
[31] Anderson, M. W., Holmes, S. M., Hanif, N., Cundy, C. S. (2000). Hierarchical Pore Structures through Diatom Zeolitization, Angew. Chem. Int. Ed., 39(15), 2707-2710.
[32] Tachibana, Y., Akiyama, H. Y., Kuwabata, S. (2007). Optical simulation of transmittance into a nanocrystalline anatase TiO_2 film for solar cell applications, Solar Energy Materials and Solar Cells, 91(2-3), 201-206.
[33] Medarevic, D., Losic, D., Ibric, S. (2016). Diatoms-nature materials with great potential for bioapplications, Hem. Ind., 70(6), 613-627.
[34] Gordon, R., Losic, D., Tiffany, M. A., Nagy, S. S., Sterrenburg, F. A. (2009). The Glass Menagerie: diatoms for novel applications in nanotechnology, Trends Biotechnol., 27(2), 116-127.
[35] Weatherspoon, M. R., Allan, S. M., Hunt, E., Cai, Y., Sandhage, K. H. (2005). Sol-gel synthesis on self-replicating single-cell scaffolds: applying complex chemistries to nature's 3-D nanostructured templates, Chem. Commun., 42(5), 651-653.
[36] Zhao, J., Gaddis, C. S., Cai, Y. (2005). KH Sandhage, 20, 282-287.
[37] Ernst, E. M., Church, B. C., Gaddis, C. S., Snyder, R. L., Sandhage, K. H. (2007). Enhanced hydrothermal conversion of surfactant-modified diatom microshells into barium titanate replicas, J. Mater. Res., 22(05), 1121-1127.
[38] Losic, D., Triani, G., Evans, P. J., Atanacio, A., Mitchell, J. G., Voelcker, N. H. (2006). Controlled pore structure modification of diatoms by atomic layer deposition of TiO_2, J. Mater. Chem., 16(41), 4029-4034.
[39] Losic, D., Yu, Y., Aw, M. S., Simovic, S., Thierry, B., Addai-Mensah, J. (2010). Surface functionalization of diatoms with dopamine modified iron-oxide nanoparticles: toward magnetically guided drug microcarriers with biologically derived morphologies, Chem. Commun., 46(34), 6323-6325.
[40] Toster, J., Iyer, K. S., Xiang, W., Rosei, F., Spiccia, L., Raston, C. L. (2013). Diatom frustules as light traps enhance DSSC efficiency, Nanoscale, 5(3), 873-876.
[41] Jantschke, A., Herrmann, A.-K., Lesnyak, V., Eychmüller, A., Brunner, E. (2012). Decoration of diatom Biosilica with noble metal and semiconductor nanoparticles (<10 nm): assembly, characterization, and applications, Chem. Asian J., 7(1), 85-90.
[42] Rosi, N. L., Thaxton, C. S., Mirkin, C. A. (2004). Control of nanoparticle assembly by using dnamodified diatom templates, Angew. Chem. Int. Ed., 43(41), 5500-5503.
[43] Cai, Y., Sandhage, K. H. (2005). Zn_2SiO_4-coated microparticles with biologically-controlled 3D shapes, phys. stat. sol., 202(10), R105-R107.
[44] Schoenwaelder, M. E. A. (2002). Novel, bioclastic route to self-assembled, 3D, chemically tailored meso/nanostructures: Shape-preserving reactive conversion of biosilica (diatom) microshells. Advanced Materials, 14(6), 429.
[45] Unocic, R. R., Zalar, F. M., Sarosi, P. M., Cai, Y., Sandhage, K. H. (2004). Anatase assemblies from algae: coupling biological self-assembly of 3-D nanoparticle structures with synthetic reaction chemistry, Chem. Commun., 796-797.
[46] Cai, Y., Dickerson, M. B., Haluska, M. S., Kang, Z., Summers, C. J., Sandhage, K. H. (2007). Manganese-Doped Zinc Orthosilicate-Bearing Phosphor Microparticles with Controlled Three-Dimensional Shapes Derived from Diatom Frustules, J. American Ceramic Society, 90(4), 1304-1308.
[47] Bao, Z., Weatherspoon, M. R., Shian, S., Cai, Y., Graham, P. D., Allan, S. M. et al. (2007). Chemical reduction of three-dimensional silica micro-assemblies into microporous silicon replicas, Nature, 446

(7132), 172-175.
[48] Shen, L., Wang, Z., Chen, L. (2014). Carbon-coated hierarchically porous silicon as anode material for lithium ion batteries, RSC Adv., 4(29), 15314-15318.
[49] Dahn, J. R., Zheng, T., Liu, Y., Xue, J. S. (1995). Mechanisms for lithium insertion in carbonaceous materials, Science, 270(5236), 590-593.
[50] Etacheri, V., Marom, R., Elazari, R., Salitra, G., Aurbach, D. (2011). Challenges in the development of advanced Li-ion batteries: a review, Energy Environ. Sci., 4(9), 3243-3262.
[51] Poizot, P., Laruelle, S., Grugeon, S., Dupont, L., Tarascon, J. M. (2001). J. Power Sources, 97-8, 235-239.
[52] Wen, C. J., Huggins, R. A. (1981). Thermodynamic study of the lithium-tin system, J. Electrochem. Soc., 128(6), 1181-1187.
[53] Losic, D., Pillar, R. J., Dilger, T., Mitchell, J. G., Voelcker, N. H. (2007a). Atomic force microscopy (AFM) characterisation of the porous silica nanostructure of two centric diatoms, J. Porous Mater., 14(1), 61-69.
[54] Jeffryes, C., Campbell, J., Li, H., Jiao, J., Rorrer, G. (2011). The potential of diatom nanobiotechnology for applications in solar cells, batteries, and electroluminescent devices, Energy Environ. Sci., 4(10), 3930-3941.
[55] Baranauskas, V., Chang, D. C., Li, B. B., Peterlevitz, A. C., Trava-Airoldi, V. J., Corat, E. J. et al. (2000). Journal of Porous Materials, 7(1/3), 401-405.
[56] Campbell, B., Ionescu, R., Tolchin, M., Ahmed, K., Favors, Z., Bozhilov, K. N. et al. (2016). Sci Rep-Uk, 6.
[57] Wang, M.-S., Fan, L.-Z., Huang, M., Li, J., Qu, X. (2012). Conversion of diatomite to porous Si/C composites as promising anode materials for lithium-ion batteries, J. Power Sources, 219, 29-35.
[58] Kang, S. M., Ryou, M.-H., Choi, J. W., Lee, H. (2012). Mussel-and diatom-inspired silica coating on separators yields improved power and safety in li-ion batteries, Chem. Mater., 24(17), 3481-3485.
[59] Liu, J., Kopold, P., van Aken, P. A., Maier, J., Yu, Y. (2015). Energy storage materials from nature through nanotechnology: a sustainable route from reed plants to a silicon anode for lithiumion batteries, Angew. Chem. Int. Ed., 54(33), 9632-9636.
[60] Lisowska-Oleksiak, A., Nowak, A. P., Wicikowska, B. (2014). Aquatic biomass containing porous silica as an anode for lithium ion batteries, RSC Adv., 4(76), 40439-40443.
[61] Wu, X., Shi, Z.-qiang., Wang, C.-yang., Jin, J, Shi, Z. Q, Wang, C. Y. (2015). Nanostructured SiO_2/C composites prepared via electrospinning and their electrochemical properties for lithium ion batteries, Journal of Electroanalytical Chemistry, 746, 62-67.
[62] Chen, J., Lu, X., Sun, J., Xu, F. (2015). Si@C nanosponges application for lithium ions batteries synthesized by templated magnesiothermic route, Mater. Lett., 152, 256-259.
[63] Liang, J., Li, X., Zhu, Y., Guo, C., Qian, Y. (2015). Hydrothermal synthesis of nano-silicon from a silica sol and its use in lithium ion batteries, Nano Res., 8(5), 1497-1504.
[64] Conway, B. E. (1999). Kluwer-Plenum Publishing Corp.
[65] Brezesinski, T., Wang, J., Tolbert, S. H., Dunn, B. Ordered mesoporous alpha-MoO_3 with isooriented nanocrystalline walls for thin-film pseudocapacitors. Nat. Mater., 9(2), 146-151, 2010.
[66] Hou, Y., Cheng, Y., Hobson, T., Liu, J. (2010). Design and synthesis of hierarchical MnO_2 nanospheres/carbon nanotubes/conducting polymer ternary composite for high performance electrochemical electrodes, Nano Lett., 10(7), 2727-2733.
[67] Chen, W., Rakhi, R. B., Hu, L., Xie, X., Cui, Y., Alshareef, H. N. (2011). High-performance nanostructured supercapacitors on a sponge, Nano Lett., 11(12), 5165-5172.
[68] Rakhi, R. B., Chen, W., Cha, D., Alshareef, H. N. (2012). Substrate Dependent Self-Organization of Mesoporous Cobalt Oxide Nanowires with Remarkable Pseudocapacitance, Nano Lett., 12(5), 2559-2567.
[69] Simon, P., Gogotsi, Y. (2008). Materials for electrochemical capacitors, Nat. Mater., 7(11), 845-854.
[70] Wang, G., Zhang, L., Zhang, J. (2012). A review of electrode materials for electrochemical supercapacitors, Chem. Soc. Rev., 41(2), 797-828.
[71] Wang, J.-G., Yang, Y., Huang, Z.-H., Kang, F. (2013). Effect of temperature on the pseudo-capacitive behavior of freestanding MnO_2@ carbon nanofibers composites electrodes in mild electrolyte, J. Power Sources, 224, 86-92.

[72] Wang, F., Xiao, S., Hou, Y., Hu, C., Liu, L., Wu, Y. (2013). Electrode materials for aqueous asymmetric supercapacitors, RSC Adv., 3(32), 13059-13084.

[73] Zhang, Y. X., Huang, M., Kuang, M., Liu, C. P., Tan, J. L., Dong, M., Yuan, Y., Zhao, X. L., and Wen, Z. Q. (2013). Facile Synthesis of Mesoporous CuO Nanoribbons for Electrochemical Capacitors Applications, International Journal of Electrochemical Science 8, 1366-1381.

[74] Santhanagopalan, S., Balram, A., Meng, D. D. (2013). Scalable high-power redox capacitors with aligned nanoforests of crystalline MnO_2 nanorods by high voltage electrophoretic deposition, ACS Nano, 7(3), 2114-2125.

[75] Huang, M., Li, F., Dong, F., Zhang, Y.X., Zhang, L.L. (2015). MnO_2-based nanostructures for high-performance supercapacitors, J. Mater. Chem. A, 3(43), 21380-21423.

[76] Aravindan, V., Suresh Kumar, P., Sundaramurthy, J., Ling, W. C., Ramakrishna, S., Madhavi, S. (2013). Electrospun NiO nanofibers as high performance anode material for Li-ion batteries, J. Power Sources, 227, 284-290.

[77] Wang, Z., Ma, C., Wang, H., Liu, Z., Hao, Z. (2013). Facilely synthesized Fe_2O_3-graphene nanocomposite as novel electrode materials for supercapacitors with high performance, J. Alloys Compd., 552, 486-491.

[78] Liang, R., Cao, H., Qian, D. (2011). MoO_3 nanowires as electrochemical pseudocapacitor materials, Chem. Commun., 47(37), 10305-10307.

[79] Qu, Q., Zhu, Y., Gao, X., Wu, Y. (2012). Core-shell structure of polypyrrole grown on V_2O_5 nanoribbon as high performance anode material for supercapacitors, Adv. Energy Mater., 2(8), 950-955.

[80] Xiong, G., Hembram, K. P. S. S., Reifenberger, R. G., Fisher, T. S. (2013). MnO_2-coated graphitic petals for supercapacitor electrodes, J. Power Sources, 227, 254-259.

[81] Wu, Q., Liu, Y., Hu, Z. (2013). Flower-like NiO microspheres prepared by facile method as supercapacitor electrodes, J. Solid State Electrochem., 17(6), 1711-1716.

[82] Wei, W., Cui, X., Chen, W., Ivey, D. G. (2011). Manganese oxide-based materials as electrochemical supercapacitor electrodes, Chem. Soc. Rev., 40(3), 1697-1721.

[83] Peng, L., Peng, X., Liu, B., Wu, C., Xie, Y., Yu, G. (2013). Ultrathin two-dimensional MnO_2/graphene hybrid nanostructures for high-performance, flexible planar supercapacitors, Nano Lett., 13(5), 2151-2157.

[84] Yang, P., Ding, Y., Lin, Z., Chen, Z., Li, Y., Qiang, P. et al. (2014). Low-cost high-performance solid-state asymmetric supercapacitors based on MnO_2 nanowires and Fe_2O_3 nanotubes, Nano Lett., 14 (2), 731-736.

[85] Kai, K., Kobayashi, Y., Yamada, Y., Miyazaki, K., Abe, T., Uchimoto, Y. et al. (2012). Electrochemical characterization of single-layer MnO_2 nanosheets as a high-capacitance pseudocapacitor electrode, J. Mater. Chem., 22(29), 14691-14695.

[86] Li, Q., Wang, Z.-L., Li, G.-R., Guo, R., Ding, L.-X., Tong, Y.-X. (2012). Design and synthesis of MnO_2/Mn/MnO_2 sandwich-structured nanotube arrays with high supercapacitive performance for electrochemical energy storage, Nano Lett., 12(7), 3803-3807.

[87] Su, M., Zhang, Y., Song, X., Ge, S., Yan, M., Yu, J. et al. (2013). Three-dimensional nanoflowerlike MnO_2 functionalized graphene as catalytically promoted nanolabels for ultrasensitive electrochemiluminescence immunoassay, Electrochim. Acta, 97, 333-340.

[88] Liu, M., Gan, L., Xiong, W., Xu, Z., Zhu, D., Chen, L. (2014a). Development of MnO_2/porous carbon microspheres with a partially graphitic structure for high performance supercapacitor electrodes, J. Mater. Chem. A, 2(8), 2555-2562.

[89] Zhang, Y.X., Li, F., Huang, M., Xing, Y., Gao, X., Li, B. et al. (2014). Hierarchical NiO moss decorated diatomites via facile and templated method for high performance supercapacitors, Mater. Lett., 120, 263-266.

[90] Li, F., Xing, Y., Huang, M., Li, K.L., Yu, T.T., Zhang, Y.X. et al. (2015). MnO_2 nanostructures with three-dimensional (3D) morphology replicated from diatoms for high-performance supercapacitors, J. Mater. Chem. A, 3(15), 7855-7861.

[91] Guo, X.L., Kuang, M., Li, F., Liu, X.Y., Zhang, Y.X., Dong, F. et al. (2016). Engineering of three dimensional (3-D) diatom @ TiO_2 @ MnO_2 composites with enhanced supercapacitor performance, Electrochim. Acta, 190, 159-167.

[92] Wen, Z. Q., Li, M., Li, F., Zhu, S. J., Liu, X. Y., Zhang, Y. X., Kumeria, T., Losic, D., Gao, Y., Zhang, W., and He, S. X. (2016) Morphology-controlled MnO_2-graphene oxide-diatomaceous earth 3-dimensional (3D) composites for high-performance supercapacitors, Dalton Transactions 45, 936–942.

[93] Zhang, Y. X., Huang, M., Li, F., Wang, X. L., Wen, Z. Q. (2014). One-pot synthesis of hierarchical MnO_2-modified diatomites for electrochemical capacitor electrodes, J. Power Sources, 246, 449–456.

[94] Hu, L., Qu, B., Chen, L., Li, Q. (2013). Low-temperature preparation of ultrathin nanoflakes assembled tremella-like NiO hierarchical nanostructures for high-performance lithium-ion batteries, Mater. Lett., 108, 92–95.

[95] Mai, Y. J., Tu, J. P., Xia, X. H., Gu, C. D., Wang, X. L. (2011). Co-doped NiO nanoflake arrays toward superior anode materials for lithium ion batteries, J. Power Sources, 196(15), 6388–6393.

[96] Marcinauskas, L., Kavaliauskas, Žydrunas., Valinčius, V. (2012). Carbon and nickel oxide/carbon composites as electrodes for supercapacitors, Journal of Materials Science & Technology, 28(10), 931–936.

[97] Bhatta, H., Kong, T. K., Rosengarten, G. (2009). Diffusion through Diatom Nanopores, JNanoR, 7(7), 69–74.

[98] Bruchez, M., Hotz, C. (2007). Humana.

[99] Hernández-Alonso, M. D., Fresno, F., Suárez, S., Coronado, J. M. (2009). Development of alternative photocatalysts to TiO_2: Challenges and opportunities, Energy Environ. Sci., 2(12), 1231–1257.

[100] Calzaferri, G. (2010). Artificial photosynthesis, Top. Catal., 53(3–4), 130–140.

[101] Chen, X., Mao, S. S. (2007). Titanium dioxide nanomaterials: synthesis, properties, modifications, and applications, Chem. Rev., 107(7), 2891–2959.

[102] Inoue, Y. (2009). Photocatalytic water splitting by RuO_2-loaded metal oxides and nitrides with d0-and d10-related electronic configurations, Energy Environ. Sci., 2(4), 364–386.

[103] Gonçalves, L. M., de Zea Bermudez, V., Ribeiro, H. A., Mendes, A. M. (2008). Dye-sensitized solar cells: A safe bet for the future., Energy Environ. Sci., 1(6), 655–667.

[104] Nazeeruddin, M. K., Kay, A., Rodicio, I., Humphry-Baker, R., Mueller, E., Liska, P. et al. (1993). Conversion of light to electricity by cis-X2bis (2,2′-bipyridyl-4,4′-dicarboxylate) ruthenium (II) charge-transfer sensitizers (X=Cl-, Br-, I-, CN-, and SCN-) on nanocrystalline titanium dioxide electrodes, J. Am. Chem. Soc., 115(14), 6382–6390.

[105] O'Regan, B., Grätzel, M. (1991). A low-cost, high-efficiency solar cell based on dye-sensitized colloidal TiO_2 films, Nature, 353(6346), 737–740.

[106] Zhang, Q., Dandeneau, C. S., Zhou, X., Cao, G. (2009). Zno nanostructures for dye-sensitized solar cells, Adv. Mater., 21(41), 4087–4108.

[107] Duong, T.-T., Choi, H.-J., He, Q.-J., Le, A.-T., Yoon, S.-G. (2013). Enhancing the efficiency of dye sensitized solar cells with an SnO_2 blocking layer grown by nanocluster deposition, J. Alloys Compd., 561, 206–210.

[108] Barea, E., Xu, X., González-Pedro, V., Ripollés-Sanchis, T., Fabregat-Santiago, F., Bisquert, J. (2011). Origin of efficiency enhancement in Nb_2O_5 coated titanium dioxide nanorod based dye sensitized solar cells, Energy Environ. Sci., 4(9), 3414–3419.

[109] Grätzel, M. (2003). Dye-sensitized solar cells, Journal of Photochemistry and Photobiology C: Photochemistry Reviews, 4(2), 145–153.

[110] Pan, L., Sander, M. B., Huang, X., Li, J., Smith, M., Bittner, E. et al. (2004). Microporous metal organic materials: promising candidates as sorbents for hydrogen storage, J. Am. Chem. Soc., 126(5), 1308–1309.

[111] Fuhrmann, T., Landwehr, S., El Rharbi-Kucki, M., Sumper, M. (2004). Diatoms as living photonic crystals, Appl. Phys. B, 78(3–4), 257–260.

[112] Huang, D.-R., Jiang, Y.-J., Liou, R.-L., Chen, C.-H., Chen, Y.-A., Tsai, C.-H. (2015). Enhancing the efficiency of dye-sensitized solar cells by adding diatom frustules into TiO_2 working electrodes, Appl. Surf. Sci., 347, 64–72.

[113] Chandrasekaran, S., Sweetman, M. J., Kant, K., Skinner, W., Losic, D., Nann, T. et al. (2014). Silicon diatom frustules as nanostructured photoelectrodes, Chem. Commun., 50(72), 10441–10444.

[114] Rosi, N. L., Eckert, J., Eddaoudi, M., Vodak, D. T., Kim, J., O'Keeffe, M. (2003). Hydrogen storage in microporous metal-organic frameworks, Science, 300(5622), 1127–1129.

[115] Mu, S.C., Pan, M., Yuan, R.Z. (2005). A new concept: Hydrogen storage in minerals, MSF, 475–479, 2441–2444.

[116] Karatepe, N., Erdoğan, N., Ersoy-Meriçboyu, A., Küçükbayrak, S. (2004). Preparation of diatomite/$Ca(OH)_2$ sorbents and modelling their sulphation reaction, Chem. Eng. Sci., 59(18), 3883–3889.

[117] Jin, J., Zheng, C.H., Yang, H.M. (2014). Funct Mater Lett, 7.

[118] Dincer, I. (2002). On thermal energy storage systems and applications in buildings, Energy and Buildings, 34(4), 377–388.

[119] Khudhair, A.M., Farid, M.M. (2004). A review on energy conservation in building applications with thermal storage by latent heat using phase change materials, Energy Conversion and Management, 45(2), 263–275.

[120] Tyagi, V.V., Kaushik, S.C., Tyagi, S.K., Akiyama, T. (2011). Development of phase change materials based microencapsulated technology for buildings: A review, Renewable and Sustainable Energy Reviews, 15(2), 1373–1391.

[121] Liu, C.-P., Seeds, A. (2010). Wireless-over-fiber technology-bringing the wireless world indoors, Opt. Photonics News, 21(11), 28–33.

[122] Zhou, D., Zhao, C.Y., Tian, Y. (2012). Review on thermal energy storage with phase change materials (PCMs) in building applications, Appl. Energy, 92, 593–605.

[123] Regin, A.F., Solanki, S.C., Saini, J.S. (2008). Heat transfer characteristics of thermal energy storage system using pcm capsules: A review, Renewable and Sustainable Energy Reviews, 12(9), 2438–2458.

[124] Memon, S.A. (2014). Phase change materials integrated in building walls: A state of the art review, Renewable and Sustainable Energy Reviews, 31, 870–906.

[125] Sarier, N., Onder, E. (2012). Organic phase change materials and their textile applications: An overview, Thermochim. Acta, 540, 7–60.

[126] Tyagi, V.V., Kaushik, S.C., Tyagi, S.K., Akiyama, T. (2011). Development of phase change materials based microencapsulated technology for buildings: A review, Renewable and Sustainable Energy Reviews, 15(2), 1373–1391.

[127] Zhang, Z., Shi, G., Wang, S., Fang, X., Liu, X. (2013). Thermal energy storage cement mortar containing n-octadecane/expanded graphite composite phase change material, Renewable Energy, 50, 670–675.

[128] Kenisarin, M.M., Kenisarina, K.M. (2012). Form-stable phase change materials for thermal energy storage, Renewable and Sustainable Energy Reviews, 16(4), 1999–2040.

[129] Nomura, T., Okinaka, N., Akiyama, T. (2009). Impregnation of porous material with phase change material for thermal energy storage, Mater. Chem. Phys., 115(2-3), 846–850.

[130] Lafdi, K., Mesalhy, O., Elgafy, A. (2008). Graphite foams infiltrated with phase change materials as alternative materials for space and terrestrial thermal energy storage applications, Carbon N Y, 46(1), 159–168.

[131] Sarı, A., Biçer, A. (2012). Thermal energy storage properties and thermal reliability of some fatty acid esters/building material composites as novel form-stable PCMs, Solar Energy Materials and Solar Cells, 101, 114–122.

[132] Li, M., Wu, Z., Kao, H., Tan, J. (2011). Experimental investigation of preparation and thermal performances of paraffin/bentonite composite phase change material, Energy Conversion and Management, 52(11), 3275–3281.

[133] Zhang, D., Zhou, J., Wu, K., Li, Z. (2005). Granular phase changing composites for thermal energy storage, Solar Energy, 78(3), 471–480.

[134] Jiao, C., Ji, B., Fang, D. (2012). Preparation and properties of lauric acid-stearic acid/expanded perlite composite as phase change materials for thermal energy storage, Mater. Lett., 67(1), 352–354.

[135] Li, M., Wu, Z., Chen, M. (2011). Preparation and properties of gypsum-based heat storage and preservation material, Energy and Buildings, 43(9), 2314–2319.

[136] Li, M., Kao, H., Wu, Z., Tan, J. (2011a). Study on preparation and thermal property of binary fatty acid and the binary fatty acids/diatomite composite phase change materials, Appl. Energy, 88(5), 1606–1612.

[137] Karaman, S., Karaipekli, A., Sarı, A., Biçer, A. (2011). Polyethylene glycol (PEG)/diatomite composite as a novel form-stable phase change material for thermal energy storage, Solar Energy Materials

and Solar Cells, 95(7), 1647-1653.
[138] Karaipekli, A., Sarı, A. (2009). Capric-myristic acid/vermiculite composite as form-stable phase change material for thermal energy storage, Solar Energy, 83(3), 323-332.
[139] Karaipekli, A., Sarı, A. (2010). Preparation, thermal properties and thermal reliability of eutectic mixtures of fatty acids/expanded vermiculite as novel form-stable composites for energy storage, Journal of Industrial and Engineering Chemistry, 16(5), 767-773.
[140] Li, M., Wu, Z., Kao, H. (2011b). Study on preparation, structure and thermal energy storage property of capric-palmitic acid/attapulgite composite phase change materials, Appl. Energy, 88(9), 3125-3132.
[141] Dong, Z., Li, Z.J., Zhou, H.M., Wu, K. (2004). Cement Concrete Res, 34, 927-934.
[142] Chen, Z., Su, D., Qin, M., Fang, G. (2015). Preparation and characteristics of composite phase change material (CPCM) with SiO_2 and diatomite as endothermal-hydroscopic material, Energy and Buildings, 86, 1-6.
[143] Jeong, S.-G., Jeon, J., Lee, J.-H., Kim, S. (2013). Optimal preparation of PCM/diatomite composites for enhancing thermal properties, Int. J. Heat Mass Transf., 62, 711-717.
[144] Sun, Z., Zhang, Y., Zheng, S., Park, Y., Frost, R.L. (2013). Preparation and thermal energy storage properties of paraffin/calcined diatomite composites as form-stable phase change materials, Thermochim. Acta, 558, 16-21.
[145] Xu, B., Li, Z. (2014a). Performance of novel thermal energy storage engineered cementitious composites incorporating a paraffin/diatomite composite phase change material, Appl. Energy, 121, 114-122.
[146] Xu, B., Li, Z. (2014b). Paraffin/diatomite/multi-wall carbon nanotubes composite phase change material tailor-made for thermal energy storage cement-based composites, Energy, 72, 371-380.
[147] Li, X., Sanjayan, J.G., Wilson, J.L. (2014). Fabrication and stability of form-stable diatomite/paraffin phase change material composites, Energy and Buildings, 76, 284-294.
[148] Werner, D. (1977). The Biology of Diatom. 1st ed. 0632000678. California, USA: University of California Press. ISBN.
[149] Noyes, J., Sumper, M., Vukusic, P. (2008). Light manipulation in a marine diatom, J. Mater. Res., 23, (No. 12), 3229-3235.
[150] Roselli, L., Stanca, E., Paparella, F., Mastrolia, A., Basset, A. (2013). Determination of coscinodiscus cf. Granii biovolume by confocal microscopy: Comparison of calculation models, J. Plankton Res., 35,, (No. 1), 135-145.
[151] Subhash, G., Yao, S., Bellinger, B., Gretz, M.R. (2005). Investigation of mechanical properties of diatom frustules using nanoindentation, J. Nanosci. Nanotechnol., 5(1), 50-56.
[152] Hamm, C.E., Merkel, R., Springer, O., Jurkojc, P., Maier, C., Prechtel, K. et al. (2003). Architecture and material properties of diatom shells provide effective mechanical protection, Nature, 421 (4925), 841-843.
[153] Gebeshuber, I.C., Kindt, J.H., Thompson, J.B., Del Amo, Y., Stachelberger, H., Brzezinski, M.A. et al. (2003). Atomic force microscopy study of living diatoms in ambient conditions, J. Microsc., 212 (3), 292-299.
[154] Gebeshuber, I.C., Stachelberger, H., Drack, M. (2005). Diatom bionanotribology—biological surfaces in relative motion: their design, friction, adhesion, lubrication and wear, J. Nanosci. Nanotechnol., 5 (1), 79-87.
[155] Arghir, M., Roucou, N., Helene, M., Frene, J. (2003). Theoretical Analysis of the Incompressible Laminar Flow in a Macro-Roughness Cell, J. Tribol., 125(No. 2), 309-318.
[156] Brajdic-Mitidieri, P., Gosman, A.D., Ioannides, E., Spikes, H.A. (2005). CFD analysis of a low friction pocketed pad bearing, J. Tribol., 127(4), 803-812.
[157] Sahlin, F., Glavatskih, S.B., Almqvist, Torbjörn., Larsson, R. (2005). Two-dimensional CFDanalysis of micro-patterned surfaces in hydrodynamic lubrication, J. Tribol., 127, (0742-4787), 96-102.

第 5 章
硅藻瓣膜形态发生的细胞机制

Yekaterina D. Bedoshvili，Yelena V. Likhoshway

5.1 简介

硅藻二氧化硅壳是自然界最奇妙的奥秘之一。有相当多的生物能够从二氧化硅中产生某种结构(例如海绵针状体、硅鞭毛骨架、金藻的包囊和鳞片)，但只有硅藻壳在微米和纳米级结构上表现出这样的多样性——各种瓣膜对称性、中缝的存在与否(瓣膜上的狭缝)、孔、刺和其他结构的复杂图案等。有趣的是,硅藻壳的一般组织在进化过程中是保守的:它们由两个与环带系统相连的瓣膜组成，它们的瓣膜装饰有穿孔图案。一方面,硅藻壳已经进化为保护硅藻细胞不被消费者吃掉或可能的酶攻击[1]，而另一方面,它允许与环境更有效的相互作用。这表明瓣膜的某些结构具有明确的功能:例如,在基质上移动所需的黏液由中缝排出[2-3];独立的刺与连接刺不同,让 *Aulacoseira* Twaites 这样的物种在不利的环境中分裂[4-5];中央支持突分泌黏液线连接浮游中心硅藻中的细胞[6]，边缘支持突排泄支持浮游中心浮选的毛藻(如 *Cyclotella*[7])。由于硅藻壳的结构具有物种特异性,因此可以合理地假设形态发生受遗传控制,并且每个物种在所涉及的基因中都有某种突变或调控变异,但此类数据目前还无法获得。此外,目前尚不清楚是哪些基因和细胞机制发生了变化,使硅藻能够如此成功地进化，分布在不同的生态位并在地球生物圈中占一席之地。

硅藻壳元素在细胞内的特殊细胞器中——二氧化硅沉积囊泡(SDV)产生[8]。因此,很明显,它们的形态发生是由细胞系统和机制控制的。在本章中,我们将硅藻瓣形态发生视为细胞学现象。

5.2 瓣膜对称性

硅藻瓣的对称性是在形态发生的最早阶段确定的,瓣膜的发展始于中心硅藻和羽状硅藻的不同结构[9-11]。中心硅藻中的瓣膜形态发生始于环的形成(见图 5.1)，肋骨从该环离心辐射,稍后,环状开口被二氧化硅填充[9-10,12-14]，在肋骨之间形成孔隙和各种过程。

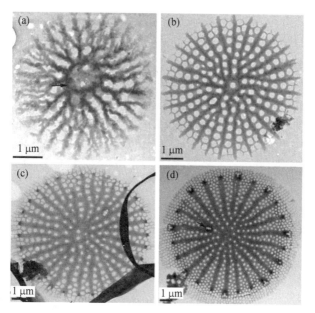

图 5.1 中心瓣膜形成的顺序阶段的 TEM 图

(a)中箭头表示环空,当肋骨之间形成开口时,环被二氧化硅填充(b、c),当地幔形成时(d),棱孔体(箭头)呈成熟的形状

一些硅藻具有特定的环状结构,如图 5.2 所示。例如,在浮游生物直链藻属的中心硅藻中,环状结构是一个巨大的环,占据了所有瓣膜的表面部分并环绕着刺[15]。

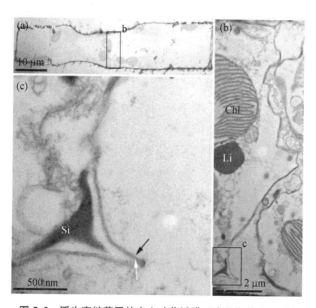

图 5.2 浮生直链藻属的中心硅藻瓣膜形成早期的 TEM 图

主要形成刺(Si),在 SDV 中间的面部部分看起来并不奇怪,黑色箭头表示质膜,白色箭头表示硅质膜;Chl—叶绿体、Li—脂滴、Si—形成瓣膜

在形态发生的最早阶段,刺开始在环上发育。类似羽状的 *Toxarium* Bailey 属具有细长的瓣膜,但没有表现出羽状硅藻的其他形态特征。研究者认为它可能有扩大和变形的环[16]。根据核小亚基 rDNA 系统发育,该属位于双极和多极中心。

羽状硅藻的对称性由称为胸骨的纵向骨架决定[17-23],一阶分支的基础命名为附属肋骨[9]或杆状连锁沟[20,24],反过来,那些会产生二级分支,即维米奈分支[20,24]。在这样布置好瓣膜的大体形状之后,孔隙和其他瓣膜元件开始形成。无壳缝和真晶体羽状硅藻的胸骨结构不同:前者是简单的线状细丝[21-22],后者的胸骨呈 π 形。在单缝物种中,这种结构在两个瓣膜上都形成,但其中一个瓣膜后来填充了二氧化硅[17],如图 5.3 所示。

图 5.3 *Achnanthidium sibiricum* 瓣膜形成的顺序的 TEM 图

(a)~(d) 为中缝瓣的形态发生变化(箭头指示为生长中的肋骨呈 π 状结构之间的间隙);(e) 二氧化硅填充的瓣膜

也有证据表明形态发生是从唇形突形成位置开始的[25-26],这种以海链藻为中心的对称性似乎是一个例外而不是规则,并且与这些物种中海链藻的中心位置有关。更多的情况下,环和海链藻是独立产生的,有时在形态发生的后期阶段从椭圆形开口形成海链藻[14-15,21]。

5.3 瓣膜硅化顺序

有几种方法可以描述瓣膜的硅化过程。第一种是分化形态发生呈水平和垂直生长,基于对两个中心物种的分析[27-28],描述了三个连续阶段:近端水平分化;纵向分化;远端水平分化。考虑到现代形态发生数据[13,21-23],我们可以将水平和垂直分化的概念扩展到所有具有或多或少厚瓣膜的硅藻,不仅是海链藻或圆筛藻的室瓣膜。根据数据,任何瓣膜的形态发生都始于"水平生长",即未来瓣膜的总体形状正在布局中。随着 SDV 的生长,形成带有放射状肋骨的环或带有侧肋的胸骨,直到覆盖层形成。在此阶段会沉积一些二氧化硅以增加成形结构的厚度("垂直生长"),但通常它仍然相当薄,并且垂直生长远不如水平生长明显。随着水平生长停止,形成的瓣膜变厚并且其精细结构正在分化。我们应该注意,这些阶段不会同时发生在整个瓣膜上,而是离心式扩散:瓣膜中心区域在其边缘之前变厚,因此可以在单个样本上同时看到几个连续的硅化阶段。

微观和宏观形态发生的概念[29]扩展了形态发生的早期分析[10],允许考虑纳米级和微米级的硅化过程。

5.4 SDV 中的二氧化硅

SDV 内容物的成分仍然未知,特别是 SDV 中的二氧化硅形式尚未完全确定。SDV 中的二氧化硅问题已在许多著作中讨论过[30-33]。通过高浓度二氧化硅能够形成三维凝胶[34]。一些研究者认为二氧化硅可以与 SDV 内的有机成分结合[30]或可能作为凝胶[32]。众所周知,形成瓣膜的超薄切口在形态发生的早期阶段显示出不致密的结构[见图 5.4(a)],后来它被分成"生长"和"压实"区[27-28]。奇怪的是,面向细胞质的瓣膜内侧的"生长"区通常比质膜更厚[见图 5.4(b)]。有时可能会看到生长中的瓣膜由大小约为 40 nm 的球状结构组成[35]。在其他情况下,可以看到直径约 10~20 nm 的细丝[见图 5.4(b)]或几纳米的小球和通道(见图 5.5)。

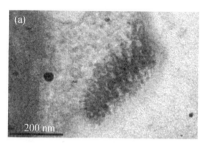
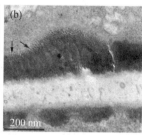

图 5.4　腹侧胞膜硅化瓣超薄片的 TEM 图

在早期形态发生阶段形成 a 的瓣膜碎片;具有"生长"(箭头)和"压实"(星号)区域的 B 型瓣膜;箭头指向瓣膜生长区上的灯丝结构

图 5.5　超薄瓣膜的 TEM 图

黑色箭头表示二氧化硅瓣膜内部的颗粒,白色箭头表示细通道

5.5 宏观形态发生控制

在分析各种细胞器对形态发生的影响期间,Pickett-Heaps 等[10]描述了膜介导的

形态发生和宏观形态发生,他们假设孔隙场和细缝缘膜等瓣膜特征的形成受膜介导的形态发生控制,并受 SDV 含量、硅膜和邻近质膜的影响。另一方面,术语"宏观形态发生"用来表示控制 SDV 和一些较大瓣膜结构的形状和位置的细胞质、细胞器或细胞骨架的更明显影响。例如,一些羽状体中缝的位置取决于微管中心的位置[10,36-37]。在某些物种中唇形突起的发展类似地由唇形突起器控制,其在本体论上与分裂纺锤体和有丝分裂后微管系统相关[36]。微管中心与唇突装置之间的相似性甚至导致了中缝和唇突之间存在个体发育联系的假设。

5.6 形态发生的细胞骨架控制

在抑制剂实验中获得了关于细胞骨架在硅藻瓣膜形态发生中的作用的信息。最值得注意的是,微管抑制剂正在阻碍宏观形态发生。例如,秋水仙碱会导致瓣膜形状的各种异常,如 *Pinnularia* sp. Ehrenberg 的中缝横向位移。将 *Hantzschia amphioxys* 中缝通道合并到瓣膜表面,以及 *Surirella Robusta* 瓣膜翅膀的异常发育[38]。秋水仙碱和鬼臼毒素导致 *Navicula saprophila*[39-40] 中缝结构异常。奥利唑啉和细胞松弛素 D 后来被证明会导致细长翼鼻状藻中的弯曲长鼻和刚毛根管藻中的唇形突起[41-42]。

使用超薄切片的透射电子显微镜和荧光显微镜观察硅藻细胞骨架。微管沿着尖头舟形藻[43-44])和 *Hantzscia amphioxys* 的细胞中缝以及沿着粗粒梭菌的中缝通道密集堆积[44-45])。在硅质毛藻形成期间,与肌动蛋白相似的细丝存在于 *Chaetoceros decipiens* Cleve 的 SDV 膜下[46],使用带有肌动蛋白和微管蛋白特异性染料的荧光显微镜,后来的工作表明,长鼻和根茎线虫中的长鼻形成是由微管"袖子"支撑的,而瓣膜直径同时受到肌动蛋白丝环的调节[41-42]。

细胞骨架抑制剂的作用并不局限于宏观的形态发生,精细的形态发生过程也受到影响。例如,将 *Cyclotella cryptica* Reimann、Lewin & Guillard 与秋水仙碱和细胞松弛素 D 一起孵育会影响边缘和中央大孔的位置,以及改变瓣膜边缘的亚微米结构[47]。

有证据表明微管和肌动蛋白丝都参与了孔隙的形成。用肌动蛋白特异性荧光染料染色硅藻细胞显示细丝模式与形成瓣膜的结构之间存在相关性[48]。在孔隙开口形成期间,蚜虫羽状 *Synedra acus* subsp. *radians* 的同步培养处理会导致瓣膜的重要区域完全缺乏孔隙[49]。另一方面,用紫杉醇(一种微管解聚抑制剂)处理类似的细胞会产生具有非特征性帆结构的异常孔隙,有时甚至是细胞壁上的大孔(见图 5.6)[50]。比较硅藻细胞同步培养中抑制微管聚合和解聚的效果,可以发现形态发生不仅受稳定微管的控制而且受动态微管的控制。

众所周知,在许多生物体中,细胞结构的形态发生是由细胞骨架控制的。例如,神经元生长锥受肌动蛋白和微管蛋白控制[51]。肌动蛋白和微管的重要性不仅限于

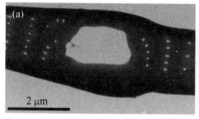
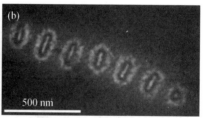

图 5.6 *Synedra acus* subs. *radians* 在紫杉醇影响下的放射性瓣膜异常图

它们在囊泡运输中的作用；在神经元的生长分支过程中，它们还充当基础或底层。研究表明肌动蛋白在拟南芥毛状体的生长和分枝过程中起主导作用[52]，而微管在该过程的早期阶段更为重要。结果证明，微管抑制剂（紫杉醇、安磺灵、丙吡胺）能够导致弯曲和扁平的毛状体，而不是它们正常的细长形状[53]。根据拟南芥毛状体形态发生的拟议模型，微管在毛状体发育过程中的作用减弱，而肌动蛋白丝对于正确形状变得越来越重要。尽管进化距离很大（硅藻和高等植物属于不同的界国，它们最后的共同祖先早于叶绿体的获得），但控制这些生物体形态发生的细胞机制有共同的特性。

5.7　囊泡在形态发生中的作用

自然界中有几种依赖肌动蛋白和微管蛋白的囊泡运输系统[54]，并且许多生物体显示出此类系统的显著多样性。然而，在硅藻中，它们尚未得到充分研究[55-56]。由于大多数硅藻基因组中不存在 Arp2 和 Arp3 复合物[57]，因此假设某些硅藻肌动蛋白依赖性囊泡运输系统[58]已经减少或显著改变是有道理的。羽纹硅藻（*Craspedostauros australis* Cox，*Nitzschia* sp. Hassal，*Pinnularia* sp.，*Craticula* spp. Grunow）运动性的研究表明[55]，硅藻中存在依赖肌动蛋白-肌球蛋白的囊泡运输。

20 世纪 70 年代的研究成果引起了人们对硅藻细胞内囊泡的关注[27]，这些囊泡将二氧化硅输送到 SDV 以沉积在成形瓣膜上[27-28,37,59]。不过，这个想法仍然存在争议：根据 Thamatrakoln 和 Kustka 的观点[60]，这些囊泡可以运输的二氧化硅量不足以用于整个瓣膜合成。事实上，SDV 的扩张很可能是由于囊泡运输所介导的微管，这些囊泡与二氧化硅膜融合的顺序能够形成二氧化硅瓣膜的确定模式[61]。

在细胞切片中观察到的最早结构是一个小的扁平 SDV，与质膜紧密相连。它包含一个刚刚开始形成的年轻瓣膜的薄板（见图 5.7）[62]。SDV 的扩展意味着增加硅膜表面，这需要某种膜材料来源。因此，如果没有囊泡的参与，这显然是无法完成的。在几个物种中，SDV 的生长是通过融合较小的囊泡来发生的[27,59]，尽管这仅在一些研究的物种中观察到[10]。

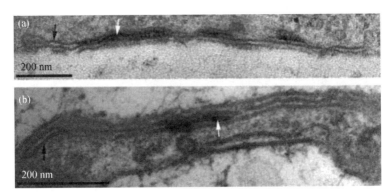

图 5.7　*Encyonema ventricosum*(a)和 *Synedra acus* subsp. *Radians* (b)瓣膜形态形成的早期阶段的 SEM 图

5.8　瓣膜胞吐作用和 SDV 起源

尽管研究者花了半个世纪的时间研究 SDV 中的二氧化硅沉积,但该隔间的起源仍有争议。有一些假说认为它是从高尔基体[59,62-64]、内质网[59,62,65]或高尔基体、内质网和质膜之间的复杂相互作用[66]演变而来的。

我们建议再次回到这个问题上来。质膜和硅质膜之间存在着明显的个体发育联系,最终,一部分硅质膜在成熟瓣膜从细胞(显然是从 SDV)排泄时加入质膜。我们必须承认,当囊泡接近质膜并发生膜融合时,瓣膜排泄并不完全遵循经典的胞吐机制。因此,囊泡内容物在细胞外排泄,最后质膜表面随着囊泡外膜的增加而增加。硅藻中存在这种胞吐现象,当粘液被排泄到细胞外空间时(见图 5.8),但这不是成熟瓣膜离开细胞的方式。观察到硅质膜和质膜在新瓣膜的覆盖层附近融合;在这种情况下,上硅膜和旧质膜的碎片夹在两个子细胞之间[59,67-71]。后来,它们或者降解,或者通过某种未知的机制,成为瓣膜有机涂层和硅藻土的一部分[10]。基本上,质膜在硅藻分裂期间形成两次:在胞质分裂之后和瓣膜形态发生期间,以及 SDV 的生长。考虑到质膜和硅膜之间紧密的结构连接,以及质膜在瓣膜吐出后形成深套陷的能力,这似乎是第一环带的初始 SDV 的形成(见图 5.9),我们认为相反的过程是可能的。质膜可能通过内陷形成未来 SDV 的初级囊泡。

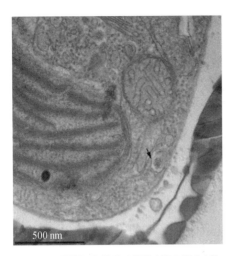

图 5.8　质膜细胞的胞吐囊泡(箭头指向)的 TEM 图

这种囊泡可能仍然附着在质膜样小窝上[72]，并通过高尔基体或内质网囊泡的同化而生长(如许多瓣膜形态发生研究所示)[27]。

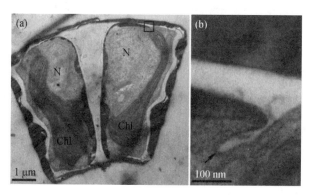

Chl—叶绿体；N—细胞核。

图 5.9　瓣膜胞吐后的细胞(a)和具有第一环带形态发生的初始阶段(b)的 TEM 图箭头指出形成了主要的 SDV

因此，我们认为 SDV 是古硅藻祖先细胞内体溶酶体系统的一部分，这一假设得到了质膜和硅膜之间的密切联系、瓣膜胞吐过程中的连接、SDV 内的酸性 pH 值[73]以及生长中的瓣膜被溶酶体的重要荧光染料 Lysotracker Yellow 主动染色这一事实的支持(见图 5.10)。

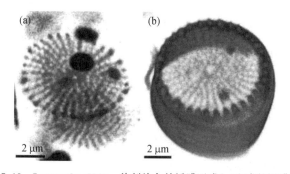

图 5.10　Lysotracker Yellow 染料染色的瓣膜形成(a)和成熟瓣膜(b)

5.9　结论

目前收集的数据表明，某些细胞机制参与了硅藻形态发生的控制。一般情况如下所示。首先，在细胞被肌动蛋白丝环隔开后[56]，微管中心向形成 SDV 的位置移动以控制其对称性。SDV 的起源仍有争议，但它与微管中心的联系似乎在形态发生的起始中起着重要作用。在形态发生的某些阶段，细胞骨架(具有微管网络和肌动蛋白丝的微管中心)通过其成分的相互作用来控制 SDV 的准确位置、形成瓣膜的形状,以及可能的精细结构。SDV 由可能依赖于微管的囊泡运输提供[61]。微管还可以确定水通道蛋白分子的特定模式,这些模式可以去除二氧化硅聚合产生的多余水分[74],或控制参与硅化作用的其他蛋白质或有机化合物的位置[75]。硅质囊膜和 SDV 含量控制微形态发生,即生化硅化本身。

当在实验中将变化引入这些机制时,它们会导致形成不同的硅质结构,因此在进化过程中有些相似的变化应该创造了当前的硅藻多样性。例如,在硅藻进化早期,环带相关基因的可能突变可能是硅藻类分离的原因,因为环带结构决定了瓣膜的形状和对称性。另一方面,调节细胞骨架和囊泡运输的蛋白质突变可能会导致瓣膜更精细的模式发生变化。

因此,硅藻形态发生背后的细胞机制大多已确定,但仍有必要检测控制硅藻中物种特异性形态发生的分子遗传机制。

参考文献

[1] Hamm, C.E., Merkel, R., Springer, O., Jurkojc, P., Maier, C., Prechtel, K., et al. (2003). Architecture and material properties of diatom shells provide effective mechanical protection, Nature, 421(6925), 841-843.

[2] Edgar, L.A., Pickett-Heaps, J. (1983). Mucilage secretions of moving diatoms, Protoplasma, 118(1), 44-48.

[3] Higgins, M.J., Molino, P., Mulvaney, P., Wetherbee, R. (2003). The structure and nanomechanical properties of the adhesive mucilage that mediates diatom-substratum adhesion and motility, J. Phycol., 39(6), 1181-1193.

[4] Davey, M.C., Crawford, R.M. (1986). Filament formation in the diatom Melosira granulata, J. Phycol. 22(2), 144-150.

[5] Bedoshvili, Ye., Bondarenko, N., Sakirko, M., Khanayev, I., Likhoshvay, Ye. (2007). The change in the length of colonies of the planktonic diatom Aulacoseira baicalensis in various stages of the annual cycle in Lake Baikal, Hydrob. J. 43(5), 81-89.

[6] Boje, R., Elbrächter, M. (1978). On the Ecological Significance of Thalassiosira partheneia in the Northwest African Upwelling Area. In Upwelling Ecosystems. (eds.). R Boje and M Tomczak. Berlin, Heidelberg: Springer.

[7] Popovskaya, G., Genkal, S., Likhoshway, Ye. (2016). Diatoms of the plankton of Lake Baikal: Atlas and Key. Novosibirsk, Nauka.

[8] Reimann, B.E. (1964). Deposition of silica inside a diatom cell, Exp. Cell Res., 34, 605-608.

[9] Mann, D.G. (1984). An ontogenetic approach to diatom systematics. In Proceedings of the 7th International Diatom Symposium. (ed.). D.G Mann. Koenigstein, Germany: O. Koeltz. pp.113-144.

[10] Pickett-Heaps, J.D., Schmid, A.-M., Edgar, L. (1990). The cell biology of diatom valve formation. In: Progress in phycological research, Vol. 7. (eds.)., F.E Round and D.J Chapman. Biopress, Bristol.

[11] Round, F., Crawford, R., Mann, D. (1990). The diatoms: biology and morphology of the genera. Bath: Cambridge Univ Press.

[12] Tiffany, M.A., Hernández-Becerril, D. (2005). Valve development in the diatom family Asterolampraceae H. L Smith 1872, Micropaleontology, 51(3), 217-258.

[13] Kaluzhnaya, O. (2008). Valve morphogenesis in the centric diatom Cyclotella baicalensis. In Proceeding of the 19th International Diatom Symposium. Y.V. Likhoshway. (ed.) Bristol: Biopress Limited. pp.31-38.

[14] Sato, S. (2010). Valve and girdle band morphogenesis in a bipolar centric diatom Plagiogrammopsis vanheurckii (Cymatosiraceae, Bacillariophyta), Eur. J. Phycol., 45(2), 167-176.

[15] Bedoshvili, Ye., Kaluzhnaya, O., Likhoshway, Ye. (2012). The frustule morphogenesis of Aulacoseira

baicalensis in the natural population, Journal of Advanced Microscopy Research. 7, 218-224.
[16] Kooistra, W. H. C. F., De Stefano, M., Mann, D. G., Salma, N., Medlin, L. K. (2003). Phylogenetic position of Toxarium, a pennate-like lineage within centric diatoms (Bacillariophyceae), J. Phycol., 39(1), 185-197.
[17] Boyle, J. A., Czarnecki, D. B., Pickett-Heaps, J. D. (1984). Valve morphogenesis in the pennate diatom Achnanthes coarctata, J. Phycol., 20(4), 563-573.
[18] Chiappino, M. L., Volcani, B. E. (1977). Studies on the biochemistry and fine structure of silicia shell formation in diatoms VII. Sequential cell wall development in the pennateNavicula pelliculosa, Protoplasma, 93(2-3), 205-221.
[19] Cox, E., Kennaway, G. (2004). Studies of valve morphogenesis in pennate diatoms: investigating aspects of cell biology in a systematic context. In Proceedings of the 17th International Diatom Symposium, Ottawa, Canada. (ed.). M Poulin. Bristol, UK: Biopress Ltd. pp.35-48.
[20] Cox, E. J. (1999). Variation in patterns of valve morphogenesis between representatives of six biraphid diatom genera (Bacillariophyceae), J. Phycol., 35(6), 1297-1312.
[21] Kaluzhnaya, O. V., Likhoshway, Y. V. (2007). Valve morphogenesis in an araphid diatom Synedra acus subsp. radians, Diatom Research, 22(1), 81-87.
[22] Sato, S., Watanabe, T., Nagumo, T., Tanaka, J. (2011). Valve morphogenesis in an araphid diatom Rhaphoneis amphiceros (Rhaphoneidaceae, Bacillariophyta), Phycol. Res., 59(4), 236-243.
[23] Tiffany, M. A. (2002). Valve morphogenesis in the marine araphid diatom Gephyria media (Bacillariophycea), Diatom Research, 17(2), 391-400.
[24] Cox, E. J. (2012). Ontogeny, homology, and terminology—wall morphogenesis as an aid to character recognition and character state definition for pennate diatom systematics, J. Phycol., 48(1), 1-31.
[25] Li, C., Volcani, B. (1985a). Studies on the biochemistry and fine structure of silica shell formation in diatoms. VIII. Morphogenesis of the cell wall in a centric diatom, Ditylum brightwellii, Protoplasma, 124, 10-29.
[26] Li, C., Volcani, B. (1985b). Studies on the biochemistry and fine structure of silica shell formation in diatoms. IX. Sequential valve formation in a centric diatom, Chaetoceros rostratum, Protoplasma, 124, 30-41.
[27] Schmid, A.-M. M., Schulz, D. (1979). Wall morphogenesis in diatoms: deposition of silica by cytoplasmic vesicles, Protoplasma, 100(3-4), 267-288.
[28] Schmid, A.-M. M., Volcani, B. E. (1983). Wall morphogenesis in Coscinodiscus wailesii. I. Valve morphology and development of its architecture, J. Phycol., 19(4), 387-402.
[29] Hildebrand, M., Wetherbee, R. (2003). Components and control of silicification in diatoms. In Progress in molecular and subcellular biology. 33. (ed.). W. E. G Müller. pp.11-57.
[30] Sullivan, C. (1986). Silicification by diatoms. In Silicon biochemistry. (eds.) D Evered and M O'Connor. Chichester: Wiley. pp.59-89.
[31] Martin-Jézéquel, V., Hildebrand, M., Brzezinski, M. A. (2000). Silicon metabolism in diatoms: implication for growth, J. Phycol., 36(5), 821-840.
[32] Grachev, M. A.., Annenkov, V. V., Likhoshway, Ye. V(2008). Silicon nanotechnologies of pigmented heterokonts, Bioessays, 30(4), 328-337.
[33] Ehrlich, H., Witkowski, A. (2013). Biomineralization in diatoms: the organic templates, Biologically-Inspired Systems, 6, 39-58.
[34] Iler, R. (1979). The chemistry of silica: solubility, polymerization, colloid and surface properties, and biochemistry. New York: Wiley.
[35] Crawford, S. A., Higgins, M. J., Mulvaney, P., Wetherbee, R. (2001). Nanostructure of the diatom frustule as revealed by atomic force and scanning electron microscopy, J. Phycol., 37(4), 543-554.
[36] Pickett-Heaps, J. D., Wetherbee, R., Hill, D. R. A. (1988). Cell division and morphogenesis of the labiate process in the centric diatom Ditylum brightwellii, Protoplasma, 143(2-3), 139-149.
[37] Schmid, A.-M. M., Eberwein, R. K., Hesse, M. (1996). Pattern morphogenesis in cell walls of diatoms and pollen grains: a comparison, Protoplasma, 193(1-4), 144-173.
[38] Cohn, S. A., Nash, J., Pickett-Heaps, J. D. (1989). The effects of drugs on diatom valve morphogenesis, Protoplasma, 149(2-3), 130-143.
[39] Blank, G. S., Sullivan, C. W. (1983a). Diatom mineralization of silicon acid. VI. The effects of

microtubule inhibitors on silicic acid metabolism in Navicula saprophila, J. Phycol., 19(1), 39-44.
[40] Blank, G. S., Sullivan, C. W. (1983b). Diatom mineralization of silicon acid. VII. Influence of microtubule drugs on symmetry and pattern formation in valves of Navicula saprophila during morphogenesis, J. Phycol., 19(3), 294-301.
[41] Van de Meene, A. M. L., Pickett-Heaps, J. D. (2002). Valve morphogenesis in the centric diatom Proboscia alata Sundstrom, J. Phycol., 38(2), 351-363.
[42] Van de Meene, A. M. L., Pickett-Heaps, J. D. (2004). Valve morphogenesis in the centric diatom Rhizosolenia setigera (Bacillariophyceae, Centrales) and its taxonomic implications, Eur. J. Phycol., 39(1), 93-104.
[43] Edgar, L. A., Pickett-Heaps, J. D. (1984). Valve morphogenesis in the pennate diatom Navicula cuspidata, J. Phycol., 20(1), 47-61.
[44] Pickett-Heaps, J. D., Kowalski, S. E. (1981). Valve morphogenesis and the microtubule center of the diatom Hantzschia amphioxysis, Eur. J. Cell Biol. 25(1), 150-170.
[45] Pickett-Heaps, J. D. (1989). Morphogenesis of the labiate process in the araphid pennate diatom Diatom vulgare, J. Phycol. 25(1), 79-85.
[46] Pickett-Heaps, J. D. (1998). Cell division and morphogenesis of the centric diatom Chaetoceros decipiens. II. Electron microscopy and a new paradigm for tip growth, J. Phycol., 34(6), 995-1004.
[47] Tesson, B., Hildebrand, M. (2010a). Extensive and intimate association of the cytoskeleton with forming silica in diatoms: control over patterning on the meso- and micro-scale, PLoS ONE, 5(12), e14300.
[48] Tesson, B., Hildebrand, M. (2010b). Dynamics of silica cell wall morphogenesis in the diatom Cyclotella cryptica: substructure formation and the role of microfilaments, J. Struct. Biol., 169(1), 62-74.
[49] Kharitonenko, K. V., Bedoshvili, Y. D., Likhoshway, Y. V. (2015). Changes in the micro- and nanostructure of siliceous valves in the diatom Synedra acus under the effect of colchicine treatment at different stages of the cell cycle, J. Struct. Biol. 190(1), 73-80.
[50] Bedoshvili, Ye., Gneusheva, K. V., Likhoshway, Y. V (2017). Changing of silica valves of diatom Synedra acus subsp. radians influenced by paclitaxel, Tsitologiia, 59(1), 53-61.
[51] Pacheco, A., Gallo, G. (2016). Actin filament-microtubule interactions in axon initiation and branching, Brain Res. Bull., 126. 300-310.
[52] Mathur, J., Spielhofer, P., Kost, B., Chua, N. (1999). The actin cytoskeleton is required to elaborate and maintain spatial patterning during trichome cell morphogenesis in Arabidopsis thaliana, Development, 126(24), 5559-5568.
[53] Mathur, J., Chua, N. H. (2000). Microtubule stabilization leads to growth reorientation in Arabidopsis trichomes, Plant Cell, 12(4), 465-478.
[54] Lodish, H., Berk, A., Kaiser, C., Krieger, M., Scott, M., Bretscher, A., et al. (2007). Molecular cell biology. 6th edn. New York: W. H. Freeman and Company.
[55] Poulsen, N. C., Spector, I., Spurck, T. P., Schultz, T. F., Wetherbee, R. (1999). Diatom gliding is the result of an actin-myosin motility system, Cell Motil. Cytoskeleton, 44(1), 23-33.
[56] Tanaka, A., De Martino, A., Amato, A., Montsant, A., Mathieu, B., Rostaing, P., et al. (2015). Ultrastructure and membrane traffic during cell division in the marine pennate diatom Phaeodactylum tricornutum, Protist, 166(5), 506-521.
[57] Aumeier, C., Polinski, E., Menzel, D. (2015). Actin, actin-related proteins and profilin in diatoms: a comparative genomic analysis, Mar. Genomics, 23, 133-142.
[58] Khaitlina, S. Y. (2014). Intracellular transport based on actin polymerization, Biochemistry Mosc., 79(9), 917-927.
[59] Dawson, P. (1973). Observations on the structure of some forms of Gomphonema parvulum Kütz. III. Frustule formation, J. Phycol., 9, 353-365.
[60] Thamatrakoln, K., Kustka, A. B. (2009). When to say when: can excessive drinking explain silicon uptake in diatoms? Bioessays, 31(3), 322-327.
[61] Parkinson, J., Brechet, Y., Gordon, R. (1999). Centric diatom morphogenesis: a model based on a DLA algorithm investigating the potential role of microtubules, Biochim. Biophys. Acta, 1452(1), 89-102.
[62] Schmid, A.-M., Borowitzka, M., Volcani, B. (1981). Morphogenesis and biochemistry of diatom cell walls. In Cytomorphogenesis in plants. 8. (eds.). O Kiermayer. Berlin Heidelberg, New York: Springer. pp. 63-97.

[63] Reimann, B.E. (1964). Deposition of silica inside a diatom cell, Exp. Cell Res., 34, 605-608.
[64] Schnepf, E., Deichgraber, G., Drebes, G. (1980). Morphogenetic processes in Attheya decora (Bacillariophyceae, Biddulphiineae), Plant Syst. Evol., 135(3-4), 265-277.
[65] Floyd, G. L., Hoops, H. J. (1979). Ultrastructure of the centric diatom, Cyclotella menenginiana: vegetative cell and auxospore development, Phycologia, 18(4), 424-435.
[66] Chiappino, M. L., Volcani, B. E. (1977). Studies on the biochemistry and fine structure of silicia shell formation in diatoms VII. Sequential cell wall development in the pennate Navicula pelliculosa, Protoplasma, 93(2-3), 205-221.
[67] Stoermer, E. F., Pankratz, H. S., Bowen, C. C. (1965). Fine structure of the diatom Amphipleura pellucida. II. Cytoplasmic fine structure and frustule formation, Am. J. Bot., 52(10), 1067-1078.
[68] Reimann, B.E., Leivin, J.C., Volcani, B.E. (1966). Studies on the biochemistry and fine structure of silica shell formation in diatoms. II. The structure of the cell wall of Navicula pelliculosa (Breb.) Hilse, J. Phycol., 2(2), 74-84.
[69] Crawford, R. (1981). The siliceous components of the diatom cell wall and their morphological variation. In Silicon and Siliceous structures in biological systems. (eds.). T. L Simpson and B. E Volcani. New York: Springer-Verlag. pp.129-156.
[70] Li, C.-W., Volcani, B.E. (1984). Aspects of silicification in wall morphogenesis of diatoms, Phil. Trans. R. Soc. Lond. B, 304(1121), 519-528.
[71] Crawford, R., Schmid, A.-M. (1986). Ultrastructure of silica deposition in diatoms. In Biomineralization in lower plants and animals System Soc. 30. (eds.). B. S Leadbeater and R Riding. pp.291-314.
[72] Thomas, C.M., Smart, E.J. (2008). Caveolae structure and function, J. Cell. Mol. Med., 12(3), 796-809.
[73] Vrieling, E.G., Gieskes, W.W.C., Beelen, T.P.M. (1999). Silicon deposition in diatoms: control by the pH inside the silicon deposition vesicle, J. Phycol., 35(3), 548-559.
[74] Grachev, M.A.., Annenkov, V.V., Likhoshway, Ye. V(2008). Silicon nanotechnologies of pigmented heterokonts, Bioessays, 30(4), 328-337.
[75] Ehrlich, H., Witkowski, A. (2013). Biomineralization in diatoms: the organic templates, Biologically-Inspired Systems, 6, 39-58.

第6章
聚焦离子束技术在硅藻超微结构分类学研究中的应用

Andrzej Witkowski1, Tomasz Płociński, Justyna Grzonka,
Izabela Zgłobicka, Małgorzata Bąk, Przemysław Dąbek,
Ana I. Gomes, Krzysztof J. Kurzydłowski

6.1 简介

聚焦离子束(FIB)技术是在20世纪后期开发的,用于对半导体制造设施中的集成电路进行修改和修复[1-2]。多年来,FIB技术取得了重大发展,并开始在透射电镜样品的制备中发挥重要作用。在1990年代后期,FIB仪器的成像分辨率可与SEM相媲美,当时它在材料科学界引起了极大的关注[1,3]。此外,FIB技术开始应用于材料科学、生命科学和纳米技术。在生物学的背景下,这种技术具有以下优点:在检查样品的精确定位部位进行精确的横截面和薄片切割,以及能够保证切片的均匀厚度。此外,FIB与SEM相结合可以被视为一种基于离子连续切片和电子成像的断层扫描技术,因为所获得的图像堆栈提供了有关所研究样品结构的3D信息。

先进的电子/离子显微镜方法在硅藻研究中的主要应用是为硅质外骨骼的复杂功能和特性的3D建模获得可靠的输入。尽管计算机化纳米断层扫描的初步结果非常有希望,但它仍远未得到常规使用[4]。等待计算机纳米断层扫描的更广泛的可及性,FIB仍然是解决硅藻壳硅质元素之间空间关系的最有利技术。特别是在具有复杂超微结构的硅藻类群中,例如具有蜂窝状瓣的硅藻类群,中缝位于高架结构上,或具有复杂的条纹结构或特殊过程。

据我们所知,Suzuki等对硅藻进行了第一次纳米切割[5-7]。Suzuki等展示了假性精氨酸球菌变种中间型的经尖端切割[5],他们还展示了利用FIB进行的菱形藻切割[6]。此后不久,Lowe在Manoylov用FIB切割并成像了耗散菱形藻[8]。Sato等对*Grammatophora*进行了纳米切割[7],目的是揭示其复杂腰带的真实结构,他是第一个沿顶端轴切割硅藻壳体的人[7]。

自2010年以来,FIB技术已被华沙理工大学材料科学与工程学院用作检查硅藻超微结构的常规工具。然而,迄今为止只有一小部分结果已发表[4,8-12]。因此,我们在此提供了更多 FIB 辅助研究的示例，旨在展示该技术在硅藻研究中的全部潜力。基于 FIB 的观察使我们能够解决和说明硅藻超微结构的基本问题：(1) *Nitzschia*、*Surirella* 和 *Simonsenia* 的中缝结构[9]；(2) *Luticola* 的纵向管道和柱头[11]；(3) *Didymosphenia geminata* (Lyngbye) M. Schmidt 的瓣膜超微结构和精确测量[4]。然而，最有趣的结果是在整个硅藻瓣的 3D 建模中使用基于 FIB 的硅藻顺序切割。标准 SEM/TEM 检查不允许解析特定瓣膜特征的空间位置。这涉及胸骨内中缝位置和形状的横截面，或纵向管的位置和形状，如 *Diploneis*。Witkowski 等发表了 *Simonsenia aveniformis* Witkowski, Gomes & Gusev 硅藻中 FIB 在 3D 瓣膜建模中的应用示例[9]。*Simonsenia* raphe 超微结构的建模能够区分芽孢杆菌目中的第三种中缝类型：除了众所周知的 *nitzschioid* 和 *surirelloid* 外，还正式发表了一种 *simonsenioid* 中缝类型[9]。

我们在此展示了沿横切面和纵切面的纳米切片，这两种硅藻之间的亲缘关系相当遥远的硅藻物种：*Biremis* 属于 *Scolitropidaceae* (Neidiaceae)，*Olifantiella* 属于 *Diadesmidaceae*[13]，他们都具有复杂的纹状超微结构。纹状超微结构在来自 *B. lucens* 野生标本和 *O. mascarenica* 的一系列纳米切割中进行了说明[14]。

6.2 材料与方法

基于 FIB 的样品制备是通过 FIB/SEM 日立 NB5000 集成系统完成的。该系统由超高性能聚焦镓离子束枪（40 kV）和高分辨率场发射扫描电子显微镜（30 kV）组成。这种双光束系统实现了高通量样品制备、高分辨率成像和分析以及精密纳米切片。FIB 样品操作是在选择离子束条件的参数下进行的，以尽量减少其损伤。

硅藻来自野生样品，因为在培养基中生长的硅藻具有硅化程度较低的瓣膜，并且缺少野生样品的同一物种中观察到的一些特征。*Biremis-lucens* 和 *O. mascarenica* 瓣膜在瓣膜外部和内部沿着横心轴和根尖轴进行纳米切割。在进行纳米切割之前，对 *B. lucens*（见图 6.1 和图 6.2）进行了鉴定，并确定其发育良好。

6.3 结果

6.3.1 复杂条纹超微结构

1. *Biremis lucens* (Hustedt) Sabbe[15]

从外部看，瓣膜显示出由两排细隙组成的条纹（见图 6.1，因此该属的名称为 *Biremis*）。在内部，沿瓣膜套观察到一个管状腔室。两条管道几乎横跨整个瓣膜套内部的长度，只有一条与顶端直接相关的非常窄的瓣膜带没有这些腔室（见图 6.2）。在外部和内部切口中，腔室显示的空间范围从 $0.2\ \mu m$（靠近顶点）到 $0.4\ \mu m$（在瓣膜中心)。垂直于心尖轴的瓣膜切口帮助我们解决了两排细隙和相应腔室之间的几何关系。沿着瓣膜套出现的两排孔隙从外部穿透硅质外骨骼，并在下面的腔室中打开。每对多孔体沿顶轴定位，对应于一个腔室。下一排靠近地幔边缘，在腔室中心打开，而上一排位于瓣膜和地幔之间的过渡处，在腔室内靠近其远端边缘打开。在瓣膜的外部和内部切口中都观察到了细隙的这种位置。

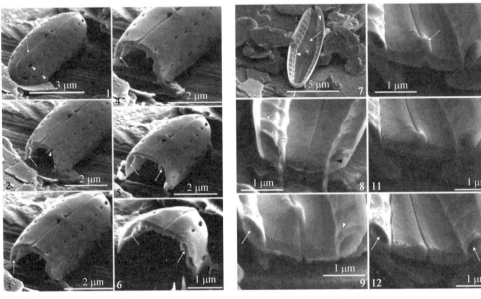

图 6.1　通过连续的纳米孔过程的外部视图显示　　图 6.2　通过连续的纳米孔过程显示的内部视图

　　$B.lucens$ 瓣膜的内部视图可能暗示管状腔内的细分。这些管子在外部显示出位于腔室高度约一半处的纵向（顶部定向）凹槽，给人一种管子被细分为两个顶部截面的印象。

　　此外，在经顶平面上，管状腔室由一系列褶边状高程细分，这些高程由轻微凹陷隔开。褶皱的存在可能意味着腔室之间的内部分隔。瓣膜内部的纳米切口显示腔室已连接。在装配（切割）瓣膜的瓣膜套内侧观察到类似的褶边结构，即在腔室内部。这些振幅较小的内部饰件遵循位于腔室外侧的形状饰件。皱褶之间沿根尖轴的距离约为 $0.4~\mu m$。

　　$Biremis$ 属于一组分类群，显示中心结节内部存在双螺旋藻。经心尖的切口显示了在心尖和中央结节之间的中缝裂缝。在 $Biremis~lucens$ 的内部和外部切割中，中缝显示为简单的直缝，而不是横截面的常见形状。穿过中央结节和双螺旋舌的切口显示中央结节具有坚实的结构。到目前为止，用 FIB 切割的双蚜虫硅藻的观察结果是正确的。

　　在 Szczecin 硅藻培养物收集中生长的 $B.~lucens$ 菌株 SZCZCh1605 中，也观察到了上述基于 $B.~lucens$ 野生标本的带有腔室的复杂条纹超微结构。

2. *Olifantiella mascarenica*[13]

与 *Biremis* 相似,另一个最近建立的 *Olifantiela* 属显示出精细的纹超微结构,带有漏斗状室(见图 6.3 和图 6.4)。与 *Biremis* 不同,*Olifantiella* 的腔室沿整个地幔长度延伸,在瓣膜顶端周围的尺寸有所减小。

图 6.3 *Olifantiella mascarenica* 瓣膜的外部切口

图 6.4 *Olifantiella mascarenica* 的内部切口

Olifantiella 的纹是复杂的,以大开口的形式出现,称为大细隙,被大的多孔膜堵塞,并沿内瓣膜套有管状腔(见图 6.3 和图 6.4)。从瓣膜到瓣膜套的过渡以固体边缘脊的存在为标志,该脊部分覆盖漏斗状腔室(见图 6.3)。位于瓣膜外表面的窠状突起瓣膜的外表面有一排单列的窠状突起,这些突起被多孔的透明膜封闭。*O. mascarenica* 细隙纹通常为大细隙状,但位于瓣膜侧的中间纹除外,中间纹被分成两个较小的多孔体(见图 6.3)。较小的多孔体位于中心结节附近,第二个较大的多孔体位于瓣膜边缘(见图 6.3 和图 6.4)。这两个多孔体由一条窄的固体硅胶条隔开。常规瓣膜面大细隙闭塞从透明瓣边缘终止约 $0.15 \sim 0.2~\mu m$(见图 6.3),并留下一个自由空间,腔室通过该空间向外部环境开放。这些开口可见于位于外部的瓣膜横切向纳米切口中。经尖端切口显示存在高度约为 $0.20~\mu m$、宽度约为 $0.5~\mu m$ 的漏斗状腔室。虽然外部切口显示了管状腔室的一般位置、形状和尺寸,但可以对经尖端的内部切口进行详细分析。在 *Olifantiella* 中,腔室由单管状结构组成,该结构沿经顶轴进行精细的外部剖切。特定的部分用模糊的横切凹槽标记,这些凹槽连接着地幔内缘和阻塞特定纹的膜,横切切口显示了腔室内部的间隙位置。

经心尖方向的切口显示存在一个特殊的过程,精确定位在瓣膜中心,名为 *buciniportula*。它位于瓣膜侧中心结节附近(见图 6.4 和图 6.5)。瓣膜侧是中缝胸骨的这一侧,它首先形成,瓣膜另一侧发展较晚,这种特殊的过程只发生在 *Olifantiella*,其高度约为 $0.4~\mu m$,外径约为 $0.2~\mu m$。在经头顶的内外切口中均可看到 *buciniportula*。我们的外部切口显示,颊孔有一个单独的开口穿透瓣膜外表面,并与第二个经心尖方向、稍大的开口相关。我们的纳米切口显示,第二个开口将腔室与瓣膜外部连接起来,就像位于瓣膜表面边缘的每个剩余开口一样。然而,这个特殊的开口有第二个分支,一根短管朝着颊孔向下倾斜,但不是笔直的,而是管本身。在 *Olifantiella mascarenica* 的瓣膜内部可以清楚地看到门和腔室之间的短连接(见图 6.4)。

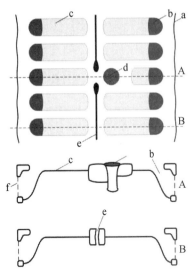

a—瓣面边缘;b—孔;c—阴影区表示大穿孔膜(菌膜)闭塞;d—外部开口;e—中缝;f—瓣膜上的孔

图 6.5 基于 FIB 纳米圈的瓣膜结构图

上图为外瓣膜面俯视图,下图是沿 A 线和 b 线的横截面,白色区域是固体二氧化硅

外部顶端切口显示中缝的形状,它不仅仅是比雷米斯中的一个简单狭缝,而且具有小于号的形状。这与传统 SEM 观察结果相矛盾,传统 SEM 观察结果显示存在简单的狭缝状中缝[16]。在经尖端的切口中,也对中心结节进行了成像,它由固体二氧化硅层组成。在内部,中心结节显示出一个小的、高硅结构的压力,一个双螺旋状结构(见图 6.4)。

6.4 讨论

FIB 纳米切割已经在一定程度上用于记录硅藻的总体形态,报告在硅藻壳体建模中使用 FIB 的论文数量正在增加[4,9,17];Zgłobicka 等[4]在 *Didymosphenia geminata* 壳体的研究中使用了一种新的强大的 3D 高分辨率成像技术——纳米 X 射线层析成像技术。目前,这项技术只能在少数专门实验室中使用,其在硅藻研究中的全部潜力仍有待检验。

FIB 纳米切割在 *B. lucens* 和 *O. mascarenica* 上的应用,显示了 FIB 在硅藻超微结构研究中的巨大潜力。可用的分子系统发育学将这两种硅藻分为不同的科:*Biremis* 属于脊柱侧凸科[18],*Olifantiella* 属于盾叶硅藻科[19]。尽管瓣膜超微结构和纹复杂性的差异表明这两个属属于不同的分类单位,但当系统发育仅基于外部形态观察时,这并不总是那么明显。最新的分子系统发育将 *Olifantiella* 和 *Luticola* 归为重链虫科[19],虽然这两个属的大体形态几乎没有相似之处,但腐蚀/断裂试样或 FIB 纳米切割显示的超微结构特征可能表明它们之间存在某种关系。在 Bąk 等的研究中[11],*L. galapagoensis* Witkowski, Bąk, Kociolek, Lange-Bertalot & Seddon 的瓣膜沿着横轴和根尖轴切割,在经心尖切口中,*L. galapagoensis* 表明沿着瓣膜边缘存在通道,这可能是 *Luticola* 和 *Olifantiella* 的衍生特征,然而该特征在不同的时间内可能有所不同[16,20-21]。一些物种有简单的、强烈减少的边缘通道,即 *L. acidoclinata* Lange Bertalot 和 *L. galapagoensi*,其他的则有相当复杂的通道,具有结构化的外表面,在某种程度上类似于在 *Olifantiella* 中观察到的通道,即 *L. submobiliensis* Levkov、Metzeltin & Pavlov。然而,在一些 *Luticola* 物种中,如 *L. mutica* (Kützing) D. G. Mann,边缘通道可能被膜堵塞[20-21]。根据我们对 Bąk 等[11]进行 *Olifantiella* 和 *Luticola* 切割以及文献检索,似乎 *Luticola* 和 *Olifantiella* 对瓣膜的超微结构有着更为普遍的规划[22]。这涉及到许多分类群中由固态透明二氧化硅组成的瓣缘和外套膜之间的接触,在瓣面上方略高,瓣套上有一排单独的细隙,以及由多个穿孔的系带组成的束带[20,22]。

Biremis 和 *Olifantiella* 中的复杂条纹显示出它们之间的显著差异,包括 *Olifantiella* 中存在 *buciniportula*。此外,在 *Biremis* 中,瓣膜和地幔之间的过渡是渐进的,而在 *Olifantiella* 中,从瓣膜到地幔的过渡是突然的,并有明显的透明硅石条带,*Biremis* 瓣膜不具有突起或柱头[15,18,20],而 *Olifantiella* 中 *buciniportula* 存在于瓣膜中心并与漏斗结构相连,尽管漏斗状结构与 *buciniportula* 之间存在连接,但两个结构之间似乎不存在直接连接[16]。在 *Biremis* 腔中,尽管漏斗状腔的内腔是完全开放的,但在经顶面和心尖面上,腔明显分为两部分,漏斗内部仅观察到短的皱褶。同样,*Olifantiella* 的漏斗结构在心尖平面和经心尖平面上被分割。在这两个属中,亚段均遵循经顶纹模式,然而这两个属之间的空间范围是不同的。在 *Biremis* 中,漏

斗状结构附着在瓣膜盖内部,并在顶部下方短距离终止,而在 *Olifantiella* 中,漏斗构成一个连续的带,仅在顶部变窄。

6.4.1 培养标本与野生标本

大多数含 FIB 的硅藻纳米切割都是在野生标本上进行的[5,7,9,17],到目前为止,有关培养硅藻切片的唯一可用数据可以在文献中找到[6]。在目前的研究中,我们研究了绿僵菌培养瓣膜的纳米切割,这在很大程度上完全符合野生标本的特征。野生和培养的绿僵菌瓣之间的唯一区别是后者通常较少硅化。因此,由于离子束切割,培养的瓣膜更脆弱,更容易变形,并且通常图像分辨率比野生标本差。在 *Simonsenia* sp. 菌株 SZCZCh839 的培养瓣膜中观察到了这一点(见图 6.6)。培养的 *Simonsenia* 的瓣膜已弱硅化,很难找到显示完整瓣膜形态发生的标本。弱硅化瓣膜,当用 FIB 切割时,向内或向外弯曲,也弯曲管缝。在一些培养的物种中,这两部分沿着中缝裂缝分开(未发表的观察)。在具有弱硅化瓣膜的 *Olifantiella mascarenica* 野生标本中也观察到了后一种现象。

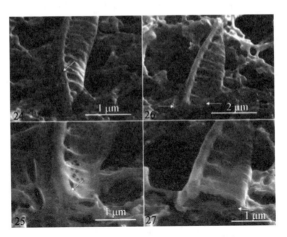

图 6.6　*Simonsenia* sp. SZCZCh839 菌株的外部切割

6.5　结论

在过去的二十年里,FIB 在硅藻分类学中得到了广泛的应用。涉及 FIB 的研究侧重于环带结构,能够解决复杂的条纹结构和硅藻壳对称性,特别是在带中缝的硅藻中,例如 *Nitzschia*、*Surirella* 或 *Simonsenia*。然而,根据我们未发表的结果,我们可以指出 FIB 在解析硅藻壳超微结构方面的众多应用主要包括以下几方面。

(1) 解析中心型、节状、节状硅藻各类工艺的超微结构,其中一些过程具有复杂的形态,需要纳米级切割才能揭示其真实的空间结构,包括 *Didymosphenia*、

Labellicula、*Luticola*、*Olifantiella* 和 *Proschkinia*。

（2）蜂窝状硅藻的瓣膜超微结构分辨率，其瓣膜面由 3D 条纹组成，类似于蜂窝。这种瓣膜结构存在于许多属中，包括 *Actinocucclus*、*Coscinodiscus*、*Caloneis*、*Pleurosigma* 和 *Psammodictyon*。FIB 的巨大优势在于它不仅可以切割硅藻瓣以显示横向或纵向横截面，而且还可以对瓣膜表层进行切片，从而揭示特定细隙的精细结构。FIB 允许对瓣膜表面进行纳米级切割，并在特别复杂的空间设置中揭示其结构。

（3）属于 *Proschkinia* 或 *Haslea* 的针状硅藻的空间结构分辨率。属于这些属的分类群具有三明治型 3D 结构。瓣膜表面的特征在于位于外部瓣膜表面上的经心尖（横向条纹）和纵向（顶端）条纹。

参考文献

[1] Phaneuf, M. W. (1999). Applications of focused ion beam microscopy to materials science specimens, Micron, 30(3), 277-288.

[2] Orloff, J., Utlaut, M., Swanson, L. (2003). High Resolution Focused Ion Beams: FIB and its Applications. Kluwer/Plenum, New York.

[3] Vasile, M. J., Nassar, R., Xie, J., Guo, H. (1999). Microfabrication techniques using focused ion beams and emergent applications, Micron, 30(3), 234-244.

[4] Zgłobicka, I., Li, Q., Gluch, J., Płocińska, M., Noga, T., Dobosz, R. et al. (2017). Visualization of the internal structure of Didymosphenia geminata frustules using nano X-ray tomography, Scientific Reports 7(1), 9086, 9086.

[5] Suzuki, H., Tanaka, J., Nagumo, T. (2001). Morphology of the marine diatom Cocconeis pseudomarginata Gregory var. intermedia Grunow, Diatom Research, 16(1), 93-102.

[6] Suzuki, H., Hanai, T., Nagumo, T., Tanaka, J. (2009). Morphology of Marine Benthic Diatom Nitzschia amabilis Hide, Suzuki (Bacillariophyceae). The Journal of Japanese Botany, 84(5), 273-278.

[7] Sato, S., Nagumo, T., Tanaka, J. (2010). Morphological study of three marine araphid diatom species of Grammatophora Ehrenberg, with special reference to the septum structure, Diatom Research, 25(1), 147-162.

[8] Manoylov, K. (2010). Diatoms of the United States. Nitzschia dissipata., http://westerndiatoms.colorado.edu/taxa/species/nitzschia_dissipata.

[9] Witkowski, A., Gomes, A., Mann, D. G., Trobajo, R., Li, C., Barka, F. et al. (2015). Simonsenia aveniformis sp. nov. (Bacillariophyceae), molecular phylogeny and systematics of the genus, and a new type of canal raphesystem, Scientific Reports, 5, 17115.

[10] Li, C. L., Ashworth, M. P., Witkowski, A., Lobban, C. S., Zgłobicka, I., Kurzydłowski, K. J. et al. (2016). Ultrastructural and molecular characterization of diversity among small araphid diatoms all lacking rimoportulae. I. five new genera, eight new species, Journal of Phycology, 52(6), 1018-1036.

[11] Bąk, M., Kociolek, J. P., Lange-Bertalot, H., Łopato, D., Witkowski, A., Zgłobicka, I. et al. (2017). Novel diatom species (Bacillariophyta) from the freshwater discharge site of Laguna Diablas (Island Isabela =Albemarle) from the Galapagos, Phytotaxa, 311(3), 201-224.

[12] Kaleli, A., Krzywda, M., Witkowski, A., Riaux-Gobin, C., Nadir Solak, C., Zgłobicka, I. B. et al. (2018). A new sediment dwelling and epizoic species of Olifantiella (Bacillariophyceae), with an account on the genus ultrastructure based on Focused Ion Beam nanocuts, Fottea, 18(2), 212-226.

[13] Riaux-Gobin, C., Compère, P. (2009). Olifantiella mascarenica gen. & sp. nov., a new genus of pennate diatom from Réunion Island, exhibiting a remarkable internal process, Phycological Research, 57(3), 178-185.

[14] Desrosiers, C. (2014). Les diatomées benthiques des zones côtières de Martinique: taxonomie, écologie et capacité bioindicatrice. PhD Thesis, Universite de Toulouse.

[15] Sabbe, K., Witkowski, A., Vyvermann, W. (1995). Taxonomy, morphology and ecology of Biremis lucens comb. nov. (Bacillariophyta): a brackish-marine, benthic diatom-species comprising different

morphological types, Botanica Marina, 38(1-6), 379-391.

[16] Riaux-Gobin, C. (2015). The elusive genus Olifantiella (Bacillariophyta): South Pacific assemblage and Indo-Pacific biogeography, Botanica Marina, 58(4), 251-266.

[17] Xing, Y., Yu, L., Wang, X., Jia, J., Liu, Y., He, J. et al. (2017). Characterization and analysis of Coscinodiscus genus frustule based on FIB-SEM, Progress in Natural Science: Materials International, 27(3), 391-395.

[18] Witkowski, A., Barka, F., Mann, D.G., Li, C., Weisenborn, J.L., Ashworth, M.P. et al. (2014). A Description of Biremis panamae sp. nov., a New Diatom Species from the Marine Littoral, with an Account of the Phylogenetic Position of Biremis D.G. Mann et E.J. Cox (Bacillariophyceae), PLoS ONE, 9(12), e114508, e114508.

[19] Han, J., Zhang, L., Wang, P., Yang, G., Wang, S., Li, Y. et al. (2018). Heterogeneity of intron presence/absence in Olifantiella sp. (Bacillariophyta) contributes to the understanding of intron loss, Journal of Phycology, 54(1), 105-113.

[20] Round, F.E., Crawford, R.D., Mann, D.G. (1990). The Diatoms. Biology and morphology of the genera Cambridge University Press.

[21] Levkov, Z., Metzeltin, D., Pavlov, A., Lange-Bertalot, H. (2013). Luticola and Luticolopsis. In Diatoms of Europe. Diatoms of the European inland waters and comparable habitats. 7. pp.1-698. H. Lange-Bertalot (eds.), Königstein: Koeltz Scientific Books.

[22] Lowe, R.L., Kociolek, P., Johansen, J.R., Vijver, B.V.D., Lange-Bertalot, H., Kopalová, K. (2014). Humidophila gen. nov., a new genus for a group of diatoms (Bacillariophyta) formerly within the genus Diadesmis: species from Hawai'i, including one new species, Diatom Research, 29(4), 351-360.

第 7 章
光与硅藻：光子学和光生物学综述

Mohamed M Ghobara，Nirmal Mazumder，Vandana Vinayak，
Louisa Reissig，Ille C Gebeshuber，Mary Ann Tiffany，Richard Gordon

7.1　简介

虽然有大量关于硅藻叶绿体及其特殊光合作用的文献[1-4]，但这篇综述侧重于一些光学效应归因于它们独特的华丽二氧化硅壳（硅藻壳光子学）以及光与活细胞之间相互作用产生的现象（光生物学），许多研究者预测硅藻将作为光学元件得到应用[5-15]。

7.2　硅藻壳独特的多尺度结构

细胞壁可以说是硅藻中最有趣的组成部分。它主要由无定形水合二氧化硅组成，被称为它的"壳"或"硅藻壳"。二氧化硅生物矿化是一种普遍现象[16]，这种现象存在于硅藻和许多其他生物[17]，一些细菌[18]，甚至可以追溯到 LECA（最后的真核生物共同祖先，见图 7.1），它包括植物中作为植硅体的二氧化硅沉积[19-21]和海绵中的针状体[22]，此外，最近在哺乳动物中发现了一种硅转运蛋白[23]，证实了早期的营养研究[24]。虽然所有这些二氧化硅生物矿化生物都值得关注，但硅藻因其精致多样的硅藻壳结构[25]、对称性[26]、漂亮的外观[27]和庞大的物种数量而成为被研究最多的生物之一[28-30]。虽然硅藻已经存在了至少 1.5 亿年[31-32]，但当前硅藻占地球初级生产力的 20%～25%[33-36]，硅藻种类的数量可能高达 200 000 种[35-36]，为研究人员提供了一个巨大的纳米技术工具箱[37]，还有许多新物种需要探索，截至 1996 年，已经有 12 000 个硅藻物种被"认可"[38]，即在分类文献中进行了讨论。此外，最近开始对硅藻进行基因改造[39-40]，并且有可能将硅藻进化成所需的形状或二氧化硅质地[41]。因此，近年来，无论是通过使用或模仿硅藻的形态方面，硅藻已开始被重新发现为用于纳米技术应用广泛且廉价的工具[42-59]。

每个硅藻壳由两部分组成，称为小壳和上皮，每个鞘膜由一个瓣膜和一个或多个环带组成，瓣膜通过环带相互连接并形成的培养皿状结构，如图 7.2 所示，较小和较新的小囊套在较大、较老的外皮囊内。在下一次细胞分裂过程中，形成了两个新的

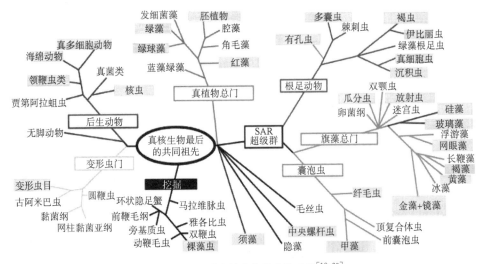

图 7.1　硅通过真核生物的生物矿化[18,60]

囊,使原来的小囊成为两个子细胞中较小的一个的新上皮,反之亦然。单个硅藻的长度或直径为 1.5 μm～5 mm(例如南极海藻)[47,61-62]。

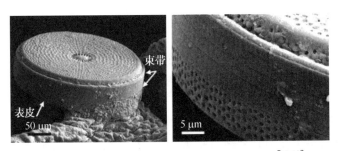

图 7.2　中心硅藻的一个完整的果状颗粒的 SEM 图[67-68]

硅藻壳不能被视为块状二氧化硅,而是由紧密结合的、烧结的二氧化硅纳米球[59,63],在微纳米长度尺度上具有或多或少的规则特征,且具有特定物种的模式和周期性。这种重复结构已在低至 3 nm 甚至可能更小的壳层中得到证实[64]。此外,细隙(即硅藻壳中的孔或穿孔)的直径范围从几百纳米到几微米,具有圆形、多边形或细长形状。如果我们包括一维菌落的形成(见表 7.2),则周期性可以跨越 8 个数量级(见表 7.1 和图 7.3)。我们可以人为地通过在规则阵列中"打印"硅藻来进一步扩展这个尺寸范围[65]。虽然在跨越"空间域大约 10 个数量级和时域大约 15 个数量级"的活细胞中可以很容易地找到如此大的动态范围[66],但所涉及的不同大小的结构在化学成分上是多种多样的。二氧化硅硅藻壳以单一、几乎化学纯的结构覆盖了如此宽的动态范围,使其真正独一无二。

表 7.1 硅藻的特征和周期性

级别	特征	报道的大小范围	最小值/nm	最大值/nm
1	单硅酸	0.3 nm[64]	0.3	0.3
2	膜（封闭物）孔径	3~100 nm[64-65]	3	100
3	一般孔隙	3 nm~2 μm[57,64]	3	2 000
4	多孔板厚度和包覆硅质骨架的有机涂层厚度	5~10 nm[16,59]；<50 nm[92]	5	50
5	硅胶	10~180 nm[59]	10	180
6	初生壁厚度	20 nm[59]	20	20
7	肋/脊	20 nm[59,95]	20	20
8	筛网孔径	43~45 nm[92]	43	45
9	层厚度	<50 nm[92]	50	50
10	孔距	68 nm[92]	68	68
11	纹/脊间距	100 nm~7.5 μm[59]	100	7 500
12	孔径	192 nm[92]	192	192
13	穹顶间距	200 nm[92]	200	200
14	孔径	200~900 nm 或>1.3 μm[92]	200	1 300
15	孔距	418~644 nm[92]	418	644
16	成熟壁厚度	700 nm[140]~1.2 μm[92]	700	1 200
17	Foramen 孔径	732 nm~1.28 μm[92]	732	1 280
18	Areolae 孔距	1 μm[92]；2.6~2.7 μm[140]	1 000	2 700
19	褐藻宽度/横轴	3.5~36 μm[86]，1.2~40 μm[70,72]	1 200	40 000
20	Cribrum 穹顶直径	1.62 μm[92]	1 620	1 620
21	Cribellum 穹顶直径	2 μm[92]	2 000	2 000
22	Pervalvar 轴	2~180 μm[65,70]	2 000	180 000
23	中心直径	2 μm~5 mm[58,62]；3 μm~3 mm[81,85]	2 000	5 000 000
24	褐藻长度	4.9~600 μm[53,81]	4 000	600 000
25	视斑直径	10~15 μm[133]	10 000	15 000

(续表)

级别	特征	报道的大小范围	最小值/nm	最大值/nm
26	刚毛长度	20～46 μm[29]	20 000	46 000
27	最大链长度	35 μm～5 cm	35 000	50 000 000

表 7.2 群落硅藻链长度的范围[69-82]

物种	细胞数范围（最小至最大）	细胞连接	细胞在链上的间距/μm	细胞直径/μm	最大群体长度/μm	Pervalvar轴/μm	群体形状
中心型硅藻							
伪角毛藻[45,83,92]	1～18	连接刺	20	23	400		直的
平滑透明盘藻[13,92]	2～4	重叠的刺	5	22～53	65～70	15～17	直的
火热骨条藻[101]	?～18	重叠的融合边缘刺环	3～5	2～6	143	3～3.5	直的或稍微波动的
角条骨藻[15,25,34,42,62,76,78,101]	2～60	重叠的融合边缘刺环	3～4	2～21	172	2～61	直的或稍微波动的
多尔尼角条骨藻[25,98]	3～94	重叠的融合边缘刺环	4～10	4～12	2 570	12	直的或稍微波动的
格雷塔角条骨藻[72]	9～111	重叠的融合边缘刺环	7～18	2～10.5	3 960	10～18	直的或稍微波动的
日本角条骨藻	3～64	重叠的融合边缘刺环	3～4	2～10	1 144	12～14	直的或稍微波动的
马里诺角条骨藻[18,25,72,80,92]	1～45	重叠的融合边缘刺环	3～9	2～12	1 197	3～18	直的或稍微波动的
伪角条骨藻[72-73]	1～34	重叠的融合边缘刺环	4～8	2～9	694	12～13	直的或稍微波动的
亚盐度角条骨藻[25,72]	7～60	重叠的融合边缘刺环	1～3	4～8	414	3～4	直的或稍微波动的
热带角条骨藻	6～93	重叠的融合边缘刺环	3～6	5.3～10	1 662	7～12	直的或稍微波动的
夏季海洋管藻[28,52,71]	2～50	连接相邻细胞的细中束	18～28	7～56	2 049	15	波动的
单形海洋管藻[21,22,26]	O(100)～25 000	中央纤维	24	4.4～15	50 000	6.4～7.5	形成直径可达1厘米的束状群体

(续表)

物种	细胞数范围（最小至最大）	细胞连接	细胞在链上的间距/μm	细胞直径/μm	最大群体长度/μm	Pervalvar 轴/μm	群体形状
圆盘海洋管藻[13,20,30,62,71,83]	1~600	连接相邻细胞的粗中束	—	8~60	—	5~20	直的或稍微波动的
羽纹细藻				顶轴			
冰岛小星藻[19,23,35,74]	2~75	通过扩展的足极（基底极）的阀面连接	0.1~0.3	15.73	—	85.11	星形（长链时呈螺旋形）
细槽状双簇藻[40]		相互锁定的刺	0.6~1	7~19	—	7~33	带状
凯尔盖朗脆笔藻[46]	14	相互锁定的刺	0.500	10~76	>35	5~7	带状
异脊藻及相关物种[27,35,51,60,68,78,102]	2~42	导纹纤毛	4~8	29~250	伸展时顶端约为4 000个	4~8	呈锯齿状至直线，具有轻微的螺旋性，细胞相互滑动

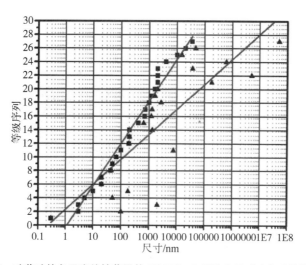

图 7.3 硅藻硅的多尺度特性范围从 $Si(OH)_4$ 分子的直径到硅藻群落的长度

为了对这种多尺度硅藻瓣膜的令人难以置信的范围给出一些看法，请考虑最近发明的用于石英玻璃的 3D 打印机，它具有"几十微米的分辨率"[83]，例如大约 30 μm。

要在玻璃雕刻雕塑中忠实地复制 5 mm 直径的硅藻瓣膜直至其 3 nm 的子结构,需要直径 50 m 的 3D 打印,即 10 000∶1 的比例。为了使肉眼可见的最小结构,例如 3 mm,比例必须为 1 000 000∶1,或直径为 5 km。

自然界中发现的硅藻在大小、对称性、表面形貌、微米和纳米结构、孔隙率、厚度以及整体硅藻壳形状方面存在很大差异[84-85]。它们的瓣膜表面可以是平面的、拱形的、圆顶形的,以及横向、纵向、同心或切向波浪形[86]或在它们的一个或两个边缘凸起,此外,还存在可被视为手性结构的扭曲硅藻壳(见图 7.4)。所有这些几何因素都可能影响硅藻壳的光学特性,因此可能导致许多不同的潜在应用。例如,可以想象印刷的手性硅藻阵列会产生雕刻薄膜[87-88](见图 7.5)。这些可能具有作为宽带圆偏振器、手性检测传感器或具有负折射率的元件的潜在应用[89]。

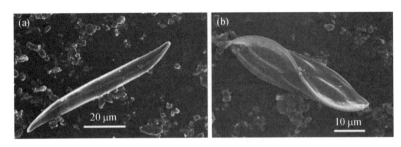

图 7.4　手性羽状硅藻的 SEM 图
(a) 二维扭曲瓣膜;(b) 三维扭曲瓣膜

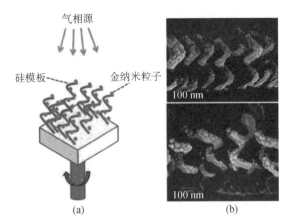

图 7.5　如何制作用金纳米颗粒装饰的手性雕刻薄膜的示意图(a)和等离子体手性纳米结构(b)[89]

此外,考虑到它们与光的相互作用,细隙的空间排列(细隙孔间距)可能是一个重要因素。光与人造类似结构(例如光子晶体纤维)的相互作用取决于周期性晶格的空间顺序[90-91]。多尺度孔隙率通常出现在同一瓣膜内的不同层中,多孔层的数量是一种特定于物种的特征,它也可能影响硅藻壳的光子特性(见图7.6)。

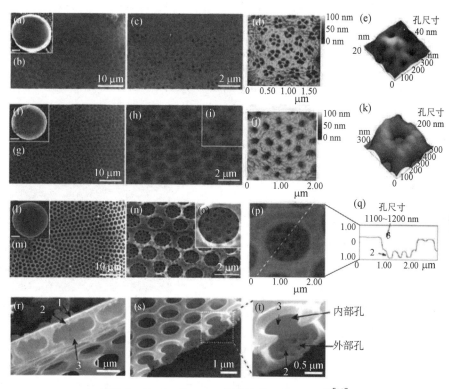

图 7.6 *Coscinodiscus* sp. 瓣膜的 SEM 和 AFM 图[92]

最适合光学和光电设备的是具有所谓薄片形状的硅藻壳[58],即对于羽状晶胞长度和宽度至少是整个硅藻壳厚度的三倍,以及中心直径至少是硅藻壳厚度的三倍。因此,这种孤立的薄片状硅藻壳通常位于它们的瓣膜上,因此垂直于施加到承载它们的设备上的光的方向。

7.3 硅藻壳的光学性质

如前所述,硅藻壳由几乎纯的无定形二氧化硅制成[93],二氧化硅会改变电磁波(例如光)的传播速度,因为它的折射率(对于波长约为 500 nm 的光为 1.46)与空气(约为 1)和水(20℃时为 1.33)的折射率不同。已知折射率会随波长增加而减小(正常色散),如图 7.7 所示,与块状二氧化硅相比,硅藻二氧化硅的折射率为 1.40~1.457[94-95],这可能是由于初始纳米颗粒生物合成引起的二氧化硅冷凝程度相对较低,它捕获各种有机分子,这也会发生在海绵骨针中[59,96-100]。折射率差异也可以通过瓣膜内二氧化硅水合程度的变化来解释,这可以通过在硅藻壳清洁过程中加热对其折射率的影响来测试,硅藻二氧化硅的水合可能在单个硅藻壳内变化,但尚未对此进行研究,与活动硅藻中的大块瓣膜二氧化硅相比,中缝衬里的二氧化硅对蚀刻的抵抗力可能伴随着折射率随瓣膜的变化(见图 7.8)[101]。圆筛藻的单个隔离瓣中的折射率的三维值被重建[46,102-103],但它们在二氧化硅内的空间变化没有相关研究,也没有观察到在此研究中使用全息方法[104],这将需要各向同性断层扫描分辨率,通过旋转安装在光纤或毛细管上的样品来实现[102,105-106]。很明显,折射率的位置依赖性变化可能会影响硅藻壳的光学特性,正如玻璃海绵所显示的那样[100]。

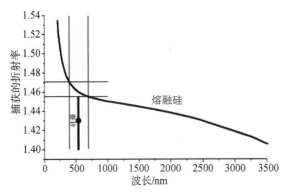 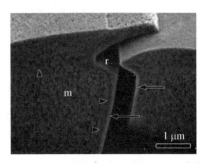

图 7.7　20℃下的熔融石英折射率曲线[107-110]　　图 7.8　已蚀刻的羽状硅藻瓣膜的横截面[46]

有趣的是,尽管硅藻壳由紧密结合的、可能是烧结的二氧化硅纳米球组成[59,63],但在扫描电镜和透射电镜图像中中硅藻的二氧化硅表面的平滑度是显而易见的,表明平滑度可能低于微米级[85]。然而,当检查瓣膜的二氧化硅横截面时,情况就不同了,它的化学和物理特性可能很不均匀(见图 7.8),可能会影响其局部折射率。然而,迄今为止,所有理论研究都假设二氧化硅是均质的,因此可能不得不重新考虑他们的结果。

此外,由于光与硅藻壳的微米和纳米结构的相互作用,硅藻通常在干涉光显微镜下出现彩虹图案[111],研究者们清楚地表示了通过某些硅藻物种的瓣膜传输的光的干扰[112],硅藻壳的多层透明性质会导致相移,并具有强烈的衍射和干涉图案,这取决于瓣膜的结构[112],并导致某些频率的放大或衰减[46]。光入射到某些硅藻壳上的角度会改变这种干涉图案,因此可以观察到取决于显微镜轴角度的颜色变化[111-114]。

Jamin-Lebedeff 显微镜可用于有意增强通过硅藻瓣膜的光与参考光束之间的干涉对比度,以产生壮观的硅藻壳彩色图像(见图 7.9)[115]。因此,它也可能是一种合适的工具,用于观察单个瓣膜内厚度的可能差异或折射率的变化。

据称,硅藻壳可以传输大约 90% 的入射光,并且可以充当 2D 亚波长衍射光栅[116]。最近的研究对这一说法进行了辩论,发现透射率和反射率模式以及透射光的百分比可能会根据光的波长、入射角、物种的瓣膜结构和周围介质而有所不同[117-119]。应该注意的是,在估计的 200 000 个物种中[120],迄今为止只有少数物种的光学特性被研究过,因此需要更多的研究来充分了解硅藻壳的光学特性。

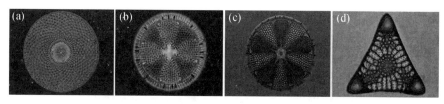

图 7.9　中心硅藻的数字增强 Jamin-Lebedef 显微图[115]

(a) *Craspedodiscus coscinodiscus* Ehrenberg（已灭绝的海洋硅藻）；(b) 和 (c) 放线菌；(d) *Triceratium morlandii* Grove & Sturt（以海洋为中心的硅藻化石）

7.3.1　作为光子晶体壁盒的壳

光子晶体（PC）是具有空间有序和周期性电介质或金属电介质微结构或纳米结构的材料，其表面折射率发生显著变化，可以控制电磁波的传播[121-123]。这些结构之间间隔的周期性决定了受 PC 影响的电磁波谱的波长。当 PC 导致在特定波长范围内完全或部分阻止光传播时，它们分别具有完全或不完全的带隙，具体带隙的特性取决于 PC 的参数[121-124]。根据周期性的空间分布，PC 可以是一维（1D）、二维（2D）或三维（3D）（见图 7.10）[124]。有时会故意将缺陷引入此类 2D 或 3D PC，以通过这些缺陷诱导光的波导（或弯曲），从而改变光束的方向和位置，这在将光聚焦到芯片上时很重要，例如光电应用（见图 7.11）[124]。在考虑 PC 板的光子特性时，应考虑波导及其光子能带结构[108]。这样，PC 可以被认为是控制电磁波行为和特性的强大工具。

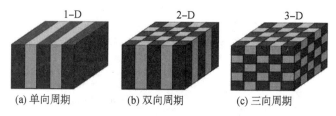

图 7.10　PC 的几何形状示意图[125]

(a) 1D；(b) 2D；(c) 3D

在自然界中，可以找到许多 PC 示例（2D 和 3D），它们以对生物体有用的方式有效地操纵光[127]。例如，Lopez-Garcia 等[128]最近在褐藻欧囊链藻中发现了由脂质制成的细胞内蛋白石状 PC，这赋予了它壮观的色彩。然而，迄今为止发现的所有天然 PCs 都具有不完整的带隙，这可能是由于它们的折射率对比度相对较低[129]。在查看硅藻的几何参数时，可以看到与 2D 或 3D 光子结构的惊人相似之处。那么某些硅藻是否出于任何目的使用 PC 结构构建硅藻壳？然而，在能够回答这样的问题之前，必须使用所有可用的技术，包括 SEM、FIB-SEM、TEM 和 AFM，对硅藻壳的超微结构

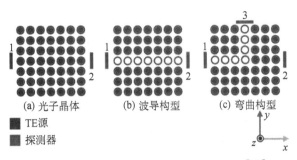

图 7.11 等离子体 PC(光子晶体)阵列示意图[126]

(a)具有完整晶体;(b)具有线性缺陷的晶体(波导构型);(c)具有两个未完成的交叉线性缺陷(弯曲构型)

进行彻底研究[130-133]。通过完整的超微结构研究,可以获得硅藻壳的晶格参数。然后,我们将能够从这些参数中预测是否有任何硅藻壳部件具有潜在的 PC 属性。理论和实验研究的结合对于获得完整的理解至关重要。一旦了解了硅藻中 PC 特性的物理特性,我们就可以询问有关它们的生物学相关性的问题。

图 7.12 显示了 *Coscinodiscus wailesii* 瓣膜横截面的简化图[134],准周期柱(充满空气、水或有机物质)存在于无定形二氧化硅介电基板,孔的典型堆积是正方形或六边形。虽然二氧化硅的折射率 $n_a=1.45$,但在实际硅藻壳二氧化硅结构中可能会有所不同(如前所述),但折射率对比度很大程度上取决于填充间隙的材料,并且对于空气($n_b=1.0$)和水($n_b=1.33$)[135],后者是硅藻的自然环境[136]。

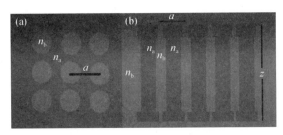

图 7.12 俯视图的原理图[134]

近年来,一些研究考虑了硅藻壳作为 PCs 的潜力。Fuhrmann[106]研究了 *Coscinodiscus granii* 硅藻壳的超微结构和光子特性。他们发现整个硅藻壳的孔隙率并不均匀,但可以识别出两种不同的孔隙模式:瓣膜中具有大晶格常数的六边形孔阵列,以及孔的方形阵列环带中的一个小的晶格常数。通过理论计算,他们得出结论,*Coscinodiscus granii* 的瓣膜和束带确实可以通过耦合到具有不同光子晶体模式的波导中来影响光在空气中的传播。已经发现这些波导模式在空气中处于蓝光区域,并且受二氧化硅厚度的影响。然而,在水中,模态的数量显著减少,但仍有一些模态存在。他们的实验研究,通过在显微镜下观察透射光的同时照亮一个垂直于其平面的束带,为这种波导特性提供了进一步的证据,因为发现只有蓝绿色从边缘发出,而红光是在束带的焦点处观察到。最近,这些结果在同一物种上得到了进一步调查和证实[118-119]。

还发现 Melosira varians 硅藻壳的内部纳米结构具有光子晶体特性,可导致空气中的蓝光吸收[136]。在另一个例子中,Romann 等[137-138]在 Coscinodiscus centralis 的清洁瓣膜中观察到其他几种光子晶体特性:光谱修改、定向光聚焦和光阻挡效应。考虑到经典的衍射现象,这些功能无法解释,因此强烈支持从瓣膜独特架构中出现的 PC 模式的存在。此外,在将相干宽带超连续谱激光聚焦在 Coscinodiscus wailesii 硅藻壳的瓣膜上时,由于硅藻壳中的准周期六边形孔阵列而产生了与位置相关的光学衍射图案[139]。

在另一项研究中,De Stefano 等[140]使用有限元分析对 Coscinodiscus wailesii 的瓣膜进行建模,发现当没有光学带隙的介电材料二氧化硅($n=1.45$)被二氧化钛($n=2.4$)取代时,一个完整的带隙可以生成。因此,由于存在具有特征色散关系的晶格周期光子模式,光的传播以及光与物质的相互作用可以在某些硅藻壳中被改变。表面改性可以增强这些特性,从而使硅藻可用作无源波导中的光子元件、太阳能电池的衍射元件或用作光子晶体激光器等有源器件[141]。

最后,还应该注意的是,在许多 PC 结构和特定类群的某些硅藻壳之间可以观察到一些有趣的结构相似性。例如,在许多物种中观察到类似于光子晶体[142]的结构,例如 Cyclotella meneghiniana(见图 7.13)。此外,一些硅藻壳部分在其原本规则的晶格结构中显示出晶格缺陷(见图 7.14 和图 7.15)。PCMLs 的聚焦特性是由一系列晶格参数引起的[143],它们可以被视为具有光聚焦特性的 2D PCs 的示例[144]。因此,未来的研究应该关注特征晶格图案的相关性以及它们对某些物种的典型缺陷及其对

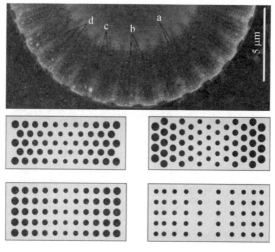

图 7.13 一个显示波状肋骨孔隙度瓣膜外部的 SEM 图

由于外部的波动,在不同层次上划分为具有不同孔隙度尺度和模式(a、b、c、d 区)的重复单元

光传播的影响。

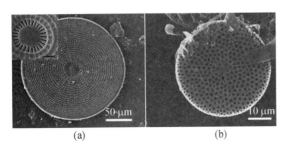

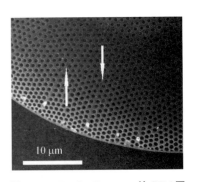

图 7.14 未成熟瓣膜的外部(a)和内部(b)的 SEM 图

图 7.15 *Coscinodiscus* sp. 的 SEM 图
没有光晕的线性二氧化硅(可以认为是线性光学缺陷)

7.3.2 光聚焦现象

在研究光通过 *Coscinodiscus* 瓣膜的透射率期间,最引人注目的观察之一是其聚焦光的能力,尤其是入射光焦距对波长的强依赖性。785 nm 二极管激光束的焦斑直径已从 100 μm 减小到 8.1 μm,距离单个瓣膜中心 104 μm。中心对应于瓣膜中心的均匀且无乳晕的透明区域(直径 15 μm),如图 7.16(c)所示。所用实验装置如图 7.17 所示。De Stefano 等将瓣膜称为"微透镜",并且可以将聚焦效应归因于由准规则散射的光的未聚焦波前的相干叠加,这些实验结果与数值模拟相结合表明,这种聚焦效应不符合 Rayleigh-Sommerfield 衍射理论。随后,在通过 *Coscinodiscus wailesii* 的单个瓣膜传输部分相干或非相干光时获得了类似的结果。然而,在非相干光的情况下,已发现重新分布模式与远距离的相干光有很大不同,并显示出环状照明。此外,这种聚焦效应不仅发生在空气中,还发生在蒸馏水或细胞质中,尽管透射光的强度会降低。Romann 等观察到透射光的方向依赖聚焦,如图 7.18 所示。

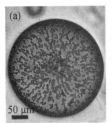  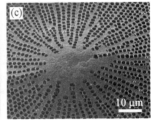

图 7.16 *Coscinodiscus wailesii* 完整活细胞的光学显微图(显示大量小的叶绿体)
(a)瓣膜视图;(b)腰带视图;(c) SEM 的中央透明区域

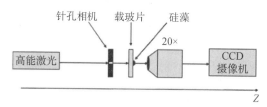

图 7.17　用相干光源和 20 倍显微镜物镜观察单个硅藻圆柱的聚焦效果的实验光学装置

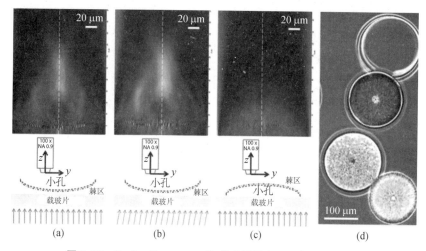

图 7.18　*Coscinodiscus centralis* 单个瓣膜在不同方向下的超图

Ferrara 等也重复了这些研究中的观察结果,研究了 *Arachnoidiscus* sp. 的单个瓣膜的光学特性。在空气中的紫外线可见区域,发现了一种聚焦效应,然而,其模式与在 *Coscinodiscus wailesii* 瓣膜中观察到的模式不同。他们的研究结果表明,瓣膜的边缘、其中心法兰、径向肋和孔隙有助于入射光的衍射和干涉图案的出现,从而产生一系列具有强光放大的斑点,他们称为"热点"。这一系列热点位于硅藻壳内部,用于红光波长,并通过将它们聚焦在硅藻壳外来减弱破坏性的紫外线波长。在另一项研究中,De Tommasi 等发现 *Arachnoidiscus* sp. 的单硅藻瓣。使用结构化激光照明和光学本征模(OEi)分解,能够在远场高效地将光限制在一个微小的亚衍射限制焦点上。

因此,假设所有中心硅藻都可以表现出这种光限制特性,该特性取决于入射光束的波长以及硅藻壳对光的方向,并且可能与光的相干程度无关。如果是这样的话,这将给这种硅藻带来进化优势,使它们能够通过硅藻壳设计来影响细胞内光的位置、强度和波长。此外,这种基于硅藻的微透镜可以为光子器件、光学显微镜、曲面光栅和纳米技术开辟新的机遇。

有趣的是,观察到的聚焦效应类似于所谓的 Talbot 效应,其中直径小于光波长

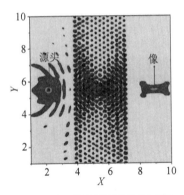

图 7.19 二维 PC(光子晶体)中的异常聚焦效应示意图

的规则孔阵列诱导光聚焦,与菲涅耳透镜的聚焦效果完全不同。瓣膜的曲率也可能导致入射光的聚焦。

另一方面,这种现象也可能与瓣膜结构类似于 2D 或 3D PCs 有关,如图 7.19 所示,已发现某些 2D PC 板中光的异常聚焦与其有效负折射率有关。此外,发现具有与硅藻中相似的周期性的 3D 二氧化硅反蛋白石 PCs 可以聚焦光。还观察到,即使在存在高吸收水分的情况下,这种反蛋白石也可以获得负折射率。在改变其周期性参数后二氧化硅反蛋白石反射的颜色显示出变化。

7.3.3 光致发光特性

光致发光(PL)是物质吸收光子后发光的现象。就硅藻而言,已发现它们的硅藻壳的 PL 特性与熔融二氧化硅以及无定形二氧化硅纳米线的类似。如果用紫外线照射,观察到的发光主要位于蓝色区域,在一些样品中可以看到绿色和红色的其他小峰。这些绿色条带可能类似于先前在嵌入无定形二氧化硅纳米线的结晶纳米二氧化硅结构中观察到的 PL 峰。已发现硅藻壳的 PL 显著依赖于物种、硅藻壳的结构以及周围环境。

由于由无定形二氧化硅制成,硅藻的硅藻壳表面可能含有大量的硅醇基(Si—OH)和硅氧烷桥(Si—O—Si),前者 PL 的出现主要与二氧化硅的蓝光区域有关,而绿色和红色带可能是由于结构缺陷或硅氧烷桥。620 nm 的波段可能与非桥氧空穴中心相关。使用傅里叶变换红外(FTIR)光谱进一步证实了 Si—OH 拉伸模式的存在。

Mazumder 等发现,在 300 nm 或 370 nm 光的激发下,中心淡水硅藻壳 (*Cyclotella* sp.)在 440 nm 处显示出强烈的蓝色光致发光。宽 PL 光谱可归因于与二氧化硅纳米结构相关的缺陷或其微环境的微小变化。Goswami、Choudhury 和 Buragohain 还研究了 *Cyclotella meneghiniana* 的发光,他们在 387 nm 和 452 nm 附近发现了两个光致发光峰,前者类似于无定形二氧化硅纳米线、氧化多孔硅和退火二氧化硅。此外,Arteaga-Larios 等发现 *Amphora coffeaeformis* 和 *Gomphocymbella* sp. 的光致发光发射的依赖于培养物的时期和清洁方案。对时期的依赖可以解释为二氧化硅可用性的变化导致实际厚度或在瓣膜上观察到的详细图案的差异。还应该注意的是,在静止培养阶段生长的成体硅藻壳中会出现孔内堵塞,这会增加二氧化硅的表面积,从而增加可能的 Si—OH 键合的数量位点,这种现象可能与光致发光发射的增强相关。

7.3.4 硅藻壳在硅藻光生物学中的可能作用

硅藻壳本身对活硅藻有几个重要作用，包括机械保护、流体动力学控制以及对细胞营养状态的控制。据推测，硅藻壳可以调节光、重新分配、优化活细胞内的照明、减少有害波长的影响，并将光合活跃的透射光与光合装置耦合。De Stefano 和他的同事还提出了硅藻壳对光合作用的重要性，光合作用发生在活细胞的叶绿体内部。

为了更详细地研究这一假设，Noyes 等通过测量单个瓣膜和整个硅藻壳的光相互作用，研究了 *Coscinodiscus wailesii* 的光操纵（见图 7.16）。在这项研究中，他们发现了一种不同的颜色传输模式，蓝色和绿色波长相对较低，但红色波长较高。如果在自然环境中被证实为活细胞，这将有助于硅藻保护自己免受过度蓝光强度、在水下观察到的主要波长和紫外线辐射的影响。Noyes 等得出结论，差异颜色传输可能是由瓣膜外壁和内壁之间的光子相互作用引起的。还应该注意的是，透射光的不同取决于入射光的位置和方向。此外，Yamanaka 等指出"蓝光可以分散在硅藻壳中，以均匀地照射硅藻内部的叶绿素有效增强光合作用"。

Romann 等通过 PC 特性和光聚焦特性的结合进行了研究，进一步表明 *Coscinodiscus centralis* 硅藻壳的结构确实是为光调节功能以及增强叶绿体附近光合活性辐射的吸收而设计的。此外，还观察到了一种高效的光捕获机制，可以防止入射光的反向散射，从而增强细胞的光吸收。最近，Goessling 等根据光合作用过程证实了硅藻壳在光调制中的作用，同时注意到早些时候叶绿体运动不参与。

此外，正如我们所讨论的，通过硅藻壳的透射光的焦平面取决于入射光的波长。它随着波长的增加而降低，这可能导致细胞内叶绿体位置附近的光合有效辐射（PAR）聚焦，同时有害的紫外线辐射缺乏聚焦。

在光保护方面，所有提到的研究表明，除了硅藻叶绿体中存在的其他分子机制外，硅藻壳可能有利于保护细胞免受紫外线波长的影响。这可以通过吸收紫外线并随后以更长的波长（即更低的能量）重新发射（紫外线诱导的 PL）来实现，通过在嵌入的类真菌氨基酸的帮助下吸收它，通过不将这些波长聚焦到细胞中，或通过反射大量此类辐射。硅藻壳的一个重要进化作用可能是保护 DNA 免受有害紫外线辐射的降解，从而降低突变的风险。

总之，我们可以总结出硅藻壳可以实现的三个潜在的重要光学特性，这些特性可能与相关的光生物学作用相关：

（1）透射光和反射光的数量取决于入射光的波长、瓣膜超微结构、入射角和周围介质；

（2）某些物种的瓣膜所传输的光的焦平面相对于瓣膜表面的位置取决于入射光的波长；

（3）透射模式显示出空间差异，具体取决于照明区域、物种和瓣膜的方向。

7.4 硅藻光生物学

光是硅藻生命周期中的关键因素。各种研究调查了活硅藻和光之间的相互作用。这个研究领域被称为硅藻光生物学。除了其他复杂的相互作用外,硅藻光生物学还有两个主要的具体方面,即光保护和光感知。硅藻可以在暴露于多种光照变化的区域中找到,并且对不同的环境条件表现出很高的耐受性。为了实现这一点,必须存在高水平的适应,包括在特定的光合机制内、羽状体的趋光性、多质体中心的叶绿体迁移,以及可以说是硅藻壳光子学,其程度取决于物种。

7.4.1 水下光场

水是所有硅藻的家园(其中一些生活在潮湿的表面)。这就是为什么目前的综述侧重于水下光场的研究,其中光强度、空间和时间相干性以及光谱特性与空气中的不同。发生这种情况的原因有很多:一天中的时间、水体的吸收和散射过程、硅藻上方的水柱高度以及有色溶解有机物和/或悬浮颗粒的存在。随着深度的增加,水强烈吸收红光和红外光,导致其蓝绿色光谱成分(400~500 nm)占主导地位。随后,在这种光照变化下观察到浮游植物生态型的深度依赖性生态位分离,因为并非所有的光合有效辐射都通过水面传播并到达水生栖息地的深处。

7.4.2 细胞周期光调节

已发现三个因素与某些硅藻物种的光照状态和细胞周期调节相关:光的波长、明暗周期的长度和光强度。硅藻使用光感受器,例如称为金色素的蛋白质来检测蓝光,蓝光控制着硅藻生理学的许多方面,包括细胞周期活动。它们存在于细胞核和细胞质中(见图 7.20)。Huysman 等观察到暴露于蓝光后细胞信号的表达增加,这反过来又导致细胞周期激活。此外,通过控制培养过程中的光照和黑暗时期,已经证明了细胞周期时间的交替。此外,光的强度在细胞周期中发挥作用,有时会触发生长孢子的形成。虽然低光强度已被证明会触发某些物种的生长孢子形成,例如 *Skeletonema costatum*、*Chaetoceros* spp. 和 *Aulacoseira subarctica*,像 *Aulacoseira baikalensis* 这样的物种在阳光明媚的日子和较高的光照强度下形成生长孢子。

7.4.3 羽纹纲中的趋光现象

另一种光调节机制已在一些羽纹纲中发现,称为"趋光效应"。这种现象可能与光的波长和强度有关。Cohn, Dunbar, Skoczylas 和 Mucha 发现所研究的四种物种中的每一种,即 *Craticula cuspidata*,*Stauroneis phoenicenteron*,*Nitzschia linearis* 和 *Pinnularia viridis*,在它们对可见光和紫外线辐射的趋光反应方面表现出不同的模式。*Craticula* 与绿光具有正向趋光相互作用,而 *Stauroneis* 与红光具有正趋光反

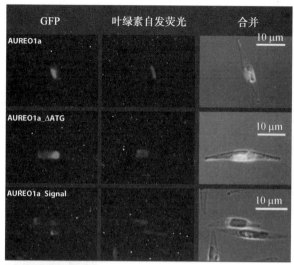

图 7.20 显示了不同金黄色色素 1a-GFP 融合蛋白[AUREO1a(49116)、AUREO1a_△ATG(49116) 和 AUREO1a_Signal(56684)]的荧光显微镜图

应,而与绿光具有负趋光反应。从这一点,本综述的作者得出结论,这种效应可能有助于分离存在于混合种群中的趋光物种,因为光点可以刺激物种分布的积极变化,正如 Cohn 所证明的那样,其中 *Craticula* sp. 绿点的比例增加,而 *Stauroneis* sp. 红点的比例增多。

    Cohn,Spurck 和 Pickett-Heaps 进一步研究了高能辐照对羽状藻类运动的影响。他们的数据(结合 DNA 分析)表明存在一个假定的光检测系统,该系统对中缝远端附近的强光和弱光条件作出响应,如图 7.21 所示。在这里,光响应触发了硅藻运动方向的变化,从强光到弱光条件。Depauw 等的分子研究进一步证实了这种光电检测系统的存在。他回顾了海洋硅藻对光反应的分子基础,并描述了可能存在的假定光传感器。远端光电探测系统可能涉及金色素(见图 7.20)。

    除了对照明特性的依赖之外,Cohn 等发现当温度升高超过 30℃ 或 35℃ 时,细胞的运动活性急剧下降。他们得出的结论是,黏液分泌或黏液系链的附着/易位可能与潜在的肌动蛋白有关,它们都参与了这一调节,因为硅藻物种 *Nitzschia palea* 和 *Navicula grimmei* 的细胞滑行期、滑行细胞百分比和速度,取决于培养物是在黑暗中还是在低光子通量(2 μmol·m^{-2}·s^{-1} 或 10 μmol·m^{-2}·s^{-1})下发生在不同的水平。

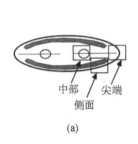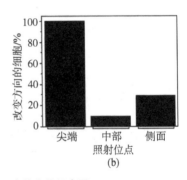

图 7.21　光反应位点的定位示意图

用 545 nm 光在细胞的尖端、中间或侧面对表皮细胞照射 1 s；(a) 表示每个辐照部位的近似位置；(b) 表示各部位照射后细胞改变方向的百分比

7.4.4　叶绿体迁移（核营养不良）

叶绿体是硅藻中另一个有趣的超微结构。它们在硅藻属和种之间的数量、排列和形态各不相同。大多数中心硅藻是多质体的,这意味着它们有不只一种叶绿体（见图 7.16）。Mann 说明了这种多质体细胞的超微结构。有趣的是,早在 1862 年,Lüders 就观察到了一个相当有趣的现象：当细胞暴露在强光下时,有时会发现叶绿体向内移动,并在细胞核周围聚集。50 多年后,这个过程被 Senn 称为"核营养不良"。进一步研究发现,机械刺激也能产生同样的效果。

Karsten 和 Mann 的综述表明,核营养不良发生在广泛的不相关硅藻中。它发生在明暗循环的黑暗阶段开始时,需要 7～12 min,而分散（周转）较慢。此外,Furukawa、Watanabe 和 Shihira-Ishikawa 发现 Pleurosira laevis 中的叶绿体迁移还取决于单色光的波长和强度。在 540 nm 附近具有单峰的绿光导致叶绿体分散,而在 450 nm 处具有主峰的蓝光导致叶绿体组装。

在其他三种中心硅藻中也研究了核营养不良：*Biddulphia pellucida*、*Ditylum brightwellii* 和 *Lauderia borealis*,但与 *Pleurosira laevis* 不同,在这些硅藻中,已发现黑暗是细胞核周围叶绿体组合的信号。对于"光感受和光反应之间的有效交流",运动的羽状体具有较少叶绿体的推测没有考虑到时间尺度的巨大差异和运动与核营养不良的光谱响应。

7.4.5　蓝光及其对细胞微管的影响

众所周知,当高能辐射照射具有相对自由价电子的原子时,会发生光电效应。这些电子基本上被高能光从原子上击落,从而产生正电荷。在细胞中,当暴露于包含蓝光或更高频率的广谱光时,这种效应可能发生在带电离子中,例如在微管中发现的。虽然靛蓝、紫色和紫外线甚至更高能量的辐射都可能导致这种效果,但蓝光最不容易被大气或水阻挡,如图 7.22 所示。此外,蓝光包含在植物的光合作用有效波长中。

这种微管蛋白的静电图显示了两个相隔 2 nm 的正电荷区域,在 β 亚基附近进一步被负电位包围。因此,微管相关蛋白(MAP)或其他微管相关分子的功能可能会受到蓝光的影响。例如,已经表明,当将光敏素(一种光敏剂)添加到人内皮细胞中时,荧光会导致细胞中的微管解聚。解

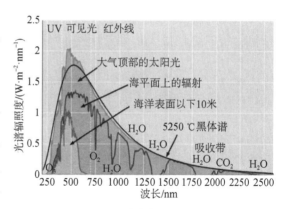

图 7.22 地球表面的太阳能

聚以剂量依赖性方式发生,在 5 分钟暴露于 60 mJ/cm² 辐射后,细胞没有留下任何有组织的微管。然而,这种解聚也可能是由于细胞内钙水平的增加。以硅藻为例,蓝光频率已被证明被许多光颜料使用(光合作用)和阻挡(光保护)。这些感光颜料也可能影响细胞行为——可能重新排列微管的化学反应。或者,已知钙离子通道可重新排列微管的构象。这些二价钙离子可以很容易地稳定通过光电效应形成的松散电子。一般来说,钙离子通道通过细胞膜的去极化而被激活,微管方向的变化可能导致它们的打开/关闭。

蓝光对硅藻生长的影响因硅藻物种而异,其中一些经历了生长增加,而另一些则没有。已发现硅藻对少量的蓝光、红光或远红外线照射有反应。然而,发现的大多数受体对蓝光有反应,在硅藻中发现了几种不同类型的感光蛋白:隐花色素、金色素(见图 7.20)和光敏色素。此外,一些硅藻壳本身具有纳米结构,可增强对蓝光频率的吸收以及它们在此类硅藻内的保留。此外,*Coscinodiscus wailesii* 叶绿体在蓝光照射下向细胞核迁移,推测这一事件也受到钙渗透通道的支持,因为叶绿体的运动也使用细肌动蛋白丝。

对微管与蓝光辐射相互作用的一些见解可以来自对植物的研究,因为已经表明,当植物中的微管暴露于蓝光时,它们会重组。例如,*Ceratopteris richardii* 配子体的生长伸长部分的皮质微管在 453±31 nm 左右的 2 watts/m² 照射下显示出从横向变为纵向的方向。光导致原叶区域弯曲,因为受影响的细胞比没有被蓝光照射的周围细胞短。此外,在多种拟南芥中,微管的运动已经被绘制出来,表明光抑制了微管从纵向阵列到微管星形阵列到横向阵列的转变速度。这种横向排列之后是更快的伸长率。此外,光还抑制了它们的微管聚合速度,如图 7.23 所示。在拟南芥叶子的情况下,蓝光已显示重新排列肌动蛋白,随后影响细胞核的位置。在这些细胞的光响应中已经显示了向光素(即在高等植物中介导响应于光的方向的光感受器)的参与。将来,需要对硅藻的细胞骨架对蓝光或任何产生光电效应的光的反应进行系统研究。

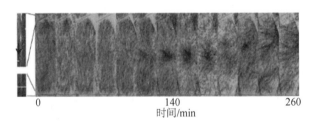

图 7.23 *Arabidopsis thaliana* Ler-hy1-EB1a 中的恒星形成过程

在恒星形成后,微管变得更加横向排列,帧之间的间隔为 20 min,细胞长度约为 50 μm

7.4.6 高光强度下的光调节策略

当硅藻细胞暴露于高光照水平时,会发生取决于暴露时间的变化。在短时间曝光的情况下,即几秒钟到几十分钟,基因表达模式不会改变,光合分子机器的灵活性可以减少损害,例如,从光收集状态切换到光保护状态。这可以通过激活允许光系统Ⅱ分散过度吸收的光能的两种机制来实现:非光化学荧光猝灭(NPQ)的激活以及光系统Ⅱ和/或光系统Ⅰ周围的电子循环。但如果曝光持续时间超过几十分钟,基因表达模式就会开始受到影响,并开始改变光合装置的组成和光化学。对于硅藻三角褐指藻,这些高光驯化机制可分为初始响应阶段、中间驯化阶段和后期驯化阶段。虽然初始阶段包括对编码参与光合作用、色素代谢和活性氧(ROS)清除系统的蛋白质的基因进行强有力和快速的调控,但随后的阶段会导致光捕获色素减少,并且在后期适应过程中光合能力强烈增加阶段。

7.4.7 紫外线辐射照射下的光调节策略

紫外线(UV)辐射对生物体,尤其是光合微生物的有害影响很普遍,主要对 DNA 造成损害。由于紫外线辐射对硅藻生产力、光合作用过程和光合作用装置的影响,已经进化出特殊的生理反应来调整活细胞以适应其自然栖息地中紫外线辐射水平的变化。例如,持续暴露在 UV-B 辐射下会导致海洋硅藻 *Coscinodiscus gigas* 的色素细胞发生变化,从而导致一种称为 UV-B 诱导的同养现象,这种现象不会显著改变光合 O_2 的释放。

硅藻对紫外线辐射的适应策略因物种而异,因此可以被认为是一种物种特异性的行为,这意味着一些物种如肋骨骨架对紫外线辐射敏感并高度影响,而另一些物种则具有更高的耐受性,如咖啡形双眉藻。尽管如此,Scholz、Rúa 和 Liebezeit 认为硅藻对短期和长期紫外线照射存在一般生理反应,包括脯氨酸、总碳水化合物和脂质的显著增加。此外,除了众所周知的光合作用装置的光化学变化外,长时间的紫外线暴露与叶黄素循环的激活有关,然后合成类真菌孢子素,这可以被认为是紫外线防护方面的重要代谢物用于活硅藻细胞,以及其他吸收紫外线的化合物的活化,这些化合物在某些硅藻物种中被鉴定,例如菌孢碱-甘氨酸和类菌胞素氨基酸。在暴露于 UV-A 辐射的情况下,硅藻显示出苯丙氨酸、天冬氨酸和饱和脂肪酸的产生水平增加,而暴露于 UV-B 辐射则增加了半乳糖、甘露糖和不饱和脂肪酸的产量。

还应注意的是,一些因素能够减少有害紫外线辐射引起的抑制,例如高温、高海洋酸化或高 Zn^{2+} 浓度。硅藻可能对海洋酸化具有很强的抵抗力。

7.4.8 硅藻和弱光

在光线不足的情况下,不同物种之间观察到的硅藻反应差异很大,这取决于它们的原始栖息地。一些硅藻物种因其自然倾向于弱光条件而闻名,其中一些是底栖、附生、北极或在洞穴和温泉中发现的。发现包括 *Pseudo-nitzschia multiseries*、*Aulacoseira subarctica*、*Phaeodactylum tricornutum* 和 *Achnanthes exigua* 在内的这些物种在低光照条件下有效控制其代谢成本,并且因此生存。在规划其在各个领域的应用时,这些方面可能是一个重要的考虑因素,因为它降低了大规模种植所需的功率。有趣的是,一些 *Coscinodiscus* spp. 适应低光强度的动物可以在大约 100 m 的深度生存。最近,在深海(海面以下约 4 km)中发现了健康的硅藻细胞,这被认为是快速下沉机制的结果。

在低光照条件下可能会引起硅藻细胞的关键变化:

(1) 在 *Pseudo-nitzschia multistriata* 中看到的毒素产生升高;
(2) 形成发育孢子;
(3) 三角褐指藻中的脂质生成;
(4)细胞大小和光系统Ⅰ和Ⅱ反应中心数量的变化以捕获最大数量的光子,如三角褐指藻;
(5) 在羽状中看到的运动性降低;
(6) 增加细胞二氧化硅含量,以增强硅藻壳的机械性能;
(7) 细胞内硝酸盐积累增加和/或细胞体积减少。

7.4.9 硅藻和无光

虽然大多数硅藻物种不能在异养条件下生长,但某些硅藻可以在黑暗缺氧的沉积物层中存活数月至数十年。这种硅藻存于沉积物、泥浆或高密度藻垫中,在那里它们接收到很少或没有光,激活了从周围吸收更多有机基质的机制,并且当再次受到光照进行光合作用时,可以关闭这些机制。此外,据报道,与其他硅藻物种不同,*Nitzschia palea* 甚至在黑暗中偶尔出现细胞分裂。此外,最近对深海硅藻浮游植物群落的研究表明,它们可以在大深度(2 000～4 000 m)存活一个月或两个月,同时保持完整和存活。一些硅藻在深色沉积物中存活至少 100 年,可能以孢子的形式存在。孢子的形成也会使硅藻在长达 6 个月的极地冬季存活。

7.4.10 光导管与细胞视觉

部分生活在弱光条件下的长硅藻菌落（例如，在苔藓上，部分菌落在苔藓内）可能进一步进化出光管道特性。一个例子是 *Ellerbeckia arenaria*。硅藻研究员 Richard 将这个物种命名为 *Ellerbeckia*，以纪念 Ellerbeck 家族，该家族拥有一座墙壁非常厚实的房子。如此厚的墙壁会使光衰减太多，无法进行适当的光合作用，未来研究的任务是评估该物种是否还可能进化出特定的增强光导特性。

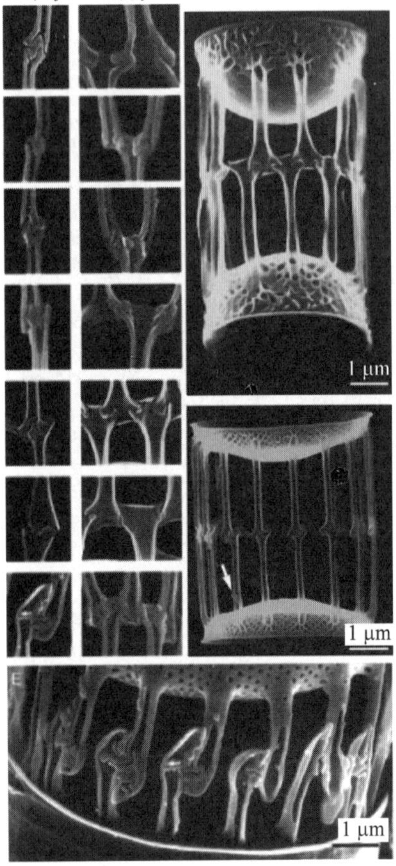

光导已在各种其他生物体中显示出来。一个例子是北极熊的毛皮，它们有效地工作并激发了纺织太阳能集光器的相关技术发展。此外，一些玻璃海绵的针状体，以及生活在高海拔的阿尔卑斯山花雪绒花的毛发，具有类似玻璃纤维的结构，出现典型的光子水晶包层，过滤掉有害的紫外线，可以在雪绒花中显示出来，将光合作用所需的光直接引导到叶子表面。从这些有希望的研究中，硅藻研究人员可以了解这些生物如何将光耦合到它们的纤维中，并尝试在硅藻菌落实验中进行类似的研究，该实验首先由 Gordon 等提出，用于同步奇异芽孢杆菌的游动菌落。

链中硅藻之间的连接可分为有机纤维、熔融二氧化硅、互锁二氧化硅或通过中缝原纤维（见表 7.2）。在这里，我们强调这些联动机制所赋予的潜在光学特性，而不是它们的机械特性。链状硅藻之间的链接全景图如图 7.24 和图 7.25 所示。为了比较，图 7.26 显示了人造包层光纤的多种结构。

图 7.24 *Skeletonema* 藻属物种的各种连锁连接点

硅黄藻连接点分别为 1∶1 或 1∶2。右上：两个阀门连接 1∶2。当所有的有机物被去除后，这些都没有分解，硅烧结融合可能是连接这些结构的部分原因

当今的包层光纤和硅藻链之间的根本区别在于，后者沿光纤长度呈现周期性。由于光纤中两种传播模式之间的相位差，"强度分布的形状会周期性地变化"。因此，我们可以预期，沿着链状硅藻的周期性将选择一对特定模式的传播，类似于弯曲损耗，并且它们的净强度可能与叶绿体或其他色素的位置同相相关。由于叶绿体可以在每个细胞内移动（核营养不良），因此可以在光管道条件下预测细胞内叶绿体沿链的相关位置。据说沿其轴具有不同折射率的人造光纤包含"光纤布拉格光栅"（见图 7.26），这导致除了窄波段之外的所有波长都具有透射率。在更复杂的结构中会

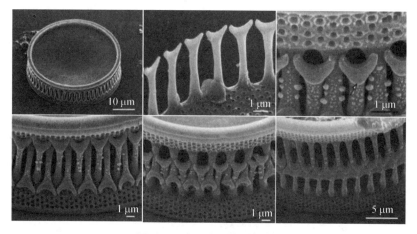

图 7.25　中心链硅藻的联锁棘

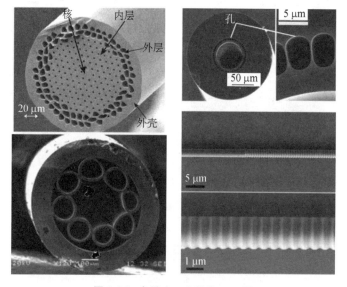

图 7.26　各种人工光纤的 SEM 图

发生什么，例如硅藻链，需要进一步研究。未来，通过铭刻方法制造光纤布拉格光栅可能会被"生长纳米技术"所取代，即培养链状硅藻。

7.5　硅藻和光的应用

7.5.1　在光催化中的应用

每克硅藻瓣膜的超大表面积（$80 \sim 211 \text{ m}^2 \cdot \text{g}^{-1}$）已被证明对催化具有潜在吸引力。虽然二氧化硅本身没有最高的催化潜力，但可以将其还原为硅，同时保留原始硅藻壳的大部分纳米结构。例如,这种纳米结构的硅可以用来将水分解成 H_2 和 O_2,从而制备氢气作为燃料。此外,用 CdS 涂覆硅增强了这种光催化效应。Chandrasekaran 等证实通过镁热转换制备的硅藻壳具有光催化和脉冲光电流产生特性。因此,由硅藻壳构建纳米结构的半导体可以改善光催化氢的产生。

7.5.2 基于生物的紫外线过滤器

最近,来自三种不同物种(格氏圆筛藻、点刺海链藻和假微型海链藻)的单层硅藻壳已均匀地分布在透明基板(SiO_2)上,以研究紫外范围内的透射率和反射率。令人惊讶的是,三个单层膜的紫外线透射率和反射率值不同(见图 7.27)。点刺海链藻获得了最佳反射率。这是将硅藻壳引入紫外线过滤器行业的有希望的做法。

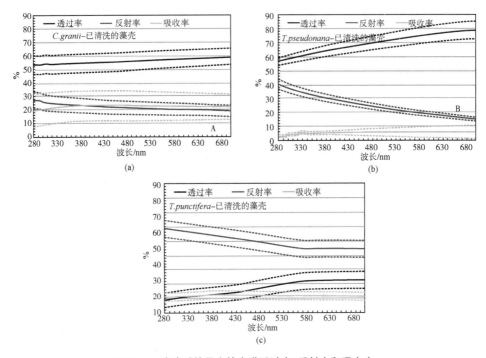

图 7.27 冲洗后的果实的光谱透过率、反射率和吸光度

虚线表示平均值±为标准差;参考光谱为紫外透明石英硅显微镜玻璃板,所有的值都被标准化为100%的覆盖率,种类为 *Coscinodiscus granii*,*Thalassiosira punctifera*,*Thalassiosira pseudonana*

7.5.3 在太阳能电池中的应用

硅藻的大表面积可以通过表面涂层,或者甚至通过生物方式将氧化锗或氧化钛插入硅藻壳本身,很容易地用光电活性物质例如二氧化钛,从而使其成为提高太阳能电池板效率的合适材料,从而使它们成为提高太阳能电池板效率的合适材料。将染料敏化太阳能电池(DSSC)中的光敏层夹在透明电极(入射光方向)和二氧化钛涂层硅藻壳之间,已显示通过增加光子反向散射的程度来提高该层中捕获的光量,这使得能够使用更薄的光敏层,从而降低制造成本。薄的光敏层还减少了染料和导电玻璃电极之间的电子扩散距离,降低了电子-电解质复合的风险,从而提高了效率。Jeffryes 等使用这种二氧化钛涂层的硅藻壳层使 DSSC 的能量转换效率提高了三倍。当加入主要由覆有二氧化钛纳米粒子的硅藻壳组成的硅藻土时,观察到类似 30% 的增幅。从绝对值来看,能量转换效率从 3.5% 上升到 4.6%。

另一种方法是使用硅藻壳作为传统无机硅太阳能电池板中的光捕获涂层。这个想法是基于这样一个事实,即与生活在阳光充足的地方的硅藻相比,生活在低光照水平(例如,在北欧海)的硅藻可能会使用它们的硅藻来提高光收集效率(一个值得测试的想法),这意味着它们能够捕获大部分入射光子,并且只反射极少量。使用这个想法确实实现了至少 4% 的硅太阳能电池板效率提升,即从 16% 到 16.6%。

7.5.4 基于发光特性的应用

De Stefano 及其同事在证明硅藻壳的光致发光光谱受到周围大气影响后,提出了将硅藻壳用作光学化学传感器的潜力。他们研究了几种中心硅藻(如圆海链藻和威氏圆筛藻)的硅藻壳的光致发光光谱对一系列气体和挥发性物质的响应。基于这些硅藻壳的传感器在二氧化氮的情况下具有高灵敏度,检测限为百万分之几浓度的十分之几。此外,基于海洋硅藻 *Coscinodiscus concinnus* 的化学改性硅藻壳的光学生物传感器已被制造出来,具有高灵敏度。此外,非均质硅藻壳或清洁的均质新鲜栽培硅藻壳已被研究作为有前途的光活性材料来制造聚合物复合随机激光器,这可能通过增加激光效率来潜在地提高激光效率基质的散射,从而降低达到饱和所需的激光瓣膜。

为了增强光致发光发射,无定形二氧化硅可以通过磁热过程随后化学处理转化为硅,同时保留硅藻壳的结构。此类方法可能会导致将蛋白质或抗体固定在硅藻壳表面上以制造光电生物传感器。

7.5.5 隐形硅藻

1978 年,在伍兹霍尔海洋生物实验室参加 Allen 和 Inoué 教授的生物医学科学光学显微镜和显微摄影课程时,RG 和一名显微镜推销员偷偷地打破了一张硅藻片,这个想法是,通过匹配活硅藻硅藻壳的折射率,人们可以清楚地看到里面。2015 年的一天下午,Andrejic 和 RG 在 Kalina Manoylov 实验室用活硅藻进行了同样的实验。最初的结果令人惊讶:瓣膜的表面看起来更加锐利。

根据 Fuhrmann 等的说法,格氏圆筛藻的二氧化硅的折射率为 1.43,即比水高 0.10,比本实验中使用的浸油($n=1.52$)低 0.09,解释了观察到的结果。事实上,使用油时图像质量的提高仅仅是去除硅藻上方水层的效果。这意味着从油浸物镜到硅藻瓣膜没有光学接口。然而,这种油对硅藻本身有毒,导致叶绿体明显降解。

因此我们得出两个结论：

(1) 如果可以找到折射率为 1.52 的无毒介质，则图像使用光学显微镜可以提高硅藻瓣膜的质量。

(2) 如果我们能找到一种折射率为 1.43 的无毒介质，硅藻可以在其中茁壮成长，那么在隐藏二氧化硅壳的同时进行内部成像（即使硅藻壳不可见）的新可能性将会打开。

甲基纤维素是一种无毒物质，不会引起细胞的皱褶，通常用于增加水溶液的粘度以降低快速微生物（如草履虫）的速度。由于甲基纤维素的聚合物长度范围很广，因此通常可以达到所需的溶液粘度。据报道，纯甲基纤维素的折射率为 1.497 0。因此，应该可以用水稀释甲基纤维素以获得硅藻瓣膜的折射率，即 1.43，因为它介于两者之间。

隐蔽的硅藻壳将为活硅藻细胞的内部提供前所未有的清晰视图。例如，人们可以观察到不被母瓣或腰带扭曲的瓣膜形态发生，必须仔细观察。细胞骨架的实时荧光成像也是可能的，叶绿体中的石油生产和运输，液滴的融合，以及叶绿体和中缝原纤维的运动也可以被可视化。

回想起来，最初的 Woods Hole 实验成功的原因可能是载玻片来自旧制剂，导致硅藻二氧化硅脱水，从而将其折射率提高到 1.43 以上。

当然，不应假定硅藻的单一值 1.43 是普遍的。有必要在各种制备方案下对各种活的和清洁的硅藻壳的折射率进行适当的研究。这可以使用折射率匹配液体来完成。

我们假设硅藻壳的光子特性不仅对我们很重要，对硅藻也很重要。硅藻的掩蔽可以提供对后一个概念的关键测试，因为掩蔽应该消除硅藻壳的大部分光子特性，除非这种情况可能发生在硅藻壳与细胞质的不规则内部边界处。

与隐身相反，上面第(1)点将更难实现。要在无毒的水性介质中获得 1.52 的折射率，与浸油相当，可能需要一些化学方法。例如，甲基纤维素上的一些侧基可以被修饰以包含高原子量的原子，这应该会提高折射率。

7.6 结论

硅藻可能是复杂的微米或纳米制造光学元件（如光子晶体）的替代来源。硅藻不是通过常见的光刻技术逐层构建光子晶体，而是可以一步构建同样复杂的结构，其分辨率挑战当前的 3D 打印技术。然而，迄今为止，仅对少数属和物种的光子特性进行了研究，尽管它们可能有 200 000 种在形态、孔径、孔间距、层厚度和多尺度孔隙率方面不同，这些因素可能会影响硅藻壳的光学特性。这个巨大的品种等待探索。例如，许多中心硅藻是多边形而不是圆柱形，其中一些是非常对称的。他们的光学器件可能与具有多边形包层、纤芯或空腔的光纤有关。可以为最丰富的属准备一个光学特性的目录。此外，许多基于硅藻壳光子特性的应用将依赖于阵列的成功实现，因此应特别注意优化形成同质取向硅藻壳单层阵列的技术。此外,有必要详细研究遗传和物理化学对每个物种的超微结构的影响,看看我们是否可以改变硅藻壳的孔隙率模式,从而直接控制/影响它们的光子特性。通过对硅藻光生物学的研究,我们可以进一步利用弱光硅藻和在没有光的情况下幸存下来的硅藻类群作为生产硅藻二氧化硅和硅藻生物柴油的更便宜的来源,以及降低通常需要的光能成本,通过光管进行的殖民交流也可能激发有希望的未来发展,包括诸如细胞视觉效应是否在硅藻中发生等基本问题,对弱光硅藻光合作用的量子方面的研究可能有助于拓展量子生物学和量子效应在量子计算机系统等新技术发展中的应用。

参考文献

[1] Bedoshvili, Y.D., Popkova, T.P., Likhoshway, Y.V. (2009). Chloroplast structure of diatoms of different classes, Cell Tiss. Biol., 3(3), 297-310.

[2] Schoefs, B., Hu, H., Kroth, P.G. (2017). The peculiar carbon metabolism in diatoms. Philosophical Transactions of the Royal Society B: Biological Sciences 372(1728, The peculiar carbon metabolism in diatoms, Eds. Benoît Schoefs, Hanhua Hu & Peter G. Kroth). #20160405.

[3] Scarsini, M., Marchand, J., Schoefs, B. (2018). Photosynthesis in diatoms. In Diatoms: Fundamentals & Applications [DIFA, Volume 1 in the series: Diatoms: Biology & Applications, series editors: Richard Gordon & Joseph Seckbach]. J. Seckbach and R. Gordon (eds.). Beverly, MA, USA: Wiley-Scrivener: 189-210.

[4] Nowak, R.D., Baraniuk, R.G. (1996). Optimally weighted highpass filters using multiscale analysis, Proceedings of the IEEE Southwest Symposium on Image Analysis and Interpretation, 224-229.

[5] Fedor, A., Freeman, M.O. (1992). Optical multiscale morphological processor using a complexvalued kernel, Appl. Opt., 31(20), 4042-4050.

[6] Morales, A., Ko, S.J. (1996). Finding optimal convex grayscale structuring elements for morphological multiscale representation, Proc. SPIE, 2662, 129-141.

[7] Boucher, V., Leblond, H., Phu, X.N. (2002). Multiscale theory of nonlinear wavepacket propagation in a planar optical waveguide, J. Opt. A Pure Appl. Opt., 4(5), 514-520.

[8] Christensen, M.P., Milojkovic, P., McFadden, M.J., Haney, M.W. (2003). Multiscale optical design for global chip-to-chip optical interconnections and misalignment tolerant packaging, IEEE J. Select. Topics Quantum Electron., 9(2), 548-556.

[9] Deumié, C., Richier, R., Dumas, P., Amra, C. (1996). Multiscale roughness in optical multilayers: atomic force microscopy and light scattering, Appl. Opt., 35(28), 5583-5594.

[10] Agarwal, A., Banwell, T., Jackel, J., Toliver, P., Woodward, T.K. (2009). Multiscale sampling for wide dynamic range electro-optic receivers OFC: 2009 Conference on Optical Fiber Communication, Vol. 1-5, pp.941-943. IEEE.

[11] McLeod, E., Arnold, C.B. (2008). Optical analysis of time-averaged multiscale Bessel beams generated by a tunable acoustic gradient index of refraction lens, Appl. Opt., 47(20), 3609-3618.

[12] Sheng, C., Norwood, R.A., Wang, J., Thomas, J., Steeves, D., Kimball, B., et al. (2009). Nonlinear optical transmission of lead phthalocyanine-doped nematic liquid crystal composites for multiscale nonlinear switching from nanosecond to continuous wave, Appl. Opt., 48(14), 731-2734.

[13] Noek, R., Knoernschild, C., Migacz, J., Kim, T., Maunz, P., Merrill, T., et al. (2010). Multiscale optics for enhanced light collection from a point source, Opt. Lett., 35(14), 2460-2462.

[14] Liu, Y., Meng, C., Zhang, A.P., Xiao, Y., Yu, H., Tong, L. (2011). Compact microfiber Bragg gratings with high-index contrast, Opt. Lett., 36(16), 3115-3117.

[15] Zerrad, M., Sorrentini, J., Soriano, G., Amra, C. (2011). Multiscale spatial depolarization of light: Electromagnetism & statistical optics, Physical Optics, 8171, #81710c.

[16] Gebeshuber, I.C., Stachelberger, H., Drack, M. (2005). Diatom bionanotribology—biological surfaces in relative motion: their design, friction, adhesion, lubrication and wear, J. Nanosci. Nanotechnol., 5(1),

79-87.

[17] Raven, J. A., Giordano, M. (2009). Biomineralization by photosynthetic organisms: Evidence of co-evolution of the organisms and their environment? Geobiology, 7(2), 140-154.

[18] Marron, A. O., Ratcliffe, S., Wheeler, G. L., Goldstein, R. E., King, N., Not, F., et al. (2016). The evolution of silicon transport in eukaryotes, Mol. Biol. Evol., 33(12), 3226-3248.

[19] Coradin, T., Lopez, P. J. (2003). Biogenic silica patterning: simple chemistry or subtle biology? ChemBioChem, 4(4), 251-259.

[20] Neethirajan, S., Gordon, R., Wang, L. (2009). Potential of silica bodies (phytoliths) for nanotechnology, Trends Biotechnol., 27(8), 461-467.

[21] Exley, C. (2015). A possible mechanism of biological silicification in plants, Front. Plant Sci., 6(112), 853.

[22] Müller, W. E. G., Wang, X., Cui, F.-Z., Jochum, K. P., Tremel, W., Bill, J., et al. (2009). Sponge spicules as blueprints for the biofabrication of inorganic-organic composites and biomaterials, Appl. Microbiol. Biotechnol., 83(3), 397-413.

[23] Ratcliffe, S., Jugdaohsingh, R., Vivancos, J., Marron, A., Deshmukh, R., Ma, J. F., et al. (2017). Identification of a mammalian silicon transporter, Am. J. Physiol. Cell. Physiol., 312(5), C550-C561.

[24] Carlisle, E. M. (1981). Silicon: A requirement in bone formation independent of vitamin D1, Calcif. Tissue Int., 33(1), 27-34.

[25] Round, F. E., Crawford, R. M., Mann, D. G. (1990). The Diatoms, Biology & Morphology of the Genera. Cambridge: Cambridge University Press.

[26] Pappas, J. L., Tiffany, M. A., Gordon, R. (2020). The uncanny symmetry of some diatoms and not of others: A new quantitative characteristic for taxonomy and a puzzle for morphogenesis. In Diatom Morphogenesis, V. Annenkov, J. Seckbach and R. Gordon (eds.). Beverly, MA, USA: Wiley-Scrivener. Submitted.

[27] Tiffany, M. A., Nagy, S. S. (2019). The beauty of diatom cells in light and scanning electron microscopy. In Diatoms: Fundamentals & Applications [DIFA, Volume 1 in the series: Diatoms: Biology & Applications, series editors: Richard Gordon & Joseph Seckbach]. J. Seckbach and R. Gordon (eds.). Beverly, MA, USA: Wiley-Scrivener, 31-40.

[28] Haeckel, E. (1862). Die Radiolarien (Rhizopoda Radiaria). Eine Monographie/The Radiolarians (Rhizopoda Radiaria). A Monograph. Berlin: Reimer.

[29] Anderson, O. R. (1983). Radiolaria. New York, NY, USA: Springer-Verlag.

[30] Piper, J. (2011). A review of high-grade imaging of diatoms and radiolarians in light microscopy optical- and software-based techniques, Diatom Research, 26(1), 57-72.

[31] Kooistra, W. H. C. F., Medlin, L. K. (1996). Evolution of the diatoms (Bacillariophyta). IV. A reconstruction of their age from small subunit rRNA coding regions and the fossil record, Mol. Phylogenet. Evol., 6(3), 391-407.

[32] Medlin, L. K. (2016). Evolution of the diatoms: major steps in their evolution and a review of the supporting molecular and morphological evidence, Phycologia, 55(1), 79-103.

[33] Kumar, S., Baweja, P., Sahoo, D. (2015). Diatoms: Yellow or golden brown algae. In Algae World. 26. pp. 235-258D Sahoo and J Seckbach (eds.). Dordrecht: Springer.

[34] van den Hoek, C., Mann, D., Jahns, H. M. (1996). Algae: An Introduction to Phycology. New York: Cambridge University Press.

[35] Mann, D. G. (1996). Chloroplast morphology, movements and inheritance in diatoms. In Cytology, Genetics and Molecular Biology of Algae. pp. 249-274B. R Chaudhary and S. B Agrawal (eds.). Amsterdam, The Netherlands: SPB Academic Publishing.

[36] Mann, D. G., Vanormelingen, P. (2013). An inordinate fondness? The number, distributions, and origins of diatom species, J. Eukaryot. Microbiol., 60(4), 414-420.

[37] Gordon, R., Hoover, R. B., Tuszynski, J. A., de Luis, J., Camp, P. J., Tiffany, M. A., et al. (2007). Diatoms in space: testing prospects for reliable diatom nanotechnology in microgravity, Proc. SPIE, 6694, V1-V15.

[38] Norton, T. A., Melkonian, M., Andersen, R. A. (1996). Algal biodiversity*, Phycologia, 35(4), 308-326.

[39] Dunahay, T. G., Jarvis, E. E., Roessler, P. G. (1995). Genetic transformation of the diatoms Cyclotella

cryptica and Navicula saprophila, J. Phycol., 31(6), 1004-1012.
[40] Kira, N., Ohnishi, K., Miyagawa-Yamaguchi, A., Kadono, T., Adachi, M. (2016). Nuclear transformation of the diatom Phaeodactylum tricornutum using PCR-amplified DNA fragments by microparticle bombardment, Mar. Genomics, 25, 49-56.
[41] Gordon, R. (1996). Computer controlled evolution of diatoms: design for a compustat, Nova Hedwigia, 112, 213-216. Festschrift for Prof. T. V. Desikachary.
[42] De Stefano, L., De Stefano, M., De Tommasi, E., Rea, I., Rendina, I. (2011). A natural source of porous biosilica for nanotech applications: the diatoms microalgae, Phys. Status Solidi C., 8(6), 1820-1825.
[43] Dolatabadi, J. E. N., de la Guardia, M. (2011). Applications of diatoms and silica nanotechnology in biosensing, drug and gene delivery, and formation of complex metal nanostructures, TracTrends. Anal. Chem., 30(9), 1538-1548.
[44] Drum, R. W., Gordon, R. (2003). Star Trek replicators and diatom nanotechnology, Trends Biotechnol., 21(8), 325-328.
[45] Gordon, R., Aguda, B. D. (1988). Diatom morphogenesis: natural fractal fabrication of a complex microstructure. In Proceedings of the Annual International Conference of the IEEE Engineering in Medicine and Biology Society, Part 1/4: Cardiology and Imaging, 4-7 Nov. 1988, New Orleans, LA, USA. pp. 273-274G. Harris and C. Walker (eds.). New York: Institute of Electrical and Electronics Engineers.
[46] Gordon, R., Losic, D., Tiffany, M. A., Nagy, S. S., Sterrenburg, F. A. (2009). The Glass Menagerie: diatoms for novel applications in nanotechnology, Trends Biotechnol., 27(2), 116-127.
[47] Gordon, R., Witkowski, A., Gebeshuber, I. C., Allen, C. S. (2010). The diatoms of Antarctica and their potential roles in nanotechnology. In Antarctica: Time of Change. pp. 84-95. Edited by M Masó. Barcelona: Editions ACTAR.
[48] Gordon, R., Sterrenburg, F. A. S., Sandhage, K. H. (2005). A Special Issue on Diatom Nanotechnology, J. Nanosci. Nanotechnol., 5(1), 1-4.
[49] Hildebrand, M. (2005). Prospects of manipulating diatom silica nanostructure, J. Nanosci. Nanotechnol., 5(1), 146-157.
[50] Jeffryes, C., Campbell, J., Li, H., Jiao, J., Rorrer, G. (2011). The potential of diatom nanobiotechnology for applications in solar cells, batteries, and electroluminescent devices, Energy Environ. Sci., 4(10), 3930-3941.
[51] Jiang, Y., Fu, J., Pan, J., An, Z., Zhang, D., Cai, J. (2014). Biopattern transfer using diatom frustules for fabrication of functional micro/nano-structures, Journal of Micromechanics and Microengineering, 14(1), 014502.
[52] Kröger, N., Poulsen, N. (2008). Diatoms: from cell wall biogenesis to nanotechnology, Annu. Rev. Genet., 42(1), 83-107.
[53] D. Kurkuri, M., Saunders, C., J. Collins, P., Pavic, H., Losic, D. (2011). Combining micro and nanoscale structures: emerging applications of diatoms, Micro and Nanosystems, 3(4), 277-283.
[54] Losic, D. (ed.) (2018). Diatom Nanotechnology: Progress and Emerging Applications, Royal Society of Chemistry, London.
[55] Losic, D., Mitchell, J. G., Voelcker, N. H. (2009). Diatomaceous lessons in nanotechnology and advanced materials, Adv. Mater., 21(29), 2947-2958.
[56] Parkinson, J., Gordon, R. (1999). Beyond micromachining: the potential of diatoms, Trends Biotechnol., 17(5), 190-196.
[57] Yang, W., Lopez, P. J., Rosengarten, G. (2011). Diatoms: Self assembled silica nanostructures, and templates for bio/chemical sensors and biomimetic membranes, Analyst (Lond)., 136(1), 42-53.
[58] Zhang, D., Wang, Y., Cai, J., Pan, J., Jiang, X., Jiang, Y. (2012). Bio-manufacturing technology based on diatom micro-and nanostructure, Chin. Sci. Bull., 57(30), 3836-3849.
[59] Gordon, R., Drum, R. W. (1994). The chemical basis of diatom morphogenesis, Int. Rev. Cytol., 150(243-372), 421-422.
[60] Adl, S. M., Simpson, A. G., Lane, C. E., Lukeš, J., Bass, D., Bowser, S. S., et al. (2012). The revised classification of eukaryotes, J. Eukaryot. Microbiol., 59(5), 429-514.
[61] Collier, A., Murphy, A. (1962). Very small diatoms: Preliminary notes and description of Chaetoceros galvestonensis, Science, 136(3518), 780-781.

[62] Smetacek, V. (2000). Oceanography: The giant diatom dump, Nature, 406(6796), 574-575.
[63] Crawford, S. A., Higgins, M. J., Mulvaney, P., Wetherbee, R. (2001). Nanostructure of the diatom frustule as revealed by atomic force and scanning electron microscopy, J. Phycol., 37(4), 543-554.
[64] Willis, L., Page, K. M., Broomhead, D. S., Cox, E. J. (2010). Discrete free-boundary reaction-diffusion model of diatom pore occlusions, Plecevo, 143(3), 297-306.
[65] Li, A., Cai, J., Pan, J., Wang, Y., Yue, Y., Zhang, D. (2014). Multi-layer hierarchical array fabricated with diatom frustules for highly sensitive bio-detection applications, J. Micromech. Microeng., 24(2), ♯025014.
[66] Tozzini, V. (2010). Multiscale modeling of proteins, Acc. Chem. Res., 43(2), 220-230.
[67] Wikipedia. (2017a). ImageJ, http://en.wikipedia.org/wiki/ImageJ.
[68] Tiffany, M. A. (2011). Epizoic and epiphytic diatoms. In The Diatom World. pp.195-209. J. Seckbach and J. P Kociolek (eds.). Dordrecht, The Netherlands: Springer.
[69] Bergkvist, J., Thor, P., Jakobsen, H. H., Wängberg, Sten-Åke., Selander, E. (2012). Grazer-induced chain length plasticity reduces grazing risk in a marine diatom, Limnol. Oceanogr., 57(1), 318-324.
[70] Marsot, P., Leclerc, M., Fournier, R. (1983). Aspect morphologique et composition chimique-de Skeletonema costatum (Bacillariophyceae) croissant en milieu nutritif naturel l'aide d'un système de culture fibres dialysantes [The morphological aspect and chemical-composition of Skeletonema costatum (Bacillariophyceae) grown in a natural nutritional medium aided by a culture fiber dialyzing system], Can. J. Microbiol., 29(10), 1235-1240.
[71] Sjöqvist, C., Kremp, A., Lindehoff, E., Båmstedt, U., Egardt, J., Gross, S., et al. (2014). Effects of grazer presence on genetic structure of a phenotypically diverse diatom population, Microb. Ecol., 67(1), 83-95.
[72] Thomas, W. H., Hollibaugh, J. T., Seibert, D. L. R. (1980). Effects of heavy metals on the morphology of some marine phytoplankton*, Phycologia, 19(3), 202-209.
[73] Schöne, H. K. (1972). Experimentelle Untersuchungen zur Ökologie der marinen Kieselalge Thalassiosira rotula. I. Temperatur und Licht [Experimental] investigations on the ecology of the marine diatom Thalassiosira rotula. I. Temperature and light, Mar. Biol., 13(4), 284-291.
[74] Mullin, M. M. (1963). Some factors affecting the feeding of marine copepodsof the genus Calanus, Limnol. Oceanogr., 8(2), 239-250.
[75] Deason, E. E. (1980b). Potential effect of phytoplankton colony breakage on the calculation of zooplankton filtration rates, Mar. Biol., 57(4), 279-286.
[76] Martin, J. H. (1970). Phytoplankton-zooplankton relationships in Narragansett Bay. IV. Seasonal importance of grazing, Limnol. Oceanogr., 15(3), 413-418.
[77] Tiffany, M. A., Nagy, S. S. (2019). The beauty of diatom cells in light and scanning electron microscopy. In Diatoms: Fundamentals & Applications [DIFA, Volume 1 in the series: Diatoms: Biology & Applications, series editors: Richard Gordon & Joseph Seckbach]. J. Seckbach and R. Gordon (eds.). Beverly, MA, USA: Wiley-Scrivener, 31-40.
[78] Pahlow, M., Riebesell, U., Wolf-Gladrow, D. A. (1997). Impact of cell shape and chain formation on nutrient acquisition by marine diatoms, Limnol. Oceanogr., 42(8), 1660-1672.
[79] Srajer, J., Majlis, B. Y., Gebeshuber, I. C. (2009). Microfluidic simulation of a colonial diatom chain reveals oscillatory movement, Acta Bot. Croat., 68(2), 431-441.
[80] Fryxell, G. A., Miller, W. I. (1978). Chain-forming diatoms: three araphid species, Bacillaria, 1, 113-136. III.
[81] Gordon, N. K., Gordon, R. (2016). Embryogenesis Explained. Singapore: World Scientific Publishing.
[82] Komárek, J., Anagnostidis, K. K. (1989). Modern approach to the classification system of Cyanophytes 4-Nostocales, Arch. Hydrobiol., 82(3), 247-345.
[83] Kotz, F., Arnold, K., Bauer, W., Schild, D., Keller, N., Sachsenheimer, K., et al. (2017). Threedimensional printing of transparent fused silica glass, Nature, 544(7650), 337-339.
[84] Pappas, J. L., Tiffany, M. A., Gordon, R. (2020). The uncanny symmetry of some diatoms and not of others: A new quantitative characteristic for taxonomy and a puzzle for morphogenesis. In Diatom Morphogenesis, V. Annenkov, J. Seckbach and R. Gordon (eds.). Beverly, MA, USA: Wiley-Scrivener. Submitted.
[85] Round, F. E., Crawford, R. M., Mann, D. G. (1990). The Diatoms, Biology & Morphology of the

Genera. Cambridge: Cambridge University Press.

[86] Gordon, R., Tiffany, M. A. (2011). Possible buckling phenomena in diatom morphogenesis. In The Diatom World. pp. 245–272, eds.: J. Seckbach and J. P. Kociolek. Dordrecht, The Netherlands: Springer.

[87] Podraza, N. J., Pursel, S. M., Chen, C., Horn, M. W., Collins, R. W. (2008). Analysis of the optical properties and structure of serial bi-deposited TiO_2 chiral sculptured thin films using Mueller matrix ellipsometry, J. Nanophotonics, 2(1), #021930.

[88] Lakhtakia, A., Geddes, J. B. (2010). Thin-film metamaterials called sculptured thin films. In Trends in Nanophysics: Theory, Experiment and Technology. pp. 59–71, eds.: A Aldea and V Barsan. Berlin, Germany: Springer-Verlag. III.

[89] Nair, G., Singh, H. J., Ghosh, A. (2015). Tuning the chiro-plasmonic response using high refractive index-dielectric templates, J. Mater. Chem. C, 3(26), 6831–6835.

[90] Huang, F. M., Zheludev, N., Chen, Y., Javier Garcia de Abajo, F. (2007). Focusing of light by a nanohole array, Appl. Phys. Lett., 90(9), 091119.

[91] Josten, G., Weber, H. P., Luethy, W. (1989). Lensless focusing with an array of phase-adjusted optical fibers, Appl. Opt., 28(23), 5133–5137.

[92] Losic, D., Rosengarten, G., Mitchell, J. G., Voelcker, N. H. (2006). Pore architecture of diatom frustules: potential nanostructured membranes for molecular and particle separations, J. Nanosci. Nanotechnol., 6(4), 982–989.

[93] Vrieling, E. G., Beelen, T. P. M., Sun, Q., Hazelaar, S., van Santen, R. A., Gieskes, W. W. C. (2004). Ultrasmall, small, and wide angle X-ray scattering analysis of diatom biosilica: interspecific differences in fractal properties, J. Mater. Chem., 14(13), 1970–1975.

[94] Aas, E. (1996). Refractive index of phytoplankton derived from its metabolite composition, J. Plankton Res., 18(12), 2223–2249.

[95] Hodgson, R. T., Newkirk, D. D. (1975). Pyridine immersion: a technique for measuring the refractive index of marine particles, Proc. SPIE, 64, 19–20.

[96] Bridoux, M. C., Ingalls, A. E. (2010). Structural identification of long-chain polyamines associated with diatom biosilica in a Southern Ocean sediment core, Geochim. Cosmochim. Acta, 74(14), 4044–4057.

[97] De Sanctis, S., Wenzler, M., Kröger, N., Malloni, W. M., Sumper, M., Deutzmann, R., et al. (2016). PSCD domains of pleuralin-1 from the diatom Cylindrotheca fusiformis: NMR structuresand interactions with other biosilica-associated proteins, Structure, 24(7), 1178–1191.

[98] Ingalls, A. E., Whitehead, K., Bridoux, M. C. (2010). Tinted windows: The presence of the UV absorbing compounds called mycosporine-like amino acids embedded in the frustules of marine diatoms, Geochim. Cosmochim. Acta, 74(1), 104–115.

[99] Lechner, C. C., Becker, C. F. W. (2013). Modified silaffin R5 peptides enable encapsulation and release of cargo molecules from biomimetic silica particles, Bioorg. Med. Chem., 21(12), 3533–3541.

[100] Aizenberg, J., Sundar, V. C., Yablon, A. D., Weaver, J. C., Chen, G. (2004). Biological glass fibers: Correlation between optical and structural properties, Proc. Natl. Acad. Sci. U. S. A., 101(10), 3358–3363.

[101] Crawford, S. A., Chiovitti, A., Pickett-Heaps, J., Wetherbee, R. (2009). Micromorphogenesis during diatom wall formation produces siliceous nanostructures with different properties, J. Phycol., 45(6), 1353–1362.

[102] Debailleul, M., Simon, B., Georges, V., Haeberlé, O., Lauer, V. (2008). Holographic microscopy and diffractive microtomography of transparent samples, Meas. Sci. Technol., 19(7), #074009.

[103] Simon, B., Debailleul, M., Georges, V., Lauer, V., Haeberlé, O. (2008). Tomographic diffractive microscopy of transparent samples, Eur. Phys. J. Appl. Phys., 44(1), 29–35.

[104] Di Caprio, G., Coppola, G., De Stefano, L. D., De Stefano, M. D., Antonucci, A., Congestri, R., et al. (2014). Shedding light on diatom photonics by means of digital holography, J. Biophotonics, 7(5), 341–350.

[105] Simon, B., Debailleul, M., Houkal, M., Ecoffet, C., Bailleul, J., Lambert, J., et al. (2017). Tomographic diffractive microscopy with isotropic resolution, Optica, 4(4), 460–463.

[106] King, G. M., Gordon, R., Karmali, K., Biberman, L. J. (1982). A new method for the immobilisation of teleost embryos for time-lapse studies of development, J. Exp. Zool., 220(2), 147–151.

[107] Malitson, I. H. (1965). Interspecimen comparison of the refractive index offused silica, J. Opt. Soc. Am., 55(10), 1205-1209.

[108] Fuhrmann, T., Landwehr, S., El Rharbi-Kucki, M., Sumper, M. (2004). Diatoms as living photonic crystals, Appl. Phys. B, 78(3-4), 257-260.

[109] Stramski, D., Reynolds, R. A. (1993). Diel variations in the optical properties of a marine diatom, Limnol. Oceanogr., 38(7), 1347-1364.

[110] Hodgson, R. T., Newkirk, D. D. (1975). Pyridine immersion: a technique for measuring the refractive index of marine particles, Proc. SPIE, 64, 19-20.

[111] Gebesbuber, I. C., Lee, D. W. (2012). Nanostructures for coloration (organisms other than animals). In Springer Encyclopedia of Nanotechnology. pp. 1790-1803. B. Bhushan, (ed.). Springer.

[112] Maibohm, C., Friis, S. M. M., Su, Y., Rottwitt, K. (2015). Comparing optical properties of different species of diatoms, Proc. SPIE, 9360, #93600b.

[113] Kieu, K., Li, C., Fang, Y., Cohoon, G., Herrera, O. D., Hildebrand, M., et al. (2014). Structurebased optical filtering by the silica microshell of the centric marine diatom Coscinodiscus wailesii, Opt. Express, 22(13), 15992-15999.

[114] Townley, H. E. (2011). Diatom frustules: physical, optical, and biotechnological applications. In The Diatom World. pp. 273-289. J. Seckbach and J. P. Kociolek (eds.). Dordrecht, The Netherlands: Springer.

[115] Stavenga, D. G., Leertouwer, H. L., Wilts, B. D. (2013). Quantifying the refractive index dispersion of a pigmented biological tissue using Jamin—Lebedeff interference microscopy, Light Sci. Appl., 2(9), #e100.

[116] Noren, A. K. (2011). Characterization of Structure and Optical Properties of Diatoms for improved Solar Cell Efficiency [Masters Thesis]. Trondheim: Norwegian University of Science and Technology, Norwegian University of Science and Technology.

[117] Aguirre, L. E., Ouyang, L., Elfwing, A., Hedblom, M., Wulff, A., Inganäs, O. (2018). Diatom frustules protect DNA from ultraviolet light, Sci. Rep., 8(1), #5138.

[118] Goessling, J. W., Frankenbach, S., Ribeiro, L., Serôdio, J., Kühl, M. (2018a). Modulation of the light field related to valve optical properties of raphid diatoms: implications for niche differentiation in the microphytobenthos, Mar. Ecol. Prog. Ser., 588, 29-42.

[119] Su, Y., Lenau, T. A., Gundersen, E., Kirkensgaard, J. J. K., Maibohm, C., Pinti, J., et al. (2018). The UV filtering potential of drop-casted layers of frustules of three diatom species, Sci. Rep., 8(1), #959.

[120] Mann, D. G. (1996). Chloroplast morphology, movements and inheritance in diatoms. In Cytology, Genetics and Molecular Biology of Algae. pp. 249-274B. R Chaudhary and S. B Agrawal (eds.). Amsterdam, The Netherlands: SPB Academic Publishing.

[121] Hall, N., Ozin, G. (2003). The photonic opal—the jewel in the crown of optical information processing, Chem. Commun 21, 2639-2643.

[122] Lourtioz, J.-M., Benisty, H., Berger, V., Gérard, J.-M., Maystre, D., Tchelnokov, A. (2008). Photonic Crystals: Towards Nanoscale Photonic Devices. Berlin: Springer, 2nd edition.

[123] Yablonovitch, E. (1987). Inhibited spontaneous emission in solid-state physics and electronics, Phys. Rev. Lett., 58(20), 2059-2062.

[124] Joannopoulos, J. D., Johnson, S. G., Winn, J. N., Meade, R. D. (2011). Photonic Crystals: Molding the Flow of Light. Princeton, New Jersey, USA: Princeton University Press. 2nd, revised.

[125] Robinson, S., Nakkeeran, R. (2013). Photonic crystal ring resonator based optical filters. In Advances in Photonic Crystals. 1. 1-26. Edited byV. M. N Passaro. Rijeka: InTech.

[126] Wang, B., Cappelli, M. A. (2016). Waveguiding and bending modes in a plasma photonic crystal bandgap device, AIP Adv., 6(6), #065015.

[127] Wilts, B. D., Vignolini, S. (2019). Living light: Optics, ecology and design principles of natural photonic structures, Interface Focus., 9(1), #20180071.

[128] Lopez-Garcia, M., Masters, N., O'Brien, H. E., Lennon, J., Atkinson, G., Cryan, M. J., et al. (2018). Light-induced dynamic structural color by intracellular 3D photonic crystals in brown algae, Sci. Adv., 4(4), #eaan8917.

[129] Vigneron, J. P., Simonis, P. (2012). Natural photonic crystals, Physica B: Condensed Matter, 407

(20), 4032-4036.
[130] Abo-Shady, A. M., Zalat, A. A., Al-Ashkar, E. A., Ghobara, M. M. (2019). Nanoporous silica of some Egyptian diatom frustules as a promising natural material., Nanoscience & NanotechnologyAsia, 8(2), doi:10.2174/2210681208666180321113834.
[131] Losic, D., Rosengarten, G., Mitchell, J. G., Voelcker, N. H. (2006). Pore architecture of diatom frustules: potential nanostructured membranes for molecular and particle separations, J. Nanosci. Nanotechnol., 6(4), 982-989.
[132] Losic, D., Short, K., Mitchell, J. G., Lal, R., Voelcker, N. H. (2007). AFM nanoindentations of diatom biosilica surfaces, Langmuir, 23(9), 5014-5021.
[133] Zalat, A. A., Abo-Shady, A. M., Al-Ashkar, E. A., Ghobara, M. E. (2017). Characterization of silica nanoporous structures of some diatom frustules and its applications, Scientific J. Sci, 38, 194-201.
[134] Kucki, M., Landwehr, S., Rühling, H., Maniak, M., Fuhrmann-Lieker, T. (2006). Light-emitting biological photonic crystals—the bioengineering of metamaterials, Proc. SPIE, 6182, #61821S.
[135] Quan, X., Fry, E. S. (1995). Empirical equation for the index of refraction of seawater, Appl. Opt., 34 (18), 3477-3480.
[136] Yamanaka, S., Yano, R., Usami, H., Hayashida, N., Ohguchi, M., Takeda, H., et al. (2008). Optical properties of diatom silica frustule with special reference to blue light, J. Appl. Phys., 103(7), #074701.
[137] Romann, J., Valmalette, J.-C., Chauton, M. S., Tranell, G., Einarsrud, M.-A., Vadstein, O. (2015a). Wavelength and orientation dependent capture of light by diatom frustule nanostructures, Sci. Rep., 5(1), #17403.
[138] Romann, J., Valmalette, J.-C., Røyset, A., Einarsrud, M.-A. (2015b). Optical properties of single diatom frustules revealed by confocal microspectroscopy, Opt. Lett., 40(5), 740-743.
[139] Kieu, K., Li, C., Fang, Y., Cohoon, G., Herrera, O. D., Hildebrand, M., et al. (2014). Structurebased optical filtering by the silica microshell of the centric marine diatom Coscinodiscus wailesii, Opt. Express, 22(13), 15992-15999.
[140] De Stefano, L., Maddalena, P., Moretti, L., Rea, I., Rendina, I., De Tommasi, E., et al. (2009a). Nano-biosilica from marine diatoms: A brand new material for photonic applications, Superlattices and Microstructures, 46(1-2), 84-89.
[141] Lourtioz, J.-M., Benisty, H., Berger, V., Gérard, J.-M., Maystre, D., Tchelnokov, A. (2008) Photonic Crystals: Towards Nanoscale Photonic Devices. Berlin: Springer, 2nd edition.
[142] Mikaelian, A. L. (1951). The use of medium properties to focus waves, Doklady of the USSR Academy of Sciences, 81, 569-571.
[143] Baghdasaryan, T., Geernaert, T., Thienpont, H., Berghmans, F. (2013). Photonic crystal Mikaelian lenses and their potential use as transverse focusing elements in microstructured fibers, IEEE Photonics J., 5(4), #7100512.
[144] Triandaphilov, Y. R., Kotlyar, V. V. (2008). Photonic crystal Mikaelian lens, Optical Memory and Neural Networks, 17(1), 1-7.

第 8 章
硅藻的光合作用

Matteo Scarsini，Justine Marchand，
Kalina M Manoylov，Benoît Schoefs

8.1 简介

  光是地球上最重要和最重要的能源,也是生物体用来收集周围环境信息的最重要信号之一[1-2]。海洋光合作用主要是由蓝藻和微藻组成的微观浮游植物进行的。这些生物是几乎所有海洋生物赖以生存的有机生物量的主要生产者。它们为一系列重要的生物地球化学过程提供燃料,从 O_2 的产生到重要营养物质的循环,如氮、硅、碳和磷。地球表面一半以上是水生的,其中微小的藻类和蓝藻种群是主要的初级生产者。它们的寿命短且人口众多,它们以快速的速度回收必需的营养物质。随着人口增长,农业中人为氮和磷的径流进入全球水循环,可能会进一步增强这些过程。微藻类占全球初级产量的近 45%,而仅占地球估计光合生物量的 1%。这种令人印象深刻的效率是由于每个细胞都具有光合作用活性这一事实,这与陆地植物的情况形成鲜明对比[3]。在水生环境中,藻类和蓝藻不需要像树干那样投资建造巨大的机械支撑结构。被水包围,它们消除了细胞和结构的复杂性,用于抵抗重力输送水。水生环境中的水损失也可以忽略不计,并且取决于溶质交换,而不是由于负水势而导致的大气损失。作为所有这些优势的最终结果,与其他初级生产者相比,藻类和蓝藻每单位或生物体产生更多的生物量。

  在所有真核浮游植物中,硅藻以及鞭毛藻和颗石藻是优势物种[3]。从进化的角度来看,硅藻被认为是两个连续的内共生事件的结果(见图 8.1)。大约 15 到 12 亿年前[4],一种原型真核生物吞噬了一种蓝细菌,从而产生了第一个光合真核生物,今天主要以红藻和绿藻为代表。发生了第二次内共生事件,其中红藻或绿藻被非光合作用的真核生物吞噬。出于这个原因,许多硅藻基因/蛋白质不仅与绿藻和红藻的序列对齐,而且与在动物、真菌和蓝藻中发现的同源物对齐[5-6]。共生体后来通过大量减少细胞结构而进化[7]。这产生了许多门,包括哺乳动物脑寄生虫和硅藻[3,8]。这种复杂的起源导致了硅藻叶绿体的特殊结构的进化,不同于在绿藻和陆地植物中发现的结构。这可能是这些生物体对快速变化的环境具有高适应能力和生存能力的主要原因之一[9-10]。这些次级内共生体的叶绿体被四个膜包围,两个来自初级内共生事件,两

个来自第二个。值得注意的是,最外层的膜与核包膜和内质网形成一个连续体(见图 8.2)[11],有关综述请参见文献[12],次生质体的超微结构将在下一段中详细描述。

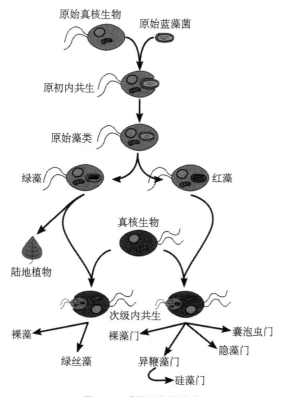

图 8.1 质体进化的代表

一种古老的蓝藻被一种古代的非光合真核生物吞噬,就产生了一种古代的藻类;被吞噬的蓝藻细菌在绿色和红色的叶绿体中进化,产生了绿藻和红藻以及它们各自的谱系,这些含有质体的细胞进一步被异养真核生物吞噬,产生了绿色谱系的次级内共生体,最终进化成陆地植物,红色谱系的次级内共生体分化为几个单系分支,包括硅藻分支

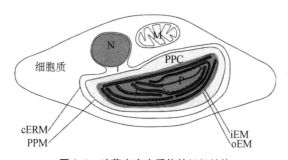

图 8.2 硅藻中次生质体的组织结构

质体受到四层膜的限制,叶绿体内质网膜(cERM),它与外核膜融合,周质膜(PPM),描绘周质膜室(PPC),外环和内膜(oEM 和 iEM);在叶绿体内部(C),存在类囊体;蛋白核(P)由虚线分隔;C—叶绿体,I—峡部,M—线粒体,N—细胞核

由于海洋中的湍流,硅藻与其他浮游植物一样,可能会暴露在变化很大的水下光照气候中。曝光的其他变化是由于吸收和散射过程。

硅藻已经进化到在不同的条件下生存和繁殖,在冰冷的水中茁壮成长,并主导上升流系统和地下层。他们被认为是动荡环境中最适合的群体,硅藻能够利用复杂的机制[2]应对快速的环境变化,以保护无定形二氧化硅细胞壁,硅藻壳的构建几乎不需要单个细胞的能量投资[13],因此细胞可以在多变的水生环境中生长得更快。此外,一些物种可以应对高度可变的光照条件,这表明硅藻能够感知、响应甚至可能预测光的变化,并且它们拥有合适的分子系统来调节光响应以及蓝藻和其他藻类物种[2,14]。

总而言之,硅藻是具有丰富物种多样性的关键生物,表现出非凡的适应性[15-17],无论环境条件如何,它们都能以最佳速率保持光合作用。这些特征在生物技术领域很重要,其中压力条件通常用于触发次级化合物的产生[18-21]。

本章概述了硅藻的光合作用机制,从细胞组织到光合作用和光保护所涉及的分子结构和过程。

8.2 叶绿体结构反映了内共生的两个步骤

硅藻起源于次级内共生事件,该事件包括早期内共生事件导致的细胞吞噬。在共生体摄取后,发生了大量的基因丢失和基因转移到细胞核中。后一种现象导致宿主细胞核基因组中出现大量基因。共生体不再是一个完全自主的有机体,而是新生成的细胞的一个组成部分。即使质体基因组仍然存在,这种基因转移在两种内共生事件中都可以观察到。初级内共生体包括数百个基因从内共生体转移到宿主基因组。hcf136、psbO 和 tic22 是编码参与光系统(PS)II 装配、PSII 功能和蛋白质输入的蛋白质的基因,是这种转移的众所周知的例子。在次生内共生事件之后,产生了囊泡藻类(包括异配子体、隐藻、甲藻和定鞭金藻)。囊泡藻类的单一系仍在争论中[22-23](见图 8.1),这些基因可能位于被吞噬的真核生物的细胞核中。Armbrust 等[5]发表的假微型海链藻基因组揭示了 hcf136、psbO 和 tic22 以及许多物质再次被重新定位到宿主的细胞核中。此外,一些基因,例如编码 PSII 的另一种蛋白质的 psbW,保留在第一个宿主的质体基因组中,并在第二个事件后转移到核基因组中。这些基因的产物靶向质体[24]。

两步内共生行为的结果是硅藻叶绿体结构的复杂化,它被细胞内高度堆积的四层膜包围(见图 8.2),最外层的膜称为叶绿体内质网膜(cERM),推测来源于非光合真核细胞的吞噬膜。在硅藻中,cERM 直接连接到内质网(ER)和最外层核包膜(oNE)。在 cERM 的正下方有被认为源自初级内共生体的质膜的质体周膜(PPM)。包围叶绿体基质和类囊体的两个最内层膜对应于源自单一内共生的质体的叶绿体包膜。外包膜和内包膜(分别为 oEM 和 iEM)描绘了简单的叶绿体,并源自革兰氏阴性蓝藻祖先类型的两个限制膜[25]。在 PPM 和 oEM 之间存在与共生体细胞质相对应的质体周室,叶绿体结构的复杂化使得代谢物和蛋白质在细胞质和叶绿体基质之间以及叶绿体亚室内的转运更加困难[26-28]。这反映在蛋白质输入机制中,在囊泡藻类中,共生体特异性内质网相关蛋白质降解(ERAD)样机制(SELMA)促进了这一机制。SELMA 源自 ER 相关降解机制,不存在于绿色谱系中,可追溯到 ER 相关降解的共生体特定运输机制,现在在 PPM 完成蛋白质运输任务[29]。在三角褐指藻中,大多数质体蛋白在细胞核中编码,并包含由 N 端信号肽和叶绿体样转运肽组成的二分拓扑信号。存在另一个短氨基酸序列(ASAFAP 基序),它是信号肽切割所必需的。含有二分靶向序列(BTS)信号的进口蛋白质涉及三个主要系统。第一个系统位于 cERM 上,称为 sec61 复合体。这是蛋白质携带信号通过 cERM 的共翻译介导,蛋白质在 cERM 的管腔中展开释放。该蛋白质现在被定向到称为 SELMA 的第二个系统,该系统之前已经介绍过。在 SELMA 中,Derlin 蛋白(推定的通道蛋白)与 N 端 BTS 的肽靶向信号相互作用,以便将蛋白质转运到内共生体细胞质中[30]。第三个系统是经典的叶绿体易位子 TOC 和 TIC,分别位于 oEM 和 iEM 上[25,31]。

Gibbs 对棕鞭藻进行的电子显微镜研究揭示了 PPC 中存在的水泡网络,称为最初是质体周围网(PPR)。PPR 是由小管、水池和囊泡组成的复杂组织,似乎是蛋白质进入叶绿体的一种输入方式,据观察,该网在整个 PPC,但存在于靠近细胞核的峡部水平的受限区域(见图 8.2)[11]。在硅藻中,这种水泡网络是物种特异性的,类核蛋白可以有或没有特定的膜[32]。

在硅藻叶绿体中,类囊体松散地堆叠在一起,三个一组,沿着它们的整个长度有一些吻合,吻合的数量取决于物种[33]。在硅藻中,没有像绿藻和陆地植物叶绿体那样在拓扑上分化为颗粒和基质薄片[12,34]。然而,最近的研究表明,结构更加复杂。最近对 P. tricornutum 叶绿体的三维重建表明存在复杂的类囊体网络[35]。叶绿体的中心部分被核糖体占据,核糖体是一种蛋白质体,通常被两个类囊体膜穿透(见图 8.2)。这种存在于硅藻和其他藻类中的叶绿体亚区室在核酮糖 1,5 二磷酸羧化酶/加氧酶(RuBisCO)附近的 CO_2 浓度中起重要作用,避免 RuBisCO 的加氧酶活性[36]。

8.3 光合色素

光合作用需要来自光的能量才能发挥作用,硅藻是具有独特色素成分的生物,在大多数方面与高等植物中存在的色素不同。叶绿素 a(Chl a)、Chl c 和岩藻黄质(Fx)是与 Fx-Chl 结合蛋白(FCP)相关的主要色素,FCP 是在硅藻中发挥光捕获功能的光捕获复合物(LHC)。

8.3.1 叶绿素

在光合生物中可以发现几种 Chl,但在硅藻中只出现两种类型:Chl a 和 Chl c,前者在其他藻类中发现,后者在捕光过程中一起作用。与其他叶绿素不同,叶绿素 c 是镁-植物卟啉而不是镁-二氢卟吩[37]。从化学的观点来看，叶绿素 c 是原叶绿素（PChlide），但由于它在光收集方面具有功能活性，所以保留了叶绿素这个术语来称呼这种类型的色素。硅藻中存在三种主要的不同 Chl c：Chl c_1、Chl c_2 和 Chl c_3，Chl c_1 和 Chl c_2 最为丰富。叶绿素 c 之间的结构差异带来了光合机构吸收光谱的变化[37-38]。虽然叶绿素 a 在硅藻和所有光合生物的光能利用中起着最重要的作用，但叶绿素 c 是一种辅助色素（就像高等植物中的叶绿素 b 一样）。当硅藻细胞暴露于高光照水平时，会发生取决于暴露时间的变化。

8.3.2 类胡萝卜素

在自然界发现的 700 多种类胡萝卜素中，硅藻中发现了 7 种。通常，类胡萝卜素在 400 到 500 之间表现出强烈的吸收，由于官能团的附着而发生微小的变化。硅藻中类胡萝卜素的主要成分是岩藻黄素，它在光收集中起重要作用。岩藻黄质具有等位基因键、共轭羰基、5,6-单环氧化物和乙酰基，它们有助于分子的独特结构和光谱特性。事实上，它的吸收光谱转移到 460～570 nm 并赋予藻类典型的棕黄色。这种转变也很重要，因为它允许吸收更多由叶绿素在绿色区域留下的间隙[38]。其他类胡萝卜素在光保护机制中至关重要[39-40]。最重要的三种是 β-胡萝卜素（β-car）、二甲黄素（Ddx）和二甲黄素（Dtx）。此外，紫黄质（Vx）、花药黄质（Ax）和玉米黄质（Zx）似乎参与了光保护过程，但仅在长光和强光条件下才起作用[38,41]。

8.4 光合器官的组织

硅藻具有典型的光合器官，由四种主要的多亚基蛋白复合物组成，即 PSII、细胞色素（Cyt）b_6f、PSI 和 ATP 合酶 CF_0F_1（见图 8.3）。它们位于类囊体膜上。前三个复合物参与光吸收，从水中提取电子并转移到 $NADP^+$，而第四个复合物使用跨类囊体膜建立的 H^+ 梯度为 ATP 合成提供动力。每个 PS 分为两个主要功能单元：光捕获

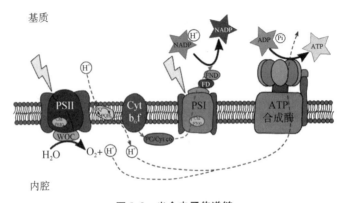

图 8.3 光合电子传递链

天线复合体(LHC)和发生光化学反应的反应中心。简而言之,光子被 LHCII 的色素吸收,与光子相关的能量被转移到反应中心,其中一个 Chl a 分子的特殊 Chl 对 PSII (P680)被激发,并经历电荷分离。产生的电子依次转移到质体醌池(PQ)。还原的 PQ 被 Cyt b_6f 氧化,然后还原电子从其转移到 PSI 的移动电子传输器。在 PSI 中,来自 LHCI 的能量在反应中心(P700)的 Chl 对水平触发电荷分离。产生的电子转移到铁氧还蛋白(Fd),然后转移到 Fd-$NADP^+$-氧化还原酶,该酶催化 $NADP^+$ 还原为 NADPH,用于碳固定反应。在 P680 水平产生的电子间隙充满了源自水氧化复合物(WOC)的水氧化的电子。该反应还产生 O_2。电子转移和水氧化通过类囊体膜产生的质子梯度通过 ATP 合酶的催化活性产生 ATP。

除了构成 WOC 的亚基之外,光合作用生物的 PS 组织结构在全球范围内高度一致[42]。作为 Cyt b_6f 和 PSI 之间的移动电子载体,硅藻使用质体蓝素或 Cyt c_6。这两种可能性之间的"选择"可能取决于铜的可用性。海洋硅藻(如 *Thalassiosira oceanica* Hasle)使用质体蓝素,而沿海物种(如 *T. pseudonana* 和 *P. tricornutum*)则使用细胞色素 c_6[43]。众所周知,由线性电子流产生的 ATP/NADPH 比率(已描述)不足以为 CO_2 进入质粒并在卡尔文循环中同化提供燃料。Viridiplantae 中存在的替代途径主要包括循环电子流(CEF)和水-水循环[44-45]。编码大多数这些替代途径成分的基因似乎存在于硅藻中,但 ATP/NADH 平衡的机制仍然未知[46-48]。最近的一项研究表明,硅藻中光合作用的优化是通过将光合作用流重新路由到线粒体呼吸而不是 CEF 来完成的[49]。

硅藻的 PSI 和 PSII 接受天线在色素-蛋白质复合物中没有特定的组成,这在陆生植物和绿藻中不太可能。天线或光捕获复合物主要由 Fx-Chl a/c 结合蛋白(FCP)组成,其中蛋白质部分与植物 Chl a/b 结合蛋白(CAB)具有共同特征。FCP 由一个多基因家族编码。例如,6 个 fcp 基因首先在三角褐指藻中,8 个在 *Sketetonema costatum* (Greville) Cleve 和 *Cyclotella cryptica* Reimann 中被描述[50-51]。由于硅藻基因组的测序,现在人们认为它存在 20 到 30 个 fcp 基因和相关基因[5]。除了叶绿素 a 外,FCP 还含有与高等植物不同的辅助色素:叶绿素 c,作为相关叶绿素;Fx,作为主要的叶黄素和较低量的 Ddx 和 Dtx。Guglielmi 等[52]和 Büchel[53]分别对 *P. tricornutum* 和 *Cyclotella meneghiniana* Kützing 进行了两项不同的研究,证明 LHC 低聚物(LHCo)由两部分组成。第一部分 LHCo-1,主要且更稳定,主要由 18 kDa 和 19 kDa 多肽组成。它携带大部分叶绿素 a 和大部分类胡萝卜素 Fx[52]。第二部分 LHCo-2 仅由 18 kDa 多肽组成,负责吸收光谱中 486 nm 的肩峰。LHCo-2 含有 10%~15% 的总天线叶绿素,只有少量叶绿素 c,并且富含 Ddx[52]。这两个级分的多肽组成不同,并且存在两个 LHCo 级分的三个主要成分。最重要的是 Lhcf,它是主要的采光色素蛋白复合物。第二个成分 Lhcx 与绿藻的 LI818 和 LhcSR 以及陆生植物的 PsbS 具有高度相似性[54]。第三个成分 Lhcr 蛋白被认为是硅藻中的 PSI 天线,因为它与红藻对应物高度相似[55-57]。Gundermann 等[56]报道的几种 Lhc 蛋白存在于硅藻中,这些蛋白结合形成 FCP 复合物。根据调节光收集能力或激活光保护机制的必要性,存在不同类型的 FCP 复合物并结合不同量的色素。事实上,不同配合物的丰度取决于光照条件。

在陆地植物和绿藻中,PSII 和 PSI 都拥有自己的集光复合物(分别为 LHCII 和 LHCI)组装成超级复合物。在硅藻中,缺乏小的 Chl 结合蛋白[48]可能是造成绿藻和陆地植物典型的大型 PSII 天线超复合物缺失的原因。在 P. tricornutum 和 Cyclotella meneghiniana 中,FCP 采用三聚体组织[58]。还分离出更高的寡聚形式[53]。PSI 核心复合物的多肽组成在不同的硅藻物种中是可变的[59]。此外,与 PSI 超复合体是三聚体的蓝藻不同,硅藻的 PSI 以单体形式存在[60]。PSI 的反应中心与单排或两排 FCP 不对称相关[17,48,60-61]。

在陆生植物和绿藻中,PSII-LHCII 超级复合物位于堆积颗粒中,而 PSI-LHCI 超级复合物位于暴露的基质薄片中[62-64]。由 Flori 等[35]进行的 PS 的免疫定位,证明 PSI 位于面向类囊体的外周基质膜中,而 PSII 位于核心定位的类囊体膜中。这种分离通过热力学有利的能量转移,即所谓的能量溢出[35],防止了两个 PS 之间的能量提取。在硅藻中,类囊体中不存在明显的结构域有利于 PS 的随机分布,即 FCP 均匀分布在类囊体膜中并为 PSs 服务[65-66],或者,最大光合的线性组织链[48],仅报告了 PS 本地化的适度差异[66]。数据表明,硅藻类囊体堆叠的外层将富含 PSI 和 ATP 合酶[57]。然而,这种 PS 的分离可能与类囊体膜的贴伏特性无关,因为"硅藻"贴壁的类囊体发生在其他原生藻中。事实上,这种分离与外部类囊体在带负电荷的饱和磺基喹诺糖基二酰基甘油(SQDG)中的富集有关,而中等薄片将在单半乳糖基二酰基甘油(MGDG)中更富集[19,57]。事实上,在红光下,PSI-LHC 超复合物分离成膜域[67],这也将富含 SQDG,这些聚合体的作用仍有待检验。

8.5 非光化学猝灭

与其他浮游植物物种一样,硅藻暴露在不断变化的光照条件下。由于光照不足会损害光合作用过程,包括硅藻在内的光合作用生物可能已经制定了对这种条件作出反应的策略。这包括增加色素细胞配额,以便在阴影或低光照的情况下更有效地收集可用光子[68],相反,应该消散多余的能量以避免形成活性氧(ROS)[69]。简而言之,当光子被叶绿素分子吸收时,相关的能量被用来形成单线态叶绿素(^{1}Chl*),这是这种色素的高能状态,^{1}Chl* 的能量可以以多种方式使用[70]。光合作用的基础是将这种能量用于光化学,即在反应中心进行电荷分离。当不可能时,例如,在饱和光照下,多余的能量会在荧光和热量的作用下消散在环境中。或者,^{1}Chl* 可以转化为 Chl 三重态(^{3}Chl*),这是一种更稳定的 Chl 状态,可以与 O_2 反应形成 ROS。^{3}Chl* 形成的概率取决于 ^{1}Chl* 的寿命,而后者又取决于前面列出的三个反应的速率。光合生物能够通过建立快速光保护机制来维持低水平的 ^{3}Chl*,从而最大限度地减少 ROS

的形成[71]。这些机制中最重要的是叶绿素荧光的非光化学猝灭(NPQ)。

在陆地植物和绿藻中,NPQ 分为三个主要成分:qE、qI 和 qT。qE 是最重要的组成部分,是依赖于类囊体膜之间产生的 ΔpH 和叶黄素循环(XC)活性的能量依赖性猝灭。不为人所知的 qI 成分与光合作用的光抑制有关。qT 是从 PSII 到 PSI 的能量重定位。如前所述,NPQ 最重要的组成部分是 qE,原则上它是由跨类囊体膜 ΔpH 的积累激活的,因此管腔酸化,激活去环氧化物酶,使无环氧化物的叶黄素积累。这些叶黄素诱导LHCs 的构象变化,将它们转化为能量耗散形式而不是感光形式[70,72]。

在硅藻中,已通过其松弛动力学详细分析了 NPQ[73]。与绿藻和陆地植物一样,已经确定了形状、半衰期和振幅的不同三种成分:慢速成分和快速成分,分别为 qNs 和 qNf,具有指数形状。中间和主要成分(qNi)具有 S 形。这三个组成部分很可能在 NPQ 的建立中有对应的部分[73-75]。然而,硅藻中这三种成分的生物学和机械基础与绿藻和陆地植物中发现的成分有部分不同。

qNf 成分可能与松弛过程开始时类囊体膜内发生的快速构象变化有关[73-76]。与其他藻类相比,硅藻中的这种构象变化较少被理解。NPQ 的建立需要体外聚集[77],尽管缺乏实验证据,但预计在体内也会发生类似的机制[78]。Lhcx-1 的作用可能对此至关重要,因为它参与了类囊体膜的结构稳定性[79-80]。实际上,类囊体膜对于在光胁迫下微调光捕获能力至关重要。据报道,在三角褐指藻中,色素-蛋白质复合物的手性有组织的大结构域的修饰发生在光胁迫下[81],这些修饰也反映了类囊体多细胞组织的变化[76]。

qNs 成分与 pH 梯度的光抑制和/或部分消散有关[73],当 T. pseudonanais 暴露于 HL 超过 60 min 时,该成分会增加[80]。qNi 成分与通过类囊体的质子梯度的形成和 Ddx 去环氧化有关。Ddx 去环氧化是 XC 的一部分,即叶黄素的去环氧化物形式酶促转化为环氧化物形式(Dtx),反之亦然。参与 XC 的两种色素通常存在于细胞中,与 FCP 结合,它们的数量取决于入射光。在高光下,Ddx 被 Ddx 去环氧化物酶(DDE,也称为 VDE)转化为 Dtx,该酶通过类囊体腔的酸化(低于 6 的 pH 值)被激活。与绿藻和陆地植物相比,硅藻类囊体腔 pH 值的降低要快得多,并导致 Dtx 环氧化反应的激活。此外,ΔpH 诱导 FCP 的解聚,随后这些复合物的质子化导致蛋白质结构的构象变化,从而使 Dtx 具有淬灭活性[73]。一旦 NPQ 建立,它的进一步存在很大程度上取决于 Dtx 的存在而不是 ΔpH[82]。

在弱光下,Dtx 环氧酶(DEP,也称为 ZEP)进行逆反应,导致 Ddx 中的 Dtx 重新转化。与去环氧化反应相反,环氧化反应总是活跃的。环氧化反应的速度比去环氧化反应慢,Ddx 会累积[82]。在 P. tricornutum 中,已鉴定出三种 DDE/VDE 和 DEP/ZEP 同工型,但尚不清楚哪种同工型参与 Ddx 循环[83]。硅藻和高等植物中的酶结构、底物和辅因子相似,但在硅藻中,光依赖性激活的机制尚不清楚。

据报道,长时间且非常强烈的 HL 应力可以触发硅藻中的另一个 XC。由于 VDE 的活动和 ZEP 将 Zx 重新转换为 Vx,第二个 XC 包括通过中间 Ax 将 Vx 转换为 Zx。这个循环主要不涉及光保护[84]。与陆地植物和绿藻相比,硅藻中 Ddx 向 Dtx 转化的激活可能高 4~5 倍。因此,在过饱和光照条件下,硅藻可以将高达 90% 的吸收能量向 NPQ 消散[85-86]。

8.6 碳吸收和固定

作为主要的初级生产者,硅藻中的碳固定机制对于生物地球化学非常重要。在海洋环境中,碱度(pH 值高于 8)和盐度非常高,改变了水性 CO_2 和 CO_3^{2-} 之间的平衡,形成碳酸氢盐 HCO_3^-[87]。溶解的无机碳(DIC)是这三种形式的 CO_2 的通用术语。在海水中,CO_2 浓度为 10~25 mM[88],即远低于使硅藻 RuBisCO(CO_2 固定的关键酶)饱和所需的 50 mM。在硅藻中,在最保守的范围内并考虑到可变性,据报道浓度在 20~70 μM[89-90]。实际上,CO_2 与 O_2 竞争独特的酶活性位点,并且根据酶周围的主要气体分压,RuBisCO 催化核酮糖-1,5-二磷酸的羧化或氧化。前一种反应允许光合细胞获得碳,而后一种反应触发 CO_2 的释放,两种反应都在竞争同一个反应中心[36],为了避免在 RuBisCO 催化位点发生加氧酶反应,它必须是饱和的[36,91-92]。作为蓝藻和绿藻,硅藻具有碳浓缩机制(CCM)以积累 DIC,从周围吸收它。藻类细胞基于类核糖体进化出不同的 CCM,但它的存在并不意味着藻类具有 CCM[93-94]。叶绿体的这个区域应该作为 CCM 的中心点。据报道,硅藻中存在两种类型的 CCM。一种是"生物物理 CCM",其中 CO_2 和 HCO_3^- 由碳酸酐酶(CA)和碳酸氢盐转运蛋白运输[95-96]。另一种 CCM 被称为"生化 CCM",它涉及将无机碳前缀到 C4 化合物中,就像在 C4 陆地植物中一样,然后在 RuBisCO 附近转化回 C3 化合物和 CO_2,有利于其羧化活性。这种机制的证据已在硅藻 *T. weissflogii* (Grunow)中获得[97-98],但仍有待在其他物种中得到证实[99-101]。

淡水和海洋硅藻都表现出对 CO_2 和碳酸氢盐的积极吸收。Matsuda 等[102]发现不同 DIC 从周围介质中吸收碳是物种特异性的。CA 是参与碳吸收的关键酶,它们的活性取决于 pH 值及其在细胞区室中的定位。最近的研究表明,核糖体中内部 CA 和 RuBisCO 的共定位[103]。这种拟核定位的 CA 在低 CO_2 浓度下被激活,并允许在与 CO 具有低亲和力的 RuBisCO 附近有效地形成 CO_2。

外部 CA 也存在。如前所述,碱性和咸味环境倾向于将平衡向碳酸氢盐的形成移动,迫使细胞将能量投入到碳酸氢盐的泵送中,但外部 CA 的存在抵消了周质未搅拌层中的这种平衡转变,使细胞能够接触到 CO_2。在硅藻中,已经表明外部 CA 的出现可能不一致,因此不是必需的[104]。在 *P. tricornutum* 中,由于使用了乙氧基唑胺(可渗透膜的 CA 抑制剂),已证明内部 CA 在生理上对 CCM 是必需的[105]。在中心硅藻和羽状硅藻中,几个 CA 已定位在不同的细胞室中。在叶绿体包膜、核蛋白和线粒体中存在定位于具有不同功能的 CA,从上述对 CCM 的影响到完全不同的过程[91,103,106]。Allen 等[107]在 *P. tricornutum* 中证明,果糖二磷酸与核糖体中的 CA 共定位,并且由于 RuBisCO 也共定位在核糖体中,这意味着在硅藻中,卡尔文循环的第一步发生在叶绿体的这个区域中。

功能性生化 CCM 需要羧化酶,例如磷酸烯醇丙酮酸羧化酶(PEPC),它催化磷酸烯醇丙酮酸(PEP)与 HCO_3^- 的羧化,形成 C4 碳化合物(草酰乙酸、苹果酸、天冬氨酸)。然后,这种 C4 化合物被脱羧酶(苹果酸酶 ME 和磷酸烯醇丙酮酸羧激酶 PEPCK)裂解,在 RuBisCO 附近产生 CO_2[108]。C4 途径所需的这些酶存在于 C3 植物中,在其中发挥不同作用[109]。C4 代谢似乎仅限于特定物种[98]。事实上,它已在硅藻 *T. weissflogii* 中得到报道[97-98],但仍有待在其他物种中得到证实。两种硅藻 *T. pseudonana* 和 *P. tricornutum* 具有操作功能性 C4 CCM 的遗传潜力,因为所需的所有基因都存在[5,110-111]。然而,这种生化 CCM 的功能尚未在这些物种中得到证实,仍然存在争议[99-100,112-114]。Kustka 等[100]认为 *T. pseudonana* 可能已经进化出一种非常规的基于 C4 的 CCM,该 CCM 涉及丙酮酸羧化酶反向工作以从叶绿体中的草酰乙酸产生 HCO_3^-。该模型需要在质膜和 ER 膜中存在碳酸氢盐转运蛋白[96]。

卡尔文循环在光合生物中的复杂性和核心作用需要精细调节。在陆地植物和绿藻中,它通过光、pH 值、氧化还原状态、代谢和蛋白质相互作用,特别是硫氧还蛋白(Trx)介导的二硫键可逆还原/氧化/半胱氨酸残基的巯基进行了很好的研究和精细调节[36]。硅藻也有许多不同的 Trx,位于不同的隔间,包括叶绿体,但关于它们的功能的知识很少。硅藻中质体 Trx 的功能似乎是有限的,据观察,除了果糖 1,6-二磷酸酶[103]外,没有一种参与卡尔文循环的酶以这种方式受到调节。

8.7 结论

尽管硅藻作为初级生产者很重要,而且它们具有显著的适应能力(见表 8.1),但它们的光合机制的细节,包括类囊体膜的结构以及这些膜对光合机制的功能的重要性,仍然有待部分地被发现。对光合作用过程的更深入了解不仅对于理解光合作用过程本身很重要,而且对于理解海洋中物种的分布和硅藻组合的修饰也很重要。因为微藻生产感兴趣的分子是基于微藻捕获的 CO_2,这些数据将影响基于微藻的生物技术过程的优化。

表 8.1　硅藻和绿藻中光合作用的主要差异

	硅藻	绿藻
进化起源	次级内共生	原初内共生
包围类囊体的膜数目	4	2
类囊体中是否存在类粒	否	是
光系分布的侧向异质性	弱	强

(续表)

	硅藻	绿藻
非光化学淬灭现象	非常强烈且快速	强烈但缓慢
碳浓缩机制	是	是,但如藻类属(绿藻门)中的 Caulerpa 等有例外情况[37,56,61,72]

参考文献

[1] Darko, E., Heydarizadeh, P., Schoefs, B., Sabzalian, M. R. (2014). Photosynthesis under artificial light: the shift in primary and secondary metabolites, Philos. T. Roy. Soc. B., 369, 20130243.

[2] Depauw, F. A., Rogato, A., Ribera d'Alcala, M., Falciatore, A. (2012). Exploring the molecular basis of responses to light in marine diatoms, J. Exp. Bot., 63(4), 1575-1591.

[3] Benoiston, A.-S., Ibarbalz, F. M., Bittner, L., Guidi, L., Jahn, O., Dutkiewicz, S., et al. (2017). The evolution of diatoms and their biogeochemical functions, Philos. Trans. R. Soc. Lond. B Biol. Sci., 372 (1728), 20160397.

[4] Keeling, P. J., Burger, G., Durnford, D. G., Lang, B. F., Lee, R. W., Pearlman, R. E. et al. (2005). The tree of eukaryotes, Trends Ecol. Evol. (Amst.)., 20(12), 670-676.

[5] Armbrust, E. V., Berges, J. A., Bowler, C., Green, B. R., Martinez, D., Putnam, N. H., et al. (2004). The genome of the diatom Thalassiosira pseudonana: Ecology, evolution, and metabolism, Science, 306 (5693), 79-86.

[6] Thiriet-Rupert, S., Carrier, G., Chénais, B., Trottier, C., Bougaran, G., Cadoret, J.-P. et al. (2016). Transcription factors in microalgae: genome-wide prediction and comparative analysis, BMC Genomics, 17 (1), 282.

[7] McFadden, G. I. (2014). Origin and evolution of plastids and photosynthesis in eukaryotes, Cold Spring Harb. Perspect. Biol., 6(4), a016105.

[8] Reyes-Prieto, A., Weber, A. P. M., Bhattacharya, D. (2007). The origin and establishment of the plastid in algae and plants, Annu. Rev. Genet., 41(1), 147-168.

[9] Spetea, C., Rintamäki, E., Schoefs, B. (2014). Changing the light environment: chloroplast signalling and response mechanisms, Philos. T. Roy. Soc. B., 369(1640), 20130220.

[10] Schoefs, B., Hu, H., Kroth, P. G. (2017). The peculiar carbon metabolism in diatoms, Phil. Trans. R. Soc. B., 372(1728), 20160405.

[11] Gibbs, S. P. (1979). The route of entry of cytoplasmically synthesized proteins into chloroplasts of algae possessing chloroplast ER, J. Cell Sci., 35(1), 253-266.

[12] Solymosi, K. (2012). Plastid structure, diversification and interconversions. I. Algae, Curr. Chem. Biol., 6(3), 167-186.

[13] Raven, J. A. (1983). The transport and function of silicon in plants, Biol. Rev., 58(2), 179-207.

[14] Grossman, A. R., Bhaya, D., He, Q. (2001). Tracking the light environment by cyanobacteria and the dynamic nature of light harvesting, J. Biol. Chem., 276(15), 11449-11452.

[15] Beauger, A., Serieyssol, K., Schoefs, B. (2015). Recent progress in diatom's taxonomy and freshwater ecology, Cryptogam. Algol., 36, 241-244.

[16] Morin, S., Rosberry, J., Van de Vijver, B., Schoefs, B. (2016). Advances in diatom biodiversity and ecology, Botany Letters, 163(2), 69-70.

[17] Herbstová, M., Bína, D., Koník, P., Gardian, Z., Vácha, F., Litvín, R. (2015). Molecular basis of chromatic adaptation in pennate diatom Phaeodactylum tricornutum, Biochim. Biophys. Acta. Bioenerg.,, 1847(6-7), 534-543.

[18] Vinayak, V., Manoylov, K. M., Gateau, H., Blanckaert, V., Herault, J., Pencreac'h, G. et al. (2015). Diatom milking: a review and new approaches, Mar. Drugs, 13(5), 2629-2665.

[19] Heydarizadeh, P., Poirier, I., Loizeau, D., Ulmann, L., Mimouni, V., Schoefs, B. et al. (2013). Plastids of marine phytoplankton produce bioactive pigments and lipids, Mar. Drugs, 11(9), 3425-3471.

[20] Sayanova, O., Mimouni, V., Ulmann, L., Morant-Manceau, A., Pasquet, V., Schoefs, B. et al. (2017). Modulation of lipid biosynthesis by stress in diatoms, Phil. Trans. R. Soc. B., 372(1728), 20160407.

[21] Schoefs, B., Hu, H., Kroth, P. G. (2017). The peculiar carbon metabolism in diatoms, Phil. Trans. R. Soc. B., 372(1728), 20160405.

[22] Katz, L. A., Grant, J. R. (2015). Taxon-rich phylogenomic analyses resolve the eukaryotic tree of life and reveal the power of subsampling by sites, Syst. Biol., 64(3), 406-415.

[23] Cavalier-Smith, T., Chao, E. E., Lewis, R. (2015). Multiple origins of heliozoa from flagellate ancestors: New cryptist subphylum corbihelia, superclass corbistoma, and monophyly of haptista, cryptista, hacrobia and chromista, Mol. Phylogenet. Evol., 93, 331-362.

[24] Nisbet, R. E. R., Kilian, O., McFadden, G. I. (2004). Diatom genomics: genetic acquisitions and mergers, Curr. Biol., 14(24), R1048-R1050.

[25] Petroutsos, D., Amiar, S., Abida, H., Dolch, L.-J., Bastien, O., Rebeille, F. et al. (2014). Evolution of galactoglycerolipid biosynthetic pathways—From cyanobacteria to primary plastids and from primary to secondary plastids, Prog. Lipid Res., 54, 68-85.

[26] Berger, H., De Mia, M., Morisse, S., Marchand, C. H., Lemaire, S. D., Wobbe, L., et al. (2016). A light switch based on protein S-nitrosylation fine-tunes photosynthetic light-harvesting in the microalga Chlamydomonas reinhardtii, Plant Physiol., 171, 821-832.

[27] Marchand, J., Heydarizadeh, P., Schoefs, B., Spetea, C. (2019). Chloroplast ion and metabolite transport in algae. In: Photosynthesis in Algae, Vol. 2, A. W. D. Larkum, A. Grossman and J. Raven, (eds.). Springer.

[28] Marchand, J., Heydarizadeh, P., Schoefs, B., Spetea, C. (2018b). Ion and metabolite transport in the chloroplast of algae: lessons from land plants, Cell. Mol. Life Sci., 75(12), 2153-2176.

[29] Grosche, C., Hempel, F., Bolte, K., Zauner, S., Maier, U. G. (2014). The periplastidal compartment: a naturally minimized eukaryotic cytoplasm, Curr. Opin. Microbiol., 22, 88-93.

[30] Lau, J. B., Stork, S., Moog, D., Schulz, J., Maier, U. G. (2016). Protein-protein interactions indicate composition of a 480 kDa SELMA complex in the second outermost membrane of diatom complex plastids, Mol. Microbiol., 100(1), 76-89.

[31] Botte, C. Y., Marechal, E. (2014). Plastids with or without galactoglycerolipids, Trends Plant Sci., 19(2), 71-78.

[32] Drum, R. W., Pankratz, H. S. (1964). Pyrenoids, raphes, and other fine structure in diatoms, Am. J. Bot. 51(4), 405-418.

[33] Bedoshvili, Y. D., Popkova, T. P., Likhoshway, Y. V. (2009). Chloroplast structure of diatoms of different classes, Cell Tiss. Biol., 3(3), 297-310.

[34] Solymosi, K., Keresztes, Á. (2013). Plastid structure, diversification and interconversions. II. Land plants, Curr. Chem. Biol., 6(3), 187-204.

[35] Flori, S., Jouneau, P. H., Bailleul, B., Gallet, B., Estrozi, L. F., Moriscot, C. et al., (2017). Plastid thylakoid architecture optimizes photosynthesis in diatoms, Nat. Commun., 8, 15885.

[36] Jensen, E., Clement, R., Maberly, S. C., Gontero, B. (2017). Regulation of the calvin-benson-bassham cycle in the enigmatic diatoms: biochemical and evolutionary variations on an original theme, Phil. Trans. R. Soc. B, 372(1728), 20160401.

[37] Zapata, M., Garrido, J. L., Jeffrey, S. W. (2006). Chlorophyll c pigments: current status. In: Chlorophylls and Bacteriochlorophylls, Vol. 25, B. Grimm, R. J. Porra, W. Rüdiger and H. Scheer, (eds.). pp. 39-53. Kluwer Academic Publishers: The Netherlands.

[38] Kuczynska, P., Jemiola-Rzeminska, M., Strzalka, K. (2015). Photosynthetic Pigments in Diatoms, Mar. Drugs, 13(9), 5847-5881.

[39] Bertrand, M. (2010). Carotenoid biosynthesis in diatoms, Photosyn. Res., 106(1-2), 89-102.

[40] Moulin, P., Lemoine, Y., Schoefs, B. (2011). Modification of the carotenoid metabolism in plastids: a response to stress conditions. In: Handbook of Plant and Crop Stress, Vol. 3, M. Pessarakli (ed.), pp. 407-434. CRC Press, Boca Raton, London, New York.

[41] Lohr, M., Wilhelm, C. (1999). Algae displaying the diadinoxanthin cycle also possess the violaxanthin cycle, P. Natl. A. Sci., 96(15), 8754-8789.

[42] Okumura, A., Nagao, R., Suzuki, T., Yamagoe, S., Iwai, M., Nakazato, K., et al. (2008). A novel protein in photosystem II of a diatom chaetoceros gracilis is one of the extrinsic proteins located on lumenal

side and directly associates with PSII core components, Biochimica et Biophysica Acta (BBA)-Bioenergetics 1777(12), 1545-1551.

[43] Masmoudi, S., Nguyen-Deroche, N., Caruso, A., Ayadi, H., Morant-Manceau, A., Tremblin, G., et al. (2013). Cadmium, copper, sodium and zinc effects on diatoms: from heaven to hell—a review, Cryptogamie Algol., 34(2), 185-225.

[44] Shikanai, T., Yamamoto, H. (2017). Contribution of cyclic and pseudo-cyclic electron transport to the formation of proton motive force in chloroplasts, Mol. Plant, 10(1), 20-29.

[45] Asada, K. (2000). The water-water cycle as alternative photon and electron sinks, Philos. Trans. R. Soc. Lond. B Biol. Sci., 355(1402), 1419-1431.

[46] Bowler, C., Allen, A. E., Badger, J. H., Grimwood, J., Jabbari, K., Kuo, A. et al. (2008). The Phaeodactylum genome reveals the evolutionary history of diatom genomes, Nature, 456(7219), 239-244.

[47] Prihoda, J., Tanaka, A., de Paula, W. B. M., Allen, J. F., Tirichine, L., Bowler, C. (1093). Chloroplastmitochondria cross-talk in diatoms, J. Exp. Bot., 63(4), 1543-1557.

[48] Grouneva, I., Rokka, A., Aro, E. M. (2011). The thylakoid membrane proteome of two marine diatoms outlines both diatom-specific and species-specific features of the photosynthetic machinery, J. Proteome Res., 10(12), 5338-5353.

[49] Bailleul, B., Berne, N., Murik, O., Petroutsos, D., Prihoda, J., Tanaka, A. et al. (2015). Energetic coupling between plastids and mitochondria drives CO_2 assimilation in diatoms, Nature, 524(7565), 366-369.

[50] Bhaya, D., Grossman, A. R. (1993). Characterization of gene clusters encoding the fucoxanthin chlorophyll proteins of the diatom Phaeodactylum tricornutum, Nucleic Acids Res., 21(19), 4458-4466.

[51] Janssen, M., Bathke, L., Marquardt, J., Krumbein, W. E., Rhiel, E. (2001). Changes in the photosynthetic apparatus of diatoms in response to low and high light intensities, Int. Microbiol., 4(1), 27-33.

[52] Guglielmi, G., Lavaud, J., Rousseau, B., Etienne, A.-L., Houmard, J., Ruban, A. V. (2005). The light-harvesting antenna of the diatom phaeodactylum tricornutum. Evidence for a diadinox anthin-binding subcomplex, Febs J., 272(17), 4339-4348.

[53] Büchel, C. (2003). Fucoxanthin-chlorophyll proteins in diatoms: 18 and 19 kDa sub units assemble into different oligomeric states, Biochemistry, 42(44), 13027-13034.

[54] Peers, G., Truong, T. B., Ostendorf, E., Busch, A., Elrad, A., Grossman, A. R. et al. (2009). An ancient light-harvesting protein is critical for the regulation of algal photosynthesis, Nature, 462(7272), 818-821.

[55] Büchel, C., Wilhelm, C., Wagner, V., Mittag, M. (2017). Functional proteomics of light-harvesting complex proteins under varying light-conditions in diatoms, J. Plant Physiol., 217, 38-43.

[56] Gundermann, K., Schmidt, M., Weisheit, W., Mittag, M., Büchel, C. (2013). Identification of several sub-populations in the pool of light harvesting proteins in the pennate diatom Phaeodactylum tricornutum, Biochim. Biophys. Acta, 1827(3), 303-310.

[57] Lepetit, B., Goss, R., Jakob, T., Wilhelm, C. (2012). Molecular dynamics of the diatom thylakoid membrane under different light conditions, Photosyn. Res., 111(1-2), 245-257.

[58] Joshi-Deo, J., Schmidt, M., Gruber, A., Weisheit, W., Mittag, M., Kroth, P. G., et al. (2010). Characterization of a trimeric light-harvesting complex in the diatom Phaeodactylum tricornutum built of FcpA and FcpE proteins, J. Exp. Bot. 61(11), 3079-3087.

[59] Ikeda, Y., Komura, M., Watanabe, M., Minami, C., Koike, H., Itoh, S., et al. (2008). Photosystem I complexes associated with fucoxanthin-chlorophyll-binding proteins from a marine centric diatom, Chaetoceros gracilis, Biochim. Biophys. Acta, 1777(4), 351-361.

[60] Veith, T., Büchel, C. (2007). The monomeric photosystem I-complex of the diatom phaeodactylum tricornutum binds specific fucoxanthin chlorophyll proteins (FCPs) as light-harvesting complexes, Biochim. Biophys. Acta, 1767(12), 1428-1435.

[61] Ikeda, Y., Yamagishi, A., Komura, M., Suzuki, T., Dohmae, N., Shibata, Y., et al. (2013). Two types of fucoxanthin-chlorophyll-binding proteins I tightly bound to the photosystem I core complex in marine centric diatoms, Biochim. Biophys. Acta, 1827(4), 529-539.

[62] Chow, W. S., Kim, E. H., Horton, P., Anderson, J. M. (2005). Granal stacking of thylakoid membranes in higher plant chloroplasts: the physicochemical forces at work and the functional consequences that ensue,

Photochem. Photobiol. Sci., 4(12), 1081-1090.
[63] Dekker, J. P., Boekema, E. J. (2005). Supramolecular organization of thylakoid membrane proteins in green plants, Biochim. Biophys. Acta, 1706(1), 12-39.
[64] Anderson, J. M., Chow, W. S., De Las Rivas, J. (2008). Dynamic flexibility in the structure and function of photosystem II in higher plant thylakoid membranes: the grana enigma, Photosyn. Res., 98(1-3), 575-587.
[65] Owens, T. G. (1986). Light-harvesting function in the diatom phaeodactylum tricornutum. II. Distribution of excitation energy between the photosystems, Plant Physiol., 80(3), 739-746.
[66] Pyszniak, A. M., Gibbs, S. P. (1992). Immunocytochemical localization of photosystem I and the fucoxanthin-chlorophyll a/c light-harvesting complex in the diatom phaeodactylum tricornutum, Protoplasma, 166(3-4), 208-217.
[67] Bína, D., Herbstová, M., Gardian, Z., Vácha, F., Litvín, R. (2016). Novel structural aspect of the diatom thylakoid membrane: lateral segregation of photosystem I under red-enhanced illumination, Sci. Rep. 6, #25583.
[68] Heydarizadeh, P., Boureba, W., Zahedi, M., Huang, B., Moreau, B., Lukomska, E., et al. (2017). Response of CO_2-starved diatom Phaeodactylum tricornutum to light intensity transition, Phil. Trans. R. Soc. B., 372(1728), 20160396.
[69] Schoefs, B., Franck, F. (2003). Protochlorophyllide reduction: mechanisms and evolution, Photochem. Photobiol., 78(6), 543-557.
[70] Müller, P., Li, X.-P., Niyogi, K. K. (2001). Non-photochemical quenching. A response to excess light energy, Plant Physiol., 125(4), 1558-1566.
[71] Krieger-Liszkay, A. (2005). Singlet oxygen production in photosynthesis, J. Exp. Bot., 56(411), 337-346.
[72] Rochaix, J.-D. (2014). Regulation and dynamics of the light-harvesting system, Annu. Rev. Plant Biol., 65(1), 287-309.
[73] Roháček, K., Bertrand, M., Moreau, B., Jacquette, B., Caplat, C., Morant-Manceau, A., et al. (2014). Relaxation of the non-photochemical chlorophyll fluorescence quenching in diatoms: kinetics, components and mechanisms, Philos. T. Roy. Soc. B., 369(1640), 20130241.
[74] Grouneva, I., Jakob, T., Wilhelm, C., Goss, R. (2008). A new multicomponent NPQ mechanism in the diatom Cyclotella meneghiniana, Plant Cell Physiol., 49(8), 1217-1225.
[75] Grouneva, I., Jakob, T., Wilhelm, C., Goss, R. (2009). The regulation of xanthophyll cycle activity and of non-photochemical fluorescence quenching by two alternative electron flows in the diatoms phaeodactylum tricornutum and cyclotella meneghiniana, Biochim. Biophys. Acta. Bioenerg., 1787(7), 929-938.
[76] Nagy, G., Szabó, M., Ünnep, R., Káli, G., Miloslavina, Y., Lambrev, P. H., et al. (2012). Modulation of the multilamellar membrane organization and of the chiral macrodomains in the diatom phaeodactylum tricornutum revealed by small-angle neutron scattering and circular dichroism spectroscopy, Photosyn. Res. 111(1), 71-79.
[77] Gundermann, K., Büchel, C. (2008). The fluorescence yield of the trimeric fucoxanthin-chlorophyll-protein FCPa in the diatom cyclotella meneghiniana is dependent on the amount of bound diatoxanthin, Photosyn. Res., 95(2-3), 229-235.
[78] Miloslavina, Y., Grouneva, I., Lambrev, P. H., Lepetit, B., Goss, R., Wilhelm, C. et al. (2009). Ultrafast fluorescence study on the location and mechanism of non-photochemical quenching in diatoms, Biochimica et Biophysica Acta (BBA)-Bioenergetics, 1787(10), 1189-1197.
[79] Bailleul, B., Rogato, A., de Martino, A., Coesel, S., Cardol, P., Bowler, C., et al. (2010). An atypical member of the light-harvesting complex stress-related protein family modulates diatom responses to light, Proc. Natl. Acad. Sci. U.S.A., 107(42), 18214-18219.
[80] Zhu, S.-H., Green, B. R. (2010). Photoprotection in the diatom Thalassiosira pseudonana: role of LI818-like proteins in response to high light stress, Biochim. Biophys. Acta, 1797(8), 1449-1457.
[81] Szabó, M., Lepetit, B., Goss, R., Wilhelm, C., Mustárdy, L., Garab, G. (2008). Structurally flexible macro-organization of the pigment-protein complexes of the diatom Phaeodactylum tricornutum, Photosyn. Res., 95(2), 237-245.
[82] Goss, R., Ann Pinto, E., Wilhelm, C., Richter, M. (2006). The importance of a highly active and ΔpH-

regulated diatoxanthin epoxidase for the regulation of the PS II antenna function in diadinoxanthin cycle containing algae, J. Plant Physiol., 163(10), 1008-1021.

[83] Kuczynska, P., Jemiola-Rzeminska, M., Strzalka, K. (2015). Photosynthetic Pigments in Diatoms, Mar. Drugs, 13(9), 5847-5881.

[84] Lohr, M., Wilhelm, C. (1999). Algae displaying the diadinoxanthin cycle also possess the violaxanthin cycle, P. Natl. A. Sci., 96(15), 8754-8789.

[85] Ruban, A., Lavaud, J., Rousseau, B., Guglielmi, G., Horton, P., Etienne, A.-L. (2004). The superexcess energy dissipation in diatom algae: comparative analysis with higher plants, Photosyn. Res., 82(2), 165-175.

[86] Lavaud, J. (2007). Fast regulation of photosynthesis in diatoms: mechanisms, evolution and ecophysiology, Functional Plant Science and Biotechonology, 1(2), 267-287.

[87] Matsuda, Y., Hara, T., Colman, B. (2001). Regulation of the induction of bicarbonate uptake by dissolved CO_2 in the marine diatom, Phaeodactylum tricornutum, Plant Cell Environ., 24(6), 611-620.

[88] Clement, R., Jensen, E., Prioretti, L., Maberly, S. C., Gontero, B. (2017). Diversity of CO_2-concentrating mechanisms and responses to CO concentration in marine and freshwater diatoms, J. Exp. Bot., 68(14), 3925-3935.

[89] Badger, M. R., Andrews, T. J., Whitney, S. M., Ludwig, M., Yellowlees, D. C., Leggat, W., et al. (1998). The diversity and coevolution of Rubisco, plastids, pyrenoids and chloroplast-based CO_2-concentrating mechanisms in algae, Can. J. Bot., 76(6), 1052-1071.

[90] Young, J. N., Hopkinson, B. M. (2017). The potential for co-evolution of CO_2-concentrating mechanisms and rubisco in diatoms, J. Exp. Bot., 68(14), 3751-3762.

[91] Hopkinson, B. M., Dupont, C. L., Matsuda, Y. (2016). The physiology and genetics of CO_2-concentrating mechanisms in model diatoms, Curr. Opin. Plant Biol., 31, 51-57.

[92] da Silva, J. M., Cruz, S., Cartaxana, P. (2017). Inorganic carbon availability in benthic diatom communities: photosynthesis and migration, Philos. T. Roy. Soc. B., 372(1728), #20160398.

[93] Ratti, S., Giordano, M., Morse, D. (2007). CO_2-concentrating mechanisms of the potentially toxic dinoflagellate Protoceratium reticulatum (Dinophyceae, Gonyaulacales), J. Phycol., 43(4), 693-701.

[94] Genkov, T., Meyer, M., Griffiths, H., Spreitzer, R. J. (2010). Functionalhybrid rubisco enzymes with plant small subunits and algal largesubunits: engineered rbcs cDNA for expression in Chlamydomonas reinhardtii, J. Biol. Chem., 285(26), 19833-19841.

[95] Kikutani, S., Nakajima, K., Nagasato, C., Tsuji, Y., Miyatake, A., Matsuda, Y. (2016). Thylakoid luminal θ-carbonic anhydrase critical for growth and photosynthesis in the marine diatom Phaeodactylum tricornutum, Proc. Natl. Acad. Sci. USA., 113(35), 9828-9833.

[96] Nakajima, K., Tanaka, A., Matsuda, Y. (2013). SLC4 family transporters in a marine diatom directly pump bicarbonate from seawater, P. Natl. A. Sci., 110(5), 1767-1772.

[97] Reinfelder, J. R., Kraepiel, A. M. L., Morel, F. M. M. (2000). Unicellular C4 photosynthesis in a marine diatom, Nature, 407(6807), 996-999.

[98] Roberts, K., Granum, E., Leegood, R. C., Raven, J. A. (2007). Carbon acquisition by diatoms, Photosyn. Res., 93(1-3), 79-88.

[99] Clement, R., Dimnet, L., Maberly, S. C., Gontero, B. (2016). The nature of the CO_2-concentrating mechanisms in a marine diatom, Thalassiosira pseudonanale, New Phytol., 209(4), 1417-1427.

[100] Kustka, A. B., Milligan, A. J., Zheng, H., New, A. M., Gates, C., Bidle, K. D., et al. (2014). Low CO_2 results in a rearrangement of carbon metabolism to support C4 photosynthetic carbon assimilation in Thalassiosira pseudonana, New Phytol., 204(3), 507-520.

[101] Haimovich-Dayan, M., Garfinkel, N., Ewe, D., Marcus, Y., Gruber, A., Wagner, H., et al. (2013). The role of C4 metabolism in the marine diatom Phaeodactylum tricornutum, New Phytol., 197(1), 177-185.

[102] Matsuda, Y., Nakajima, K., Tachibana, M. (2011). Recent progresses on the genetic basis of the regulation of CO acquisition systems in response to CO concentration, Photosyn. Res., 109(1), 191-203.

[103] Matsuda, Y., Kroth, P. G. (2014). Carbon fixation in diatoms. In: The structural basis of biological energy generation, Vol. 39, M. F. Hohmann-Marriott (ed.), pp.335-362. Springer Science.

[104] John-McKay, M. E., Colman, B. (1997). Variation in the occurrence of external carbonic anhydrase among strains of the marine diatom Phaeodactylum tricornutum (Bacillariophyceae), J. Phycol., 33(6),

988-990.

[105] Satoh, A., Kurano, N., Miyachi, S. (2001). Inhibition of photosynthesis by intracellular carbonic anhydrase in microalgae under excess concentrations of CO_2, Photosyn. Res., 68(3), 215-224.

[106] Giordano, M., Norici, A., Forssen, M., Eriksson, M., Raven, J. A. (2003). An anaplerotic role for mitochondrial carbonic anhydrase in Chlamydomonas reinhardtii, Plant Physiol., 132(4), 2126-2134.

[107] Allen, A. E., Moustafa, A., Montsant, A., Eckert, A., Kroth, P. G., Bowler, C. (2012). Evolution and functional diversification of fructose bisphosphate aldolase genes in photosynthetic marine diatoms, Mol. Biol. Evol., 29(1), 367-379.

[108] Sage, R. F. (2004). The evolution of C4 photosynthesis, New Phytol., 161(2), 341-370.

[109] Aubry, S., Brown, N. J., Hibberd, J. M. (2011). The role of proteins in C3 plants prior to their recruitment into the C4 pathway, J. Exp. Bot., 62(9), 3049-3059.

[110] Bowler, C., Allen, A. E., Badger, J. H., Grimwood, J., Jabbari, K., Kuo, A. et al. (2008). The phaeodactylum genome reveals the evolutionary history of diatom genomes, Nature, 456 (7219), 239-244.

[111] Shen, C., Dupont, C. L., Hopkinson, B. M. (2017). The diversity of CO_2-concentrating mechanisms in marine diatoms as inferred from their genetic content, J. Exp. Bot., 68(14), 3937-3948.

[112] Raven, J. A. (2010). Inorganic carbon acquisition by eukaryotic algae: four current questions, Photosyn. Res., 106(1), 123-134.

[113] Reinfelder, J. R., Milligan, A. J., Morel, F. M. M. (2004). The role of the C4 pathway in carbon accumulation and fixation in a marine diatom, Plant Physiol., 135(4), 2106-2111.

[114] Tanaka, R., Kikutani, S., Mahardika, A., Matsuda, Y. (2014). Localization of enzymes relating to C4 organic acid metabolisms in the marine diatom, Thalassiosira pseudonana, Photosyn. Res., 121(2), 251-263.

第 9 章
硅藻中的铁

John A. Raven

9.1 简介

Fe 是迄今为止研究的几乎所有生物体生长的必需微量营养素元素[1-2],在氧化还原反应中具有特殊作用[3]。当地表水中的 O_2 非常少时,Fe 已融入早期生物的新陈代谢中,因此 Fe^{2+} 的可用性很高。在由含氧光合作用和有机碳封存引起的全球氧化后,Fe^{2+} 被氧化成 Fe^{3+},沉淀为 Fe_2O_3 和 Fe_3O_4[3]。与其他必需元素相比,Fe 的缺乏在当今海洋的高营养(N,P)低叶绿素(HNLC)区域最为明显[4-5]。

有大量关于硅藻中铁的吸收、代谢用途和储存的信息。基本上所有的工作都是关于海洋浮游硅藻的,因此对海洋底栖或淡水硅藻知之甚少。然而,有趣的是,基因组最先被测序的两种海洋硅藻,分别是中心假微型海链藻和羽状三角褐指藻,最先发现起源于淡水[6-7],并出现在淡水以及半咸水和海洋生境中[7]。由于硅藻起源是海洋硅藻,因此至少假微型海链藻可能是次海洋硅藻[8-9]。

9.2 硅藻对铁的获取

含氧水中的 Fe 主要处于 Fe^{3+} 氧化还原状态,或呈胶态或溶解为有机螯合物。一些溶解的 Fe^{2+} 是通过太阳紫外线的光还原产生的,溶解的 O_2 再将 Fe^{2+} 氧化成 Fe^{3+}[10-11]。扩散到质膜表面的 Fe^{3+} 可以通过 Fe^{3+} 转运蛋白转移到胞质溶胶中,类似地,扩散到质膜表面的 Fe^{2+} 可以通过 Fe^{2+} 转运蛋白转移到胞质溶胶中。质膜上的 Fe^{3+} 也可能受到表面还原酶活性的影响,然后使用含铜氧化酶转运至胞质溶胶并氧化为 Fe^{3+}[12]。

铁载体是由一些细菌、真菌和草产生的有机化合物,它们将 Fe^{3+} 从环境中隔离出来,带 Fe^{3+} 的铁载体可以被生产者生物体或其他一些生物体吸收。硅藻梅尼小环藻可以产生类似铁载体的化合物,但并没有证明这与 Fe 得有关,同时也不清楚培养环境是否是无菌的[13]。硅藻也与细菌有关,有时与硅藻和细菌之间具有特异性有关。一些细菌可以产生铁载体,但尚不清楚铁载体是否有助于硅藻吸收铁[14]。这种关联还可以为藻类提供维生素[15],这也可能与 Fe-维生素 B_{12} 相互作用有关[16]。吞噬铁的获取不能在硅藻中发生[17-18]。

9.3 硅藻含铁蛋白质与铁利用的经济性

Ho 等和 Quigg 等提供了 29 种蓝细菌和真核微藻的营养丰富培养物的元素组成信息。为了考虑到细胞体积和液泡化程度的差异,元素的细胞定额是相对于 P 来表示的。Fe 和 P 的摩尔比的最高值出现在蓝藻、灰胞藻和具有叶绿素 b 的微藻中,即叶绿体的"绿色"线。在缺乏叶绿素 b 的微藻中发现较低的 Fe:P 值,即那些具有通过次生内共生从红藻衍生的叶绿体(叶绿体的"红"线)的微藻。硅藻处于"红"线,平均 F 和 P 的摩尔比值低于任何其他"红"线生物,因此低于任何测试的生物[19-20]。四种硅藻中 Fe 和 C 的摩尔比分别为 0.3、1.7、3.1 和 6.7[19],与 Marchetti 等研究的数据相比,Fe 和 C 的摩尔比分别为 6.7、21、84 和 148。6 种硅藻的 Fe 和 C 的摩尔比分别为 156.5~161.5、248.3、186.1、176.4、221.5 和 114.6[21]。沿海硅藻和大洋硅藻的 Fe 和 C 的摩尔比没有明显差异。应该强调的是,这些值是针对营养丰富的培养物中的铁含量,而不是铁需求量;当然,储存的 Fe 也可能超过要求,但因为其含量是以细胞 P 为标准的,因此也可能存在储存 P 的情况。特别是考虑到外部铁的可用性(在这种情况下,铁处于饱和状态)对红藻和类蓝藻中的 P 储存的影响[22]。很少有数据集将硅藻中的铁含量与特定含铁蛋白质的含量联系起来,但血红素 b 是一个例外,线粒体中的细胞色素 b 和质体的 Cyt b_{559} 和 Cyt b_6[23]。

在催化的含铁蛋白质和蛋白质复合物中存在恒定的铁:蛋白质比率。这些在硅藻中或与类囊体膜相关的催化复合物是 PSII、Cyt $b_6f Fe_{nh}$、PSI、Cyt c_6、Fd、PTOX(见表 9.1)、抗坏血酸过氧化物酶(可能在水-水循环中发挥作用)和超氧化物歧化酶。对于线粒体内膜,Fe 蛋白是复合物 I、II、III、IV 和 AOX(见表 9.2)。Fe-蛋白质复合物是 NR、NiR 和亚硫酸还原酶、乌头酸酶[24]和硫氧还蛋白(酶活性调节,尤其是明暗变化)[25-27]。

表 9.1 硅藻光合线性电子流中蛋白质或蛋白质复合物的铁

蛋白质或蛋白质复合物	每个复合体的铁
PS II	3;2 Cyt b_{559},1 Fe_{nh}
Cyt $b_6 f Fe_{nh}$	5:2 Cyt b_6,1 Cyt f,2 Fe_{nh}
Cyt c_6(可以被替换为基因型由不含铁元素的质体蓝素)	1:1 Cyt c_6
PS I	12:3×4 Fe_{nh}
铁氧化还原蛋白(可以在表型上被不含铁的黄酮素所取代)	2:2 Fe_{nh}
PTOX	2

表 9.2　硅藻呼吸电子流中蛋白质或蛋白质复合物的铁

蛋白质或蛋白质复合物	每个复合物的铁
复合物Ⅰ	30 Fe_{nh}
复合物Ⅱ	9 Fe
复合物Ⅲ	5：2 Cyt b, 1 Cyt c_1, 2 Fe_{nh}
Cyt c	1 per Cyt c
复合物Ⅳ	2：1 Cyt a; 1 Cyt a
AOX	2

根据目前的证据,硅藻中不存在将 PSI 中的电子和溶液中的 H^+ 转化为 H_2O 的黄素二铁蛋白,但其在光合作用中可能具有水-水循环的功能。同样,硅藻缺乏过氧化氢酶,将 $2H_2O_2$ 转化为 $2H_2O$ 和 $1O_2$ [28-29]。所有真核生物过氧化氢酶都含有 Fe[30],而过氧化氢酶缺乏 Fe 的情况可以通过与含有过氧化氢酶的细菌的结合来补偿[28]。一些细菌具有 Mn 过氧化氢酶[30-31],因此硅藻与细菌的相互作用可能会总体上节省 Fe。硫酸盐还原酶是一种 APR-B,它不使用参与维管植物 APR-A 中的硫化亚铁簇(或任何其他铁催化剂)[32]。对于硅藻超氧化物歧化酶,如果不是唯一的话,其主要形式使用 Mn 而不是 Fe[33-35];假微型海链藻缺 Fe 使 Mn 的配额增加了三倍,部分原因可能与超氧化物歧化酶对 Mn 的需求量增加有关[34-35]。硅藻在循环电子流(CEF)中也缺乏高含铁 H^+ 泵送 NDHdh 复合物[36]。

APR-B 而非 APR-A 的出现,以及缺铁时用 Mn 代替 Fe,被认为与低铁生态环境中硅藻的出现有关,最显著的地方是公海 HNLC(高营养[P,结合 N],低叶绿素)栖息地。硅藻的高 PsII:PsI 和 Cyt b_6fFe_{nh}:PSI 比率也节省了 Fe,因为配合物的 Fe 含量增加了 PSII:Cyt b_6fFe_{nh}:PSI[37]。PsII:PsI 和 cyt b_6fFe_{nh}:PSI 的高比率尤其适用于在公海、低铁环境(例如大洋海链藻)而不是来自高铁沿海环境的那些(例如威氏海链藻)中生长的硅藻[37](见表 9.3)。相对于高(饱和)光下的铁补充,在低铁限制培养物下的比率尤其高,在低(限制)光下生长时的比率较低[37](见表 9.3)。Allen 等使用沿海(即高铁适应环境)硅藻三角褐指藻,在低铁培养条件下生长时 PSI:PSII 比率也低于高铁培养物,但对 Cyt b_6fFe_{nh}:PSII 的影响更大。Szrzepek 和 Harrison 指出,大洋海链藻的高 PSII:高 PSI 限制了对低辐照度的适应。Poncel 等发现,缺铁培养的三角褐指藻不仅光合电子传递能力下降[38-39],而且也显示出对高辐照度更敏感。更令人惊讶的是,由于某些限制氧化损伤的机制需要铁,缺铁硅藻赋予了对氧化损伤的瞬时抵抗力[40]。

表 9.3 在低铁(限制)条件下(饱和)条件下(低铁)和高铁(高铁沿海海洋海生境)中含铁光合配合物的比例比较

Fe 复合物	大洋海链藻低光	大洋海链藻高光	威氏海链藻低光	威氏海链藻高光
PS Ⅱ	8.7	10.5	2.2	4.4
Cyt c_6 f	1.1	3.2	2.0	1.9
Cyt c_6	0	0	1.5	1.5
PS Ⅰ	1	1	1	1

高铁叶绿体 NDHdh 的缺乏以 CEF 的低 $H^+:e^-$ 和 $ATP:e^-$ 为代价节约了铁[36]。但是,如果测量的通过 CEF 的低电子通量[41]对应于低 CEF 容量,则 Fe 的节省是有限的。此外,CEF 在硅藻中的有限作用与线粒体呼吸的重要作用是平行的,在光照下,每单位 ATP 生产对 Fe 的需求很高[41]。目前尚不清楚这是否涉及光周期中的线粒体容量比暗周期中维持和生物合成所需的更多,线粒体铁的需求相应增加。然而,Roncel 等表明,三角褐指藻中的铁缺乏对暗呼吸 O_2(83%)吸收的抑制作用比对光合 O_2 产生(56%)和比生长率(30%)的抑制作用更大。

Groussman 等在几乎所有受检的硅藻中都发现了 Cyt b_6 转录物,其中质体蓝素的分布更受限制,在几个物种中两种转录物共存,而质体蓝素的表达更受限制;因此,质体蓝素的功能性表达是质体蓝素蛋白差异表达的结果。质体蓝素已在大洋海链藻中进行了功能性验证,它栖息在铁含量低的环境中[34,42],不含 Cyt b_6[37],并且比来自铁含量较高的栖息地的硅藻对铜的需求量更大,这种需要可以完全由质体蓝素的含量来解释[42-43]。

已在许多硅藻的转录组中检测到铁氧还蛋白和黄素氧还蛋白基因[44];在硅藻的低铁环境中,铁氧还蛋白在功能上被黄素氧还蛋白取代,硅藻保留了黄素氧还蛋白基因的铁响应拷贝。在沿海(即铁利用率高的栖息地)假微型海链藻,该基因的铁响应拷贝已经丢失,黄素氧还蛋白表达与铁可用性无关[45]。相比之下,在微氏海链藻的公海(低铁可用性)菌株中,黄素氧还蛋白基因的铁敏感拷贝被保留,黄素氧还蛋白在低外部铁浓度下取代铁氧还蛋白[45]。Haitori-Saito 等表明,随着西亚北极太平洋几种硅藻物种的铁可用性降低,铁氧还蛋白与铁氧还蛋白的比例也随之降低;然而,并没有显示在所有的采样站且都是相同的基因型,因此这些效应是驯化性的,而非适应性的。

低铁(HNLC)环境以 NO_3^- 作为无机 N 源,因此需要含铁的 NR 和 NiR 来供应用于 NH_4^+ 同化途径的 NH_4^+,从而增加生长对所铁的需求;NO_3^- 同化途径酶,尤其对铁的需求高于 NR 的 NiR 下降,导致 NH_4^+ 的产生和 NO_2^- 的积累减慢。对于具有固氮营养的蓝藻共生体的硅藻,发现对 N 同化的 Fe 需求甚至更高。Fe 的消耗量按 $N_2 > NO_3^- > NO_2^- > NH_4^+$、尿素的顺序降低[46-48]。蓝藻共生体包括海洋硅藻半管藻和有根管藻的硅藻壳内但在原生质体之外的 Richelia,海洋硅藻角毛藻的硅藻壳表面的胶须藻,以及棒杆藻(光亮窗纹藻和棒杆藻)中淡水硅藻的非光合蓝藻内共生体[49-52]。

在适应低铁环境的硅藻基因型中,存在质子泵视紫红质。这使得硅藻在非常低的铁消耗下实现光能转换成为可能[53-55]。然而,基于质子泵视紫红质的能量转换每吸收一个光子泵出的质子数之比为 1.0,这与仅涉及 PSII 和 PTOX 的水-水循环(1.0)相同,但低于同时涉及 PSI 和 PSII 的线性电子流水-水循环(1.5),同时低于在硅藻中缺少质子泵 NADdh 的循环电子流(即 H^+:光子吸收比率为 2.0)[55-56]。线性电子流与 $NADP^+$ 还原耦合的附加能量存储与基于叶绿素的质子泵视紫红质的能量转换无关,因为这种能量转换只是 $NADP^+$ 还原的化学计量补充。此外,由于质子视紫红质可能位于质膜上,这意味着由质子视紫红质生成的质子梯度无法驱动 ATP 的磷酸化,除了在缺乏类囊体的蓝藻胶藻菌,光合作用、呼吸作用需要基于视紫红质的质子流以及 CF_0CF_1 ATP 合酶同时存在[56]。

9.4 铁储存

铁蛋白是古细菌、细菌和真核生物中广泛但非普遍的细胞内铁储存方式[57-61]。2012 年,Marchetti 等从分子遗传学证据显示,铁蛋白存在于针状硅藻圆柱拟脆杆藻、三角褐指藻、伪南极菱形藻、拟菱形藻和公海伪菱形藻属(严格意义上的硅藻纲)中,但不存在于中心硅藻假微型海链藻(中心硅藻目)中。然而,Groussman 等利用转录组学证据,在所有四类硅藻的一些物种中发现了铁蛋白。在严格意义上的硅藻纲的 18 个物种中有 17 个,脆杆藻纲 9 个物种中有 4 个,所有 6 个中型硅藻纲,以及 22 个中心硅藻纲中的 6 个。

铁蛋白的主要功能是在生长期储存铁,使铁供应饱和,在外源铁限制生长时利用储存的铁进行生长。Marchetti 等表明,与缺乏铁蛋白的大洋海链藻相比,含有铁蛋白的伪菱形藻在提取外部铁后经历了 4 轮多的细胞分裂(二元裂变)。Marchetti 等使用富含铁和缺铁细胞的铁含量来预测 4 株伪菱形藻和两株大洋海链藻可以支持的细胞分裂数量。Raven 认为存储容量(存储元素支持的倍增次数)与临界含量成反比(刚好足以提供最小倍增时间的内容);硅藻中铁的数据对其他水生光石营养体中的氮、磷、钾、镁和钙的数据支持有限。

缺乏铁蛋白的沿海硅藻假微型海链藻和威氏海链藻液泡中的铁储存似乎是利用多磷酸盐结合的铁[62]。在高营养(N,P)-低叶绿素(HNLC)的海洋硅藻中,缺铁海洋中存在过量的 P,这可能使 P 在短暂的铁有效期内用于铁储存。在另一种原生藻菌中,褐藻长囊水云是一种主要的铁储存化合物,类似于细菌和植物铁蛋白的无定形铁磷矿物核心[63]。

铁蛋白在绿枝藻纲海洋微浮游生物牛头芽孢杆菌中的另一个作用是光周期/暗周期铁稳态[58]。虽然含铁蛋白的羽状硅藻三角褐指藻的转录组和代谢组中的昼夜变化受铁可用性的调节[64-65],但并未提及铁蛋白。虽然上面讨论的证据清楚地表明,铁蛋白显然起到了铁储存的作用,但 Pfaffen 等指出,拟菱形藻多系列中的铁蛋白在

Fe^{2+} 氧化、铁缓冲以及铁储存中起作用。而 Pfaffen 等则认为,这种铁蛋白在 Fe^{2+} 氧化方面得到了优化,而不是在铁储存方面得到了优化,因此应在伪造优化声明的情况下予以考虑。Arosio 等综述了硅藻铁蛋白与非植物铁蛋白功能的关系。

9.5 结论

硅藻有几种吸收铁的机制。硅藻在光合作用和呼吸作用中的电子传递中具有典型的 Fe 催化剂范围,尽管在来自低 Fe 栖息地的硅藻中,含 Fe 细胞色素 c_6 在基因型上被无 Fe 质体蓝蛋白取代,并且无 Fe 黄素氧还蛋白在表型上取代了含 Fe 铁氧还蛋白。硅藻缺乏循环电子流的从铁氧还蛋白到 PQ 组分的 H^+ 泵送、高 Fe 途径,因此硅藻循环电子流的 H^+:电子比低于蓝藻、一些绿藻和大多数胚芽植物。一些来自低铁栖息地的硅藻具有不含铁的 H^+ 泵视紫红质,作为光能转换的额外手段。根据有限数量的研究,羽状硅藻将 Fe 储存为铁蛋白,而中心硅藻使用 FeS 复合物。没有关于 Fe 与硅藻相互作用的数据涉及淡水硅藻。

参考文献

[1] Archibald, F. (1983). Lactobacillus plantarum, an organism not requiring iron, FEMS Microbiol. Lett., 19(1), 29-32.

[2] Troxell, B., Xu, H., Yang, X. F. (2012). Borrelia burgdorferi, a pathogen that lacks iron, encodes manganese-dependent superoxide dismutase essential for resistance to streptonigrin, J. Biol. Chem., 287(23), 19284-19293.

[3] Williams, R. J. P., Fraústa da Silva, J. J. R. (1997). Natural Selection of the Chemical Elements. Oxford: Oxford University Press. p.672.

[4] Martin, J. H., Fitzwater, S. E. (1988). Iron deficiency limits phytoplankton growth in the north-east Pacific subarctic, Nature, 331(6154), 341-343.

[5] Martin, J. H., Gordon, R. M., Fitzwater, S. E. (1990). Iron in Antarctic waters, Nature, 345(6271), 156-158.

[6] Alverson, A. J., Beszteri, B., Julius, M. L., Theriot, E. C. (2011). The model marine diatom Thalassiosira pseudonana likely descended from a freshwater ancestor in the genus Cyclotella, BMC Evol. Biol., 11, 125.

[7] Martino, A. D., Meichenin, A., Shi, J., Pan, K., Bowler, C. (2007). Genetic and phenotypic characterization of Phaeodactylum tricornutum (Bacillariophyceae) accessions, J. Phycol., 43(5), 992-1009.

[8] Benoiston, A. S., Ibarbalz, F. M., Bittner, L., Guidi, L., Jahn, O., Dutkiewicz, S., et al. (2017). The evolution of diatoms and their biogeochemical functions, Philos. Trans. R. Soc. Lond. B Biol. Sci., 372(1728), 20160397.

[9] Kooistra, W., Gersomde, R., Medlin, D. G. (2007). The origin and evolution of the diatoms: their adaptation to a planktonic existence. In Evolution of Primary Producers in the Seas. (eds.). P. G. Falkowski and A. H. Knoll. Elsevier, Amsterdam. pp.207-241.

[10] Sutak, R., Botebol, H., Blaiseau, P. L., Léger, T., Bouget, F. Y., Camadro, J. M., et al. (2012). A comparative study of iron uptake mechanisms in marine microalgae: iron binding at the cell surface is a critical step, Plant Physiol., 160(4), 2271-2284.

[11] Raven, J. A. (2013). Iron acquisition and allocation in stramenopile algae, J. Exp. Bot., 64(8), 2119-2127.

[12] Peers, G., Quesnel, S.-A., Price, N. M. (2005). Copper requirements for iron acquisition and growth of coastal and oceanic diatoms, Limnol. Oceanogr., 50(4), 1149-1158.

[13] Caboj, A., Kosakowska, A. (2005). The marine diatom Cyclotella meneghiana Kützing as a produce of siderophore-like substances, Oceanolog Hydrobiologic Studies, 34, 57-72.

[14] Amin, S. A., Parker, M. S., Armbrust, E. V. (2012). Interactions between diatoms and bacteria, Microbiol. Mol. Biol. Rev., 76(3), 667-684.

[15] Croft, M. T., Lawrence, A. D., Raux-Deery, E., Warren, M. J., Smith, A. G. (2005). Algae acquire vitamin B12 through a symbiotic relationship with bacteria, Nature, 438(7064), 90-93.

[16] Cohen, N. R., Ellis, K. A., Burns, W. G., Lampe, R. H., Schuback, N., Johnson, Z., et al. (2017). Iron and vitamin interactions in marine diatom isolates and natural assemblages of the Northeast Pacific Ocean, Limnol. Oceanogr., 62(5), 2076-2096.

[17] Maranger, R., Bird, D. F., Price, N. M. (1998). Iron acquisition by photosynthetic marine phytoplankton from ingested bacteria, Nature, 396(6708), 248-251.

[18] Flynn, K. J., Stoecker, D. K., Mitra, A., Raven, J. A., Glibert, P. M., Hansen, P. J., et al. (2013). Misuse of the phytoplankton-zooplankton dichotomy: the need to assign organisms as mixotrophs within plankton functional types, J. Plankton Res., 35(1), 3-11.

[19] Ho, T. Y., Quigg, A., Finkel, Z. V., Milligan, A. J., Wyman, K., Falkowski, P. G., et al. (2003). The elemental composition of some marine phytoplankton, J. Phycol., 39(6), 1145-1159.

[20] Quigg, A., Irwin, A. J., Finkel, Z. V. (2011). Evolutionary inheritance of elemental stoichiometry in phytoplankton, Proc. Biol. Sci., 278(1705), 526-534.

[21] Marchetti, A., Maldonado, M. T., Lane, E. S., Harrison, P. J. (2006). Iron requirements of the pennate diatom Pseudo-nitzschia: Comparison of oceanic (high-nitrate, low-chlorophyll waters) and coastal species, Limnol. Oceanogr., 51(5), 2092-2101.

[22] Nagasaka, S., Yoshimura, E. (2008). External iron regulates polyphosphate content in the acidophilic, thermophilic alga Cyanidium caldarium, Biol. Trace Elem. Res., 125(3), 286-289.

[23] Honey, D. J., Gledhill, M., Bibby, T. S., Legiret, F. E., Pratt, N. J., Hickman, A. E., et al. (2013). Heme b in marine phytoplankton and particulate material from the North Atlantic Ocean, Mar. Ecol. Prog. Ser., 483, 1-17.

[24] Nunn, B. L., Faux, J. F., Hippmann, A. A., Maldonado, M. T., Harvey, H. R., Goodlett, D. R., et al. (2013). Diatom proteomics reveals unique acclimation strategies to mitigate Fe limitation, PLoS ONE, 8 (10), article e75653.

[25] Jensen, E., Clément, R., Maberly, S. C., Gontero, B. (2017). Regulation of the Calvin-Benson Bassham cycle in the enigmatic diatoms: biochemical and evolutionary variations on an original theme, Philos. Trans. R. Soc. Lond. B Biol. Sci., 372(1728), 20160401.

[26] Kikutani, S., Tanaka, R., Yamazaki, Y., Hara, S., Hisabori, T., Kroth, P. G., et al. (2012). Redox regulation of carbonic anhydrases via thioredoxin in chloroplast of the marine diatom Phaeodactylum tricornutum, J. Biol. Chem., 287(24), 20689-20700.

[27] Weber, T., Gruber, A., Kroth, P. G. (2009). The presence and localization of thioredoxins in diatoms, unicellular algae of secondary endosymbiotic origin, Mol. Plant, 2(3), 468-477.

[28] Hünken, M., Harder, J., Kirst, G. O. (2008). Epiphytic bacteria on the Antarctic ice diatom Amphiprora kufferathii Manguin cleave hydrogen peroxide produced during algal photosynthesis, Plant Biol. (Stuttg)., 10(4), 519-526.

[29] Winkler, U., Stabenau, H. (1995). Isolation and characterisation of peroxisomes from diatoms, Planta, 195(3), 403-407.

[30] Zamocky, M., Furtmüller, P. G., Obinger, C. (2008). Evolution of catalase from bacteria to humans, Antioxid. Redox Signal., 10(9), 1527-1548.

[31] Whittaker, J. W. (2012). Non-heme manganese catalase—the "other" catalase, Arch. Biochem. Biophys., 525(2), 111-120.

[32] Takahashi, H., Kopriva, S., Giordano, M., Saito, K., Hell, R. (2011). Sulfur assimilation in photosynthetic organisms: molecular functions and regulations of transporters and assimilatory enzymes, Annu. Rev. Plant Biol., 62, 157-184.

[33] Ken, C. F., Hsiung, T. M., Huang, Z. X., Juang, R. H., Lin, C. T. (2005). Characterization of Fe/Mn superoxide dismutase from diatom Thalassiosira weissflogii: cloning, expression, and property, J. Agric. Food Chem., 53(5), 1470-1474.

[34] Peers, G., Price, N. M. (2004). A role for manganese in superoxide dismutases and growth of iron-

deficient diatoms, Limnol. Oceanogr., 49(5), 1774-1783.
[35] Wolfe-Simon, F., Starovoytov, V., Reinfelder, J. R., Schofield, O., Falkowski, P. G. (2006). Localization and role of manganese superoxide dismutase in a marine diatom, Plant Physiol., 142(4), 1701-1709.
[36] Larkum, A. W. D., Szabo, M., Fitzpatrick, D., Raven, J. A. (2017). Cyclic electron flow in cyanobacteria and eukaryotic algae. In Photosynthesis and Bioenergetics. (eds.). J. Barber and A. V. Ruban. World Scientific. pp. 305-343.
[37] Strzepek, R. F., Harrison, P. J. (2004). Photosynthetic architecture differs in coastal and oceanic diatoms, Nature, 431(7009), 689-692.
[38] Geider, R. J., Roche, J., Greene, R. M., Olaizola, M. (1993). Response of the photosynthetic apparatus of Phaeodactylum tricornutum (Bacillariophyceae) to nitrate, phosphate or iron starvation, J. Phycol., 29(6), 755-766.
[39] Petrou, K., Trimborn, S., Rost, B., Ralph, P. J., Hassler, C. S. (2014). The impact of iron limitation on the physiology of the Antarctic diatom Chaetoceros simplex., Mar. Biol., 161(4), 925-937.
[40] Graff van Creveld, S., Rosenwasser, S., Levin, Y., Vardi, A. (2016). Chronic Iron Limitation Confers Transient Resistance to Oxidative Stress in Marine Diatoms, Plant Physiol., 172(2), pp. 00840.2016-.00840.2979.
[41] Bailleul, B., Berne, N., Murik, O., Petroutsos, D., Prihoda, J., Tanaka, A., et al. (2015). Energetic coupling between plastids and mitochondria drives CO_2 assimilation in diatoms, Nature, 524(7565), 366-369.
[42] Peers, G., Price, N. M. (2006). Copper-containing plastocyanin used for electron transport by an oceanic diatom, Nature, 441(7091), 341-344.
[43] Hippmann, A. A., Schuback, N., Moon, K. M., McCrow, J. P., Allen, A. E., Foster, L. J., et al. (2017). Contrasting effects of copper limitation on the photosynthetic apparatus in two strains of the open ocean diatom Thalassiosira oceanica, PLoS ONE, 12(8), e0181753.
[44] Groussman, R. D., Parker, M. S., Armbrust, E. V. (2015). Diversity and evolutionary history of iron metabolism genes in diatoms, PLoS ONE, 10(6), e0129081.
[45] Whitney, L. P., Lins, J. J., Hughes, M. P., Wells, M. L., Chappell, P. D., Jenkins, B. D. (2011). Characterization of putative iron responsive genes as species-specific indicators of iron stress in thalassiosiroid diatoms, Front. Microbiol., 2, article 234.
[46] Hattori-Saito, A., Nishioka, J., Ono, T., McKay, R. M. L., Suzuki, K. (2010). Iron deficiency in micro-sized diatoms in the Oyashio region of the Western subarctic Pacific during spring, J. Oceanogr., 66(1), 105-115.
[47] Kustka, A., Sañudo-Wilhelmy, S., Carpenter, E. J., Capone, D. G., Raven, J. A. (2003). A revised estimate of the iron use efficiency of nitrogen fixation with special reference to nitrogen fixation, with special reference to the marine cyanobacterium Trichodesmium spp. (Cyanophyta), J. Phycol., 39, 12-25.
[48] Twining, B. S., Baines, S. B. (2013). The trace metal composition of marine phytoplankton, Ann. Rev. Mar. Sci., 5, 191-215.
[49] Hilton, J. A., Foster, R. A., Tripp, H. J., Carter, B. J., Zehr, J. P., Villareal, T. A. (2013). Genomic deletions disrupt nitrogen metabolism pathways of a cyanobacterial diatom symbiont, Nat. Commun., 4(1), article 1767.
[50] Kneip, C., Voss, C., Lockhart, P. J., Maier, U. G. (2008). The cyanobacterial endosymbiont of the unicellular algae Rhopalodia gibba shows reductive genome evolution, BMC Evol. Biol., 8(1), article 30.
[51] Nakayama, T., Kamikawa, R., Tanifuji, G., Kashiyama, Y., Ohkouchi, N., Archibald, J. M., et al. (2014). Complete genome of a nonphotosynthetic cyanobacterium in a diatom reveals recent adaptations to an intracellular lifestyle, Proceed. Natnl. Acad. Sci. USA, 111(31), 11407-11412.
[52] Wouters, J., Raven, J. A., Minnhagen, S., Janson, S. (2009). The luggage hypothesis: comparison of two phototrophic hosts with nitrogen-fixing cyanobacteria and implications for analogous life strategies for kleptoplastids/secondary symbioses in dinoflagellates, Symbiosis, 49(2), 61-70.
[53] Marchetti, A., Schruth, D. M., Durkin, C. A., Parker, M. S., Kodner, R. B., Berthiaume, C. T., et al. (2012). Comparative metatranscriptomics identifies molecular bases for the physiological responses of phytoplankton to varying iron availability, Proc. Natl. Acad. Sci. U.S.A., 109(6), E317-E325.
[54] Marchetti, A., Catlett, D., Hopkinson, B. M., Ellis, K., Cassar, N. (2015). Marine diatom proteorhodopsins and their potential role in coping with low iron availability, ISME J., 9(12), 2745-2748.

[55] Raven, J. A. (2009). Functional evolution of photochemical energy transformations in oxygenproducing organisms, Functional Plant Biol., 36(6), 505-149.
[56] Larkum, A. W. D., Ritchie, R. J., Raven, J. A. (2018). Living off the sun: Chlorophylls, Bacteriochrophylls and Rhodopsins, Photosynthetica.
[57] Arosio, P., Carmona, F., Gozzelino, R., Maccarinelli, F., Poli, M. (2015). The importance of eukaryotic ferritins in iron handling and cytoprotection, Biochem. J., 472(1), 1-15.
[58] Botebol, H., Lesuisse, E., Šuták, R., Six, C., Lozano, J.C., Schatt, P., et al. (2015). Central role for ferritin in the day/night regulation of iron homeostasis in marine phytoplankton, Proc. Natl. Acad. Sci. U.S.A., 112(47), 14652-14657.
[59] Marchetti, A., Parker, M. S., Moccia, L. P., Lin, E. O., Arrieta, A. L., Ribalet, F., et al. (2009). Ferritin is used for iron storage in bloom-forming marine pennate diatoms, Nature, 457(7228), 467-470.
[60] Masuda, T., Yamamoto, A., Toyohara, H. (2015). The iron content and ferritin contribution in fresh, dried, and toasted nori, Pyropia yezoensis, Biosci. Biotechnol. Biochem., 79(1), 74-81.
[61] Seekbach, J., Iron content and ferritin in leaves of iron treated Xanthium pensylvanicum plants. Plant Physiol., 44(6), 816-820, 1969.
[62] Nuester, J., Vogt, S., Twining, B. S. (2012). Localization of iron within centric diatoms of the genus Thalassiosira, J. Phycol., 48(3), 626-634.
[63] Böttger, L. H., Miller, E. P., Andresen, C., Matzanke, B. F., Küpper, F. C., Carrano, C. J. (2012). Atypical iron storage in marine brown algae: a multidisciplinary study of iron transport and storage in Ectocarpus siliculosus, J. Exp. Bot., 63(16), 5763-5772.
[64] Smith, S. R., Gillard, J. T. F., Kustka, A. B., McCrow, J. P., Badger, J. H., Zheng, A. R., et al. (2016). Transcriptional orchestration of the global response of a model pennate diatom to light cycling under iron limitation to diel light cycling under iron limitation, PLoS Genet., 12, e1006490.
[65] Pfaffen, S., Bradley, J.M., Abdulqadir, R., Firme, M.R., Moore, G.R., Le Brun, N.E., et al. (2015). A diatom ferritin optimised for iron oxidation but not iron storage, J. Biol. Chem., 290(47), 28416-28427.

第 10 章
硅藻与其他光合自养生物共生

Rosalina Stancheva，Rex Lowe

10.1 简介

　　de Bary 提出的术语"共生"，即"不同命名的生物生活在一起"，是一个中性词,被用于描述生物之间的各种联系,今天更狭义地用于描述那些双方都受益的关系。通常,单细胞或多细胞宿主为共生体提供了获取营养和稳定环境的途径,并反过来受益于共生体的代谢产物[1]。共生体的范围从临时的兼性联合,其中共生体在每一代宿主中重新定居到永久的专性关系,其中至少一个伙伴不是自主的,共生体垂直传递到下一代宿主,因此两种生物一起进化以适应彼此[2]。真核细胞中质体起源的内共生体理论表明了共生的进化重要性[3]。Mereschkowski 的理论在当时很大程度上被忽视了,但后来被 Margulis 重新提出。

　　硅藻是最多样化的单细胞真核藻类之一,其进化涉及捕获红藻质体,这意味着硅藻起源于中生代早期[4]。硅藻几乎分布在所有淡水和海洋生态环境中,占全球初级产品的近四分之一和海洋初级产品的 40%[5]。它们在多变的环境条件下具有很强的竞争力,部分原因在于它们与其他生物体形成关联的灵活性。一些硅藻属含有细胞内固氮蓝藻,这对它们的营养有益。从进化和生态学的角度来看,这种内共生关系是很重要。固氮的蓝藻与主要属于棒杆藻科家族的淡水硅藻和与根管藻科或半管藻科的海洋成员的细胞内共生代表了研究基因组进化的分子适应的新模型系统,这些进化伴随着从自由生活到细胞内存在的转变[6-7]。多年来,出现了许多关于这些硅藻-蓝藻共生体的生物学、进化和生态学的评论[8-12]。在这里,我们将根据其生态影响总结硅藻-蓝藻共生体,因为它们在贫营养生态系统中的氮供应中发挥着重要作用。此外,我们将重点介绍硅藻在低营养生态系统中与淡水和海洋大型藻类的附生、内生和内生"松散共生"关系中的一些生态方面。最后,将简要讨论硅藻与甲藻的三级内共生。其他有趣的硅藻共生体,例如与较大的有孔虫[13]以及硅藻与无脊椎动物和脊椎动物的表生共生[14-15]不是本次总结的主题,因为可以获得关于这些主题全面的最新的综述[16-18]。

10.2 具有固氮球状蓝藻内共生体的硅藻

Drum 和 Pankratz 首先报道内共生蓝藻和硅藻棒杆藻属的联系，他们描述了卵形夹层的细胞微粒，宽度为 4~6 μm，长为 5~7 μm，并且利用五层壁与硅藻细胞质隔开[19]。这些内含物既不是核糖体也不是油体，它们可能是独特的细胞器，其功能未知的，或可能是经过修饰的球状蓝藻。Geitler 将窗纹藻、棒杆藻和布伦齿藻中的内共生菌称为"sphäroidkörper"，怀疑它们是蓝藻[20]。Lowe 等证明了窗纹藻中蓝藻内共生体的精细结构[21]。在形态上与棒杆藻中的内共生体相似。这种共生似乎是普遍存在的，在属于 *Rhopalodia* 的棒杆藻和 *Epithemia* 属的所有硅藻物种中都存在，它们主要分布在淡水中，但在咸水和海洋环境中也有一些丰富的分类群[22]。然而，文献中蓝藻内共生体的插图仅适用于淡水类群（见表 10.1），这增加了它们在海洋物种中存在的不确定性[11]。因此，我们在这里说明了布雷比松弯棒藻 Krammer 和小肌弯棒藻(Kütz.) O. Müll. 中的内共生体（见图 10.1），它们主要栖息在半咸水和海洋沿海水域[11,23]。

表 10.1 文献中记载的蓝藻类细菌内共生体

硅藻类宿主	CE 宿主细胞	CE 尺寸范围/μm	图片
膨胀短缝藻(Kütz.) Bré	2~6 2~6	3.2~7.6×5.5~7.7 5~6×5.3~6.6	N/A 9 (LM)
萨克森附着短缝藻变种(Kütz.) R. M. Patrick	4~8 (16)	N/A 3.2~4×4~50.2	116 (D) 1 (D)
附着短缝藻变种(Kütz.) R. Ross	4~8 (16)	3.2~4×4~50.2	2 (D)
鮋短缝藻 Kütz	1~2 (4) 1~2	2.6~3.2 3~7.1×3.9~7.9	3A-D (D) 1E (LM)
膨胀果壳藻(Ehrenb.) Kütz	4~8 (16) 2~12	5×5~6 4.1~8.1×5.6~8.7	4 (D) 1F (LM)
布雷比松弯棒藻 Krammer	1~2	2.7~3.8×3.6~4.8	1~4 (LM)
凸起弯棒藻(Ehrenb.) O. Müll.	2 1~2 (4) 4 2 1~4 1~2 2 1~2	4~6×5~7 4~5×5~6 N/A N/A 4.4~6.3×5.5~13 N/A N/A 4.9~5.5×6~80.7	1, 4~7 (TEM) 3E (D) 1, 2 (LM, TEM) 3~2 (LM) 1A (LM) 1 (CLM), 2 (TEM) 1E (LM) 10 (LM)

（续表）

硅藻类宿主	CE 宿主细胞	CE 尺寸范围/μm	图片
凸起小弯棒藻（Ehrenb.）O. Müll	2	4~6×5~7	2，3 (TEM)
小肌弯棒藻（Kütz.）O. Müll.	1~2 (4)	3.2~4.8×4~60.4	5~8 (LM)
蛭形弯棒藻 O. Müll.	N/A	N/A	片状2：2，5，7 (LM)
蠕虫形弯棒藻 O. Müll.	N/A	N/A	片状3：6 (LM)
凡氏四月藻（Brun）Hamsher, Graeff, Stepanek and Kociolek	1~2 1~2	N/A 2.8~3.4×3~40.8	118 (D) N/A

CE—蓝藻内共生体；D—绘图、LM—光学显微镜显微照片；CLM—共聚焦光学显微照片；透射电子显微镜显微照片；N/A—没有

棒杆藻物种中蓝藻的存在一定是一个相对较早的进化事件，因为它似乎只出现在两个密切相关的属中。Nakayama 等比较了两种上皮藻和一种红藻属的蓝藻内共生体的 DNA 序列，并得出结论："红藻科硅藻中的内共生体是由红藻硅藻的共同祖先获得的，并在宿主物种形成过程中保留下来"[24]。此外，Hamsher 等研究了苏门答腊岛齿状体属物种的瓣膜形态，包括从爪哇描述的 D. vanheurckii[25]。他们得出的结论是，这些分类群与齿状体的模式物种显著不同，并将它们转移到棒杆藻的新属 Tetralunata。这证实了 Nakayama 的结论，即固氮蓝藻内共生体是棒杆藻属所特有的。事实上，在凡氏齿形藻（现在的四月藻属）细胞中存在"sphäroidkörper"的唯一证据是 Geitler 的一幅画。在有性生殖过程中，已经观察到

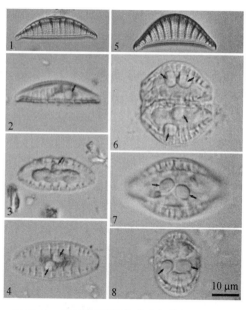

图 10.1 含有球状蓝藻内共生体的 *Rhopalodia brebissonii*（1~4）和肌状棒杆藻（5~8）的光学显微镜图像

蓝藻内共生菌垂直传播到下一代硅藻宿主的现象，这种现象发生在几种上皮细胞和棒杆藻（见图 10.2），最终发生在凡氏齿形藻[26]。

内共生体随后成为了一些生理学和分子研究的目标[27-30]。这些研究表明，内共生体与自由生活的固氮蓝藻蓝小球藻 ATCC 51142 密切相关，它能固定元素氮（N_2），尽管存在类囊体，但其固定元素氮（N_2）且光合作用无活性。在大多数其他的内共生实例中，藻类向宿主提供还原碳（C）产物[31]，但在棒杆藻科中，蓝藻是专性共生体，从

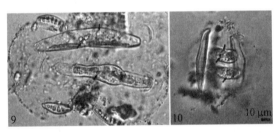

图 10.2 *Epithemia adnata*(9)和 *Rhopalodia gibba*(10)有性繁殖细胞的光学显微镜图

硅藻获得它们的能量需求,同时向硅藻宿主提供固定的氮。硅藻细胞内固氮的内共生体的数量和生物体积是可变的,并且似乎是外部无机氮源可用性的函数。DeYoe 等在生长室实验中观察到,硝酸盐和 N/P 比值的降低导致膨胀伞藻和凸腹囊藻内共生细胞数量和生物体积的增加[32]。因此,Stancheva 等记录了粘附伞藻、鮈伞藻、膨胀伞藻和凸腹囊藻随着美国南加州溪流中环境硝酸盐浓度的增加的变化[33]。此外,Stancheva 等为环境流样本中上位症内共生体中固氮酶基因的表达提供了分子证据。

R. gibba 内共生体在基因组减少的过程中显示出单个基因的丢失[29],这导致了依赖于硅藻细胞代谢物的内共生体的永久和专性关系。Adler 等分析了目前 *R. gibba* 的基因组和形态学数据[8]。结果认为,与其他研究内共生体向细胞器转化的模型系统相比,这种共生关系在分子和形态特征上处于相对较早的阶段。同样,*E.* 和凸腹囊藻揭示了他们的基因组具有明显还原性,虽然缺乏大多数参与光合作用的基因,但具有与固氮相关的保守基因[7,34]。然而,比较基因组分析表明,*R. gibberula* 在与光合相关的基因丢失方面表现出较慢的进展与内共生体的代谢途径比较。这表明,红藻藻类硅藻的专性内共生体基因组之间存在一定程度的多样性[7]。根据 Cavalier-Smith 和 Lee 的研究,共生发生不仅涉及共生基因的丢失,还包括共生基因转移到宿主的细胞核中,随后基因产物转移到细胞器中,这需要蛋白质输入系统的进化。宿主控制的细胞器从宿主的细胞质中输入细胞核中编码的大部分必需的蛋白质[35]。尽管蓝藻内共生体在藻纲硅藻中的发育是还原性的,但它们的进化仍处于细胞器形成的早期阶段[7]。

在低氮的淡水和微咸水环境中,窗纹藻和棒杆藻常常是底栖生物群落的主要组成部分,它们与固氮蓝藻细菌一起为生态系统提供氮素。根据 Howarth 等的说法,底栖动物 N_2-少营养系统中的固定是由蓝藻介导的,与中营养和富营养湖泊和河口中的异养细菌相反。底栖动物固氮是许多少营养湖泊和泻湖的主要氮来源,尽管固定率适中,但其他氮输入往往较低[36]。Fairchild 和 Lowe 通过在美国密歇根州道格拉斯湖海岸区部署营养扩散基质(NDS)来研究硅藻的营养生态位。NDS 释放出由硝酸盐、磷酸盐和硝酸盐组成的钾盐。他们发现释放磷酸盐的 NDS 刺激了以鱼腥藻、棒杆藻和窗纹藻为主的藻类组合,并得出结论,当磷酸盐水平相对于硝酸盐高时,固氮物种(包括棒杆藻成员)在底栖藻类群落中占主导地位。在美国西部的两条河流中,也对硝酸盐有效性和上皮增生之间的关系进行了实验研究。在加利福尼亚州南叉鳗鱼河的夏末孵化过程中,在环境少氮(约为 7 $\mu g \cdot L^{-1}$)河水中孵化的绿藻刚毛藻丝状体上的附生植物群落主要由附生植物物种组成(见图 10.3),而在实验中富氮(约为 300 $\mu g \cdot L^{-1}$)以非 N_2 为主的固定硅藻的室内附生植物主要由非固氮硅藻极小曲丝藻和谷皮菱形藻主导[37]。同样,美国亚利桑那州沙漠溪流中早期的演替群落也不富集,实验底物由窗纹藻属主导[38],Grimm 和 Fisher,Bahls 和 Weber 在美国蒙大拿州溪流的硅藻组合的调查中发表了确信证据[39-40]。他们记录 E. Sorex 主导的流组合中,平均 N/P 比为 3.2,远低于 Redfield 的比率(N/P=16)[41],这表明 N 对许多藻类物种限制。同样,Stancheva 等表明,在南加州河流中,在 N 梯度的低端和 N/P 比低于 15 的情况下,随着环境无机 N 浓度的增加,窗纹藻和棒杆藻分类群的相对比例下降[33]。

图 10.3 附生窗纹藻属和 E. Cladophora glomerata 的刚毛藻 SEM 图

这种共生关系对围生植物群落结构的影响及其对水生食物网的重要性已经被一些研究者记录下来[42-44]。位于加州北部的鳗鱼河,是美国西部受地中海气候影响的流域的典型河流[44]。鳗鱼河和太平洋西北部的许多水生生态系统一样,都是由最近上升的海相沉积沉积物组成的分水岭[45]。这些岩石以及太平洋西北部其他盆地的火山成因的岩石产生的土壤和径流富含磷,但有效氮相对较低,导致光自养生物的氮限制。石藻科的氮固定硅藻是这些水生生态系统中食物网的重要成员。Power 等研究了附生生活在绿藻刚毛藻上的窗纹藻对增强鳗鱼河食物网的作用[43]。在鳗鱼河上的冬季风暴过后,刚毛藻(它主导着夏季绿藻的生长)在几米长的溪流中繁殖。刚毛藻质地粗糙的细胞壁为鼠形窗纹藻和 E. turgida 提供了稳定的栖息地,每个细菌都有内共生的蓝藻细菌。他们发现,捕食有附位植物的昆虫(摇蚊科)的出现率是捕食没有附位植物优势的昆虫或其他缺乏这些附生植物的丝状藻类的昆虫的 3~25 倍。此外,来自这些藻团产生的昆虫的生物量比来自河流中缺乏密集附位种群的地区的昆虫的生物量大 8~10 倍,水生昆虫依次进入水生和陆地的食物网[46]。Furey 等进一步揭示了优势食草动物(理查氏假摇蚊)与附位症之间的密切营养关系,证明了这种资源对食草动物成功的重要性[42]。实际上,需要对硅藻细菌共生的生态方面进行更多研究,因为在生态系统层面上尚不很好地了解淡水底栖食物网中蓝细菌 N_2 固定的重要性,尤其是减轻系统范围内的氮限制的潜力[47]。

在海洋环境中,在浮游链形成的非球状硅藻中观察到类似的胞内球藻蓝藻[48]。硅藻宿主细胞包含 20~30 个直径为 2.5~3.5 μm 的蓝藻内共生细胞。系统发育分析表明,气候内共生体与自由生活的重氮营养蓝藻隐杆藻 ATCC 5114 有密切联系[48]，但由于超微结构差异，例如大淀粉颗粒的存在，不太可能与棒杆藻内共生体相同[10]。最近对蓝藻内共生菌部分 16SrRNA 序列的重新评价，显示与自由生活的重氮营养蓝藻沃氏 8501（水氏蓝藻的同义词）的完整序列一致[9]。此外，作者还证明了 N_2-由蓝藻内共生体的固定和随后的 N 转移到硅藻细胞[9]。偶尔，内共生球藻蓝藻已被报道在其他海洋链形成的浮游硅藻，像链菌和新链菌，被认为与内共生体相似[12]，但尚未进一步研究。

10.3 具有固氮丝状异细胞蓝藻内共生体的硅藻

硅藻与细胞内丝状蓝藻共生，其中一些包含异型细胞，在海洋浮游环境中进化。据报道，含有丝状蓝藻内共生体的海洋硅藻属于非羽状硅藻树更原始的分支，其分类多样性最近被总结[11]。然而，许多这样的共生关系只是偶尔被报道的，具体研究的还很少。最具特征和生态意义的共生关系是胞内丝状杂胞蓝藻之间的共生关系。胞内植生藻及其浮藻浮游宿主属于根管藻、半管藻、角毛藻、辐杆藻和几内亚藻[12]。硅藻与植生藻的共生关系通常在缺乏营养的热带和亚局部海洋中非常丰富，它们在氮循环中发挥作用，而很少在沿海水域被发现[12]。

蓝藻内共生体 R. intracellularis 细胞内有一个短的毛状体，首次在根瘤细胞内观察到[49]。典型的一到几个植生藻毛状体位于大的宿主根瘤菌细胞的顶端（见图10.4）在细胞膜和硅细胞壁之间的周质空间[50]。在半毛的小细胞中，宿主的细胞质团可以掩盖 Richelia 毛状体，用明亮的视野显微镜几乎看不见。荧光显微镜显示每个半球细胞有一个或两个植生藻内共生体（见图 10.5），但它们的细胞位置尚不完全清楚[6]。主要的硅藻-植生藻相互作用被认为是向硅藻细胞提供固定的氮，这是由于存在特殊的异质细胞，其中氮气的固定是局部的。一些证据证明了氮由内共生的植生藻进行的固定，如，在现场和培养条件下，乙炔的还原率较高[50-51]；分子遗传学研究[52]和氮通过高分辨率纳米级二级离子质谱测量固定的氮产物的固定和转移到硅藻宿主细胞的直接实验证据[9]。在后一项研究中，作者估计植生藻固定的氮比自身生长所需的多 81%~744%，并且高达 97.3% 的固定氮转移到硅藻伙伴。

图 10.4 根管藻属（12）和半管藻属（13）硅藻细胞内共生蓝藻胞内植生藻的光学显微镜图像
明视野显微镜显示硅藻宿主细胞与内共生体（箭头）；藻红蛋白荧光显示具有末端杂细胞的放大的蓝细菌毛状体

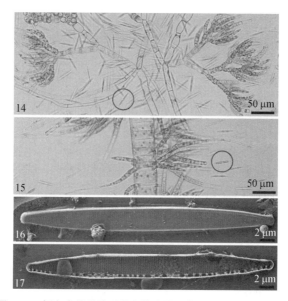

图 10.5　栖息在优美胶毛藻和簇生竹枝藻粘液中的谷皮菱形藻（圆圈细胞）的光学显微镜和 SEM 图

在宿主细胞分裂过程中,内共生体从宿主硅藻垂直转移到子细胞[50,53-54]。然而,附生[12,55-56]和自由生活的毛状体的植生藻已被观察到[9,56-58],这表明内共生体的额外水平传播。Hilton 等认为,一些关于自由生活的植生藻细丝[59-60]导致硅藻瓣膜破裂。宿主细胞内的植生藻毛状体分裂与硅藻细胞分裂不相连,即使在氮限制条件下培养也会导致根瘤细胞失衡[54]。由于有报道称根细胞只有 94%~98% 的根细胞具有氰内共生体[53],因此认为水华期间氮限制的减少可能允许非共生的根菌细胞持续一段时间[50]。同样,用荧光显微镜观察在夏威夷收集的大约 80% 的半细胞,以及来自北大西洋西南部的超过 98% 的半细胞含有含有植生藻内共生体[61-62]。因此,这种共生关系对双方来说都不是必须的。

另一种丝状杂细胞蓝藻,根顶藻,被记录为硅藻宿主上的附生植物(通常在毛藻上),包含植生藻内共生体。两种蓝藻细菌在形态上相似,导致分类学混淆。植生藻和眉藻的综合形态学和分子分析[52,63]结果表明,它们在每个硅藻组合的系统发育、毛状体形态、位置和每个硅藻伙伴的数量上存在差异,表明它们是不同的类群。最近的一项研究表明,胞内共生的植生藻和胞外根瘤菌菌株的基因组大小和含量存在很大差异,这表明细胞位置可能决定了不同的进化路径和维持共生的不同机制[3]。胞内植生藻基因组显示减少,其特征在于基因组大小、编码百分比和 GC 含量的减少,表明蓝藻依赖于宿主,这与密切相关的胞外眉藻的基因组相反,后者与自由生活的异细胞蓝藻的基因组更相似[3]。作者提出,植生藻-硅藻共生关系可能对固氮共生关系的进化和调控植物-蓝藻相互作用具有潜在意义。

浮游硅藻-杂细胞蓝藻共生在海洋表面的生物地球化学中发挥着重要的作用,但仍被低估,仅被用来支持营养不良的热带水域的初级生产。一般来说,植生藻与根菌的关联更常见于北太平洋中央环流,而与半环的共生在大西洋、加勒比海和巴哈曼群岛有更高的丰度[9]。硅藻-蓝藻共生体在世界上的公海中呈斑片状分布,这可能是因为硅藻宿主需要硅来构建它们的硅藻壳性[64]。迄今为止,这些组合的最大密度是在亚马逊河羽流中,该亚马逊河羽流向热带北大西洋输送高浓度的硅酸盐和其他营养物质[65]。尽管硅藻分布不均匀和丰富,但有证据表明,与 *Rhichelia* 和 *Calothrix* 的共生在输出碳和氮方面效率很高[9,65]。因此,它们可能是它们发生的生物泵中非常重要的组成部分[64]。在硅藻寄主根茎藻和半毛藻水华期间,这种共生关系可满足北太平洋中部环流总的浮游植物氮需求的 15%[62]。据估计,在加利福尼亚湾中部盆地,固定提供了高达 35%~48% 的浮游植物氮需求[58]。在亚热带的北大西洋,Carpenter 等遇到了广泛的植生藻组合,并估计氮的供应在共生体中的固定量超过了来自深水的硝酸盐通量[66]。在地中海缺乏营养的水域,但缺乏大规模的水华和低 N_2-固定率归因于 P 饥饿条件[67]。居住在印度洋的植生藻和眉藻也被记录在印度洋[68]、红海[69]和中国西部海洋[56],这使他们成为最广泛的海洋硅藻-蓝藻组合。

10.4　附生、内源性和内生附生硅藻

硅藻对大型藻类和其他植物基质的附生作用对定居者有许多好处,尽管这两种生物之间的化学相互作用相当复杂,包括化感作用。在淡水生态系统中(与海洋环境相反),目前还没有公认的藻类定居者只对一种特定的植物宿主有特定的偏好[70],这将藻类附生性归类为两种生物之间的非专性关联("松散共生")。然而,有证据表明,植物寄主是附生藻类的主要营养来源,特别是在少营养和中营养生境中,自然植物的附生植物群落与人工植物明显不同[71]。研究发现,大型植物可以将溶解的有机 C 和少量的组织 P 浸出附生植物[2]。相比之下,在富营养化条件下,松散附着的藻类附生植物不表现出基质偏好,因此水体可能在供应其营养物质方面发挥了主要作用[70]。

谷皮菱形藻,是所讨论的生态学区别的一个很好的例子。这种生长速度快的富营养化运动物种[72]栖息在刚毛藻附生生物膜的顶层。在环境氮和磷浓度升高的情况下,可能从水中获得营养物质[37]。有趣的是,据报道,谷皮菱形藻和其他菱形藻植物也被报道为内源藻,栖息在大型绿藻的黏液中[73],同样,我们在南加州低营养非多年生溪流的优美胶毛藻和簇生竹枝藻的黏液中也发现谷皮菱形藻(见图 10.5)。绿藻胶毛藻和竹枝藻具有位于长根尖毛细胞表面的碱性磷酸酶,它催化有机磷化合物的水解,并释放周围粘液中的正磷酸盐,可供细胞摄取。当有机底物浓度较高时,谷皮菱形藻能够在高辐照度下以异养性补充 C 固定[74]。在浮游球藻类蓝藻铜绿微囊藻的粘液菌落中也观察到谷皮菱形藻[73,75-76]。我们可以合理地假设,有机底物摄取和代谢的机制是被激活的。在环境无机营养限制下居住藻类粘液时的谷皮菱形藻。菱形藻与产生粘液的藻类(如海洋浮游褐藻棕囊藻),可能与菱形藻的异养倾向有关。

已经证明,许多在亚优光照条件下生长的藻类能够利用不同范围的有机 C 源[74]。事实上,这些有机和无机化合物,可以在低营养条件下由硅藻进行异养代谢,实际上可以被其他藻类和蓝藻细菌排出体外。在生物评价研究中没有考虑藻类营养相互作用,这可能会影响水质评价的精度。例如,Snoeijs 和 Murasi 报道了居住在波罗的海固氮蓝细菌的粘液菌落中的硅藻的共生关系[77],他们认为,硅藻可能通过保护防止掠过和物理干扰而获益,利用蓝藻粘液作为运动的基质,并作为从河流细胞排出的无机和有机营养物质的来源。须藻属成员能够在高无机磷供应时期固定大气氮,并在受磷的条件下形成长无色的多细胞毛,这是利用有机磷酸盐的磷酸单酯酶活性位点[78]。显然,营养交换和适当的光照条件有利于硅藻的硅藻黏膜,但宿主是否如此尚不清楚,研究者认为这些硅藻组合是共生的,但对硅藻和宿主来说都是暂时的和可选的。

海洋环境中硅藻和海藻之间的内生性和更专门化的联系可能为硅藻提供营养和保护作用。例如,在褐藻的容器和顶端的细胞间物质中观察到泡叶藻[79-80]。其他附生硅藻,有时记录与叶内舟形藻宿主体内的内生菌[79-80]。黄褐藻和黄褐藻在红藻的粘液中内生生长[81],但也在沉积物中被发现[23,82]。金角藻和长葵生活在红海藻腔节藻的节间。*Pseudogomphonema* sp. 在海洋红藻和 *N. decipiens* 的囊虫囊内已有记录,并分散在整个叶片的皮质层和髓质层中[83]。*Gyrosigma coelophilum* 和 *Pseudogomphonema* sp. 均部分异养,接收低光强度[83-84]。

内生硅藻与其宿主之间关系的确切性质尚不清楚。然而,上面列出的内生硅藻具有发育良好的色素叶绿体和富含硅的硅藻壳,没有结构还原的迹象。它们的细胞在宿主体内分裂,但目前尚不清楚它们是在每一代宿主中重新定植于宿主并保持它们的自主性,还是垂直传播给下一代宿主。在 *N. dumontiae* 中,由于硅藻经常被记录为在宿主外自由生活,因此似乎可以多次重新定殖宿主。建立关系的机制,和穿透表面从一个宿主菌体传播到另一个宿主是未知的。似乎居住在海藻生殖结构中的 *N. dumontiae* 和 *Pseudogomphonema* sp. 似乎可能存在垂直传播。硅藻细胞可以与孢子一起分布到新的发育中的植物中,因此,硅藻的亲缘关系可能是必然的。据报道,*Pseudogomphonema* sp. 大量内生生长对于 *N. endophytica* 而言可能是消极的,但不会导致形态变化(即宿主菌体的扭曲),也没有达到寄生水平[83]。

10.5 鞭毛藻的藻内共生体

鞭毛藻拥有各种各样的内共生体,鞭毛藻宿主和硅藻内共生体之间的三级内共生导致了一组被称为"恐龙藻"的鞭毛藻[85]。鞭毛藻具有独特的核和线粒体冗余,它们来自三级内共生事件中两个进化上不同的真核生物谱系。不同的鞭毛藻属共生菌宿主来自至少 5 个不同的真核生物门的内共生体[86]。硅藻内共生体的包含发生在许多不同的甲藻类群中。在大多数例子中,内共生硅藻被还原并修饰为一个具有功能的固碳质体,具有典型的硅藻色素,而不是鞭毛藻。Inagaki 等的研究表明,两种鞭毛藻（*Peridinium balticum* 和 *Peridinium foliaceum*）内部的硅藻内共生体相似,表明内共生体是在这些多甲藻"姐妹种"分化之前获得的[87]。这些"姐妹种"需要进一步的分析,因为它们已经被转移到不同的属。此外,内共生体是一种底栖硅藻的底栖物种[88],他们讨论了浮游鞭毛藻和底栖硅藻的栖息地并置,特别是建立了浮游鞭毛鞭毛藻和海底硅藻之间的共生关系,他们认为宿主和共生体的环境相关的生命策略阶段可能对硅藻/鞭毛藻结合的形成至关重要。*Galeidinium rugatum* M. Tamura et T. Horig 也被证明含有一种硅藻内共生体,似乎与羽状硅藻、圆柱藻属密切相关,这是一种典型的底栖硅藻物种[89]。最近的系统发育和叶绿体基因组分析再次证实了这些内共生体的羽状硅藻祖先与三角褐指藻[90]和菱形藻[91]密切相关。这进一步表明,底栖内共生体可能更容易被在底栖生境觅食的甲藻所利用。

在淡水生态系统和热带海洋中,营养物质往往是初级生产的关键限制因素,在那里经常观察到硅藻共生。营养限制似乎是硅藻的主要驱动因素,它通过不同的代谢途径与其他光致营养生物建立胞内和胞外关系,以获得营养效益。硅藻共生关系的范围从暂时共存,允许由其他藻类产生的营养化合物的异养同化,再到使细胞内氮的永久强制性积累的固氮蓝藻细菌,最终将固定的氮产物转移到硅藻宿主细胞中。因此,在硅藻与其他光自生营养生物之间建立一种共生关系可能被认为是对少营养生境中生活的一种生态适应。

参考文献

[1] Schönfeld, B. I. K. (2012). The pattern and processes of genome change in endosymbionts old and new. A thesis presented in partial fulfillment of the requirement for the degree of Doctor of Philosophy in Evolutionary Biology, Institute of Molecular BioSciences, Massey University, New Zealand., https://mro.massey.ac.nz/handle/10179/4719.

[2] Moeller, R. E., Burkholder, J. M., Wetzel, R. G. (1988). Significance of sedimentary phosphorus to a rooted submersed macrophyte (Najas flexilis) andits algal epiphytes, Aquatic Botany, 32(3), 261-281.

[3] Merezhkowsky, C. (1905). Ueber Natur und Ursprung der Chromatophoren im Pflanzenreiche, Biologisches Centralblatt, 25, 595-596.

[4] Medlin, L. K. (2016). Evolution of the diatoms: major steps in their evolution and a review of the supporting molecular and morphological evidence, Phycologia, 55(1), 79-103.

[5] Falkowski, P. G., Barber, R. T., Smetacek, V. (1998). Biogeochemical controls and feedbacks on ocean primary production, Science, 281(5374), 200-206.

[6] Hilton, J. A., Foster, R. A., Tripp, H. J., Carter, B. J., Zehr, J. P., Villareal, T. A. (2013). Genomic deletions disrupt nitrogen metabolism pathways of a cyanobacterial diatom symbiont, Nat. Commun., 4(1), 1767.

[7] Nakayama, T., Inagaki, Y. (2017). Genomic divergence within non-photosynthetic cyanobacterial endosymbionts in rhopalodiacean diatoms, Sci. Rep., 7(1), 13075.

[8] Adler, S., Trapp, E. M., Dede, C., Maier, U. G., Zauner, S. (2014). Rhopalodia gibba: The first steps in the birth of a novel organelle? In Endosymbiosis. (ed.). W Löffelhardt (Wien: SpringerVerlag. pp. 167-179.

[9] Foster, R. A., Kuypers, M. M. M., Vagner, T., Paerl, R. W., Musat, N., Zehr, J. P. (2011). Nitrogen fixation and transfer in open ocean diatom-cyanobacterial symbioses, Isme J., 5(9), 1484-1493.

[10] Janson, S. (2002). Cyanobacteria in symbiosis with diatoms. In Cyanobacteria in Symbiosis. (eds.). A. N Rai, B Bergman and U Rasmussen. Dordrecht: Kluwer Academic Publ. pp. 1-10.

[11] Kociolek, J. P., Hamsher, S. E. (2017). Diatoms: By, with and as Endosymbionts. In Algal and

cyanobacteria symbioses. (eds.). M Grube, J Seckbach and L Muggia. London: World Scientific Publishing Europe Ltd. pp. 371-399.

[12] Villareal, T. A. (1992). Marine nitrogen-fixing diatom-cyanobacteria symbioses. In Marine pelagic cyanobacteria: Trichodesmium and other diazotrophs. (eds.). E. J Carpenter, D. G Capone and J. G Rueter. Kluwer. pp. 163-175.

[13] Lee, J. J. (1994). Diatoms and their chloroplasts as endosymbiotic partners for larger foraminifera, Memoirs of the Southern California Academy of Sciences, 17, 21-36.

[14] Denys, L. (1997). Morphology and taxonomy of epizoic diatoms (epiphalaina and tursiocola) on a sperm whale (physeter macrocephalus) stranded on the coast of belgium, Diatom Research, 12(1), 1-18.

[15] Wetzel, C. E., Van de Vijver, B., Cox, E. J., Bicudo, DdeC., Ector, L. (2012). Tursiocola podocnemicola sp. nov., a new epizoic freshwater diatom species from the Rio Negro in the Brazilian Amazon Basin, Diatom Research, 27(1), 1-8.

[16] Lee, J. J. (2011). Diatoms as endosymbionts. In The diatom world. (eds.). J Seckbach and P. J Kociolek. Springer. pp. 437-465.

[17] Robinson, N. J., Majewska, R., Lazo-Wasem, E. A., Nel, R., Paladino, F. V., Rojas, L., et al. (2016). Epibiotic diatoms are universally present on all sea turtle species, PLoS ONE, 11(6), e0157011.

[18] Totti, C., Romagnoli, T., De Stefano, M., Di Camilo, C. G., Bavestrello, G. (2011). The diversity of epizoic diatoms: relationships between diatoms and marine invertebrates. In All Flesh is Grass. Cellular Origin, Life in Extreme Habitats and Astrobiology 16. (eds.). J Seckbach and Z Dubinski. Springer Science. pp. 323-343.

[19] Drum, R. W., Pankratz, S. (1965). Fine structure of an unusual cytoplasmic inclusion in the diatom genus, Rhopalodia, Protoplasma, 60(1), 141-149.

[20] Geitler, L. (1977). Zur entwicklungsgeschichte der epithemiaceen epithemia, rhopalodia und denticula (diatomophyceae) und ihre vermutlich symbiotischen Sphäroidkörper, Plant Syst. Evol., 128(3-4), 259-275.

[21] Lowe, R. L., Rosen, B. H., Fairchild, G. W. (1984). Endosymbiotic blue-green algae in freshwater diatoms: an advantage in nitrogen poor habitats, J. Phycol., 20, 24.

[22] Kociolek, J. P., Spauling, S. A., Lowe, R. L. (2015). Bacillariophyceae: The Raphid Diatoms. In Freshwater Algae of North America. Ecology and Classification (Second Edition). (eds.). J Wehr, D, R. G Sheath and J. P Kociolek. San Diego: Elsevier. pp. 709-772.

[23] Witkowski, A., Lange-Bertalot, H., Metzeltin, D. (2000). Diatom flora of marine coasts. In Iconographia Diatomologica. 7. (ed.). H Lange-Bertalot. Ruggell: Gantner Verlag. pp. 1-925.

[24] Nakayama, T., Ikegami, Y., Nakayama, T., Ishida, K., Inagaki, Y., Inouye, I. (2011). Spheroid bodies in rhopalodiacean diatoms were derived from a single endosymbiotic cyanobacterium, J. Plant Res., 124(1), 93-97.

[25] Hamsher, S., Graeff, C. L., Stepanek, J. G., Kociolek, J. P. (2014). Frustular morphology and polyphyly in freshwater Denticula (Bacillariophyceae) species, and the description of Tetralunata gen. nov. (Epithemiaceae, Rhodopalodiales), Plant Ecology and Evolution, 147(3), 346-365.

[26] Geitler, L. (1932). Der Formwechsel Der Pennaten Diatomeen (Kieselalgen, Archiv für Protistunkunde, 78, 1-226.

[27] Bothe, H., Tripp, H. J., Zehr, J. P. (2010). Unicellular cyanobacteria with a new mode of life: the lack of photosynthetic oxygen evolution allows nitrogen fixation to proceed, Arch. Microbiol., 192(10), 783-790.

[28] Floener, L., Bothe, H. (1980). Nitrogen fixation in Rhopalodia gibba, a diatom containing blue greenish inclusions symbiotically. In Endo-cytobiology, endosymbiosis and cell biology. Vol. 1. (eds.). W Schwemmler and H. E. A Schenk. Berlin: Walter de Gruyter and Co. pp. 541-552.

[29] Kneip, C., Voss, C., Lockhart, P. J., Maier, U. G. (2008). The cyanobacterial endosymbiont of the unicellular algae Rhopalodia gibba shows reductive genome evolution, BMC Evol. Biol., 8(1), 30.

[30] Prechtl, J., Kneip, C., Lockhart, P., Wenderoth, K., Maier, U. G. (2004). Intracellular spheroid bodies of Rhopalodia gibba have nitrogen-fixing apparatus of cyanobacterial origin, Mol. Biol. Evol., 21(8), 1477-1481.

[31] Taylor, D. L. (1984). Autotrophic eukaryotic marine symbionts. In Cellular Interactions. (eds.). H. F Linskens and J Heslop-Harrison. Berlin, Heidelberg: Springer-Verlag. pp. 75-90.

[32] DeYoe, H. R., Lowe, R. L., Marks, J. C. (1992). Effects of nitrogen and phosphorus on the

endosymbiont load of rhopalodia gibba and epithemia turgida (bacillariophyceae)1, J. Phycol., 28(6), 773-777.

[33] Stancheva, R., Sheath, R. G., Read, B. A., McArthur, K. D., Schroepfer, C., Kociolek, J. P., et al. (2013). Nitrogen-fixing cyanobacteria (free-living and diatom endosymbionts): their use in southern California stream bioassessment, Hydrobiologia, 720(1), 111-127.

[34] Nakayama, T., Kamikawa, R., Tanifuji, G., Kashiyama, Y., Ohkouchi, N., Archibald, J. M., et al. (2014). Complete genome of a nonphotosynthetic cyanobacterium in a diatom reveals recent adaptations to an intracellular lifestyle, Proceedings of the National Academy of Sciences, 111(31), 11407-11412.

[35] Cavalier-Smith, T., Lee, J. J. (1985). Protozoa as hosts for endosymbioses and the conversion of symbioses into organelles. 2, J. Eukaryot. Microbiol., 33, 376-379.

[36] Howarth, R. W., Marino, R., Lane, J., Cole, J. J. (1988). Nitrogen fixation in freshwater, estuarine, and marine ecosystems. 1. Rates and importance, Limnol. Oceanogr., 33, 669-687.

[37] Marks, J. C., Power, M. E. (2001). Nutrient induced changes in the species composition of epiphytes on Cladophora glomerata Kutz. (Chlorophyta), Hydrobiologia, 450(1/3), 187-196.

[38] Peterson, C. G., Grimm, N. B. (1992). Temporal variation in enrichment effects during periphyton succession in a nitrogen-limited desert stream ecosystem, Journal of the North American Benthological Society, 11(1), 20-36.

[39] Grimm, N. B., Fisher, S. G. (1986). Nitrogen Limitation in a Sonoran Desert Stream, Journal of the North American Benthological Society, 5(1), 2-15.

[40] Bahls, L. L., Weber, E. E. (1988). Ecology and distribution in Montana of Epithemia sorex Kütz., a common nitrogen-fixing diatom, Proceedings of the Montana Academy of Sciences, 48, 15-20.

[41] Redfield, A. C. (1958). The biological control of chemical factors in the environment, Am. Sci., 46, 205-221.

[42] Furey, P. C., Lowe, R. L., Power, M. E., Campbell-Craven, A. M. (2012). Midges, Cladophora, and epiphytes: shifting interactions through succession, Freshwater Science, 31(1), 93-107.

[43] Power, M., Lowe, R., Furey, P., Welter, J., Limm, M., Finlay, J., et al. (2009). Algal mats and insect emergence in rivers under Mediterranean climates: towards photogrammetric surveillance, Freshw. Biol., 54(10), 2101-2115.

[44] Power, M. E., Holomuzki, J. R., Lowe, R. L. (2013). Food webs in Mediterranean rivers, Hydrobiologia, 719(1), 119-136.

[45] Lock, J., Kelsey, H., Furlong, K., Woolace, A. (2006). Late Neogene and Quaternary landscape evolution of the northern California Coast Ranges: Evidence for Mendocino triple junction tectonics, Geol. Soc. Am. Bull., 118(9-10), 1232-1246.

[46] Power, M. E., Dietrich, W. E. (2002). Food webs in river networks, Ecol. Res., 17(4), 451-471.

[47] Scott, J. T., Marcarelli, A. M. (2012). Cyanobacteria in freshwater benthic environments Ecology of Cyanobacteria II. Their Diversity in Space and Time. (ed.). Whitton, B. A London: Springer. pp. 271-289.

[48] Carpenter, E. J., Janson, S. (2000). Iintracellular cyanobacterial symbionts in the marine diatom climacodium frauenfeldianum (bacillariophyceae), J. Phycol., 36(3), 540-544.

[49] Ostenfeld, C. H., Schmidt, J. (1902). Plankton fra det Røde Hav og Adenbugten (Plankton from the Red Sea and the Gulf of Aden, Videnskabelige Meddelelser fra Dansk Naturhistorisk Forening, 1901, 141-182.

[50] Villareal, T. A. (1990). Laboratory cultivation and preliminary characterization of the Rhizosolenia (Bacillariophyceae)-Richelia (Cyanophyceae) symbiosis, P. Z. S. N. I: Marine Ecology, 11, 117-132.

[51] Mague, T. H., Weare, N. M., Holm-Hansen, O. (1974). Nitrogen fixation in the North Pacific Ocean, Marine Biology, 24(2), 109-119.

[52] Foster, R. A., Zehr, J. P. (2006). Characterization of diatom-cyanobacteria symbioses on the basis of nifH, hetR and 16S rRNA sequences, Environ. Microbiol., 8(11), 1913-1925.

[53] Taylor, F. J. R. (1982). Symbioses in marine microplankton, Annales de l'Institutocéanographique, Paris, 58, 61-90.

[54] Villareal, T. A. (1989). Division cycles in the nitrogen-fixing Rhizosolenia (Bacillariophyceae)-Richelia (Nostocaceae) symbiosis, British Phycological Journal, 24(4), 357-365.

[55] Carpenter, E. J. (2002). Marine cyanobacterial symbioses. Biology and Environment, Proceedings of the Irish Royal Academy, 102B, 15-18.

[56] Gómez, F., Furuya, K., Takeda, S. (2005). Distribution of the cyanobacterium Richelia intracellularis as

an epiphyte of the diatom Chaetoceros compressus in the western Pacific Ocean, J. Plankton Res., 27(4), 323-330.

[57] Foster, R. A., Subramaniam, A., Mahaffey, C., Carpenter, E. J., Capone, D. G., Zehr, J. P. (2007). Influence of the Amazon River plume on distributions of free-living and symbiotic cyanobacteria in the western tropical north Atlantic Ocean, Limnol. Oceanogr., 52(2), 517-532.

[58] White, A. E., Prahl, F. G., Letelier, R. M., Popp, B. N. (2007). Summer surface waters in the Gulf of California: Prime habitat for biological N_2 fixation, Global Biogeochem. Cycles, 21(2), n/a-11.

[59] Lyimo, T. J. (2011). Distribution and abundance of the cyanobacterium Richelia intracellularis in the coastal waters of Tanzania, Journal of Ecology and The Natural Environment, 3, 85-94.

[60] Zhang, Y., Zhao, Z., Sun, J., Jiao, N. (2011). Diversity and distribution of diazotrophic communities in the South China Sea deep basin with mesoscale cyclonic eddy perturbations, FEMS Microbiol. Ecol., 78(3), 417-427.

[61] Heinbokel, J. F. (1986). Occurrence of richelia intacellularis (cyanophyta) within the diatommms hemiaulus haukii adn h. Membranaceus off hawaii, J. Phycol., 22(3), 399-403.

[62] Villareal, T. A. (1994). Widespread Occurrence of the Hemiaulus-cyanobacterial Symbiosis in the Southwest North Atlantic Ocean, Bull. Mar. Sci., 54, 1-7.

[63] Janson, S., Wouters, J., Bergman, B., Carpenter, E. J. (1999). Host specificity in the Richelia diatom symbiosis revealed by hetR gene sequence analysis, Environ. Microbiol., 1(5), 431-438.

[64] Sohm, J. A., Webb, E. A., Capone, D. G. (2011). Emerging patterns of marine nitrogen fixation, Nat. Rev. Microbiol., 9(7), 499-508.

[65] Subramaniam, A., Yager, P. L., Carpenter, E. J., Mahaffey, C., Björkman, K., Cooley, S., et al. (2008). Amazon River enhances diazotrophy and carbon sequestration in the tropical North Atlantic Ocean, Proc. Natl. Acad. Sci. U. S. A., 105(30), 10460-10465.

[66] Carpenter, E. J., Montoya, J. P., Burns, J., Mulholland, M. R., Subramaniam, A., Capone, D. G. (1999). Extensive bloom of a N_2-fixing diatom/cyanobacterial association in the tropical Atlantic Ocean, Mar. Ecol. Prog. Ser., 185, 273-283.

[67] Zeev, E. B., Yogev, T., Man-Aharonovich, D., Kress, N., Herut, B., Bejà, O., et al. (2008). Seasonal dynamics of the endosymbiotic, nitrogen-fixing cyanobacterium Richelia intracellularis in the eastern Mediterranean Sea, Isme J., 2(9), 911-923.

[68] Norris, R. E. (1961). Observations on Phytoplankton organisms collected on the N. Z. O. I. Pacific Cruise, September 1958, New Zealand Journal of Science, 4, 162-168.

[69] Kimor, B., Gordon, N., Neori, A. (1992). Symbiotic associations among the microplankton in oligotrophic marine environments, with special reference to the Gulf of Aqaba, Red Sea, J. Plankton Res., 14(9), 1217-1231.

[70] Burkholder, J. M. (1996). Interactions of benthic algae with their substrata Algal Ecology: Freshwater Benthic Ecosystems. (eds.). Stevenson, R. J., Bothwell, M. L. and Lowe, R. L. San Diego: Academic Press. pp.253-298.

[71] Burkholder, J. M., Wetzel, R. G. (1989). Microbial colonization on natural and artificial macrophytes in a phosphorus-limited, hardwater, J. Phycol., 25(1), 55-65.

[72] Marks, J. C., Lowe, R. L. (1993). Interactive effects of nutrient availability and light levels on the periphyton composition of a large oligotrophic lake, Can. J. Fish. Aquat. Sci., 50(6), 1270-1278.

[73] Round, F. E. (1984). The ecology of algae. Cambridge: Cambridge University Press.

[74] Tuchman, N. C., Schollett, M. A., Rier, S. T., Geddes, P. (2006). Differential heterotrophic utilization of organic compounds by diatoms and bacteria under light and dark conditions. In Advances in Algal Biology: A Commemoration of the Work of Rex Lowe. (eds.). R. J Stevenson, Y Pan, J. P Kociolek and J. C Kingston. Dordrecht: Springer. pp.167-177.

[75] Flower, R. J. (1982). The occurrence of an epiphytic diatom on Microcystis aeruginosa, Irish Naturalist Journal, 20, 553-555.

[76] Morales, E. A., Rivera, S. F., Wetzel, C. E., Hamilton, P. B., Bicudo, D. C., Pibernat, R. A., et al. (2015). Hypothesis: the union Microcystis aeruginosa Kütz, Nitzschia palea (Kütz.) W. Sm.-bacteria in Alalay Pond, Cochabamba, Bolivia is symbiotic. Acta Nova, 7, 122-142.

[77] Snoeijs, P., Murasi, L. W. (2004). Symbiosis between diatoms and cyanobacterial colonies, Vie Milieu, 5, 163-170.

[78] Whitton, B. A., Mateo, P. (2012). Rivulariaceae Ecology of Cyanobacteria II. Their Diversity in Space and Time. (ed.). Whitton B. A. London: Springer. pp. 561-592.

[79] Hasle, G. R. (1968). Navicula endophytica Sp. Nov., A pennate diatom with an unusual mode of existence, British Phycological Bulletin, 3(3), 475-480.

[80] Taasen, J. P. (1972). Observations on Navicula endophytica Hasle (Bacillariophyceae, Sarsia, 51(1), 67-82.

[81] Baardseth, E., Taasen, J. P. (1973). Navicula dumontiae sp. nov., an endophytic diatom inhabiting the mucilage of Dumontia incrassata (Rhodophyceae), Norwegian Journal of Botany, 20, 80-87.

[82] Álvarez-Blanco, I., Blanco, S. (2014). Benthic diatoms from Mediterranean coasts, Bibliotheca Diatomologica, 60, 1-409.

[83] Klochkova, T. A., Pisareva, N. A., Park, J. S., Lee, J. H., Han, J. W., Klochkova, N. G., Klochkova, N. G., et al. (2014). An endophytic diatom, Pseudogomphonema sp. (Naviculaceae, Bacillariophyceae), lives inside the red alga Neoabbottiella (Halymeniaceae, Rhodophyta, Phycologia, 53(3), 205-214

[84] Okamoto, N., Nagumo, T., Tanaka, J., Inouye, I. (2003). An endophytic diatom Gyrosigma coelophilum sp. nov. (Naviculales), Bacillariophyceae) lives inside the red alga Coelarthrum opuntia (Rhodymeniales, Rhodophyceae, Phycologia, 42(5), 498-505.

[85] Imanian, B., Keeling, P. J. (2014). Horizontal gene transfer and redundancy of tryptophan biosynthetic enzymes in dinotoms, Genome Biol. Evol., 6(2), 333-343.

[86] Stat, M., Carter, D., Hoegh-Guldberg, O. (2006). The evolutionary history of Symbiodinium and scleractinian hosts—Symbiosis, diversity, and the effect of climate change, Perspectives in Plant Ecology Evolution and Systematics, 8(1), 23-43.

[87] Inagaki, Y., Dacks, J. B., Doolittle, W. F., Watanabe, K. I., Ohama, T. (2000). Evolutionary relationship between dinoflagellates bearing obligate diatom endosymbionts: insight into tertiary endosymbiosis, Int. J. Syst. Evol. Microbiol., 50(6), 2075-2081.

[88] Chesnick, J. M., Kooistra, W. H., Wellbrock, U., Medlin, L. K. (1997). Ribosomal RNA analysis indicates a benthic pennate diatom ancestry for the endosymbionts of the dinoflagellates Peridinium foliaceum and Peridinium balticum Pyrrhophyta, J. Eukaryot. Microbiol., 44(4), 314-320.

[89] Tamura, M., Shimada, S., Horiguchi, T. (2005a). Galeidiniium rugatum gen. Et sp. Nov. (dinophyceae), a new coccoid dinoflagellate with a diatom endosymbiont, J. Phycol., 41(3), 658-671.

[90] Imanian, B., Pombert, J. F., Keeling, P. J. (2010). The complete plastid genomes of the two "dinotoms" Durinskia baltica and Kryptoperidinium foliaceum, PLoS ONE, 5(5), e10711.

[91] Takano, Y., Hansen, G., Fujita, D., Horiguchi, T. (2008). Serial Replacement of Diatom Endosymbionts in Two Freshwater Dinoflagellates, Peridiniopsis spp. (Peridiniales, Dinophyceae, Phycologia, 47(1), 41-53.

第 11 章
硅藻有性繁殖和生命周期

Aloisie Poulíčková，David G Mann

11.1 简介

大多数硅藻的生命周期都是在长期(数月或数年)的有丝分裂的无性生长和发育、有丝分裂的细胞分裂和有丝分裂的细胞分裂之间交替进行。这一点本身并不值得注意，因为许多多细胞植物和动物也是这样。这种变化与生物体大小的变化相结合，这也不是特别奇怪。在植物和动物中，以及在褐藻(硅藻的近亲)中，这是意料之中的。奇怪的是，然而在大多数"高等"生物体中，成熟只有经过一段时间的扩大和分化之后才会出现，而有性生殖的结果是形成一个小的胚胎个体。在硅藻中，性成熟的个体是该物种中最小的，同一物种中较大的个体通常是无能的[1-4]。硅藻的短性阶段实际上是恢复大尺寸：因为硅藻细胞分裂(新的细胞壁元素形成在现有的壁)，加上细胞壁的物理特性(每个瓣膜或环带可以弯曲但不能拉伸)导致的长营养期细胞尺寸逐渐减小，尺寸恢复是有必要的。因此，辅助孢子的形成和发育辅助孢子是硅藻的特殊构造细胞，它能膨胀以重建最大尺寸，它的形成和发展通常也是减数分裂和有性重组发生的阶段[4-5]。

鉴于硅藻生命周期的独特和反直觉的特征[6-7]，由于与性别相关的缓慢尺寸缩小和快速尺寸恢复的锯齿形模式，因此在很长一段时间后才完全了解生命周期，这也许并不奇怪。细胞的配对和其内容物的结合(融合)首先由 Thwaites 发现，之后 Griffith 和 Carter 进行了进一步的观察，但对这一过程的解释尚不清楚[8-12]，结合的产物最初被认为是产生孢子的结构，Smith 认为每个扩大的细胞迅速分裂形成许多小细胞，就像那些配对形成"孢子囊"的细胞一样，他的画图是准确的，但它们说明了变形虫原生生物的进食阶段，而不是硅藻的繁殖。直到 1869 年，硅藻的细胞分裂机制才有了明确的描述[13-15]，这导致(在大多数情况下)细胞尺寸的不可避免下降，因此显示了尺寸恢复阶段的重要性[16]。后来，倍性水平的交替，从整个营养阶段的二倍体到配子中的单倍体，被 Surirella 和 Karsten 通过双菱藻中配子发生期间减数分裂的证明所证实，并且经过多年来对底栖硅藻进行了广泛的调查，显示出形态上的同配生殖，尽管有时行为上的异配生殖[3,17]。最后，在 Thwaites 对硅藻性行为的初步观察 100 多年后，von Stosch[18]研究表明，一些硅藻，特别是浮游物种，并不像 Thwaites

等研究的底栖硅藻那样结合,而是卵合子的。然而,即使在这之后,一些人[19]依旧质疑尺寸缩小恢复周期是否像教科书中所暗示的那样是硅藻的特征。

已经有一些关于硅藻的有性繁殖、缺陷孢子形成和生命周期动力学的综述[1-4,17,20-21]以及最近的一篇论文[22]已经发表,试图解释和标准化应用于这些阶段的术语。我们不会试图重复或重新描述,因为它们中的大多数都很容易获得,而且它们所包含的大部分信息仍然有效。本章我们提出了一些可能没有得到应有重视的一般性观点。我们不讨论性行为阶段在种群结构方面对理解硅藻的物种形成和维持多样性的重要性[23-29]。

11.2 辐射型硅藻

11.2.1 生命周期和繁殖

传统上,分类学家将硅藻分为两组——中心状(在瓣膜上)和羽状,前者围绕圆形或细长环(即"环")放射状排列,后者围绕肋骨状"胸骨"左右两侧排列;这两个组通常被认为是标准[30]。这种分离并不反映系统发育,因为中心硅藻不是单系的[31-32],但它仍然是一个方便的分界,因为它是基于容易可见的瓣膜模式的特点,与栖息地有广泛的关系(大多数中心是浮游,大多数羽状是底栖),对应整体分裂基于细胞的内部组织(大多数中心有许多小叶绿体,大多数羽状有一个,两个或几个大叶绿体),最重要的是,反映了有性生殖的根本差异:中心表现出卵,而羽状是异卵(或很少有异卵),永远不会产生精子。事实上,鞭毛和中心粒从未存在于羽状硅藻的生命周期的任何阶段,而且可能在它们进化的早期就丢失了。

然而,虽然卵配生殖的孢子形成仅限于有中心的硅藻,但这并不意味着所有中心硅藻都有卵藻,也不是配子发生和卵生孢子发育的特点是统一的,也不表示一个物种或细胞谱系可以通过卵配子恢复大小。很明显,例如,一些中心种能够通过非配子生殖产生的营养孢子或通过营养体扩大而扩大(营养扩大扩张后的细胞抛弃细胞壁,没有减数分裂或受精,没有特殊的保护和塑造结构,例如,鳞片,孔隙,通常与增生孢子形成有关)。此外,在目前的分子系统发育中,至少有一个嵌套在其他中心硅藻谱系中,卵配体已经被一种非鞭毛虫同配生殖所取代,这与一些无壳羽状硅藻非常相似[33-34]。

在一个生命周期中,可能会出现非两性生殖的辅助孢子和无性繁殖。在威氏圆筛藻[35]、菱角海链藻[36]和圭亚内近海藻[37]中发现了无性或自产的缺陷孢子的例子;在念珠藻直链藻八面变种[2]、圆筛藻、玛氏骨条藻[38]角毛藻一些种群,和明威星状藻[39]发现了营养性扩大的记录。Von Stosch指出,一些明威星状藻种群似乎仅通过或主要通过营养体扩大来恢复大小(例如,在德国北海湾和美国罗德岛的纳拉甘塞特湾),但其他种群主要是性的(例如,在澳大利亚东南部的菲利普湾港)。然而,解释是困难的,因为观察自然种群布氏双尾藻的可能性是指不同的隐种,它们单独生长或一起生长[40-41]。此外,尽管 Koester 等的研究结果表明,来自美国缅因州的布氏双尾藻在培养中可以同时进行卵增孢和营养扩大,但目前尚不清楚这两种扩大模式在自然种群中的相对贡献。细胞大小恢复的混合策略最近也在玛氏骨条藻中被发现[27]。

中心硅藻生命周期的另一个仍然令人惊讶地不清楚的方面是,该循环在多大程度上表现出"基点",即一个物种的大小特征范围,在此之外没有表现出特定的性别、生理或发育特征。基点的概念嵌入到我们目前对硅藻生命周期的理解中,很大程度上是 Geitler 对羽状硅藻[3]和 von Stosch 对中心硅藻的工作。羽状硅藻中基点的例子是通过缺陷孢子扩张产生的细胞的最大尺寸(这个最大值是一个范围),细胞可以被性别化并产生配子的最大和最小细胞尺寸(这些基点是瓣膜)和生存能力的最小尺寸(也是一个瓣膜,最小存活尺寸可能与性化的最小尺寸相同,也可能不相同)。中心基点的例子是辅助孢子形成后达到的最大细胞尺寸,产生卵细胞的最大和最小尺寸,产生精子的最大和最小尺寸,以及最小的活细胞尺寸。举个例子,细胞的最大直径接近 100 μm,卵发生的最大尺寸超过 40 μm,配子发生的最小尺寸 16 μm,最小存活尺寸 10 μm;精子发生和卵子发生的瓣膜相同,但大细胞中卵子发生的相对频率更大。在移动三翼虫中,精子发生和卵子发生的最大值和最小值是不同的。其他关于"基点"的中心物种包括变异直链藻、菱状盒形藻,帕维氏镜藻硅藻的鉴定可能是值得怀疑的;显微图显示了一种类似威氏圆筛藻的鼓形物种[42]和玛氏骨条藻[27]。

然而,只有极少数的中心硅藻物种被证实存在基点(尽管中心硅藻由于其同宗配合的性质而相对容易地在培养中被保持),事实上,有几个理由可以怀疑这个概念在中心型硅藻中是否像在羽扇型硅藻中一样有用或适用。原因之一是在几个远亲谱系中显示了两或三步式增生孢子,许多藻类学家可能仍不了解这种现象。最近报道的斑点海链藻[43],其中辅助孢子形成可以通过至少两步进行。在培养过程中,直径 26~41 μm 的小细胞逐渐形成直径 83~124 μm 的初始细胞。经过一些分裂,扩大的细胞(92~120 μm)本身能够作为卵原,产生缺陷孢子膨胀至 132~153 μm。此外,Chepurnov 等推测点纹海星藻的尺寸上限为 183 μm,表明在某些情况下点纹海星藻可能会发生了第三步扩张[44]。类似的现象也在拟银币直链藻[45],串珠美洲藻[46],强氏圆筛藻和雅尼氏镜藻出现[47]。在所有这些情况下,尽管可以进行有性繁殖的细胞大小可能有一个下限,而最大的营养缺陷孢子所产生的细胞大小可能有一个最大值,但似乎最小的细胞本身不能直接产生最大的细胞,而只能产生中等大小的细胞,这与硅藻生命周期的经典模式相悖。Armbrust 和 Chisholm 报道冰河星杆藻培养中发生的意外的尺寸变化模式可能也反映了多步生孢的发生和缺乏真正的"基点"[48]。分子系统发育的检查表明,硅藻表现多步生孢并不局限于一个血统(海链藻、圆筛藻和直链藻不是姐妹组),表明这不能假定这种类型的生命周期,或 von Stosch 和其同事报告的更严格的"正统"类型的生命周期是普遍的一个。

Chepurnov 等进一步报道了冰河星杆藻一个有趣的现象,这可能提示了羽状硅藻的等配体（以及异常的中心硅藻 Ardissonea）可能已经出现[34,43]。这就是观察到的冰河星杆藻能够移动变形虫：相邻的卵原细胞相互作用，一个细胞的卵细胞推入另一个细胞的果囊内。Findlay 对圆筛藻 pavillardii 也进行了类似的观察[42]。

总的来说，最近和更早的研究结果表明，中心硅藻的生命周期可能比我们想象的更灵活和多样。从上述观察中可以得到的一个教训是，不应该对一个特定的中心物种或种群的生命周期过程做出任何预先的假设。演示广泛的大小的特定物种的种群可能表明存在着一个大小减少-恢复循环周期发生（尽管这也可能反映出存在大小不同但大小不变的不同物种，因为并非所有硅藻都不可避免地会发生大小减小）[1]但它没有说明循环如何发生。卵配生殖是一种可能性，但其他的可能性是非卵配生殖的孢子形成和营养体增大。此外，虽然精子的形成是着丝粒硅藻中卵配生殖的先决条件，但群体内精子的形成并不意味着发育缺陷孢子形成将以卵配生殖的方式发生，因为并没有已知产生精子但发育缺陷孢子通过自配生殖形成的实例[1-2]。卵配子性的严格证明必须包括直接观察精子发生和卵子发生同时发生，以及观察卵子和精子核融合；或通过遗传证据进行检测核聚变已经发生的基本原理和应用（正如 Godhe 等 2014 在玛氏骨条藻所做的那样）。事实上，异配生殖是极其难证明的；即使在培育中也是如此，在自然种群中更是如此。另外，同样重要的是要认识到，无论是在培育中还是在自然种群中，对一个物种的有性繁殖没有任何观察，并不意味着它不会发生：Moore 等最近报道了在假微型海链藻中诱导卵原细胞和精子的现象[49]，这是一种以前似乎是无性的物种[50-51]，现在也有遗传证据表明该物种的有性谱系及最近进化出一种专性无性生物，这种生物显然在几百年内全球分布[52]。

11.2.2　配子发生和配子结构

关于中心硅藻的配子体发生和配子结构的新报道很少。硅藻精子发生经典分为两类根据整个精母细胞原生质体，包括叶绿体，分配减数分裂后——这是同源的精子发生；还有精母细胞，包括叶绿体留在一个"残留体"——这是异生精子发生。Samanta 等最近记录了一种变异的生殖类型，其中叶绿体的排除一直持续到精子形成的最后阶段[53]。Drebes 和 Round 等已经描述和说明了精子发生的两种经典模式[2,4]，Idei 等最近的一篇论文说明了辐环藻的次生发生机制[54]。有时也有人声称[55]，同源精子发生和亚生精子发生具有系统发育意义，每种精子发生都表征了中心硅藻的主要谱系，但情况似乎并非如此：两者甚至可以发生在同一属内[54,56-57]。此外，很明显，如之前的综述所述，同源和亚源精子发生[2,4]有点过于简化，因为一些硅藻表现出了 Idei 等所称的"疟原虫精子发生"。在这种情况下，在未分裂的精原细胞内重复的有丝分裂导致多核疟原虫的形成；每个二倍体核在疟原虫内进行减数分裂，然后精子出芽，留下疟原虫作为一个大的无核团块[54]。这种类型的精子发生最近被 Idei 等[58]在水链藻中进行了一些详细的记录，但在光滑侧链藻，蛇目沟盘藻和菱软几内亚藻[59-60]等早期也有关于相同现象的简短报道。与经典的异生精子发生的相似之处是明显的,这两种类型都导致精子缺乏叶绿体,因此,推测可能在叶绿体基因组的母体遗传中。与同源性和经典次生发生一样,表现出疟原虫精子发生的类群并不都是密切相关的[32];相对于"正常"的生理发生,疟原虫精子发生的选择优势尚不清楚。

精子的超微结构最近被记录在由 Idei 等[61]撰写的两种硅藻湖泊海星藻和八角串珠美洲藻变种中,加入之前研究过的波状石丝藻,光滑侧链藻,威氏圆筛藻和垂缘角毛藻[56,59,62]。研究的数量仍然很少,但早期注意到的一些特征似乎仍然对硅藻很普遍,例如鞭毛轴素中心微管对的缺失(即"9+0"结构),在基体中存在九个双联体而不是三联体,并且不存在真正的鞭毛根系。相比之下,组成硅藻姐妹群的双子叶植物纲-伞形科,有正常的"9+2"结构的鞭毛。然而,"9+0"结构是否对所有现存的硅藻都是共同的,尚不清楚;考虑到环毛藻和细柱藻在某些分子系统发育中的基础位置,对它们的研究可能具有指导意义[32]。

11.2.3 产卵

当然,精子的功能是游到卵细胞并使其受精。这一过程已经在实验室中被观察了好几次,但在自然界中,卵子和精子形成和扩散的动态和空间模式是一个谜。以卵为中心的硅藻本质上是"广播产卵细胞":精子和卵细胞(无论是否脱离卵细胞)在水柱或膜中受精的地方是自由的。这似乎是一个危险的策略,因为精子和卵子可能会被比精子游泳速度快得多的湍流运动迅速"稀释"。在底栖物种的情况下,如直链藻或水链藻,配子可能暂时集中,并通过流动促进它们的接触,正如 Crimaldi 和 Zimmer 对无脊椎动物精子和卵子提出的那样,但这似乎不太可能像浮游中心生物如冰河星杆藻或波状石丝藻[63]。有性生殖与高细胞密度的紧密耦合,如在水华结束时出现的那些,可能是部分答案,也可能是通过有性生殖以外的方式恢复尺寸的能力(Godhe 等讨论了与玛氏骨条藻的关系)。有性生殖与水华形成的耦合可以通过群体感应、细胞密度之间的正反馈和信息素介导的性化之间的正反馈来实现[27]或外部因素的检测通常与峰值丰度相关(例如,低水平的限制性营养:Jewson 等表明低磷可能在贝加尔湖的斯科尔托夫弓形藻中很重要)[64]。Moore 等[49]表明,高水平的铵可能参与在某些情况下,这标志着当水华已经足够密集以支持大量的食草动物时,铵会随着食草动物进食释放出来(食草动物在水华高潮时释放的铵可能是中心硅藻有性形态的主要生态触发因素)。然而,关于中心硅藻在自然界中何时发生有性繁殖的信息还相当少。Crawford[15]记录了海洋浮游硅藻角刺藻的大规模性事件,并有一些关于淡水浮游植物的详细研究[65-67];但这样的例子很少见,可能是因为如果没有进行数月或数年的密集抽样活动,用显微镜发现性行为的机会几乎可以忽略不计。元转录组学的方法导致在了解自然种群的繁殖方面取得重大进展,例如使用有性繁殖的标记,如编码鞭毛元素的西格玛基因的转录本[68-69]。特别令人感兴趣的是,观察栖息在浮游生物和底栖生物中的中心硅藻在产卵行为上是否有重大差异,可以预期配子形成、释放和扩散的空间和时间动态在本质上更有利于繁殖成功。

11.3 羽状硅藻的生命周期和繁殖

在绝大多数的羽状硅藻中,配子在形态上是相似的。在真核生物有性生殖的更广泛背景下,从卵配体的进化是令人惊讶的,因为它们的晚期起源于中心硅藻谱系[70]。然而,在大多数羽状硅藻中发现的异性繁殖是在配子交配之前,这与它们的底栖生活方式有关。羽状硅藻不传播它们的配子,因此与中心硅藻形成鲜明的对比。尤其是针晶硅藻(但至少也有一些羽纹硅藻)[71-72]是运动的,可以在周围和上层发生,这允许一种基于形成少量相对较大的等配子的繁殖方法[70]。

近年来,我们在对羽状硅藻,特别是羽纹硅藻 *Seminavis* 的生物化学、遗传控制和有性繁殖机制的理解方面取得了显著的进展[73-78]。性化和配偶吸引已被证明是由信息素介导的[74-75,79-82],该方法可以通过趋化作用来改变交配细胞的运动模式[73]。配子的运动也可能受到羽状硅藻中的信息素的引导[81],尽管在其他情况下,搜索似乎是通过随机游动[83]。此外,在 *araphid pennates* 中发现了一些完全意想不到的配子运动和行为类型[81,83-84]。而以前唯一已知的羽状硅藻受精期间的配子运动和活动类型是肿胀和变形虫运动[3,85-86];一些羽纹硅藻的"雄性"配子现在已经被证明具有可延伸的突起,这些突起参与产生稳定或不稳定的易位或旋转运动[81,84,87-88]。这些突起的超微结构尚未被研究;它们显然含有微管,但似乎与鞭毛无关[81]。至少有一组中心硅藻(极其细长的 *Ardissonea*)产生具有相似突起的非鞭毛雄配子[33]。

对羽状硅藻有性繁殖的遗传基础的探索,首先试图确定雌雄异株硅藻半舟藻[77]的交配型位点,在三个物种(三角褐指藻,圆柱拟脆杆藻和多列拟菱形藻,以及中心硅藻假微型海链藻)和另外两个物种(半舟藻和尖刺拟菱形藻)的基因组中寻找可能参与减数分裂的基因。毫无疑问,这些研究只是我们探索硅藻的性行为和孢生缺陷的一个新阶段的开始。基因组/转录组学证据已经表明,硅藻减数分裂和"经典"减数分裂之间可能存在显著差异(如高等植物和后鞭毛生物),包括交叉机制可能存在差异[76]。他们还暗示了"模型"硅藻三角褐指藻可能具有一个性周期,尽管直到现在还没有[51]任何迹象表明它经历了尺寸缩小和增孢子。

羽状硅藻的有性繁殖和增生孢子形成的巨大差异已经存在了 150 多年。Geitler 已经能够列出 90 多条记录,根据每个配子囊产生的配子数和每个母细胞产生的生子孢子数量等特征,将其分为四大类[89]和一些子类别。随后,他开发了一个更详细的分类[17],根据配子行为、交配结构的存在/缺失、配对方法和缺陷孢子方向划分了主要类别。然而,尽管相关物种或属的特定类群有时以特定类型的有性生殖为特征,从而支持或至少不与基于形态学和分子证据的分类相矛盾,但在更高的分类学水平上,很难找到继续提供令人信服的系统发育信号的特征。Mann[70]给出的一个例子是长蓖藻属和 Biremis,它们表现出相似的配子发生、配子和缺陷孢子发育,这些属之间的密切关系得到了分子系统数据的支持[90]。一个有时使确定有性生殖的进化变得困难的问题是,由于分类学的变化,有时我们不能确切地确定哪个有性生殖的记录属于哪个物种[91-92],这就指出了凭证材料的重要性[93]。

我们对羽状硅藻有性生殖的理解有了快速的进展,但仍然没有关于大多数硅藻属的信息。此外,关于有性生殖的数据来源也发生了显著变化。直到 20 世纪 90 年代,羽状硅藻繁殖的绝大多数观察都是在自然种群上进行的[17,86,94],尽管在羽状硅藻的生命周期中,当细胞因为超过临界尺寸阈值而不能有性生殖[95]。这种依赖自然种群(甚至是 Geitler 在他 1932 年的论文中,他主要负责发起有性生殖的实验室研究)大大推迟了发现许多羽状硅藻是异性生殖,但它的优点是调查了比通常在实验室中所能获得的结果更多的克隆间交配的结果,并提供了一些关于繁殖季节性的初步信息(尽管这些数据尚未审查和合成)。尤其是 20 世纪 90 年代以来根据[1,47]的开创性工作,数据主要来自实验室实验,这些实验允许交配系统被调查,并提供了关于性化的触发器的信息[96]以及细胞相互刺激和吸引的机制[73-75,81],尽管还没有细胞识别彼此的方式,我们怀疑这是物种形成的关键阶段。能够吸引相关物种的细胞,但不识别它们为"相同"的细胞的能力[97-98]。

羽状硅藻硅藻长期以来被认为是主要的异体(双亲)性生物[1,99],这可能是真的。然而,一些较老的概括已经被证明是不太可靠的。例如,历史上,所有的硅藻都被认为是同源的,但看起来只是中心硅藻的规律(唯一已知的例外是 *Ardissone*)[33-34],许多羽状被证明是异宗配合的。最初的异宗配合记录是在三个无瓶偏瓜藻:亚得里亚海棒形藻、海洋字瓶藻和脆壳细支藻,这些是直到 Drebes 的综述之前已知道的唯一一例子[2]。然而,从那时起,异宗硅藻的例子就成倍增加了。在异宗羽状硅藻中,有性繁殖发生在属于不同交配类型的无性繁殖之间[见图 11.1(a)~(f)]。当交配类型之间的配子行为不同时("生理"或"行为"异配子),可以谈论"雄性"和"雌性",或"＋"和"－"[22],但不同交配类型产生的配子和配子之间有时没有明显或行为差异[101]。近年来,在许多鞭毛属中都发现了异种现象,包括伪尼兹希藻[102-104],拟脆杆藻属[105],简柱藻[106],长蓝藻[98],羽纹藻[99],哈斯勒藻[107-108]和裂矛形藻[33-34],加上 Chepurnov 等当时已经知道的综述。也有其他关于 *raphid* 羽状的异种繁殖的报道,如平板藻、榆叶藻和倒叶藻科[87]。有些物种不是专性异孢子的,因为它们表现出一定(通常相当有限)程度的克隆内增生孢子[109]。

交配系统可以在近缘物种之间有所不同(在同一属内)。异宗和同宗配合行为已在不同物种的鞍型藻属和拟菱形藻中被证实[110-112],以及不同物种长蓝藻,鞍型藻属,和菱形藻的异种生殖行为和自动生殖行为[110,113-115]。此外,对交配行为的复杂控制在长柄曲壳藻中很明显,克隆可以是"雌雄同株"、"双性"和"单性"[116-118]。

在鞍型藻属和菱形藻的研究中,自动(单亲)繁殖似乎是从双亲祖先反复进化而来的[110,115,119],尽管我们仍然没有足够的信息(可用的内容没有进行正式分析),我们认为,异型繁殖的广泛分布和固有的更大的发育复杂性与这种状态在羽鳃目硅藻中

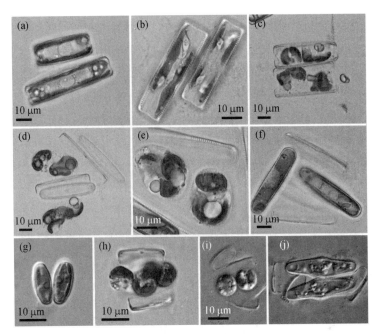

图 11.1 羽状硅藻的双亲本繁殖的异亲本繁殖

的原始状态是一致的,同型繁殖和不同类型的单亲繁殖是衍生的。

在异型硅藻和同型硅藻中,人们发现生殖隔离有时与形态差异不挂钩。在舟形藻属的一些物种中已经观察到习惯性的克隆内繁殖,例如隐头舟形藻和洛萨尔格特勒舟形藻生殖分离,形成遗传分化的姐妹支,但根据 LM 硅藻壳形态几乎无法区分[91,120]。另一个相关的和形态相似的物种是自发孤雌生殖普通舟形藻[92],其中每个细胞谱系的定义都是生殖隔离的,并包括一个单独的"微物种"。微种(由于遗传自动融合:见下文)也在菱形藻组有记录。Lanceolatae[115,121] 和 N. inconspicua 被发现由几个分支组成的复合体,其中一些分支与形态上非常不同的硅藻(双球舟形藻和克兹氏齿藻)长期以来一直被认为是独立的物种[115]。在雌雄异株的硅藻中,生殖分离、隐或假分泌物种的证据继续积累例如中国鞍型藻属[122]和拟菱形藻[112]。在 DeDecker 等研究中分析了同域的半舟藻谱系之间的不完全生殖隔离[123]。

同宗配合克隆的有性繁殖通常在接种到新培养基后的几天内就会被诱导[61],且该过程可能非常迅速且激烈,以至于如果在分离后对克隆的检查不够频繁,可能会错过这一过程[124]。当然,在理论上,这种克隆内繁殖的影响可以被细胞尺寸的变化所检测,即使有性繁殖本身所忽略。事实上,由于强烈的克隆内辅助孢子形成,培养物中小型有性能力细胞的比例经常迅速下降,在一个月甚至一周内,几乎可以完全被扩大的初始细胞和初始后细胞所取代。(未发表的观察资料包括节状小舟藻,Neidium cf. ampliatum,菱形藻和鞍型藻)[124]。然而,一旦培养物只包含新扩大的细胞,可能在数月或数年内没有进一步的有性生殖,直到再次有足够的细胞低于有性阈值。因此,很容易忽略同性和其他形式的无内有性生殖。以 Luticola 为例,它的自然栖息地可能是可以理解的——它在洞穴中生长[99]——它只是间歇潮湿,可能只提供有限和零星的机会来满足外部的性伴侣。然而,同宗配合的物种,例如 S. bisexualis,也发生在更多的环境中。

在瘤双壁藻中,观察到克隆内的有性繁殖,但一些无性系对培养基的补充无反应,并保持营养[61]。在这种情况下,交配系统可能基本上是异型的,但具有一定的系内繁殖能力,这种行为已在菱形藻,平板藻,脆杆藻,哈斯勒藻和肘杆藻[47,88,107,125],也发生在 S. blackfordensis(Chepurnov 和 Mann 未发表的观察结果)。Roshchin[47]将这种行为解释为两种性行为(雌雄同株、雌雄异株)的世代交替,但 Davidovich 等[109,125]报道了平板藻和菱形藻的雄性无性系内繁殖,并提出了"兼性雄性"一词[22,125]。兼性雄激素可能代表了栖息在新栖息地的先驱种群的适应优势,或者代表了异雄性生殖尚未开始的种群中性行为的"最后机会"。因此,它们已经接近了生存能力和/或性能力丧失时的大小阈值[109]。

与中心硅藻一样,大小缩小-恢复周期的存在并不意味着一个种群是有性的,细胞之间有交配和有性繁殖。生长孢子也可以自动形成,(即,融合发生在单个细胞产生的配子之间)或非融合。在这两种情况下,缺陷孢子都是"单亲本";通常一个单一的未配对细胞产生一个单一的缺陷孢子,尽管在卵形藻属中记录了假种群,包括细胞之间的配对[126]。单亲本繁殖以前一直被认为是罕见的[1],但最近在各种属中发表了一些新的病例,包括曲壳藻属,短缝藻属,穆勒利亚藻属,舟形藻属,长蓝藻属,菱形藻属和羽纹藻属[92,110,113-115,121,127-130]。单亲本繁殖的原则已经被解释了好几次了[1]。简单地说,已知的单亲本缺陷孢子形成的两种主要形式——无融合生殖和自动融合生殖。在无融合生殖中,减数分裂缺失或高度修饰("假减数分裂"),核没有性融合,而在自动融合过程中,正常的减数分裂发生,然后是两个减数分裂产物之间的融合。自动融合技术有两种变体[22]。在配子自动融合中[见图 11.2(a)~(h)],未配对的亲本细胞在减数分裂 I 后形成两个配子,并融合形成一个二倍体受精子[113-114]。在自配子自融合中[见图 11.2(i)~(n)],减数分裂 I 后没有胞质分裂,两个单倍体核在未分裂的细胞内融合形成一个缺陷孢子。不出所料,双藻配子中从未报道过,因为精子和卵细胞从未在同一配子囊中形成。在目前已知的所有情况下,双配子繁殖发生在相关的属中,双亲繁殖种每个配子产生两个配子[113,128,131]。相比之下,自配子发生在鞍型藻属的一些物种中,在这个属中,大多数物种是双亲本的,每个配子体只产生一个配子[110]。祖先的性状重建表明,鞍型藻属是原始的双亲本,单亲本自孢子形成进化得相当晚,可能有几次[110]。类似地,泛婚制可能在菱形藻属至少进化了两次[113]。总的来说,相对于双亲本有性繁殖,自动融合在羽状硅藻中似乎仍然不常见,并且在其他成员为异体的不相关谱系中偶尔发生[113,121]。这种形式——异配或自配——似乎在很大程度上取决于异体祖先的行为。

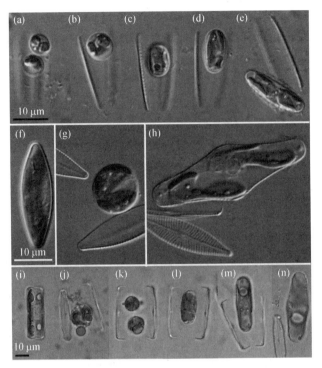

图 11.2 戊状硅藻的单亲本繁殖

11.4 生长孢子的发育与结构

除了上述所有的生命周期和性的特性外,硅藻缺陷孢子形成的一个方面也很少受到应有的关注,即缺陷孢子发展。这是生命周期中形状从头产生的阶段[132],理解它是理解硅藻形态的巨大多样性是如何产生的关键。然而,尽管缺陷孢子的 SEM 图像最早出现在 35 年前的文献中[133],以及更早的透射电镜和高质量的 LM 图像,但对受精卵和缺陷孢子超微结构的比较研究是一个相对较新的兴趣领域。

11.4.1 初期阶段

似乎受精后的第一个事件通常是由年轻的受精卵形成的多糖层(初级壁),这也可能包含二氧化硅鳞片。鳞片通常是扁平的±圆形结构,有一个类似简化的中心瓣,但有时鳞片具有二分式分支的刺[134]。一段时间以来,我们已经知道缺陷孢子的发育可能只涉及(1)鳞片的形成(例如,参见最近关于辐环藻属对辐环藻属的描述的文献)[54],或(2)鳞片与外壁一起[22]。有可能一些缺陷孢子具有完全有机壁,尽管据我们所知,这从未被证实:即使在脆弱的筒柱藻中,也存在腹膜环带[106]。在生长孢子生长之前形成的元素被称为"婴儿"(来自拉丁语的"襁褓衣服"或"摇篮"):用

Kaczmarska 等的话来说,"初级受精卵壁和随后在其作为缺陷孢子扩张之前被添加到受精卵壁的部分(次级元素)组成了'无囊'"[22]。然而,在放线菌环菌中,鳞片甚至存在于卵细胞周围[54],尽管分泌了更多的物质来覆盖受精卵和扩大缺陷孢子。在这里,卵壁、包囊壁和缺陷孢子壁之间的区别尚不清楚。然而,在其他硅藻中,有更明显的区别。例如,在塞拉藻属物种和隐形尼兹希藻中,鳞片包含圆形鳞片,在生长孢子扩张之前完成,而生长孢子壁由横向和纵向的边缘带加强[21,113]。在一些羽纹藻属和菱形藻属中,微小的元素包括精细的未形成图案的条带,它们被轻微硅化[99,121,128]。这些条带大致横向于缺陷孢子的长轴,但在一定程度上纵横交错,有时类似于一个"毛球"[121]。

针晶硅藻长蓖藻及其盟友的硅化程度特别强烈。在最著名的案例中(在长蓖藻的一种)[135]有一个由六个内骨元素组成的分化良好的系统——两个极帽和四个侧板。帽重叠的板,因此可能会首先形成,但发育似乎或多或少同时发生,并发生在受精卵仍被封闭在配子囊内。帽的超微结构表明,它们可能来源于融合的鳞片。后来,由于生长孢子的生长,囊被分成两半,每一半由一个帽和两个侧板组成;这些覆盖了缺陷孢子的末端,似乎确保了它在扩张时保持恒定的直径。类似的摇篮似乎存在于穆勒利亚藻属和旋舟藻属[20]和双桨藻[90]。

主壁和内壁在其他半部中也一分为二,尽管似乎这一半并不总是像长蓖藻属那样相等[21]。关于刺骨及其发展的细节很少被确定。在针晶羽状硅藻上的等径尺度已被多次报道,包括小杆线虫属,弓桥藻属,斑条藻属,伪条纹藻属和平板藻[136-139]。关于针晶硅藻这种规模的报告并不常见,但包括拟菱形藻[140],鞍型藻属[95]和双壁藻属[61]。鳞状和带状都可以在系统发育密切相关的类群中发现,如鞍型藻属(鳞状砧骨)和羽纹藻属(砧骨带;鞍型藻属和羽纹藻属的关系)[26,55],甚至在一个属内(例如菱形藻属)[113,121],表明摇篮发生了趋同进化。在中心硅藻水链藻属[58]和 Ardissonea[141]中出现类似于菱形藻属和羽纹藻属进一步证明了这一点。

总的来说,从硅藻中鳞片囊的广泛分布来看,这似乎是祖先类型[95];这些覆盖物在 Round 和 Crawford[142]提出的早期硅藻进化模型中发挥了重要作用,并邀请与硅藻姐妹群色藻界的比例情况进行比较[94,143]。

11.4.2 外壁

在属于不同谱系的"径向中心"中,如埃勒贝克藻[144],辐环藻属[54],海滨藻[37],和海链藻[144],扁平的圆形鳞片是唯一添加到缺陷孢子壁的二氧化硅结构。然而,在缺陷孢子扩张过程中,一些中心硅藻("极"中心,具有细长、多边形或多极瓣膜)和羽状形成了硅带和环的复杂系统,称为"胞孔藻"。它的功能似乎是创造和控制生长孢子的非等距扩张。戊状硅藻的边缘层通常由两个系列的带组成,横向和纵向[133]。一些中心硅藻(以前称为"属性硅藻")[22]的分化不太明显,而异常中心硅藻的高度细长的缺陷孢子是由一个奇怪的"鳞片带"系统支持的[141]由于其精致,其形态和精确排列难以辨别。

横向围生孢子是由扩张的生长孢子依次产生的,在其顶端添加横向带(初生和次生)[132]。在大多数情况下,纵向腹膜形成的时间仍然不清楚,因为它是在横向腹膜内部形成的,因此在 SEM 中隐藏在整个正在发育的缺陷孢子中,在 LM 中几乎不可见;只在没有横向腹膜带的物种中[145-147],纵向腹膜暴露。值得注意的是,目前还没有关于生长孢子发育的薄片研究。然而,在使用荧光色素 PDMPO 研究凯尔盖朗脆藻中,Fuchs 等表明,正如从其位置所预期的那样,纵向包囊只有在横向包囊完成后才会添加[105]。

这也是因为纵向围腹带的内部位置(但通常也因为它们的微妙),而横向腹带已被发现约 150 年[148]和它的性质被多次回顾[1,4,22,132],对纵向囊的详细形态知之甚少。它是由 von Stosch 发现的,对其结构的完整描述仅适用于小杆线虫属,弯楔藻属[133],曲壳藻[145,147,149],菱形藻属[113],双壁藻属[61]和平板藻[146]。其中 4 种(小杆线虫属,弯楔藻属,菱形藻属和双壁藻属),尽管系统发育分离广泛,但纵向腹带具有相同的基本结构,由 5 个带组成——中间的初级带两侧各有两个次级带[61]。然而,在两种已知的情况下,纵向的腹膜(曲壳藻属和平板藻)[145-147],条带的数量更高。尽管其条带的平面外观,纵向的腹膜可以是一个高度三维的结构[61]。它的功能是未知的;Idei 等推测,"它可能有助于缺陷孢子的开裂,使初始细胞通过提供一个舌状支撑,初始细胞可以按压打开横向系列,从而使其从缺陷孢子逃逸"。

11.5 性生殖诱导

硅藻在长时间的营养分裂后恢复大小需要有性繁殖。在羽状硅藻中,有性繁殖成功的一个先决条件是细胞出现低于临界大小阈值[1]。另一个问题是有兼容的性伴侣。然而,由于个体硅藻的种群物种通常主要由相对较小的细胞低于临界尺寸阈值[95],很明显,前提条件往往会满足,提供人口密度足够高,这意味着外部因素可能高度影响控制有性生殖的发生。然而,人们对这些因素的性质以及它们在特定种群中的运作方式仍然知之甚少;这种情况在其他藻类群中也类似,例如双星藻[150]。光、温度和营养物质都被证明对不同的物种和生物因素(例如,特定细菌的存在)也已被证明在特定的情况下有影响。Drebes[2]和 Chepurnov 等[1]回顾了早期的文献。

在实验室中,低光水平和短光周期有利于牡蛎哈斯勒藻的增生孢子生长[96],对应于克隆分离地点的冬季和早春条件。Davidovich 等[151]的研究表明,夏季温度也会抑制牡蛎哈斯勒藻和卡拉达根哈斯勒藻有性繁殖,但允许快速的营养生长。不幸的是,和几乎所有其他硅藻物种一样,似乎没有关于 *Haslea* 在自然环境中有性繁殖的数据。光照条件也被发现对多列拟菱形藻[152],亚得里亚海杆线藻[100]和盾卵形藻[153]的副孢子形成很重要。然而在其他硅藻中,缺陷孢子形成明显不受光的影响[154-155]。连续光照已被证明可以抑制 *Haslea* 和中心硅藻角毛藻的有性繁殖[47,96,156],但支持其他中心辅助孢子的形成(小环藻和圆筛藻)[157-159]。光周期似乎对一些硅藻的性化很重要[160]，可能是通过细胞周期的同步[96]。节状小舟藻发生了两种自配途径，它们的相对频率似乎受到光照条件的影响[128]。在早期的出版物[161]中，有报道称光质量(波长)对角毛藻的影响[161]，而红光似乎对牡蛎哈斯勒藻的性化是必要的[96]。在对光质量的反应中，硅藻类似于其他藻类，既是相对密切相关的藻类，也是属于不同的超级类群；例如海褐藻的配子释放[162]，栅藻属的产生[163]，以及裂殖藻科的性化[150]。

在其他藻类中，营养水平已被证明对引起性反应很重要，如双星藻的性诱导和克里藻属和栅藻属中游动细胞的产生[150,163-164]。如前所述(在"产卵"下)，一些中心硅藻物种也会对营养水平作出反应——对海链藻中高水平的铵做出反应[49]，冠盘藻中硝酸盐的补充[67]，以及斯科沃尔佐夫藻中明显的磷水平较低[64]，尽管Jewson[66]认为在北极新星藻地区，有性繁殖的线索是低光照水平对生长的一种抑制。另一方面，Chepurnov等[1]指出，"到目前为止，还没有发现在羽状硅藻中有性繁殖是由严重的营养(如N或P)消耗引发的案例"，自2004年以来发表的关于羽状交配系统的论文通常记录了性化是在指数增长的文化中发生的。Scalco等[165]对性化和生殖成功进行了特别详细的研究。在一种海洋浮游物种(多纹拟菱形藻)中，表现出明显的季节性变化，因此可能会对营养水平的变化作出反应。他们记录了成功的寄生菌的主要要求，除了兼容的细胞的存在在生命周期的正确阶段，是无限的增长(成功繁殖迅速下降的指数阶段)，高细胞密度(确保频繁的细胞接触)，和缺乏一个强有力的混合机制。

参考文献

[1] Chepurnov, V. A., Mann, D. G., Sabbe, K., Vyverman, W. (2004). Experimental studies on sexual reproduction in diatoms, Int. Rev. Cytol., 237, 91-154.
[2] Drebes, G. (1977). Sexuality. In The Biology of Diatoms. pp. 250-283. D Werner (ed). Oxford: Blackwells.
[3] Geitler, L. (1932). Der Formwechsel der pennaten Diatomeen, Archiv für Protistenkunde, 78, 1-226.
[4] Round, F. E., Crawford, R. M., Mann, D. G. (1990). The diatoms. Biology and morphology of the genera. Cambridge: Cambridge University Press.
[5] Mann, D. G., Crawford, R. M., Round, F. E. (2017). Bacillariophyta. In Handbook of the Protists. Second edition. J. M Archibald, A. G. B Simpson and C. H Slamovits (eds.). Dordrecht: Springer International.
[6] Beraldi-Campesi, H., Mann, D. G., Cevallos-Ferriz, S. R. S. (2015). Life cycle of 70 Ma-old nonmarine pennate diatoms, Cretaceous Research, 56, 662-672.
[7] Jewson, D. H., Harwood, D. M. (2017). Diatom life cycles and ecology in the Cretaceous, J. Phycol., 53 (3), 616-628.
[8] Thwaites, G. H. K. (1847a). On conjugation in the Diatomaceae, Annals and Magazine of Natural History, 20, 9-11.
[9] Thwaites, G. H. K. (1847b). On conjugation in the Diatomaceae, Annals and Magazine of Natural History, 20, 343-344.
[10] Smith, W. (1856). Synopsis of British Diatomaceae, vol. 2. J. London: van Voorst.
[11] Griffith, J. W. (1855). On conjugation of the Diatomaceae, Annals and Magazine of Natural History,

series, 2(16), 92-94.
[12] Carter, H. J. (1856). On the conjugation of Cocconeis, Cymbella and Amphora; together with some remarks of Amphiprora alata (?), Kg, Annals and Magazine of Natural History, 2(17), 92-94. series.
[13] Macdonald, J. D. (1869). I.—On the structure of the Diatomaceous frustule, and its genetic cycle, Annals and Magazine of Natural History, 3(3), 1-8.
[14] Pfitzer, E. (1869). Ueber den Bau und die Zelltheilung der Diatomeen, Botanische Zeitung, 27, 774-776.
[15] Crawford, R. M. (1995). The role of sex in the sedimentation of a marine diatom bloom, Limnol. Oceanogr., 40(1), 200-204.
[16] Pfitzer, E. (1871). Untersuchungen über Bau und Entwickelung der Bacillariaceen (Diatomaceen). In Botanische Abhandlungen. 1. pp. 1-189. Hanstein (ed).
[17] Geitler, L. (1973). Auxosporenbildung und Systematik bei pennaten Diatomeen und die Cytologie von Cocconeis-Sippen, Österr. Bot. Z., 122(5), 299-321.
[18] VON STOSCH, H. A, Stosch, H. Avon. (1950). Oogamy in a centric diatom, Nature, 165(4196), 531-532.
[19] Margalef, R. (1969). Size of centric diatoms as an indicator, Mitteilungen der Internationale Verein für theoretische und angewandte Limnologie, 17, 202-210.
[20] Edlund, M. B., Stoermer, E. F. (1997). Ecological, evolutionary, and systematic significance of diatom life histories, J. Phycol., 33(6), 897-918.
[21] Mann, D. G., Poulíčková, A., Sato, S., Evans, K. M. (2011). Scaly incunabula, auxospore development, and girdle polymorphism in Sellaphora marvanii sp. nov. (Bacillariophyceae), J. Phycol., 47 (6), 1368-1378.
[22] Kaczmarska, I., Poulíčková, A., Sato, S., Edlund, M. B., Idei, M., Watanabe, T., et al. (2013). Proposals for a terminology for diatom sexual reproduction, auxospores and resting stages, Diatom Research, 28(3), 263-294.
[23] Adams, N. G., Trainer, V. L., Rocap, G., Herwig, R. P., Hauser, L. (2009). Genetic population structure of Pseudo-nitzschia pungens (Bacillariophyceae) from the Pacific Northwest and the North Sea, J. Phycol., 45(5), 1037-1045.
[24] Casteleyn, G., Leliaert, F., Backeljau, T., Debeer, A.-E., Kotaki, Y., Rhodes, L., et al. (2010). Limits to gene flow in a cosmopolitan marine planktonic diatom, Proceedings of the National Academy of Sciences, 107(29), 12952-12957.
[25] Chen, G., Rynearson, T. A. (2016). Genetically distinct populations of a diatom co-exist during the North Atlantic spring bloom, Limnol. Oceanogr., 61(6), 2165-2179.
[26] Evans, K. M., Chepurnov, V. A., Sluiman, H. J., Thomas, S. J., Spears, B. M., Mann, D. G. (2009). Highly differentiated populations of the freshwater diatom Sellaphora capitata suggest limited dispersal and opportunities for allopatric speciation, Protist, 160(3), 386-396.
[27] Godhe, A., Kremp, A., Montresor, M. (2014). Genetic and microscopic evidence for sexual reproduction in the centric diatom Skeletonema marinoi, Protist, 165(4), 401-416.
[28] Harnstrom, K., Ellegaard, M., Andersen, T. J., Godhe, A. (2011). Hundred years of genetic structure in a sediment revived diatom population, Proceedings of the National Academy of Sciences, 108(10), 4252-4257.
[29] Vanormelingen, P., Evans, K. M., Mann, D. G., Lance, S., Debeer, A. E., D'Hondt, S, D'Hondt, S., et al. (2015). Genotypic diversity and differentiation among populations of two benthic freshwater diatoms as revealed by microsatellites, Mol. Ecol., 24(17), 4433-4448.
[30] Karsten, G. (1928). Abteilung Bacillariophyta (Diatomeae). In Die natürlichen Pflanzenfamilien. 2. 2. pp. 105-303. A. Engler and K. Prantl (ed). Leipzig: W. Engelmann.
[31] Medlin, L. K., Kooistra, W. H. C. F., Schmid, A.-M. M. (2000). A review of the evolution of the diatoms—a total approach using molecules, morphology and geology. In The Origin and Early Evolution of the Diatoms: Fossil, Molecular and Biogeographical Approaches. pp. 13-35. A. Witkowski and J. Sieminska (eds.). Cracow, Poland: W. Szafer Institute of Botany, Polish Academy of Sciences.
[32] Theriot, E. C., Ashworth, M. P., Nakov, T., Ruck, E., Jansen, R. K. (2015). Dissecting signal and noise in diatom chloroplast protein encoding genes with phylogenetic information profiling, Mol. Phylogenet. Evol., 89, 28-36.
[33] Davidovich, N. A., Davidovich, O. I., Podunay, Y. A., Gastineau, R., Kaczmarska, I., Poulíčková, A., et

al. (2017a). Ardissonea crystallina has a type of sexual reproduction that is unusual for centric diatoms, Sci. Rep., 7(1), 14670.

[34] Davidovich, N. A., Davidovich, O. I., Witkowski, A., Li, C., Dąbek, P., Mann, D. G., et al. (2017b). Sexual reproduction in Schizostauron (Bacillariophyta) and a preliminary phylogeny of the genus, Phycologia, 56(1), 77-93.

[35] Nagai, S., Hori, Y., Manabe, T., Imai, I. (1995). Restoration of cell size by vegetative cell enlargement in Coscinodiscus wailesii (Bacillariophyceae), Phycologia, 34(6), 533-535.

[36] Mills, K. E., Kaczmarska, I. (2006). Autogamic reproductive behavior and sex cell structure in Thalassiosira angulata (Bacillariophyta), Botanica Marina, 49(5/6), 417-430.

[37] Kaczmarska, I., Ehrman, J. M. (2015). Auxosporulation in Paralia guyana MacGillivary (Bacillariophyta) and Possible New Insights into the Habit of the Earliest Diatoms, PLoS ONE, 10(10), e0141150.

[38] Gallagher, J. C. (1983). Cell enlargement in Skeletonema costatum (Bacillariophyceae), J. of Phycology, 19(4), 539-542.

[39] Koester, J. A., Brawley, S. H., Karp-Boss, L., Mann, D. G. (2007). Sexual reproduction in the marine centric diatom Ditylum brightwellii (Bacillariophyta), Eur. J. Phycol., 42(4), 351-366.

[40] Koester, J. A., Swalwell, J. E., von Dassow, P., Armbrust, E. V. (2010). Genome size differentiates co-occurring populations of the planktonic diatom Ditylum brightwellii (Bacillariophyta), BMC Evol. Biol., 10(1), 1.

[41] Sharpe, S. C., Koester, J. A., Loebl, M., Cockshutt, A. M., Campbell, D. A., Irwin, A. J., et al. (2012). Influence of cell size and DNA content on growth rate and photosystem II function in cryptic species of Ditylum brightwellii, PLoS ONE, 7(12), e52916.

[42] Findlay, I. W. O. (1969). Cell size and spore formation in a clone of a centric diatom, Coscinodiscus pavillardii Forti, Phykos, 8, 31-41.

[43] Chepurnov, V. A., Mann, D. G., von Dassow, P., Armbrust, E. V., Sabbe, K., Dasseville, R., et al. (2006). Oogamous reproduction, with two-step auxosporulation, in the centric diatom Thalassiosira punctigera (Bacillariophyta), J. Phycol., 42(4), 845-858.

[44] Hasle, G. R. (1983). Thalassiosira punctigera (Castr.) comb, nov., a widely distributed marine planktonic diatom, Nordic Journal of Botany, 3(5), 593-608.

[45] Schreiber, E. (1931). Über Reinkulturversuche und experimentelle Auxosporenbildung bei Melosira nummuloides, Archiv für Protistenkunde, 73, 331-344.

[46] Kustenko, N. G. (1978). Obrazovanie krupnykh kletok v koloniyakh diatomei Melosira moniliformis, Biologiya Morya, 1978, 72-74.

[47] Roshchin, A. M. (1994). Zhiznennye tsikly diatomovyh vodoroslej. Kiev: Naukova Dumka.

[48] Armbrust, E. V., Chisholm, S. W. (1992). Patterns of cell size change in a marine centric diatom: variability evolving from clonal isolates, J. Phycol., 28(2), 146-156.

[49] Moore, E. R., Bullington, B. S., Weisberg, A. J., Jiang, Y., Chang, J., Halsey, K. H. (2017). Morphological and transcriptomic evidence for ammonium induction of sexual reproduction in Thalassiosira pseudonana and other centric diatoms, bioRxiv preprint.

[50] Chepurnov, V. A., Chaerle, P., Vanhoutte, K., Mann, D. G. (2012). How to breed diatoms: examination of two species with contrasting reproductive biology. In The Science of Algal Fuels. pp. 325-340. R. Gordon and J. Seckbach (eds.). Dordrecht: Springer.

[51] Chepurnov, V. A., Mann, D. G., von Dassow, P., Vanormelingen, P., Gillard, J., Inzé, D., et al. (2008). In search of new tractable diatoms for experimental biology, Bioessays, 30(7), 692-702.

[52] Koester, J. A., Berthiaume, C. T., Hiranuma, N., Parker, M. S., Iverson, V., Morales, R., et al. (2018). Sexual ancestors generated an obligate asexual and globally dispersed clone within the model diatom species Thalassiosira pseudonana, Sci. Rep., 8(1), 10492.

[53] Samanta, B., Heffell, Q., Ehrman, J. M., Kaczmarska, I. (2018). Spermatogenesis in the bipolar centric diatom Plagiogrammopsis vanheurckii (Mediophyceae), Phycologia, 57(3), 354-359.

[54] Idei, M., Osada, K., Sato, S., Toyoda, K., Nagumo, T., Mann, D. G. (2012). Gametogenesis and Auxospore Development in Actinocyclus (Bacillariophyta), PLoS ONE, 7(8), e41890.

[55] Medlin, L. K., Kaczmarska, I. (2004). Evolution of the diatoms: V. Morphological and cytological support for the major clades and a taxonomic revision, Phycologia, 43(3), 245-270.

[56] Jensen, K. G., Moestrup, Øjvind., Schmid, A.-M. M. (2003). Ultrastructure of the male gametes from

two centric diatoms, Chaetoceros laciniosus and Coscinodiscus wailesii (Bacillariophyceae), Phycologia, 42(1), 98-105.

[57] Mizuno, M. (2008). Evolution of centric diatoms inferred from patterns of oogenesis and spermatogenesis, Phycological Research, 56(3), 156-165.

[58] Idei, M., Sato, S., Nagasato, C., Motomura, T., Toyoda, K., Nagumo, T., et al. (2015). Spermatogenesis and auxospore structure in the multipolar centric diatom Hydrosera, J. Phycol., 51(1), 144-158.

[59] Heath, I. B., Darley, W. M. (1972). Observations on the ultrastructure of the male gametes of Biddulphia laevis Ehr, J. Phycol., 8, 51-59.

[60] Hoppenrath, M., Elbrächter, M., Drebes, G. (2009). Marine Phytoplankton. Selected microphytoplankton species from the North Sea around Helgoland and Sylt. E. Schweizerbart'sche. Stuttgart, Germany: Verlagsbuchhandlung.

[61] Idei, M., Sato, S., Watanabe, T., Nagumo, T., Mann, D. G. (2013). Sexual reproduction and auxospore structure in Diploneis papula (Bacillariophyta), Phycologia, 52(3), 295-308.

[62] Manton, I., Stosch, H. Avon. (1966). Observations on the fine structure of the male gamete of the marine centric diatom Lithodesmium undulatum, Journal of the Royal Microscopical Society, 85(2), 119-134.

[63] Crimaldi, J. P., Zimmer, R. K. (2014). The physics of broadcast spawning in benthic invertebrates, Ann. Rev. Mar. Sci., 6(1), 141-165.

[64] Jewson, D. H., Granin, N. G., Zhdanov, A. A., Gorbunova, L. A., Bondarenko, N. A., Gnatovsky, R. Y. (2008). Resting stages and ecology of the planktonic diatom Aulacoseira skvortzowii in Lake Baikal, Limnol. Oceanogr., 53(3), 1125-1136.

[65] Jewson, D. H., Granin, N. G. (2015). Cyclical size change and population dynamics of a planktonic diatom, Aulacoseira baicalensis, in Lake Baikal, Eur. J. Phycol., 50(1), 1-19.

[66] Jewson, D. H. (1992a). Size reduction, reproductive strategy and the life cycle of a centric diatom, Phil. Trans. R. Soc. Lond. B, 336(1277), 191-213.

[67] Jewson, D. H. (1992b). Life cycle of a Stephanodiscus sp. (Bacillariophyta)1, J. Phycol., 28(6), 856-866.

[68] Armbrust, E. V. (1999). Identification of a new gene family expressed at the onset of sexual reproduction in the centric diatom Thalassiosira weissflogii, Appl. Environ. Microbiol., 65, 3121-3128.

[69] Yamagishi, T., Motomura, T., Nagasato, C., Kawai, H. (2009). Novel proteins comprising the stramenopile tripartite mastigoneme in Ochromonas danica (Chrysophyceae), J. Phycol., 45(5), 1110-1115.

[70] Mann, D. G. (1993). Patterns of sexual reproduction in diatoms, Hydrobiologia, 269-270(1), 11-20.

[71] Hopkins, J. T. (1969). Diatom motility: its mechanism, and diatom behaviour patterns in estuarine mud, PhD thesis, University of London.

[72] Sato, S., Medlin, L. K. (2006). Motility of non-raphid diatoms, Diatom Research, 21(2), 473-477.

[73] Bondoc, K. G. V., Lembke, C., Vyverman, W., Pohnert, G. (2016). Searching for a mate: pheromone-directed movement of the benthic diatom Seminavis robusta, Microb. Ecol., 72(2), 287-294.

[74] Gillard, J., Frenkel, J., Devos, V., Sabbe, K., Paul, C., Rempt, M., et al. (2013). Metabolomics enables the structure elucidation of a diatom sex pheromone, Angew. Chem. Int. Ed., 52(3), 854-857.

[75] Moeys, S., Frenkel, J., Lembke, C., Gillard, J. T. F., Devos, V., Van den Berge, K., et al. (2016). A sexinducing pheromone triggers cell cycle arrest and mate attraction in the diatom Seminavis robusta, Sci. Rep., 6(1), 19252.

[76] Patil, S., Moeys, S., von Dassow, P., Huysman, M. J. J., Mapleson, D., De Veylder, L., et al. (2015). Identification of the meiotic toolkit in diatoms and exploration of meiosis-specific SPO11 and RAD51 homologs in the sexual species Pseudo-nitzschia multistriata and Seminavis robusta, BMC Genomics, 16(1), 930.

[77] Vanstechelman, I., Sabbe, K., Vyverman, W., Vanormelingen, P., Vuylsteke, M. (2013). Linkage mapping identifies the sex determining region as a single locus in the Pennate diatom Seminavis robusta, PLoS ONE, 8(3), e60132.

[78] Basu, S., Patil, S., Mapleson, D., Russo, M. T., Vitale, L., Fevola, C., et al. (2017). Finding a partner in the ocean: molecular and evolutionary bases of the response to sexual cues in a planktonic diatom, New Phytol., 215(1), 140-156.

[79] Frenkel, J., Vyverman, W., Pohnert, G. (2014). Pheromone signaling during sexual reproduction in algae, Plant J., 79(4), 632-644.
[80] Lembke, C., Stettin, D., Speck, F., Ueberschaar, N., De Decker, S., Vyverman, W., et al. (2018). Attraction pheromone of the benthic diatom Seminavis robusta: studies on structure-activity relationships, J. Chem. Ecol., 44(4), 354-363.
[81] Sato, S., Beakes, G., Idei, M., Nagumo, T., Mann, D. G. (2011). Novel sex cells and evidence for sex pheromones in diatoms, PLoS ONE, 6(10), e26923.
[82] Venuleo, M., Raven, J. A., Giordano, M. (2017). Intraspecific chemical communication in microalgae, New Phytol., 215(2), 516-530.
[83] Edgar, R., Drolet, D., Ehrman, J. M., Kaczmarska, I. (2014). Motile male gametes of the araphid diatom Tabularia fasciculata search randomly for mates, PLoS ONE, 9(7), e101767.
[84] Davidovich, N. A., Kaczmarska, I., Karpov, S. A., Davidovich, O. I., MacGillivary, M. L., Mather, L. (2012b). Mechanism of male gamete motility in araphid pennate diatoms from the genus Tabularia (Bacillariophyta), Protist, 163(3), 480-494.
[85] Mann, D. G., Stickle, A. J. (1991). The genus Craticula, Diatom Research, 6, 79-107.
[86] Mann, D. G., Stickle, A. J. (1995). Sexual reproduction and systematics of Placoneis (Bacillariophyta), Phycologia, 34(1), 74-86.
[87] Kaczmarska, I., Gray, B. S., Ehrman, J. M., Thaler, M. (2017). Sexual reproduction in plagiogrammacean diatoms: First insights into the early pennates, PLoS ONE, 12(8), e0181413.
[88] Podunay, Y. A., Davidovich, O. I., Davidovich, N. A. (2014). Mating system and two types of gametogenesis in the fresh water diatom Ulnaria ulna. 24. pp. 3-18. Algologiya: Bacillariophyta.
[89] Hustedt, F. (1930). Bacillariophyta (Diatomeae). In Die Süsswasser-Flora Mitteleuropas 10. pp. 1-466. A. Pascher and G. Fischer (eds.). Jena.
[90] Witkowski, A., Barka, F., Mann, D. G., Li, C., Weisenborn, J. L., Ashworth, M. P., et al. (2014). A Description of Biremis panamae sp. nov., a new diatom species from the marine littoral, with an account of the phylogenetic position of Biremis D. G. Mann et E. J. Cox (Bacillariophyceae, PLoS ONE, 9 (12), e114508.
[91] Poulíčková, A., Mann, D. G. (2006). Sexual reproduction in Navicula cryptocephala (Bacillariophyceae), J. Phycol., 42(4), 872-886.
[92] Poulíčková, A., Veselá, J., Neustupa, J., Škaloud, P. (2010). Pseudocryptic diversity versus cosmopolitanism in diatoms: a case study on Navicula cryptocephala Kütz. (Bacillariophyceae) and morphologically similar taxa, Protist, 161(3), 353-369.
[93] Mann, D. G. (2015). Unconventional diatom collections, Nova Hedwigia, Beiheft, 144, 35-59.
[94] Mann, D. G., Marchant, H. (1989). The origins of the diatom and its life cycle. In The Chromophyte Algae: Problems and Perspectives (Systematics Association Special Volume 38). pp. 305-321. J. C Green, B. S. C Leadbeater and W. L Diver (eds.). Oxford: Clarendon Press.
[95] Mann, D. G. (2011). Size and sex. In The Diatom World. pp. 147-166. J Seckbach and J. P Kociolek (eds.). Dordrecht: Springer.
[96] Mouget, J.-L., Gastineau, R., Davidovich, O., Gaudin, P., Davidovich, N. A. (2009). Light is a key factor in triggering sexual reproduction in the pennate diatom Haslea ostrearia, FEMS Microbiol. Ecol., 69 (2), 194-201.
[97] Mann, D. G., Chepurnov, V. A., Droop, S. J. M. (1999). Sexuality, incompatibility, size variation, and preferential polyandry in natural populations and clones of Sellaphora pupula (Bacillariophyceae), J. Phycol., 35(1), 152-170.
[98] Mann, D. G., Chepurnov, V. A. (2005). Auxosporulation, mating system, and reproductive isolation in Neidium (Bacillariophyta), Phycologia, 44(3), 335-350.
[99] Poulíčková, A., Hašler, P. (2007). Aerophytic diatoms from caves in central Moravia (Czech Republic), Preslia, 79, 185-204.
[100] Rozumek, K. E. (1968). Der Einfluß der Umweltfaktoren Licht und Temperatur auf die Ausbildung der Sexualstadien bei der pennaten Diatomee Rhabdonema adriaticum Kütz, Beiträge zur Biologie der Pflanzen, 44, 365-388.
[101] Mann, D. G., Poulíčková, A. (2010). Mating system, auxosporulation, species taxonomy and evidence for homoploid evolution in Amphora (Bacillariophyta), Phycologia, 49(2), 183-201.

[102] Amato, A., Orsini, L., D'Alelio, D., Montresor, M. (2005). Life cycle, size reduction pattern, and ultrastructure of the pennate planktonic diatom Pseudo-nitzschia delicatissima (Bacillariophyceae), J. Phycol., 41(3), 542-556.

[103] Chepurnov, V. A., Mann, D. G., Sabbe, K., Vannerum, K., Casteleyn, G., Verleyen, E., et al. (2005). Sexual reproduction, mating system, chloroplast dynamics and abrupt cell size reduction in Pseudo-nitzschia pungens from the North Sea (Bacillariophyta), Eur. J. Phycol., 40(4), 379-395.

[104] Scalco, E., Stec, K., Iudicone, D., Ferrante, M. I., Montresor, M. (2014). The dynamics of sexual phase in the marine diatom Pseudo-nitzschia multistriata (Bacillariophyceae), J. Phycol., 50(5), 817-828.

[105] Fuchs, N., Scalco, E., Kooistra, W. H. C. F., Assmy, P., Montresor, M. (2013). Genetic characterization and life cycle of the diatom Fragilariopsis kerguelensis, Eur. J. Phycol., 48(4), 411-426.

[106] Vanormelingen, P., Vanelslander, B., Sato, S., Gillard, J., Trobajo, R., Sabbe, K., et al. (2013b). Heterothallic sexual reproduction in the model diatom Cylindrotheca, Eur. J. Phycol., 48(1), 93-105.

[107] Davidovich, N. A., Gastineau, R., Gaudin, P., Davidovich, O. I., Mouget, J.-L. (2012a). Sexual reproduction in the newly-described blue diatom, Haslea karadagensis, Fottea, 12(2), 219-229.

[108] Davidovich, N. A., Mouget, J.-L., Gaudin, P. (2009). Heterothallism in the pennate diatom Haslea ostrearia (Bacillariophyta), Eur. J. Phycol., 44(2), 251-261.

[109] Davidovich, N. A., Kaczmarska, I., Ehrman, J. M. (2010). Heterothallic and homothallic sexual reproduction in Tabularia fasciculata (Bacillariophyta), Fottea, 10(2), 251-266.

[110] Poulíčková, A., Sato, S., Evans, K. M., Chepurnov, V. A., Mann, D. G. (2015). Repeated evolution of uniparental reproduction in Sellaphora (Bacillariophyceae, Eur. J. Phycol., 50(1), 62-79.

[111] Quijano-Scheggia, S., Garcés, E., Andree, K., Fortuño, J. M., Camp, J. (2009a). Homothallic auxosporulation in Pseudo-Nitzschia brasiliana (Bacillariophyta), J. Phycol., 45(1), 100-107.

[112] Quijano-Scheggia, S. I., Garcés, E., Lundholm, N., Moestrup, Øjvind., Andree, K., Camp, J. (2009b). Morphology, physiology, molecular phylogeny and sexual compatibility of the cryptic Pseudo-nitzschia delicatissima complex (Bacillariophyta), including the description of P. arenysensis sp. nov, Phycologia, 48(6), 492-509.

[113] Mann, D. G., Sato, S., Rovira, L., Trobajo, R. (2013). Paedogamy and auxosporulation in Nitzschia sect. Lanceolatae (Bacillariophyta), Phycologia, 52(2), 204-220.

[114] Poulíčková, A. (2008a). Pedogamy in Neidium (Bacillariophyceae, Folia Microbiol. (Praha)., 21, 125-129.

[115] Rovira, L., Trobajo, R., Sato, S., Ibáñez, C., Mann, D. G. (2015). Genetic and Physiological Diversity in the Diatom Nitzschia inconspicua, J. Eukaryot. Microbiol., 62(6), 815-832.

[116] Chepurnov, V. A., Mann, D. G. (1997). Variation in the sexual behaviour of natural clones of Achnanthes longipes (Bacillariophyta), Eur. J. Phycol., 32(2), 147-154.

[117] Chepurnov, V., Mann, D. (1999). Variation in the sexual behaviour of Achnanthes longipes (Bacillariophyta). II. Inbred monoecious lineages, Eur. J. Phycol., 34(1), 1-11.

[118] Chepurnov, V., Mann, D. (2000). Variation in the sexual behaviour of Achnanthes longipes (Bacillariophyta). III. Progeny of crosses between monoecious and unisexual clones, Eur. J. Phycol., 35(3), 213-223.

[119] Geitler, L. (1953). Allogamie und Autogamie bei der Diatomee Denticula tenuis und die Geschlechtsbestimmung der Diatomeen, Osterr. Bot. Z., 100(4-5), 331-352.

[120] Poulíčková, A., Neustupa, J., Hašler, P., Tomanec, O., Cox, E. J. (2016). A new species, Navicula lothargeitleri sp. nov., within the Navicula cryptocephala complex (Bacillariophyceae, Phytotaxa, 273(1), 23-33.

[121] Trobajo, R., Mann, D. G., Chepurnov, V. A., Clavero, E., Cox, E. J. (2006). Taxonomy, life cycle, and auxosporulation of Nitzschia fonticola (Bacillariophyta), J. Phycol., 42(6), 1353-1372.

[122] Vanormelingen, P., Evans, K. M., Chepurnov, V. A., Vyverman, W., Mann, D. G. (2013a). Molecular species discovery in the diatom Sellaphora and its congruence with mating trials, Fottea, 13(2), 133-148.

[123] De Decker, S., Vanormelingen, P., Pinseel, E., Sefbom, J., Audoor, S., Sabbe, K., et al. (2018). Incomplete reproductive isolation between genetically distinct sympatric clades of the pennate model diatom Seminavis robusta, Protist, 169(4), 569-583.

[124] Poulíčková, A. (2008b). Morphology, cytology and sexual reproduction in the aerophytic cave diatom Luticola dismutica (Bacillariophyceae), Preslia, 80, 87-99.

[125] Davidovich, N. A., Kaczmarska, I., Ehrman, J. M. (2006). The sexual structure of a natural population of the diatom Nitzschia longissima (Bréb.) Ralfs. In Proceedings of the 18th International Diatom Symposium. pp. 27-40. A. Witkowski (ed). Bristol: Biopress Limited.

[126] Geitler, L. (1982). Die infraspezifischen Sippen von Cocconeis placentula des Lunzer Seebachs, Archiv für Hydrobiologie Algological Studies, 1(30), 1-11.

[127] Edlund, M. B., Spaulding, S. A. (2006). Initial observations on uniparental auxosporulation in Muelleria (Frenguelli) Frenguelli and Scoliopleura Grunow (Bacillariophyceae). In Advances in phycological studies. Festschrift in honour of Prof. Dobrina Temniskova-Topalova. pp. 211-223. N. Ognanova-Rumenova and K. Manoylov (eds.). Sofia and Moscow: Pensoft Publishers.

[128] Poulíčková, A., Mann, D. G. (2008). Autogamous auxosporulation in Pinnularia nodosa (Bacillariophyceae), J. Phycol., 44(2), 350-363.

[129] Sabbe, K., Chepurnov, V. A., Vyverman, W., Mann, D. G. (2004). Apomixis in Achnanthes (Bacillariophyceae); development of a model system for diatom reproductive biology, Eur. J. Phycol., 39 (3), 327-341.

[130] Vanormelingen, P., Chepurnov, V. A., Mann, D. G., Sabbe, K., Vyverman, W. (2008). Genetic divergence and reproductive barriers among morphologically heterogeneous sympatric clones of Eunotia bilunaris sensu lato (Bacillariophyta), Protist, 159(1), 73-90.

[131] Geitler, L. (1970). Pädogame Automixis und Auxosporenbildung bei Nitzschia frustulum var. perpusilla, Österr. Bot. Z., 118(1-2), 121-130.

[132] Mann, D. G. (1994). The origins of shape and form in diatoms: the interplay between morphogenetic studies and systematics. In Shape and form in plants and fungi. pp. 17-38. D. S. Ingram and A. J. Hudson (eds.). London: Academic Press.

[133] Mann, D. G. (1982). Structure, life history and systematics of Rhoicosphenia (Bacillariophyta). II. Auxospore formation and perizonium structure of Rh. curvata, J. Phycol., 18(2), 264-274.

[134] Samanta, B., Kinney, M. E., Heffell, Q., Ehrman, J. M., Kaczmarska, I. (2017). Gametogenesis and auxospore development in the bipolar centric diatom Brockmanniella brockmannii (family Cymatosiraceae, Protist, 168(5), 527-545.

[135] Mann, D. G., Poulíčková, A. (2009). Incunabula and perizonium of Neidium (Bacillariophyta), Fottea, 9(2), 211-222.

[136] Sato, S., Nagumo, T., Tanaka, J. (2004). Auxospore formation and the morphology of the initial cell of the marine araphid diatom Gephyria media (Bacillariophyceae), J. Phycol., 40(4), 684-691.

[137] Sato, S., Mann, D. G., Nagumo, T., Tanaka, J., Tadano, T., Medlin, L. K. (2008a). Auxospore fine structure and variation in modes of cell size changes in Grammatophora marina (Bacillariophyta), Phycologia, 47(1), 12-27.

[138] Sato, S., Mann, D. G., Matsumoto, S., Medlin, L. K. (2008b). Pseudostriatella (Bacillariophyta): a description of a new araphid diatom genus based on observations of frustule and auxospore structure and 18S rDNA phylogeny, Phycologia, 47(4), 371-391.

[139] Sato, S., Kuriyama, K., Tadano, T., Medlin, L. K. (2008c). Auxospore fine structure in a marine araphid diatom Tabularia parva (Bacillariophyta), Diatom Research, 23(2), 423-433.

[140] Kaczmarska, I., Bates, S. S., Ehrman, J. M., Leger, C. (2000). Fine structure of the gamete, auxospore and initial cell in the pennate diatom Pseudonitzschia multiseries (Bacillariophyta) Nova Hedwigia, 71, 337-357.

[141] Kaczmarska, I., Ehrman, J. M., Davidovich, N. A., Davidovich, O. I., Podunay, Y. A. (2018). Structure and Development of the Auxospore in Ardissonea crystallina (C. Agardh) Grunow Demonstrates Another Way for a Centric to Look Like a Pennate, Protist, 169(4), 466-483.

[142] Round, F. E., Crawford, R. M. (1981). The Lines of Evolution of the Bacillariophyta. I. Origin, Proceedings of the Royal Society B: Biological Sciences, 211(1183), 237, B 211-260.

[143] Yamada, K., Yoshikawa, S., Ohki, K., Ichinomiya, M., Kuwata, A., Motomura, T., et al. (2016). Ultrastructural analysis of siliceous cell wall regeneration in the stramenopile Triparma laevis (Parmales, Bolidophyceae), Phycologia, 55(5), 602-609.

[144] Schmid, A.-M. M., Crawford, R. M. (2001). Ellerbeckia arenaria (Bacillariophyceae): formation of auxospores and initial cells, Eur. J. Phycol., 36(4), 307-320.

[145] Toyoda, K., Williams, D. M., Tanaka, J., Nagumo, T. (2006). Morphological investigations of the

frustule, perizonium and initial valves of the freshwater diatom Achnanthes crenulata Grunow (Bacillariophyceae, Phycological Res., 54(3), 173-182.

[146] Mather, L., Ehrman, J. M., Kaczmarska, I. (2014). Silicification of auxospores in the araphid diatom Tabularia fasciculata (Bacillariophyta), Eur. J. Protistol., 50(1), 1-10.

[147] Toyoda, K., Idei, M., Nagumo, T., Tanaka, J. (2005). Fine-structure of the vegetative frustule, perizonium and initial valve of Achnanthes yaquinensis (Bacillariophyta), Eur. J. Phycol., 40(3), 269-279.

[148] Lüders, J. E. (1862). Beobachtungen über die Organisation, Theilungund Copulation der Diatomeen, Botanische Zeitung 20, 65-69.

[149] Idei, M., Mizuno, M. (1996). Perizonium structure in Achnanthes javanica f. subconstricta Abstracts, 14th International Diatom Symposium. p.32. Tokyo, Japan.

[150] Zwirn, M., Chen, C., Uher, B., Schagerl, M. (2013). Induction of sexual reproduction in Spirogyra clones-does an universal trigger exist? Fottea, 13(1), 77-85.

[151] Davidovich, O. I., Davidovich, N. A., Mouget, J.-L. (2018). The effect of temperature on vegetative growth and sexual reproduction of two diatoms from the genus Haslea Simonsen, Russ. J. Mar. Biol., 44 (1), 8-13.

[152] Hiltz, M., Bates, S. S., Kaczmarska, I. (2000). Effect of light: dark cycles and cell apical length on the sexual reproduction of the pennate diatom Pseudo-nitzschia multiseries (Bacillariophyceae) in culture, Phycologia, 39(1), 59-66.

[153] Mizuno, M., Okuda, K. (1985). Seasonal change in the distribution of cell size of Cocconeis scutellum var. ornata (Bacillariophyceae) in relation to growth and sexual reproduction, J. Phycol., 21(4), 547-553.

[154] Davidovich, N. A., Chepurnov, V. A. (1993). Intensivnost auksosporoobrazovaniya u dvuh vidov Bacillariophyta v zavisimosti ot osveshchennosti i prodolzhitelnosti fotoperioda, Algologiya, 3, 34-41.

[155] Davidovich, N. A. (2002). Fotoregulyatsiya polovogo vosproizvedeniya u Bacillariophyta (Obzor), Algologiya, 12, 259-272.

[156] Furnas, M. J. (1985). Dial synchronization of sperm formation in diatom Chaetoceros curvisetum Cleve, J. Phycol., 21(4), 667-671.

[157] Schultz, M. E., Trainor, F. R. (1968). Production of male gametes and auxospores in the centric diatoms Cyclotella meneghiniana and C. cryptica, J. Phycol., 4(2), 85-88.

[158] Roshchin, A. M. (1972). Vliyanie uslovij osveshcheniya na obrazovanie auksospor i skorost deleniya kletok Coscinodiscus granii Gough, Fiziologiya Rastenij, 19, 180-185.

[159] Roshchin, A. M. (1976). Vliyanie uslovij osveshcheniya na vegetativnoe razmnozhenie kletok i polovoe vosppoizvedenie dvuh vidov tsentricheskih diatomovyh vodoposlej, Fiziologiya Rastenij, 23, 715-719.

[160] Vaulot, D., Chisholm, S. W. (1987). Flow cytometric analysis of spermatogenesis in the diatom Thalassiosira weissflogii (Bacillariophyceae), J. Phycol., 23(1), 132-137.

[161] Baatz, I. (1941). Die Bedeutung der Lichtqualität für Wachstum und Stoffproduktion planktonischer Meeres Diatomeen, Planta, 31(4), 726-766.

[162] Pearson, G. A., Serrao, E. A., Dring, M., Schmid, R. (2004). Blue-and green-light signals for gamete release in the brown alga, Silvetia compressa, Oecologia, 138(2), 193-201.

[163] Cepák, V., Přibyl, P. (2006). The effect of colour light on production of zooids in 10 strains of the green chlorococcal alga Scenedesmus obliquus, Czech Phycology, 6, 127-133.

[164] Lokhorst, G. M. (1996). Comparative taxonomic studies on the genus Klebsormidium (Charophyceae) in Europe. In Cryptogamic Studies 5G. Fischer Verlag. pp.1-135. W. Jülich (ed). Stuttgart.

[165] Scalco, E., Amato, A., Ferrante, M. I., Montresor, M. (2016). The sexual phase of the diatom Pseudo-nitzschia multistriata: cytological and time-lapse cinematography characterization, Protoplasma, 253(6), 1421-1431.

第 12 章
北极底栖生物的生态学、细胞生物学和超微结构

Ulf Karsten，Rhena Schumann，Andreas Holzinger

12.1 简介

　　硅藻在不仅在浮游领域，而且在底栖生物的栖息地，特别是在柔软的底部海洋中扮演着生态的关键作用。在这里，它们以羽状形态大量形成微型底栖植物群落，这些群落往往覆盖广泛的沉积区。这些单细胞光自养生物排泄粘性的胞外聚合物（EPS），从而稳定沉积物在水动力下作用下的侵蚀性[1]。此外，底栖硅藻会改变沉积物的表面性质，从而影响 O_2 和营养物质在沉积物-水界面上的水平交换[2]。底栖硅藻和远洋硅藻占全球固碳量的 25%[3]，占海洋初级生产总量的 45%～50%[4-5]。有很多报告指出，全球底栖硅藻的初级生产量普遍较高[6]，导致不同河口和滩涂的总产量贡献高达 50%[7]。然而，在这些相当温带的栖息地中，微藻由于沉积物基质内的垂直迁移，在沉积物表面的细胞数量有很大的变化[8-9]。目前对北极微植物底栖生物群落的研究仍然很少，同时与温带地区相比，很少有研究表明初级产量很高[10-14]。因此，底栖硅藻组合是北极沿海地区软基质上营养网功能的关键组成部分，尽管研究还很薄弱。虽然对北极较高的营养水平有很好的描述[15]，但关于底栖硅藻的生态生理学和细胞生物学的信息仍然很少。因此，我们的目标是通过研究北极硅藻的生态生理学和细胞生物学特性来填补这一空白。

12.2 北极的环境背景

　　低温、季节性波动的太阳辐射以及长时间的冰雪覆盖是控制北极初级生产的关键环境因素[15]。底栖硅藻在一年中只有很短的时间光线充足，极地日从 4 月中旬到 8 月底，春季和秋季主要为黄昏条件[16]。极夜从 10 月底到 2 月中旬持续约 4 个月，北纬 80°的年太阳辐射比中纬度地区少约三分之一。此外，这些长时间的黑暗可以进一步延长，例如，在冬季受海冰形成保护的海湾和峡湾的内部，在春季和初夏之间海冰的破裂情况各不相同[16]。如果冰也被雪覆盖，太阳辐射就会减少到表面值的 2%以下。因此，光养的底栖生物群落可能会暴露在大约 10 个月的黑暗或非常低的光照

条件下[17-18]。北极水域这种极端波动的光照条件对底栖硅藻的初级生产和季节性生长有着强烈的影响。

虽然太阳辐射在北极表现出极端的季节性定性和定量变化,但年模式(例如,日长)不受全球变化的影响。相比之下,自1981年以来,北极水块一般每十年升温0.7~1.2℃[19],对底栖硅藻和其他海洋生物的影响未知[20]。北冰洋的平均表面温度在地理上取决于北大西洋的温水入侵(例如:斯匹茨卑尔根岛西部形成)。例如,在弗拉姆海峡(78°50′N),根据2002年9月至2003年8月的测量记录,斯匹次卑尔根西海岸(9°E)的水团温度为5℃,而格陵兰东海岸(8°W)的水团温度为-2℃[21]。北冰洋持续变暖的迹象很多,反映在地表空气温度升高、海冰覆盖消退和变薄、融化季节延长和河流流量增加等方面[22-23]。在当地,2011年6月至2012年6月,康斯峡湾(斯瓦尔巴群岛)沿岸沙质沉积物遗址的年平均温度为2.4℃±2.1℃,该地点表现出多样和密集的微底栖植物群落[13]。然而,在夏季,这个站的水温可以暂时上升到8℃以上。相比之下,2011年6月至8月,康斯峡湾深水区(大于5 m)的温度相对稳定,约为1~5℃[11]。

12.3 生长随温度的函数

从生态生理学的角度来看,生长响应模式代表了描述底栖硅藻在温度梯度下表现的最相关过程,因为他们最优地整合所有积极和消极的环境影响细胞,因此反映他们的适应潜力[24]。北极的底栖硅藻优先生长在沉积物和硬底层顶部的浅水区中,温度较低,但相对恒定。对北极底栖硅藻生长的第一个生态生理学研究是在两种脆杆藻属物种上进行的,它们是在康斯峡湾(斯瓦尔巴群岛)采集的海藻上分离出来的。这两个种群在12~14℃时均表现出最佳生长速率,生长良好,但在0℃时生长速率降低,在20℃时未存活[25]。相比之下,直舟形藻是从康斯峡湾的次滨海岩石中分离出来的,在1~15℃的生长速度非常相似,因此表现出广泛的生长最优(见图12.1)。然而,暴露于20℃会导致很高的死亡率(见图12.1)。

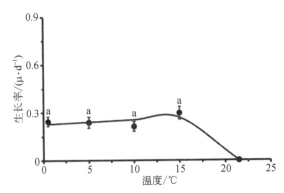

图12.1 温度升高对从斯瓦尔巴群岛康斯峡湾的潮下岩石中分离出来的定向底栖直舟形藻生长速率的影响

最近，Schlie & Karsten 评估了从冒险峡湾（斯瓦尔巴群岛）分离出来的 8 种底栖硅藻，关于它们在 1~20℃ 的控制条件下生长的温度要求。5/8 的样本在 1~20℃ 之间生长，但速率不同，2 株在 1℃ 时生长受到抑制，1 株在 20℃ 时生长受到抑制。此外，Schlie & Karsten 报道，只有 *Grammonema rostockensis* Stachura-Suchoples 和 *Nitzschia* sp. aff. *dubiiformis* Hustedt 生长表现出部分低温需求，而其他底栖硅藻在高温下表现出更高的生长潜力。因此，大多数类群的一般反应模式清楚地显示出对 6~15℃ 的偏好，表明北极底栖硅藻具有相当广热和耐冷的生长行为[26]。对于北极直舟形藻，在 0.5~4.5℃ 之间观察到生长速率的增加[27]，但更高的温度尚未被测试。

与北极硅藻相比，南极底栖硅藻的特征是生长需要非常低的温度，同时往往具有狭窄的温度耐受性。对于南极分离株，测量了布纹藻 *subsalinum* Peragallo 和 *Odontella litigiosa*（van Heurck）Hoban 的最佳生长速率，两种窄热底栖硅藻的生存温度均在 6~8℃ 之间[28]。在长期暴露（几周）期间，温度的升高导致了高死亡率，但是短期暴露是可以耐受的。同样，南极的 *Nitzschia lecointei* van Heurck 在 2~5℃ 之间最佳生长，而温度略有升高的 8℃ 则伴随着强烈的生长抑制[29]。北极和南极底栖硅藻生长的温度要求的显著差异的解释与两个极地地区的冷水历史有关[30]。与北极"年轻"的冷水地质历史相比，南极洲被认为有大约 2 300 万年[31]的冷水历史。这两种冷水系统的显著差异支持了南极洲许多海洋特有生物的发展，其中大多数对甚至略微升高的温度（高于 5℃）都很敏感[32]。

12.4　长期暗培养后的生长情况

极地硅藻在极地夜晚生存的生理状态以及潜在的生化和细胞生物学机制仍然知之甚少[33]。在这些微藻中，已经记载了长期黑暗存活的不同机制[34]。其中包括对储存能源产品的利用[33]，代谢活动的减少[35]，静息阶段的形成[36-37]和/或兼性异养生活方式[38]。能量储存产品的利用，如脂质（三酰基甘油），最近被证明为北极底栖硅藻在长时间黑暗期间的细胞维持代谢提供能量[33]。

我们还对来自斯匹次卑尔根岛的底栖硅藻直舟形藻进行了暗生存潜力的实验。该物种在模拟极地夜晚的黑暗中保存了 5 个月，并在 2、3、5 个月后，分别用 10 μmol 光子 $m^{-2} \cdot s^{-1}$ 的连续低光子通量率进行再辐照。2 和 3 个月后，直舟形藻在光照下表现出较高的生长速率（$\mu = 0.41 \sim 0.46/d$），而处理 5 个月后，生长速率明显下降至 $\mu = 0.17/d$（见图 12.2）。然而，与此同时，暗孵育时间越长，在再次建立最佳生长之前的滞后期就越长（见图 12.3）。在黑暗 2、3 和 5 个月后，滞后期分别强烈增加到 12、17 和 27 天，因此在模拟极地夜晚 5 个月后，再照射后近 1 个月直舟形藻才生长（见图 12.3）。

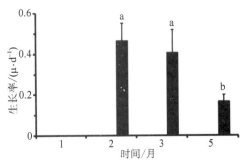

图 12.2 从康斯峡湾斯斯匹次卑尔根岛的一块次海岸岩石中分离出来的定向直舟形藻

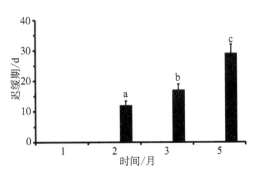

图 12.3 从康斯峡湾斯斯匹次卑尔根岛的一块亚滨海岩石中分离出来的直舟形藻恢复生长的滞后期

极地底栖硅藻的高暗生存潜力早前已有报道[25,39-40]。特别是高纬度的硅藻似乎有能力承受长时间的黑暗，这在考虑到北极和南极洲的波动和变化的辐射条件时可能是有益的。暗孵化的温带远洋硅藻的光合器官似乎在几周后已经受到受损，这反映在再照射后非常长的恢复阶段。相比之下，用于光合作用和生长的光捕获可以在极地之夜后在南极地区迅速恢复[41-42]。来自北极水域的其他底栖硅藻，如条纹脆杆藻，在 5℃暗孵育 3 个月后，叶绿体长度减少了 30%～40%，表明其通过细胞器成分分解为维持代谢补充了能量[25]。由于人们对北极底栖硅藻的黑暗生存机制通常知之甚少，因此必须进行全面和精确的评估，例如对储存脂质的再动员[33]。

最近，Kamp、deBeer、Nitsch、Lavik 和 Stief 提出了一种新的类似细菌的机制来解释底栖硅藻如何在长时间的黑暗中生存[43]。这些作者报道了暗生存潜力与细胞内积累 NO_3^- 的能力之间有很强的相关性，存储的 NO_3^- 在黑暗中被异化地还原为铵，即硅藻似乎能够呼吸硝酸盐来维持其新陈代谢。

12.5 长期暗培养后的细胞生物学性状

为了更好地了解北极底栖硅藻直舟形藻的暗生存机制的细胞生物学特征，我们使用各种荧光染料评估了细胞活力和活性。根据 Veldhuis，Kraay 和 Timmermans 的方案[44]，使用核酸染色 SYTOX，将具有完整细胞膜(质膜)的硅藻与具有渗透膜的硅藻区分开来。SYTOX 只能通过受损或受损的细胞膜，并染色细胞核，导致其在蓝光激发下的荧光增强。根据 Freese，Karsten 和 Schumann，以 5-氯甲基荧光素(GreenCMFDA)作为人工底物，通过细胞内酯酶活性测定直舟形藻的细胞活性。CMFDA 被酶解水解成绿色荧光的 5-氯甲基荧光素(CMF)，它通过其氯甲基结合到细胞内的蛋白池中。被 CMFDA 阳性染色的硅藻必须有至少两种不同的酶类型：一种水解酶(酯酶)和一种转移酶(谷胱甘肽-s-转移酶)。这种细胞被解释为(水解上的)活性硅藻。此外,荧光CTC(5-Cyano-2,3-二甲基四唑氯化铵)被活性电子传递系统还原。因此,CTC被广泛用于检测呼吸微生物[45]。

SYTOX染色的应用清楚地表明,95%以上的直舟形藻细胞显示完整的膜,甚至在黑暗5个月后,因为只有一小部分细胞核显示荧光(见图12.4)。膜完整性作为抵御所有环境影响的重要防御屏障,被认为是细胞进一步繁荣的先决条件,因此被报道为微生物对胁迫最不敏感的特性[45]。结果表明,直舟形藻在长期黑暗中没有发生细胞生物学损伤。

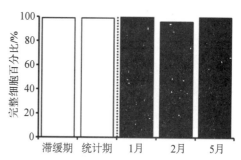

图12.4 从康斯峡湾斯斯匹次卑尔根岛的一块次海岸岩石中分离出来的直舟形藻中完整细胞的百分比($n=400$)

与细胞活力形成强烈对比的是,在黑暗中培养2个月和5个月的样品中,直舟形藻的酯酶活性从对照组的80%～90%大幅下降至3%以下(见图12.5)。这些数据清楚地表明,这种北极底栖硅藻的水解活性在黑暗2个月后被完全抑制或不表达。

水解活性的降低是与休眠期或孢子的形成有关[41],还是与异养活性的增加有关[46]仍是一个未解的问题。相比之下,其他羽纹硅藻在黑暗中孵育12 d时表现出连续不变的代谢活性,这允许在光照后保持细胞的丰度和生长能力,这是由FDA荧光测定的[47]。

将CTC作为呼吸可视化的染料清楚地表明,即使在黑暗5个月后,大多数硅藻细胞也是活跃的(见图12.6)。呼吸是极性夜间的一个关键代谢途径,因为光合作用的光缺失,它通过脂质等储存产物的降解为细胞提供能量[33]。不同储存产物的重新动员,以及有机分子(如糖、氨基酸)从外部来源的吸收,它们的代谢氧化是由呼吸活动介导的。直舟形藻的数据表明,大多数细胞在5个月后具有生物化学活性,但重点是维持代谢需要高呼吸活性(见图12.6)和较少的水解活性(见图12.5)。

舟形藻新鲜材料[见图12.7(a)]或通过H_2O_2降解制备的瓣膜可视化材料[见图12.7(b)]的光学显微镜观察结果显示了细胞结构。两个叶绿体沿壁叶排列,而细胞核位于细胞中心。在细胞核的两侧都可以观察到两个大的空泡[见图12.7(a)],中央中缝如图12.7(b)所示。

这里显示的数据首次进一步证明,在3个月的黑暗培养过程中,直舟形藻约50%的叶绿体被降解(见图12.8)。这些结果与呼吸作用(见图12.6)和在再照射后再次建立最佳生长前增加滞后期的结果一致(见图12.3)。在极地的夜晚,它利用部分叶绿体来获得能量(增强呼吸作用),因此在光线开始后,需要一段时间来再生这些细胞器的原始大小才能继续生长。这似乎是北极底栖硅藻的一种常见机制,因为在黑暗培养3个月后,条纹脆杆藻叶绿体长度也减少了30%～40%[25],而新月筒柱藻的叶绿体体积下降了约50%(未发表的结果)。因此,通过分解叶绿体成分来补充能量来维持代谢似乎是北极底栖硅藻在极地夜晚生存的关键机制。

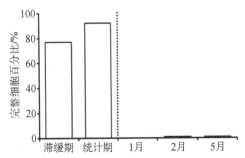

图 12.5 从康斯峡湾斯斯匹次卑尔根岛的一块次次海岸岩石中分离出的直舟形藻中活性细胞的百分比($n=400$)

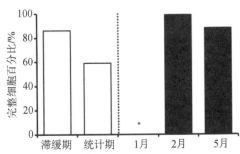

图 12.6 从康斯峡湾斯斯匹次卑尔根岛的一块次海岸岩石中分离出的直舟形藻中活性细胞的百分比($n=400$)

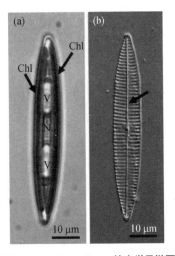

图 12.7 Navicula directa 的光学显微图

(a) 有两个叶绿体,中央核和两个叶绿体空泡;(b) 瓣的光镜图

图 12.8 直舟形藻的光学显微图

12.6 超微结构特征

根据 Holzinger,Roleda 和 Lütz 的说法[48],新鲜收获的直舟形藻细胞用标准化学固定法(2.5%戊二醛,1% OsO_4)对该细胞进行透射电镜观察。中央区域的横截面显示有一个中央核[见图 12.9(a)]。可见大面积区域被脂质体占据。叶绿体显示出完整的类囊体膜,每一个都是蛋白核。纵向截面可见广泛的脂质体,同样是顶叶的叶绿体[见图 12.9(b)]。高倍镜观察超微结构,可见高尔基体和核周围的内质网[见图 12.9(c)],伴有大量线粒体,有时可达 3 μm[见图 12.9(d)]。这些类蛋白核的直径可

达 1.8 μm,且仅含有少量的类囊体膜[见图 12.9(d)]。

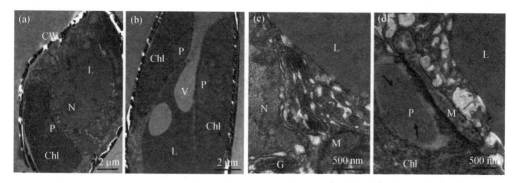

图 12.9 *Navicula directa* 的 TEM 图

12.7 结论

本研究对一种北极分离的直舟形藻的生理活动和超微结构进行了研究。可以得出结论,该物种具有广泛的生态振幅,需要生长的温度(在 1～15℃ 几乎相等的生长速率),并能耐受极夜自然发生的长期黑暗(长达 5 个月)。即使在长时间的暗孵育后,也可以观察到活性电子传递系统以及完整的膜,在此处理期间只有水解活性下降。然而,在转移到光照处后,滞后阶段在最长的黑暗暴露(5 个月)后需要 27 天,这表明需要大量的修复。从超微结构的观察可以看出,直舟形藻细胞含有大量的线粒体,并以脂质的形式积累储备,这些脂质将用于能量生产和叶绿体降解。这些特征使它能够在高纬度地区的恶劣环境条件下生存下来,而且很可能,观察到的北极水域的变暖不会轻易取代这一物种。

参考文献

[1] de Brouwer, J. F., Wolfstein, K., Ruddy, G. K., Jones, T. E., Stal, L. J. (2005). Biogenic stabilization of intertidal sediments: the importance of extracellular polymeric substances produced by benthic diatoms, Microb. Ecol., 49(4), 501-512.

[2] Petersen, N. R.-., Rysgaard, S., Nielsen, L. P., Revsbech, N. P. (1994). Diurnal variation of denitrification and nitrification in sediments colonized by benthic microphytes, Limnol. Oceanogr., 39(3), 573-579.

[3] Smetacek, V. (1999). Diatoms and the ocean carbon cycle, Protist, 150(1), 25-32.

[4] Mann, D. G. (1999). The species concept in diatoms, Phycologia, 38(6), 437-495.

[5] Yool, A., Tyrrell, T. (2003). Role of diatoms in regulating the ocean's silicon cycle, Global Biogeochem. Cycles, 17(4), 1103.

[6] Cahoon, L. B. (1999). The role of benthic microalgae in neritic ecosystems, Oceanogr. Mar. Biol. Annu. Rev., 37, 47-86.

[7] Underwood, G. J. C., Kromkamp, J. (1999). Primary production by phytoplankton and microphytobenthos in estuaries, Adv. Ecol. Res., 29, 93-153.

[8] MacIntyre, H. L., Geider, R. J., Miller, D. C. (1996). Microphytobenthos: the ecoloical role of the

"secret garden" of unvegetated, shallow-water marine habitats, I. Distribution, abundance and primary production. Estuaries, 19, 186-201.

[9] Paterson, D. M., Hagerthey, S. E. (2001). Microphytobenthos in constrasting coastal ecosystems: Biology and dynamics. In Ecological Comparisons of Sedimentary ShoresEcological Studies 151. pp. 105-125. K Reise (ed). Springer Berlin.

[10] Glud, R. N., Woelfel, J., Karsten, U., Kühl, M., Rysgaard, S. (2009). Benthic microalgal production in the Arctic: applied methods and status of the current database, Bot. Mar., 52(6), 559-571.

[11] Woelfel, J., Eggert, A., Karsten, U. (2014). Marginal impacts of rising temperature on Arctic benthic microalgae production based on in situ measurements and modelled estimates, Mar. Ecol. Prog. Ser., 501, 25-40.

[12] Glud, R. N., Kühl, M., Wenzhöfer, F., Rysgaard, S. (2002). Benthic diatoms of a high Arctic fjord (Young Sound, NE Greenland): importance for ecosystem primary production, Mar. Ecol. Prog. Ser., 238, 15-29.

[13] Sevilgen, D. S., de Beer, D., Al-Handal, A. Y., Brey, T., Polerecky, L. (2014). Oxygen budgets in subtidal arctic (Kongsfjorden, Svalbard) and temperate (Helgoland, North Sea) microphytobenthic communities, Mar. Ecol. Prog. Ser., 504, 27-42.

[14] Woelfel, J., Schumann, R., Peine, F., Kruss, A., Tegowski, J., Blondel, P., et al. (2010). Microphytobenthos of Arctic Kongsfjorden (Svalbard, Norway)-Quantification of ex situ primary production by use of incubation chambers equipped with planar optode spots and structural analyses of biomass, Polar Biol., 33, 1239-1253.

[15] Hop, H., Pearson, T., Hegseth, E. N., Kovacs, K. M., Wiencke, C., Kwasniewski, S., et al. (2002). The marine ecosystem of Kongsfjorden, Svalbard, Polar Res., 21(1), 167-208

[16] Svendsen, H., Beszczynska-Møller, A., Hagen, J. O., Lefauconnier, B., Tverberg, V., Gerland, S., et al. (2002). The physical environment of Kongsfjorden-Krossfjorden, an Arctic fjord system in Svalbard, Polar Res., 21, 133-166.

[17] Chapman, A. R. O., Lindley, J. E. (1980). Seasonal growth of Laminaria solidungula in the Canadian High Arctic in relation to irradiance and dissolved nutrient concentrations, Mar. Biol., 57(1), 1-5.

[18] Dunton, K. H. (1990). Growth and production inLaminaria solidungula: relation to continuous underwater light levels in the Alaskan High Arctic, Mar. Biol., 106(2), 297-304.

[19] Serreze, M. C., Francis, J. A. (2006). The Arctic on the fast track of change, Wea, 61(3), 65-69.

[20] Hegseth, E. N., Tverberg, V. (2013). Effect of Atlantic water inflow on timing of the phytoplankton spring bloom in a high Arctic fjord (Kongsfjorden), Svalbard, J. Mar. Syst., 113-114, 94-105.

[21] Schauer, U., Beszczynska-Möller, A., Walczowski, W., Fahrbach, E., Piechura, J., Hansen, E. (2008). Variation of measured heat flow through the Fram Strait between 1997 and 2006. In Springer, Dordrecht. pp. 65-85. A.-S. O Fluxes, P. R Dickson, J Meincke and P Rhines (eds.).

[22] Bintanja, R., van der Linden, E. C. (2013). The changing seasonal climate in the Arctic, Sci. Rep., 3, 1556.

[23] Overland, J. E., Wang, M., Walsh, J. E., Stroeve, J. C. (2014). Future Arctic climate changes: adaptation and migration time scales, Earth's Future, 2(2), 68-74.

[24] Gustavs, L., Schumann, R., Eggert, A., Karsten, U. (2009). In vivo growth fluorometry: accuracy and limits of microalgal growth rate measurements in ecophysiological investigations, Aquat. Microb. Ecol., 55, 95-104.

[25] Karsten, U., Schumann, R., Rothe, S., Jung, I., Medlin, L. (2006). Temperature and light requirements for growth of two diatom species (Bacillariophyceae) isolated from an Arctic macroalga, Polar Biol., 29(6), 476-486.

[26] Schlie, C., Karsten, U. (2017). Microphytobenthic diatoms isolated from sediments of the Adventfjorden (Svalbard): growth as function of temperature, Polar Biol. 40, 1043-1051.

[27] Torstensson, A., Chierici, M., Wulff, A. (2012). The influence of increased temperature and carbon dioxide levels on the benthic/sea ice diatom Navicula directa, Polar Biol., 35(2), 205-214.

[28] Longhi, M. L., Schloss, I. R., Wiencke, C. (2003). Effect of irradiance and temperature on photosynthesis and growth of two Antarctic benthic diatoms, Gyrosigma subsalinum and Odontella litigiosa, Bot. Mar., 46(3), 276-284.

[29] Torstensson, A., Hedblom, M., Andersson, J., Andersson, M. X., Wulff, A. (2013). Synergism

between elevated pCO$_2$ and temperature on the Antarctic sea ice diatom Nitzschia lecointei, Biogeosciences, 10, 6391-6401.

[30] Zacher, K., Rautenberger, R., Hanelt, D., Wulff, A., Wiencke, C. (2009). The abiotic environment of polar marine benthic algae, Bot. Mar. 52(6), 483-490.

[31] Sabbe, K., Verleyen, E., Hodgson, D. A., Vanhoutte, K., Vyverman, W. (2003). Benthic diatom flora of freshwater and saline lakes in the Larsemann Hills and Rauer Islands, East Antarctica, Antarctic Sci., 15 (2), 227-248.

[32] Gómez, I., Wulff, A., Roleda, M. Y., Huovinen, P., Karsten, U., Quartino, M. L., et al. (2009). Light and temperature demands of marine benthic microalgae and seaweeds in polar regions, Bot. Mar. 52(6), 593-608.

[33] Schaub, I., Wagner, H., Graeve, M., Karsten, U. (2017). Effects of prolonged darkness and temperature on the lipid metabolism in the benthic diatom Navicula perminuta from the Arctic Adventfjorden, Svalbard, Polar Biol., 40(7), 1425-1439; in press.

[34] McMinn, A., Martin, A. (2013). Dark survival in a warming world, Proc. R. Soc. Lon. B Biol. Sci., 280(1755), 20122909.

[35] Palmisano, A. C., Sullivan, C. W. (1982). Physiology of sea ice diatoms. I. Response of three polar diatoms to a simulated summer-winter transition, J. Phycol., 18(4), 489-498.

[36] Durbin, E. G. (1978). Aspects of the biology of resting spores of Thalassiosira nordenskioeldii and Detonula confervacea, Mar. Biol., 45(1), 31-37.

[37] McQuoid, M. R., Hobson, L. A. (1996). Diatom resting stages, J. Phycol., 32(6), 889-902.

[38] Tuchman, N. C., Schollett, M. A., Rier, S. T., Geddes, P. (2006). Differential heterotrophic utilization of organic compounds by diatoms and bacteria under light and dark conditions, Hydrobiol, 561(1), 167-177.

[39] Schlie, C., Woelfel, J., Ruediger, F., Schumann, R., Karsten, U. (2011). Ecophysiological performance of benthic diatoms from arctic waters. In COLE—cellular origin, life in extreme habitats and astrobiology. pp. 425-436. J Seckbach and P Kociolek (eds.). Berlin: Springer.

[40] Wulff, A., Roleda, M. Y., Zacher, K., Wiencke, C. (2008). Exposure to sudden light burst after prolonged darkness-A case study on benthic diatoms in Antarctica, Diatom Res., 23(2), 519-532.

[41] Peters, E. (1996). Prolonged darkness and diatom mortality: II. Marine temperate species, J. Exp. Mar. Biol. Ecol., 207(1-2), 43-58.

[42] Peters, E., Thomas, D. N. (1996). Prolonged darkness and diatom mortality I: Marine Antarctic species, J. Exp. Mar. Biol. Ecol., 207(1-2), 25-41.

[43] Kamp, A., de Beer, D., Nitsch, J. L., Lavik, G., Stief, P. (2011). Diatoms respire nitrate to survive dark and anoxic conditions, Proc. Natl. Acad. Sci. U. S. A., 108(14), 5649-5654.

[44] Veldhuis, M., Kraay, G., Timmermans, K. (2001). Cell death in phytoplankton: correlation between changes in membrane permeability, photosynthetic activity, pigmentation and growth, Eur. J. Phycol., 36 (2), 167-177.

[45] Freese, H. M., Karsten, U., Schumann, R. (2006). Bacterial abundance, activity, and viability in the eutrophic River Warnow, northeast Germany, Microb. Ecol., 51(1), 117-127.

[46] White, A. W. (1974). Uptake of organic compounds by two facultatively heterotrophic marine centric diatoms, J. Phycol., 10, 433-438.

[47] Jochem, F. J. (1999). Dark survival strategies in marine phytoplankton assessed by cytometric measurement of metabolic activity with fluorescein diacetate, Mar. Biol., 135(4), 721-728.

[48] Holzinger, A., Roleda, M. Y., Lütz, C. (2009). The vegetative arctic freshwater green alga Zygnema is insensitive to experimental UV exposure, Micron, 40(8), 831-838.

第 13 章
淡水硅藻的生态学
——当前的发展趋势和应用

Aloisie Poulíčková，Kalina Manoylov

13.1 简介

微生物群落的生态学仍然大多未被探索,因为我们经常应用和测试为规模较大、较长寿命和较小群落多样性的群体生态学环境。通常的假设是,最丰富的物种对当前的条件是最好的竞争对手,而低丰富度的类群是随机出现的,并被优势类群击败。在低营养条件下,期望记录具有特殊的低营养类群的低多样性群落[1],但硅藻(像所有藻类一样)将在高的营养条件下增殖。因此,在高营养条件下,需要仔细考虑生物多样性、丰富度、丰度等群落指标。当营养物质以高的浓度存在时,许多类群将增加,从而提高均匀性[1]。多样性有多大可能成为水质的一个负面指标,这一问题迄今尚未得到解决。在非常高的营养条件下,很少有类群被认为比它们的竞争对手生长得更快并支配着群落。很少有一个单一的硅藻主导着淡水群落。藻类生态学家面临的挑战仍然是理解物种身份、群落相互作用和种群密度。淡水硅藻的生态可以从许多方向接近,因为它们发生在所有的水生环境中。

淡水硅藻具有多种生态功能。作为广泛存在的初级生产者,它们对 O_2 的生产贡献显著(占世界初级生产的 20%~25%)[2],对养分循环、能量流动和更高水平的能源供应尤其重要[3-4]。据估计,物种的数量在 3 万到 10 万之间[5],与被描述的物种 12 000 个[6]相比,这意味着大多数物种都在等待它们的描述。

除高温和高盐生态环境外,硅藻是微藻中最普遍的类群[7]。它们自由地生活在静水的浮游生物中,或附着在静水和静水生态系统的底栖生物表面[8]。具有选定代表的最重要的淡水硅藻群落如图 13.1 所示。

底栖硅藻生态学比浮游硅藻生态学了解得要少得多。浮游生物研究的"黄金时代"尤其与世界各地的湖泊地区有关[9-13]。随着近年来人们的主要兴趣从局部变化转向全球变化,目前的研究主要是"间接基于浮游生物",主要基于湖泊沉积物的古沉积物重建[14-15]。底栖硅藻由于取样方面的困难,特别是定量方面的困难,在很大程度上被生态学家忽视了[16]。这一领域最近取得了巨大的进展[17-19]。在温度上升最快的北极湖泊中,由于冰覆盖动态和营养栖息地的可用性,硅藻群落发生了显著变

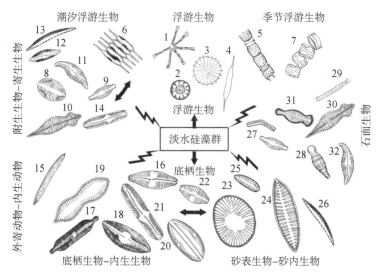

图 13.1　淡水硅藻组合

浮游生物包括适应水体中生存的真浮游硅藻(1—星杆藻,2—小环藻,3—冠盘藻,4—针状菱形藻,5—沟链藻);浅池塘或深湖泊海岸的沉积物(6—脆杆藻属,7—平板藻);附生菌(在藻类和大型植物的表面)和内生菌(在其他藻类的粘膜包膜内)[低调硅藻(8—卵形藻,9—针孔藻)、高调硅藻(10—异极藻,11—桥弯藻)和活动硅藻(12—舟形藻,13—菱形藻,14—羽纹藻)];Epizoon(在动物的表面)以 15—菱形藻为代表;(在细沉积物表面)和内膜(在细沉积物内)代表活动的物种(16—鞍型藻属,17—长篦藻,18—美壁藻,19—波缘藻,20—无槽藻,21—羽纹藻,22—平齿藻;上膜(砂粒表面)和内膜(沙质沉积物内)有 22—平扁盘藻,23—曲盘藻,24—双菱藻,25—槌棒藻,26—菱形藻;石上硅藻附着在石头上[低剖面(27—链状弯壳藻,28—扇形藻,29—硅藻)、高剖面(30—异极藻,31—双楔藻,32—桥弯藻)]

化[20]。在人类视角看来,偏远的湖泊继续向新的生态模式转变[21]。过去的十年对硅藻群落的研究是令人兴奋,因为之前没有类似于在寒冷气候中观察和预测的湖泊变化。由于人为的营养沉积和总体温度的增加,预计硅藻的生物量和食物来源将会增加[13]。具有高表体积比的硅藻保持浮力,强烈影响通过食物网的能量转移,并有助于增加碳吸收。非生物驱动因素影响浮游群落的大小结构,温暖的气候有利于小型的硅藻细胞生存[22]。在高纬度湖泊中,代表小环藻属的远洋小型硅藻物种的数量增加,可能是由于较长的无冰季节和层化增加导致[23]。这些趋势如何与结构、功能和营养水平的变化相关尚不清楚,这促使人们对大型温带湖泊采取一系列监测措施[24-26]。

硅藻对多种生态因素表现出强烈的反应,它们对污染、养分可利用性、盐度和酸度变化的敏感性使它们成为极好的生物指示剂[27-29];此外,与直接的化学分析相比,硅藻反映了水化学的长期变化[29-30],这是由于硅藻整合了跨时间和空间尺度的生态条件[31]。

水质已被确定为与人类健康有关的一个非常重要的问题。在本文中,我们提出了硅藻分析的作用,以评估和预测水生态系统的服务和商品的状态。由于人类对煤炭产生的能源、工业以及更多粮食生产的需求,水生环境中含有酸、金属和农业径流的人为污染不断增加。本书综合了目前有关硅藻多样性、分布、扩散和群落结构的知识。生态系统过程的测量,如营养吸收率、胞外酶活性和代谢,正越来越多地被用作评估工具,但相对于传统的基于群落结构的方法,仍处于起步阶段。功能措施的持续开发、测试和实施及其与硅藻群落结构指标一起的使用将进一步推进生物评估,将为管理决策提供信息,并量化实现恢复目标的进展。

13.2 硅藻分布

虽然硅藻已经被研究了近300年(第一个特定的记录是在1703年)[16],但它们的多样性、分布和生物地理学仍然鲜为人知[5]。传统的观点认为,包括硅藻在内的微生物具有无限的扩散潜力[32],缺乏任何生物地理学。由于微生物的小体型和天文数字般的种群规模,预计会有无限的扩散[33-34]。矛盾的是,一些大型生物体(如苔藓植物)比硅藻更接近于"一切都是无处不在"的模式[35],很可能是由于苔藓植物繁殖体的微观尺寸。在水生和陆地环境中,其他小生物(原生生物和多细胞生物)中也记录了身体大小和数量与分布的类似反比关系[34]。更具体地说,生态位宽、丰度高、分布广泛的物种表现出优化的体型尺寸(硅藻小,但鱼类中等)[34];这种类型的分布将受当地环境条件的控制[36]。相比之下,一些研究表明,微生物分布和生物地理模式的空间模式与已知的大型生物基本相似[37-40]。Finlay等[41]提出原生生物快速普遍扩散导致地方性比例较低。虽然我们没有硅藻中狭窄特有现象的明确例子,但新物种的描述可以从普雷斯帕湖、奥赫里德湖、贝加尔湖或夏威夷群岛得出[42-44],到目前为止,在其原始地点以外还没有任何记录。相比之下,在广泛分布的硅藻的情况下,已经记录了地理结构的种群(基于微卫星数据),例如尖刺伪菱形藻和月形短缝藻[45-47]。

13.3 硅藻扩散能力

淡水生地主要代表被陆地生地包围的离散地点(岛屿)[48-49]。水体的尺寸是影响底栖生物群落中硅藻物种丰富度的关键变量,而小岛效应是它们物种-面积关系的一个特征[50]。尽管这些场所之间缺乏连通性,但缺乏主动扩散能力的微藻可能被动地通过风、水流、动物媒介和人类活动等媒介被动地实现广泛分布[51-52]。Kristiansen[53]对所有潜在的传播媒介进行了回顾,最近很少有关于空气粉尘、压载水和非本地鱼类引入的论文发表[19,54-56]。尽管在硅藻学的各个方面都取得了进展,但我们仍然主张Round的声明,即我们没有关于硅藻运输机制的真实证据[8]。一般来说,围生类群(附着的)似乎比浮游植物物种具有较低的扩散能力[57]。

扩散能力对于理解生物多样性、入侵和适当保护的变化至关重要[58],它主要与藻类在不同类型的压力下生存的能力有关。在活的硅藻细胞的情况下,超过几米的短距离扩散应该是可能的。由于淡水硅藻对干燥和温度胁迫的耐受性较低,预计不会有常见的长距离传播[59-60]。在陆地硅藻中,可以预期有更高的扩散潜力,特别是居住在土壤中的硅藻,如北方羽纹藻和双尖菱板藻小头变种,它们对极端温度和干燥表现出更高的耐受性[59-60]。虽然在世界性分布的热门候选物种中这一优势排名为北方羽纹藻和双尖菱板藻,这两个物种在遗传上都具有高度分化,表明这些气生硅藻复合体具有隐蔽的多样性[60]。

一些微藻有望在没有特殊保护结构的动物潮湿表面生存,或以休眠状态生存[19,61]。Stoyneva 注意到,超过 90%的与鸟类长期运输相关的物种都有粘液鞘。动物类一直被认为是藻类运输的重要机制[62-64]。

藻类在鱼类消化道内抵抗消化的能力已经在过滤取食的亚洲鲤鱼中进行了测试,这些鲤鱼经常被引入世界各地的富营养化湖泊和许多河流。随着亚洲鲤鱼大规模入侵密西西比河三角洲及其支流,浮游植物群落的变化和浮游硅藻的生存情况也被记录下来[65]。在培养的后肠样本中发现了硅藻、蓝藻细菌、去丝粒和绿藻的活细胞[56]。然而,这一事实本身并不能保证硅藻在新的生物群落中的成功定植。硅藻受到它们的繁殖行为的限制,即在它们的营养分裂期间受到细胞减少(在种群的大部分内)的限制,并且通过通常伴随有性繁殖的孢子形成来恢复大小。因此,在大多数硅藻中,性别并不像其他藻类那样是兼性的[66]。由于大多数硅藻是异丘脑的[67-69],两个(兼容的)性伴侣的代表需要在运输中生存下来,这是相当不可能的。这似乎是很清楚的应用经验,在有性繁殖的实验室中,要想在有性大小范围内分离许多克隆培养物,就必须存在两个性伴侣(在异宗藻中)[67-68]。

因此,微藻作为入侵物种是如何表现出来的呢?生物入侵是由非本地物种引起的,它们可以改变群落中的优势物。入侵硅藻最著名的例子是双生双楔藻(Didymo)。Didymo 是一种底栖淡水硅藻,原产于地球周边地区和北半球的河流[70]。历史上,它被描述为一种罕见的硅藻河流,直到最近几十年[71],它才开始形成大量的生物量。目前,它正在成功入侵北美、欧洲、亚洲以及南半球的淡水地区(新西兰、阿根廷、智利、西班牙)[72-74]。最近,Didymo 在 50 多个国家被引用,并影响到除南极洲以外的所有大陆[74]。它是否有可能无处不在,但却被忽视了?最近的研究支持了其生态位正在扩大的假说[74]。不幸的是,对 Didymo 群体的遗传结构的研究正处于分子方法优化的阶段[72],因此我们对其种群的遗传多样性的了解很少,但正在取得一些进展[75]。

13.4 硅藻生态学中的功能分类

底栖硅藻组合的组成和动态取决于类群的定植能力、细胞大小和生物体积、营养需求以及对干扰的敏感性[76-79]。

硅藻生态学中的功能分类对于深入了解硅藻组合的动态和功能变得越来越重要。虽然这种对浮游藻类和蓝藻细菌的分类有着悠久的传统[80],底栖硅藻功能群直到最近才得到关注[79,81-84]。

从预测浮游植物物种沿环境梯度分布的功能方法来看,最广泛的方法是 Reynolds 等和 Padisák 等在特定条件下提出的一种方法;然而,将单个物种正确地分类为相应的功能群需要深入了解它们的个体生物学知识[85]。最近的功能分类是基于物种特征的,即生理、形态学和物候特征[83,86]。尽管许多物种缺乏信息,但功能性状方法经常被应用[87-90]。

在底栖硅藻的情况下,建立了几个功能群——集群:低轮廓、高轮廓、运动集群[81]。由于体型小和水平生长,低轮廓的集群物种适应了生物膜中的不利位置[见图 13.2(b)、(g)、(i)~(k)]。在高轮廓和可运动的集群物种可以利用它们的栖息地

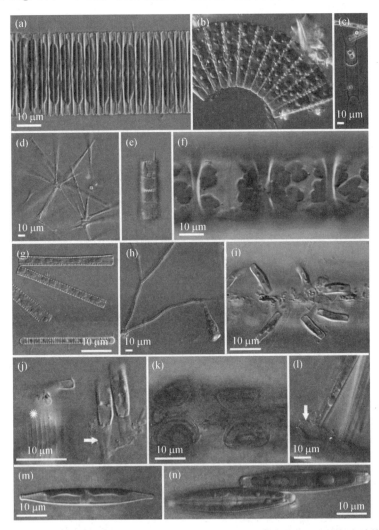

图 13.2 用 LM 图像(BF、DIC、Zeiss AxioImager)、来自捷克共和国水生栖息地的硅藻图像记录的淡水硅藻组合的代表

浮游属[(a) 脆杆藻,(c) 棘角藻,(d) 星杆藻,(e) 沟链藻];底栖生物属和低剖面生物群[(b) 扇形藻,(g) 硅藻,(i,j) 由针孔藻附着在绿色丝状藻类鞘藻的短的粘性茎上(箭头),(k) 由瓣面附着的卵形藻,(l) 由粘膜垫附着的脆杆藻];底栖生物属和高知名度行会[(f) 直链藻,(h) 具有长分枝茎的异极藻。底栖生物属];游动生物群[(m) 菱形藻,(n) 舟形藻]

获得更有益的位置。长柄[见图 13.2(h)]或运动性[见图 13.2(m)、(n)]。在这里,我们说明了一些低轮廓的集群物种的增长模式,可以允许它们被提升为附生植物植物:短柄附生植物[见图 13.2(j)]、并生附生植物[见图 13.2(k)]和带有粘液垫的针垫附生植物[见图 13.2(l)]。由于硅藻倾向于改变其策略,因此在理解生态集群分类方面仍有很多未知数。例如,在一个培养实验中,新月短缝藻[见图 13.3(c)、(d)]要么是活动的,要么是形成簇状菌落[45]。在一项竞争实验中,使用基因相同的种群(使用商业获得的培养物)极小曲丝藻(Kützing) Czarnecki[见图 13.2(i)],在低光和高营养条件下生长的种群产生短垫和柄,并附着在可用的垂直基质上[91]。形成箱形桥弯藻(Ehr.) Kirchner(高轮廓的集群成员)促进了极小曲丝藻(低轮廓的集群成员)的成长。关于初始密度,极小曲丝藻(种群在不同的光照和营养条件下生长,产生附着在基质上的波浪状粘液链。这些球状链内的硅藻壳状物呈松散或块状排列[91]。除了前面提到的三个行会[81],浮游行会[见图 13.2(a)、(c)~(e)和图 13.3(a)、(b)、(e)]已经被 Rimet 和 Bouchez[83] 添加,因为浮游物种由于沉积而经常出现在生物膜中。Berthon 等[82]提出了另一种与硅藻群落的特定功能相关时分为五种大小类的分类方法,即条纹图案、波动、中缝裂缝的大小、点状闭塞和其他结构,基于 SEM 形态特征的分类将实现更精细的尺度分类(见图 13.3)。最后 B-Beres 等[92]将这两种分类结合为生态形态功能分类,获得了灵活、精细的动态评估方法应用工具。硅藻种类和行会组成随当地环境、空间和气候变量的变化而变化,但行会分布主要由环境驱动,对历史因素的依赖性较小[4]。因此,行会组成更适合用于全球环境变化的评估(富营养化、酸化、全球变暖)。

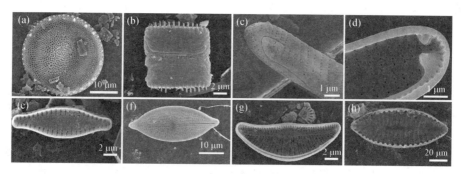

图 13.3 淡水硅藻组合的 SEM 图

浮游生物属[(a) 湖泊海链藻(Grunow) Hasle 波状瓣叶表面,具许多近中心的单孔开口,(b) 腹部十字藻(Ehrenberg) Cleve and Möller,具有许多接合部而无穿孔的截头锥体,位于瓣套上的纹之间的中空刺,(e) 短条纹假十字藻(Grunow) Williams and Round 不规则的第三排圆形乳晕,大部分被侵蚀,但可见复杂的帆];底栖物种,低剖面群[(c) 和 (d) 双月真藻,(c) 极部外观瓣面有钩状中缝,外套膜上有缩短的中缝,(d) 内部极视图,有发达的螺旋舌,一个非常接近的边缘孔和一个中断纹的光滑区域];底栖生物属、高剖面生物群[(g) 双眉藻。内部观察,背纹由数个小晕组成,腹纹由两个小晕组成(近顶处有 3 个小晕),中缝远端简单,无中央螺旋舌];底栖生物属、游动生物群[(f) 暗额藻具有三种不同形状的孔和不同类型的闭塞,(h) 双菱藻 W. Smith 单极瓣膜,窄披针形,瓣膜中心有凹陷的透明轴区]

13.5　空间生态学和集合群落

人们越来越关注微生物的空间生态学[93]和影响淡水硅藻的全球因素,如气候[94]。以湖泊沉积物及其近期和化石记录(特别是硅藻)为代表的长期的环境信息档案,可用于湖泊内以及集水区内的局部和全球影响的古矿物学重建[95-96]。

湖泊生态系统评估的准确性在很大程度上依赖于我们对关键硅藻物种生态学的理解。感觉广义环藻属应该是全球变化的早期指标,但我们需要记住,最明显的因素,比如温度,并不是它们丰度和物种组成变化的唯一驱动因素[23,97]。

环境影响预计将遵循某种层次结构:大尺度的模式影响,精细尺度的生态环境特征影响其环境内的生物。这一点在河流生态系统中尤其明显[98-99]。

Pajunen等[100]重点研究了气候对硅藻分布的影响。他们的结果表明,大尺度的气候因素显著影响了硅藻的分布,而在相对较小的区域尺度上覆盖了局部变量。它支持了一些以前对其他微生物群的研究[37,101-102];然而,这种影响似乎是物种特异性的[100]。

基于集合群落理论的生物多样性预测的空间结构最近成为生态学-集合群落生态学的分支学科[103-104]。生态学家使用四种主要的集合群落组织的理论类型[105-106]:斑块动态、种分选、质量效应和中立[107]。集合群落强烈地依赖于景观的物理结构和可用栖息地的配置。在淡水生物的情况下,集合群落作为由分散连接的离散局部群落可以用树突状生态网络表示,如河流流域[103-104,108]。河流集合群落是由局部(环境过滤)和区域(扩散)过程共同组成的结构,后者更为重要,例如在山区河流中[109]。不同地点之间的陆地距离,以前用于河流生物的空间关系模型,似乎是较差的预测因子[103]。底栖硅藻亚群落似乎局限于河流网络的边界,通过河流的传播比通过空气或通过动物媒介更重要[103,109]。同样,由于山地屏障限制了河流间的扩散,河流走廊是高山的主要扩散路径[109]。在公会的空间分布上可以发现差异(低轮廓与高轮廓和运动),后者在河网的较低区域更为常见[81,103],这是由于电流速度的差异。这些结果强调了在集合群落生态学中,基于特征的方法可以发挥重要作用[57,103,110]。

与当地环境变量相比,河流流域特征可以更好地预测群落结构,这在河流生物(硅藻、苔藓植物和浮游动物)中得到了证实,而在湖泊生物(昆虫、大型植物和鱼类)中则没有。尽管如此,这两种影响(环境和盆地效应)都导致了所有正在研究的生物群体和生境的高系数多样性[104]。不同的模式应该是由于湖泊和河流生物之间的扩散限制的差异或它们的身体/繁殖体的大小造成的[110-113]。

13.6　水生生态系统的生物监测

底栖硅藻的物种组成与河流中的环境条件有关,以开发用于评估生物条件的硅藻指标[29]。人为污染造成的生物多样性丧失被认为是我们这个时代最严重的生态危机。由于人为过度使用，改变了生态系统功能，淡水系统特别脆弱[114]。湖泊系统的生态系统功能模型已经开始考虑到自养过程在驱动河流生产中的重要性。大型河流系统中水生微生物的多样性可能对理解河流如何提供诸如 O_2 可用性、营养减少和食物生产等服务至关重要[115]。除了多样性和物种组成的变化外，污染还会导致硅藻的畸形学。是什么成分能杀死一些硅藻物种，而不是其他的物种？基于硅藻无性分裂的独特方式，在形态发生过程中，形态上的改变可能会导致个体死亡。由于晶体内的互补瓣膜是在不同的时间和条件下合成的，通常只有一个瓣膜的形态发生了改变。所有有毒物都可能引起类似的畸形吗？Lavoie 等[116]发现了物种对金属的特异性敏感性[117]和有机化合物[118]影响不成比例的双蚜虫、类和不对称双蚜硅藻[116,119]。最后，非分类参数，即脂质体的数量和大小以及类胡萝卜素与叶绿素 a 的比率，在镉和锌污染下发生了变化，而没有改变硅藻细胞壳的轮廓[120]。

物种性状通常使用加权平均方法计算，以定义物种环境最优、耐受性和回归[9,121]。分类群被定义为敏感的（如果在某些条件下消失）或耐受的（其中应激源浓度的增加并不改变分类群的相对丰度）。在美国，州和联邦项目除了分析鱼类和大型无脊椎动物外，还纳入了硅藻，而欧洲项目则基于浮游植物、大型植物和底栖植物[18]。制定的大多数指标都是为了评估反映一般水质和区域气候的生态系统健康状况[122]。也有许多研究报告了硅藻对金属污染的反应[120]，酸性矿井排水[123]和有机污染物[124-125]。

标准协议通常要求列举一定数量的瓣膜，并识别到尽可能低的分类水平。常见类群的物种特定生态作用已经被探索[93]，但有更多的罕见类群需要更多的工作。营养浓度和其他人为压力源被用于代表美国河流的损害水平的河流分类[126]。来自相同溪流的硅藻组合被用于生物条件梯度（BCG）分类。基于硅藻的 BCG 水平与总磷和森林覆盖率的相关性最强。

在最近的一项研究中，Young 等[127]研究了硅藻二磷酸核酮糖羧化酶的调控特性，发现二磷酸核酮糖羧化酶的激活状态随环境条件的响应而不同（例如温度、光、CO_2 和营养物质）。他们提出，物种特有的生理学在很大程度上是未知的。硅藻的资源利用效率与光合作用等基本生物功能同步的影响有待进一步研究。

认识到丰度的生态价值，Lavoie 等[128]剔除了 40%的已记录的硅藻物种，或所有丰度低于 2%的类群，并发现生态信号没有差异。排除更多的类群仍然可以区分受影响的地点和参考地点（大多数生物评估项目的目标），尽管生态状况的细微变化已经消失[129]。他们提出了基于存在/不存在或属级识别的排序，这导致了在细微变化方面的信息丢失，但在不同的项目中，受影响的地点和参考地点之间的大体分离仍然是可能的。

事实上，应用生态学家需要经济有效的生物监测方法，倾向于用更高的分类群或基于特征的方法取代基于物种的方法[4,130-134]、生态协会、生命形式和历史，或生存策略[79]。使用高等分类群(属、科)作为物种的替代品,经常在陆地、淡水和海洋环境中被讨论,并基于分类学充分性的概念。在硅藻中也进行了分类充分性测试,表明属水平分辨率似乎足够稳健,可以描述数据集的主要梯度[84,129,135-136]。因此,硅藻研究的最近趋势似乎遵循两个相互矛盾的方向:避免耗时的精细分类学导致物种替代[137]与基于分子方法和条形码的神秘多样性识别[7],最终允许"无分类学"的方法[138]。然而,大属内的物种在生态需求上有所不同[139],而物种复合体内的隐性物种在某些情况下是生态分化的[129,140]。因此,在生物评价中忽略精细的分类法,可能会导致精细尺度生态信息的丢失。建议保留更大的生态信息(精细的分类分辨率),特别是在容易识别的良好指示物种的情况下[129,133]。

水质监测的第一个分子方法主要是基于物种特异性的寡核苷酸序列[141-142]。一些更先进的想法包括使用微阵列(系统芯片)分析对预期的社区,并识别可能对人类健康有风险的淡水病原体[143-144]。这些分子方法允许在不栽培的情况下分析硅藻群落,有利于生物多样性和生物评估研究。在用形态学分析检测双歧杆菌具有来自多个全球来源的基因后使用环境 DNA(eDNA)来分析水华群体[75]。快速检测这种潜在入侵者的存在可以通过探索和检测系统发育标记作为 rRNA 基因的替代方法来实现,该基因有限的序列差异不允许在物种水平上准确区分硅藻。小规模的原型 DNA 芯片可以同时识别和识别多种硅藻物种。这些方法一旦完善,将有助于更深入地了解硅藻群落,并更快地返回数据,以便及时作出管理决策。

13.7 结论

在未来,了解淡水生态系统的功能多样性是重要的,目前仍未得到有效探索,需要长期的投资和承诺。我们应该努力更好地了解藻类的分类、遗传和功能多样性以及更具体的硅藻组合之间的相互作用和反馈。将如此多的参数(环境学、分类学、遗传学和生态学)整合到一个生物模型中将具有挑战性。现有的和新的建模方法都应该在必要时进行测试、验证和优化。

参考文献

[1] Manoylov, K.M., Stevenson, R. J., Wang, Y. K. (2013). Sustaining abundance and distributional patterns of benthic diatoms from streams in Kentucky, USA, WIT Transactiona ov Ecology and the Environment, 173, 111-123.

[2] Werner, D. (1977). Introduction with a note on taxonomy. In The biology of diatoms. Botanical monographs 13. D. Werner. (ed.) Oxford: Blackwell Scientific Publications. pp.1-17.

[3] Falkowski, P.G., Fenchel, T., DeLong, E. F. (2008). The microbial engines that drive earth's biogeochemical cycles, Science, 320(5879), 1034-1039.

[4] Soininen, J., Jamoneau, A., Rosebery, J., Passy, S. I. (2016). Global patterns and drivers of species and trait composition in diatoms, Global Ecol. Biogeogr., 25(8), 640-650.

[5] Mann, D.G., Vanormelingen, P. (2013). An inordinate fondness? The number, distribution and origins of diatom species, J. Eukaryot. Microbiol., 60(4), 414-420.

[6] Guiry, M.D. (2012). How many species of algae are there? J. Phycol., 48(5), 1057-1063.

[7] Zimmermann, J., Abarca, N., Enke, N., Skibbe, O., Kusber, W. H. et al. (2014). Taxonomic reference libraries for environmental barcoding: a best practice example from diatom research, PLoS ONE, 9(9), e108793, e108793.
[8] Round, F. E. (1981). The ecology of algae. Cambridge, UK: Cambridge University Press.
[9] Poulíčková, A., Duchoslav, M., Dokulil, M. (2004). Littoral diatom assemblages as bioindicators of lake trophic status: A case study from perialpine lakes in austria european, J. Phycol., 39(2), 143-152.
[10] Reynolds, C. S., Maberly, S. C., Parker, J. E., De Ville, M. M. (2012). Forty years of monitoring water quality in Grasmere (English Lake District): sensitivity of phytoplankton to environmental forcing in a sensitive area, Freshw. Biol., 57, 384-399.
[11] Talling, J. F. (2012). Freshwater phytoplankton ecology: the british contribution in retrospect, Freshwater Reviews, 5(1), 1-20.
[12] Rühland, K. M., Hargan, K. E., Jeziorski, A., Paterson, A. M., Keller, W. (Bill)., Smol, J. P. (2014). A multi-trophic exploratory survey of recent environmental changes using lake sediments in the Hudson bay lowlands, Ontario, Canada, Arctic, Antarctic, and Alpine Research, 46(1), 139-158.
[13] Rühland, K. M., Paterson, A. M., Smol, J. P. (2015). Lake diatom responses to warming: reviewing the evidence, J. Paleolimnol., 54(1), 1-35.
[14] De Laender, F., Verschuren, D., Bindler, R., Thas, O., Janssen, C. R. (2012). Biodiversity of freshwater diatom communities during 1 000 years of metal mining, land use, and climate change in Central Sweden, Environ. Sci. Technol., 46(16), 9097-9105.
[15] Reavie, E. D., Sgro, G. V., Estepp, L. R., Bramburger, A. J., Shaw Chraïbi, V. L., Pillsbury, R. W. et al. (2017). Climate warming and changes in Cyclotella sensu lato in the laurentian great lakes, Limnol. Oceanogr., 62(2), 768-783.
[16] Round, F. E., Crawford, R. M., Mann, D. G. (1990). The diatoms. Biology and morphology of the genera. Cambridge, UK: Cambridge University Press.
[17] Poulíčková, A., Dvořák, P., Mazalová, P., Hašler, P. (2014). Epipelic microphototrophs: an overlooked assemblage in lake ecosystems, Freshwater Science, 33(2), 513-523.
[18] Kelly, M. G., Birk, S., Willby, N. J., Denys, L., Drakare, S., Kahlert, M. et al. (2016). Redundancy in the ecological assessment of lakes: Are phytoplankton macrophytes and phytobenthos all necessary? Science of The Total Environment, 568, 594-602.
[19] Villac, M. C., Kaczmarska, I., Ehrman, J. M. (2013). The diversity of diatom assemblages in ships'ballast sediments: colonization and propagule pressure on canadian ports, J. Plankton Res., 35(6), 1267-1282.
[20] Griffiths, K., Michelutti, N., Sugar, M., Douglas, M. S. V., Smol, J. P. (2017). Ice-cover is the principal driver of ecological change in high arctic lakes and ponds, PLoS ONE, 12(3), e0172989.
[21] Hobbs, W. O., Telford, R. J., Birks, H. J. B., Saros, J. E., Hazewinkel, R. R., Perren, B. B. et al. (2010). Quantifying recent ecological changes in remote lakes of north america and greenland using sediment diatom assemblages, PLoS ONE, 5(4), e10026.
[22] Winder, M., Reuter, J. E., Schladow, S. G. (2009). Lake warming favours small-sized planktonic diatom species, Proc. Biol. Sci., 276(1656), 427-435.
[23] Saros, J. E., Anderson, N. J. (2015). The ecology of the planktonic diatom Cyclotella and its implications for global environmental change studies, Biol. Rev. Camb. Philos. Soc., 90(2), 522-541.
[24] Warner, D. M., Lesht, B. M. (2015). Relative importance of phosphorus, invasive mussels and climate for patterns in chlorophyll a and primary production in Lakes Michigan and Huron, Freshw. Biol., 60(5), 1029-1043.
[25] Reavie, E. D., Cai, M., Twiss, M. R., Carrick, H. J., Davis, T. W., Johengen, T. H., et al. (2016). Winter-spring diatom production in Lake Erie is an important driver of summer hypoxia, J. Great Lakes Res. 42(3), 608-618.
[26] Beall, B. F. N., Twiss, M. R., Smith, D. E., Oyserman, B. O., Rozmarynowycz, M. J., Binding, C. E. et al. (2016). Ice cover extent drives phytoplankton and bacterial community structure in a large north-temperate lake: implications for a warming climate, Environ. Microbiol., 18(6), 1704-1719.
[27] Eloranta, P., Soininen, J. (2002). Ecological status of some finnish rivers evaluated using benthic diatom communities, J. Appl. Phycol., 14(1), 1-7.
[28] Lavoie, I., Campeau, S., Zugic-Drakulic, N., Winter, J. G., Fortin, C. (2014). Using diatoms to monitor

stream biological integrity in eastern Canada: An overview of 10 years of index development and ongoing challenges, Science of The Total Environment, 475, 187-200.

[29] Stevenson, J. (2014). Ecological assessments with algae: a review and synthesis, J. Phycol., 50(3), 437-461.

[30] Smucker, N. J., Vis, M. L. (2011). Diatom biomonitoring of streams: reliability of reference sites and the response of metrics to environmental variations across temporal scales, Ecol. Indic., 11(6), 1647-1657.

[31] Virtanen, L. K., Soininen, J. (2016). Temporal variation in community-environment relationships and stream classifications in benthic diatoms: implications for bioassessment, Limnologica, 88, 11-19.

[32] Finlay, B. J. (2002). Global dispersal of free-living microbial eukaryote species, Science, 296(5570), 1061-1063.

[33] Finlay, B. J., Clarke, K. J. (1999). Ubiquitous dispersal of microbial species, Nature, 400(6747), 828.

[34] Passy, S. I. (2012). A hierarchical theory of macroecology, Ecol. Lett., 15(9), 923-934.

[35] Hájek, M., Roleček, J., Cottenie, K., Kintrová, K., Horsák, M., Poulíčková, A. et al. (2011). Environmental and spatial controls of biotic assemblages in a discrete semi-terrestrial habitat: comparison of organisms with different dispersal abilities sampled in the same plots, J. Biogeogr., 38(9), 1683-1693.

[36] Van der Gucht, K., Cottenie, K., Muylaert, K., Vloemans, N., Cousin, S., Declerck, S. J., et al. (2007). The power of species sorting: local factors drive bacterial community composition over a wide range of spatial scales, Proc. Natl. Acad. Sci. U. S. A., 104(51), 20404-20409.

[37] Martiny, J. B. H., Bohannan, J. M., Brown, J. H., Colwell, R. K., Fuhrman, J. A., Green, J. L. et al. (2006). Microbial biogeography: putting microorganisms on the map, Nat. Rev. Microbiol., 4(2), 102-112.

[38] Vyverman, W., Verleyen, E., Sabbe, K., Vanhoutte, K., Sterken, M., Hodgson, D. A. et al. (2007). Historical processes constrain patterns in global diatom diversity, Ecology, 88(8), 1924-1931.

[39] Verleyen, E., Vyverman, W., Sterken, M., Hodgson, D. A., De Wever, A., Juggins, S. et al. (2009). The importance of dispersal related and local factors in shaping the taxonomic structure of diatom metacommunities, Oikos, 118(8), 1239-1249.

[40] Astorga, A., Oksanen, J., Luoto, M., Soininen, J., Virtanen, R., Muotka, T. (2012). Distance decay of similarity in freshwater communities: do macro-and microorganisms follow the same rules? Global Ecology and Biogeography, 21(3), 365-375.

[41] Finlay, B. J., Esteban, G. F., Fenchel, T. (2004). Protist diversity is different? Protist, 155(1), 15-22.

[42] Levkov, Z., Krstic, S., Metzeltin, D., Nakov, T. (2007). Diatoms of lakes Prespa and Ohrid. About 500 taxa from ancient lake system. In: Iconographia Diatomologica, BiogeographyEcology-Taxonomy Lange-Bertalot, Vol. 16, A. R. G. Gantner and Ruggell, pp. 1-611. Liechtenstein.

[43] Kulikovskyi, M. S., Lange-Bertalot, H., Metzeltin, D., Witkowski, A. (2012). Lake Baikal: hotspot of endemic diatoms I. In: Iconographia DiatomologicaTaxonomy—Biogeography-Diversity, Vol. 23, H Lange-Bertalot, A. R. G Ganter and L Ruggell, pp. 7-607.

[44] Kociolek, J. P., Woodward, J., Graeff, C. (2016). New and endemic Gomphonema C. G. Ehrenberg (Bacillariophyceae) species from Hawaii, Nova Hedw., 102(1), 141-171.

[45] Vanormelingen, P., Chepurnov, V. A., Mann, D. G., Cousin, S., Vyverman, W. (2007). Congruence of morphological, reproductive and ITS r DNA sequence data in some australasian Eunotia bilunaris (Bacillariophyta), Eur. J. Phycol., 48, 61-79.

[46] Evans, K. M., Mann, D. G. (2009). A proposed protocol for nomenclaturally effective DNA barcoding of microalgae, Phycologia, 48(1), 70-74.

[47] Casteleyn, G., Leliaert, F., Backeljau, T., Debeer, A.-E., Kotaki, Y., Rhodes, L. et al. (2010). Limits to gene flow in a cosmopolitan marine planktonic diatom, Proceedings of the National Academy of Sciences, 107(29), 12952-12957.

[48] MacArthur, R. H., Wilson, E. O. (1967). The theory of island biogeography, Monographs in Population Biology. Princeton, NJ: Princeton University Press.

[49] Incagnone, G., Marrone, F., Barone, R., Robba, L., Naselli-Flores, L. (2014). How do freshwater organisms cross the "dry ocean"? A review on passive dispersal and colonization proces with special focus on temporary ponds, Hydrobiologia, 50, 103-123.

[50] Bolgovics, A., Ács, É., Várbíró, G., Görgényi, J., Borics, G. (2016). Species area relationship (SAR) for benthic diatoms: a study on aquatic islands, Hydrobiologia, 764(1), 91-102.

[51] Kristiansen, J. (2008). Dispersal and biogeography of silica-scaled chrysophytes, Biodivers. Conserv., 17

(2), 419-426.
- [52] Padisák, J., Vasas, G., Borics, G. (2015). Phycogeography of freshwater phytoplankton—traditional knowledge and new molecular tools, Hydrobiologia, 764(1), 3-27.
- [53] Kristiansen, J. (1996). Dispersal of freshwater algae—a review, Hydrobiologia, 336(1-3), 151-157.
- [54] Sharma, N., Singh, S. (2010). Differential aerosolization of algal and cyanobacterial particles in the atmosphere, Indian J. Microbiol., 50(4), 468-473.
- [55] Boltovskoy, D., Almada, P., Correa, N. (2011). Biological invasions: assessment of threat from ballast-water discharge in Patagonia ports, Environ. Sci. Policy, 14(5), 578-583.
- [56] Görgényi, J., Boros, G., Vitál, Z., Mozsár, A., Várbíró, G., Vasas, G. et al. (2016). The role of filterfeeding asian carps in algal dispersion, Hydrobiologia, 764(1), 115-126.
- [57] Wetzel, C. E., Bicudo, D. C., Ector, L., Lobo, E. A., Soininen, J., Landeiro, V. L., Bini, L. M. et al. (2012). Distance decay of similarity in neotropical diatom comminities, PLoS ONE, 7(9), e45071.
- [58] Wilk-Woźniak, E., Najberek, K. (2013). Towards clarifying the presence of alien algae in inland waters—can we predict places of their occurrence? Biologia, 68(5), 838-844.
- [59] Souffreau, C., Vanormelingen, P., Verleyen, E., Sabbe, K., Vyverman, W. (2010). Tolerance of benthic diatoms from temperate aquatic and terrestrial habitats to experimental desiccation and temperature stress, Phycologia, 49(4), 309-324.
- [60] Souffreau, C., Vanormelingen, P., Van de Vijver, B., Isheva, T., Verleyen, E., Sabbe, K. et al. (2013). Molecular evidence for distinct antarctic lineages in the cosmopolitan terrestrial diatoms Pinnularia borealis and Hantzschia amphioxys, Protist, 164(1), 101-115.
- [61] Padisák, J., O'Sullivan, P. E., Reynolds, C. S. (2004). Phytoplankton. In: The lakes hadbook, P. E O'Sullivan and C. S Reynolds, pp. 251-307. Blackwell Science Ltd, Oxford.
- [62] Figuerola, J., Green, J. (2002). Dispersal of aquatic organisms by waterbirds: a review of past research and priorities for future studies, Freshw. Biol., 47(3), 483-494.
- [63] Cellamare, M., Leitao, M., Coste, M., Dutartre, A., Haury, J. (2010). Tropical phytoplankton taxa in aquitaine lakes (France), Hydrobiologia, 639(1), 129-145.
- [64] Stoyneva, M. P. (2016). Allochthonous planktonic algae recorded during the last 25years in Bulgaria and their possible dispersal agents, Hydrobiologia, 764(1), 53-64.
- [65] Pongruktham, O., Ochs, C., Hoover, J. J. (2010). Observations of silver carp (Hypophthalmichthys molitrix) planktivory in a floodplain lake of the Lower Mississippi River Basin, J. Freshw. Ecol., 25(1), 85-93.
- [66] Mann, D. G. (1999). The species concept in diatoms, Phycologia, 38(6), 437-495.
- [67] Chepurnov, V. A., Mann, D. G., Sabbe, K., Vyverman, W. (2004). Experimental studies on sexual reproduction in diatoms, Internatioal Review of Cytology, 237, 61-154.
- [68] Poulíčková, A., Mayama, S., Chepurnov, V. A., Mann, D. G. (2007). Heterothallic auxosporulation, incunabula and perizonium in Pinnularia (Bacillariophyceae), European J. Phycol. 42(4), 367-390.
- [69] Vanormelingen, P., Vanelslander, B., Sato, S., Gillard, J., Trobajo, R., Sabbe, K. et al. (2013). Heterothallic sexual reproduction in the model diatom Cylindrotheca, Eur. J. Phycol., 48(1), 93-105.
- [70] Blanco, S., Ector, L. (2009). Distribution, ecology and nuisance effects of the freshwater invasive diatom Didymosphenia geminata (Lyngbye) M. Schmidt: a literature review, Nova Hedw., 88(3), 347-422.
- [71] Coste, M., Ector, L. (2000). Diatomées invasives exotiques ou rares en France: principales observations effectuées au cours des dernières décennies, Systematics and Geography of Plants, 70(2), 373-400.
- [72] Jaramillo, A., Osman, D., Caputo, L., Cardenas, L. (2015). Molecular evidence of a Didymosphenia geminata (Bacillariophyceae) invasion in chilean freshwater systems, Harmful Algae, 49, 117-123.
- [73] Montecino, V., Molina, X., Bothwell, M., Muñoz, P., Carrevedo, M. L., Salinas, F. et al. (2016). Spatio temporal population dynamics of the invasive diatom Didymosphenia geminata in central-southern Chilean rivers, Science of The Total Environment, 568, 1135-1145.
- [74] Sanmiguel, A., Blanco, S., Álvarez-Blanco, I., Cejudo-Figueiras, C., Escudero, A., Pérez, M. E. et al. (2016). Recovery of the algae and macroinvertebrate benthic community after Didymosphenia geminata mass growths in spanish rivers, Biol. Invasions, 18(5), 1467-1484.
- [75] Keller, S. R., Hilderbrand, R. H., Shank, M. K., Potapova, M. (2017). Environmental DNA genetic monitoring of the nuisance freshwater diatom, Didymosphenia geminata, in eastern north american streams, Diversity Distrib., 23(4), 381-393.

[76] Ács, É., Kiss, K. T. (1993). Colonization proces of diatoms on artificial substrate in the River Danube near Budapest (Hungary), Hydrobiologia, 269-270, 307-315.

[77] Stenger-Kovács, C., Lengyel, E., Crossetti, L. O., Üveges, V., Padisák, J. (2013). Diatom ecological guilds as indicators of temporally changing stressors and disturbances in the small Tornastream, Hungary, Ecol. Indic., 24, 138-147.

[78] B-Béres, V., Török, P., Kókai, Z., T-Krasznai, E., Tóthmérész, B., Bácsi, I. (2014). Ecological diatom guilds are useful but not sensitive enough as indicators of extremely changing water regimes, Hydrobiologia, 738(1), 191-204.

[79] Kókai, Z., Bácsi, I., Török, P., Buczkó, K., T-Krasznai, E., Balogh, C. et al. (2015). Halophilic diatom taxa are sensitive indicators of even short term changes in lowland lotic systems, Acta Bot. Croat., 74(2), 287-302.

[80] Salmaso, N., Naselli-Flores, L., Padisák, J. (2014). Functional classifications and their application in phytoplankton ecology, Freshw. Biol., 60(4), 603-619.

[81] Passy, S. I. (2007). Diatom ecological guilds display distinct and predictable behavior along nutrient and disturbance gradients in running waters, Aquatic Botany, 86(2), 171-178.

[82] Berthon, V., Bouchez, A., Rimet, F. (2011). Using diatom life-forms and ecological guilds to assess organic pollution and trophic level in rivers: a case study of rivers in south-eastern France, Hydrobiologia, 673(1), 259-271.

[83] Rimet, F., Bouchez, A. (2012a). Life-forms, cell-sizes and ecological guilds of diatoms in European rivers, Knowl. Managt. Aquatic Ecosyst., 406(01), 01.

[84] Rimet, F., Bouchez, A. (2012b). Biomonitoring river diatoms: Implications of taxonomic resolution, Ecol. Indic., 15(1), 92-99.

[85] Padisák, J., Crossetti, L. O., Naselli-Flores, L. (2009). Use and misuse in the application of the phytoplankton functional classification: a critical review with updates, Hydrobiologia, 621(1), 1-19.

[86] Litchman, E., Klausmeier, C. A. (2008). Trait-based community ecology of phytoplankton, Annu. Rev. Ecol. Evol. Syst., 39(1), 615-639.

[87] Vogt, R. J., Beisner, B. E., Prairie, Y. T. (2010). Functional diversity is positively associated with biomass for lake diatoms, Freshw. Biol., 55, 1636-1646.

[88] Weiher, E., Freund, D., Bunton, T., Stefanski, A., Lee, T., Bentivenga, S. (2011). Advances, challenges and a developing synthesis of ecological community assembly theory, Philos. Trans. R. Soc. Lond. B Biol. Sci., 366(1576), 2403-2413.

[89] Žutinić, P., Gligora Udovič, M., Kralj Borojević, K., Plenković-Moraj, A., Padisák, J. (2014). Morpho-functional classifications of phytoplankton assemblages of two deep karstic lakes, Hydrobiologia, 740, 147-166.

[90] Cellamare, M., Lançon, A. M., Leitão, M., Cerasino, L., Obertegger, U., Flaim, G. (2016). Phytoplankton functional response to spatial and temporal differences in a cold and oligotrophic lake, Hydrobiologia, 764(1), 199-209.

[91] Manoylov, K. M. (2009). Intra-and interspecific competition for nutrients and light in diatom cultures, J. Freshw. Ecol., 24(1), 145-157.

[92] B-Béres, V., Lukács, Á., Török, P., Kókai, Z., Novák, Z., T-Krasznai, E. et al. (2016). Combined eco-morphological functional groups are reliable indicators of colonisation processes of benthic diatom assemblages in a lowland stream, Ecol. Indic., 64, 31-38.

[93] Potapova, M. G., Charles, D. F. (2002). Benthic diatoms in USA rivers: distributions along spatial and environmental gradients, J. Biogeogr., 29(2), 167-187.

[94] Soininen, J. (2012). Macroecology of unicellular organisms—patterns and processes, Environ. Microbiol. Rep., 4(1), 10-22.

[95] Battarbee, R. W. (2000). Paleolimnologtical approaches to climate change with special regard to the biological record, Quat. Sci. Rev., 19(1-5), 107-124.

[96] Williamson, C. E., Saros, J. E., Schindler, D. W. (2009). Sentinels of changes, Science, 323, 887-888.

[97] Wang, Q., Yang, X., Anderson, N. J., Dong, X. (2016). Direct versus indirect climate controls on Holocene diatom assemblages in a sub-tropical deep, alpine lake (Lugu Hu, Yunnan, SW China) Quaternary Research. 86 07. pp. 1-12.

[98] Burcher, C., Valett, H., Benfield, E. (2007). The land-cover cascade: relationships coupling land and

water, Ecology, 88(1), 228-242.

[99] Atkinson, C. L., Cooper, J. T. (2016). Benthic algal community composition across a watershed: coupling processes between land and water, Aquatic Ecol., 50(2), 315-326.

[100] Pajunen, V., Luoto, M., Soininen, J. (2016). Climate is an important driver for stream diatom distributions, Global Ecology and Biogeography, 25(2), 198-206.

[101] Hillebrand, H., Watermann, F., Karez, R., Berninger, U. (2001). Differences in species richness patterns between unicellular and multicellular organisms, Oecologia, 126(1), 114-124.

[102] Passy, S. I. (2009). The relationship between local and regional diatom richness is mediated by the local and regional environment, Global Ecology and Biogeography, 18(3), 383-391.

[103] Liu, J., Soininen, J., Han, B.-P., Declerck, S. A. J. (2013). Effects of connectivity, dispersal directionality and functional traits on the metacommunity structure of river benthic diatoms, J. Biogeogr., 40(12), 2238-2248.

[104] Heino, J., Soininen, J., Alahuhta, J., Lappalainen, J., Virtanen, R. (2017). Metacommunity ecology meets biogeography: effects of geographical region, spatial dynamics and environmental filtering on community structure in aquatic organisms, Oecologia, 183(1), 121-137.

[105] Leibold, M. A., Holyoak, M., Mouquet, N., Amarasekare, P., Chase, J. M., Hoopes, M. F. et al. (2004). The metacommunity concept: A framework for multi-scale community ecology, Ecol. Lett., 7(7), 601-613.

[106] Cottenie, K. (2005). Integrating environmental and spatial processes in ecological community dynamics, Ecol. Lett., 8(11), 1175-1182.

[107] Bottin, M., Soininen, J., Alard, D., Rosebery, J. (2016). Diatom cooccurrence shows less segregation than predicted from niche modeling, PLoS ONE, 11(4), e0154581.

[108] Brown, B. L., Swan, C. M., Auerbach, D. A., Campbell Grant, E. H., Hitt, N. P., Maloney, K. O. et al. (2011). Metacommunity theory as a multispecies, multiscale framework for studying the influence of river network structure on riverine communities and ecosystems, Journal of the North American Benthological Society, 30(1), 310-327.

[109] Dong, X., Li, B., He, F., Gu, Y., Sun, M., Zhang, H. et al. (2016). Flow directionality, mountain barriers and functional traits determine diatom metacommunity structuring of high mountain streams, Sci. Rep., 6(1), 24711.

[110] De Bie, T., De Meester, L., Brendonck, L., Martens, K., Goddeeris, B., Ercken, D. et al. (2012). Body size and dispersal mode as key traits determining metacommunity structure of aquatic organisms, Ecol. Lett., 15(7), 740-747.

[111] Finkel, Z. V., Vaillancourt, C. J., Irwin, A. J., Reavie, E. D., Smol, J. P. (2009). Environmental control of diatom community size structure varies across aquatic ecosystems, Proceedings of the Royal Society B: Biological Sciences, 276(1662), 1627-1634.

[112] Lavoie, I., Grenier, M., Campeau, S., Dillon, P. J. (2010). The eastern Canadian diatom Index (IDEC) Version 2.0: Including meaningful ecological classes and an expanded coverage area that encompasses additional geological characteristics, Water Qual. Res. J. Can., 45(4), 463-477.

[113] Hájek, M., Poulíčková, A., Vašutová, M., Syrovátka, V., Jiroušek, M., Štěpánková, J., et al. (2014). Small ones and big ones: cross-taxon congruence reflects organism body size in ombrotrophic bogs, Hydrobiologia 726(1), 95-107.

[114] Dudgeon, D., Arthington, A. H., Gessner, M. O., Kawabata, Z. I., Knowler, D. J., Leveque, C. et al. (2006). Freshwater biodiversity: importance, threats, status and conservation challenges, Biol. Rev., 81(02), 163-182.

[115] Abell, R., Thieme, M. L., Revenga, C., Bryer, M., Kottelat, M., Bogutskaya, N. et al. (2008). Freshwater ecoregions of the world: A new map of biogeographic units for freshwater biodiversity conservation, Bioscience, 58(5), 403-414.

[116] Lavoie, I., Hamilton, P. B., Morin, S., Kim Tiam, S., Gonçalves, S., Falasco, E., et al. (2017). Diatom teratologies as biomarkers of contamination: are all deformities ecologically meaningful? Ecological indicators In press, 82, 539-550.

[117] Leguay, S., Lavoie, I., Levy, J. L., Fortin, C. (2016). Using biofilms for monitoring metal contamination in lotic ecosystems: The protective effects of hardness and pH on metal bioaccumulation, Environ. Toxicol. Chem., 35(6), 1489-1501.

[118] Morin, S., Proia, L., Ricart, M., Bonnineau, C., Geiszinger, A., Ricciardi, F. et al. (2010). Effects of a bactericide on the structure and survival of benthic diatom communities, Vie et Milieu, 60, 109-116.

[119] Mc Farland, B. H., Hill, B. H., Willingham, W. T. (1997). Abnormal Fragilaria spp. (Bacillariophyceae) in streams impacted by mine drainage, Journal of Fresh water Ecology, 12(1), 141-149.

[120] Pandey, L. K., Bergey, E. A. (2016). Exploring the status of motility, lipid bodies, deformities and size reduction in periphytic diatom community from chronically metal (Cu, Zn) polluted waterbodies as a biomonitoring tool, Science of The Total Environment, 550, 372-381.

[121] Schönfelder, I., Gelbrecht, J., Schönfelder, J., Steinberg, C. E. W. (2002). Relationships between littoral diatoms and their chemical environment in northeastern german lakes and rivers, J. Phycol., 38 (1), 66-89.

[122] Lavoie, I., Grenier, M., Campeau, S., Dillon, P. J. (2006a). A diatom-based index for water quality assessment in eastern Canada: an application of canonical analysis, Canadian Journal of Fisheries and Aquatic Sciences, 63, 1793-1811.

[123] Smucker, N. J., Drerup, S. A., Vis, M. L. (2014). Roles of benthic algae in the structure, function, and assessment of stream ecosystems affected by acid mine drainage, J. Phycol., 50(3), 425-436.

[124] Debenest, T., Silvestre, J., Coste, M., Delmas, F., Pinelli, E. (2008). Herbicide effects on freshwater benthic diatoms: Induction of nucleus alterations and silica cell wall abnormalities, Aquatic Toxicology, 88 (1), 88-94.

[125] Debenest, T., Silvestre, J., Coste, M., Pinelli, E. (2010). Effects of pesticides on freshwater diatoms. In Reviews of Environmental Contamination and Toxicology. 87-103. D. M Whitacre. (ed). Berlin: Springer.

[126] Hausmann, S., Charles, D. F., Gerritsen, J., Belton, T. J. (2016). A diatom-based biological condition gradient (BCG) approach for assessing impairment and developing nutrient criteria for streams, Science of The Total Environment, 562, 914-927.

[127] Young, J. N., Heureux, A. M. C., Sharwood, R. E., Rickaby, R. E. M., Morel, F. M. M., Whitney, S. M. (2016). Large variation in the Rubisco kinetics of diatoms reveals diversity among their carbon-concentrating mechanisms, J. Exp. Bot. 67(11), 3445-3456.

[128] Lavoie, I., Dillon, P. J., Campeau, S. (2009). The effect of excluding diatom taxa and reducing taxonomic resolution on multivariate analyses and stream bioassessment, Ecol. Indic., 9(2), 213-225.

[129] Poulíčková, A., Letáková, M., Hašler, P., Cox, E., Duchoslav, M. (2017). Species complexes within epiphytic diatoms and their relevance for the bioindication of trophic status, Science of The Total Environment, 599-600, 820-833.

[130] Jones, F. C. (2008). Taxonomic sufficiency: The influence of taxonomic resolution on freshwater bioassessments using benthic macroinvertebrates, Environ. Rev., 16(NA), 45-69.

[131] Mandelik, Y., Roll, U., Fleischer, A. (2010). Cost-efficiency of biodiversity indicators for mediterranean ecosystems and the effects of socio-economic factors, Journal of Applied Ecology, 47(6), 1179-1188.

[132] Mellin, C., Delean, S., Caley, J., Edgtar, G., Meekan, M., Pitcher, R. et al. (2011). Effectiveness of biological surrogates for predicting patterns of marine biodiversity: a global meta-analysis, PLoS ONE, 6 (6), e20141.

[133] Bevilacqua, S., Terlizzi, A., Claudet, J., Fraschetti, S., Boero, F. (2012). Taxonomic relatedness does not matter for species surrogacy in the assessment of community responses to environmental drivers, Journal of Applied Ecology, 49(2), 357-366.

[134] Bevilacqua, S., Claudet, J., Terlizzi, A. (2013). Best practicable aggregation of species: a step forward for species surrogacy in environmental assessment and monitoring, Ecol. Evol., 3(11), 3780-3793.

[135] Raunio, J., Soininen, J. (2007). A practical and sensitive approach to large river periphyton monitoring: comparative performance of methods and taxonomic levels, Boreal Environmenal Research, 12, 55-63.

[136] Chen, X., Bu, Z., Stevenson, M. A., Cao, Y., Zeng, L., Qin, B. (2016). Variations in diatom communities at genus and species level in peatlands (central China) linked to microhabitats and environmental factors, Science of The Total Environment, 568, 137-146.

[137] Terlizzi, A., Bevilacqua, S., Fraschetti, S., Boero, F. (2003). Taxonomic sufficiency and the increasing insufficiency of taxonomic expertise, Mar. Pollut. Bull., 46(5), 544-560.

[138] Apothéloz-Perret-Gentil, L., Cordonier, A., Straub, F., Iseli, J., Esling, P., Pawlowski, J. (2017). Taxonomy-free molecular diatom index for high-throughput eDNA biomonitoring, Mol. Ecol. Resour., 17

(6), 1231-1242.

[139] Heino, J., Soininen, J. (2007). Are higher taxa adequate surrogates for species-level assemblage patterns and species richness in stream organisms? Biological Conservation, 137, 78-89.

[140] Poulíčková, A., Špačková, J., Kelly, M.G., Duchoslav, M., Mann, D.G. (2008). Ecological variation within Sellaphora species complexes (Bacillaryophyceae): specialists or generalists? Hydrobiologia, 614 (1), 373-386.

[141] Nguyen, T.N.M., Berzano, M., Gualerzi, C.O., Spurio, R. (2011). Development of molecular tools for the detection of freshwater diatoms, J. Microbiol. Methods, 84(1), 33-40.

[142] Manoylov, K.M. (2014). Taxonomic identification of algae (morphological and molecular): species concepts, methodologies, and their implications for ecological bioassessment, J. Phycol., 50(3), 409-424.

[143] Baudart, J., Guillebault, D., Mielke, E., Meyer, T., Tqandon, N., Fischer, S.A. et al. (2016). Microarray (phylochip) analysis of freshwater pathogens at several sites along the northern german coast transecting both estuarine and freshwaters, Appl. Microbiol. Biotechnol., 101(2), 871-886.

[144] Cimarelli, L., Saurabh Singh, K., Mai, N.T.N., Dhar, B.C., Brandi, A., Brandi, L. et al. (2015). Molecular tools for the selective detection of nine diatom species biomarkers of various water quality levels, Int. J. Environ. Res. Public Health, 12(5), 5485-5504.

第 14 章

法医学中的硅藻：
法医学中硅藻检验的分子方法

Vandana Vinayak, S. Gautam

14.1 简介

多年来,硅藻一直被用于法医调查以帮助确定个人是否溺水以及案件是死前还是死后溺水。在医学法律系统中,溺水的诊断刚从水中提取的尸体主要是基于溺水迹象,如口腔、鼻孔和呼吸通道的细白色泡沫、肺气肿和肺部肋骨压痕;但腐烂时,这些明显的溺水症状往往会消失[1-2]。Rushton 认为,硅藻分析是法医科学在溺水诊断方面的众多发现之一,但仅仅在受害者组织中发现硅藻被认为仅是一种支持性证据,而不是一种完整的证明方法。虽然硅藻测试对溺水的诊断价值一直存在争议,但大多数法医专家仍然认为它是法医实践中相对可靠和有用的诊断工具[3-8]。溺水和硅藻测试背后的机制是当一个人在水介质中溺水时,硅藻与其他细菌、病毒、真菌、浮游植物一起进入气道并到达肺部。由于硅藻细胞壁是由二氧化硅组成的,它会抵抗身体的酸性环境,并将存活很长时间,而其他微生物将在适当的时间内死亡。在溺水的过程中,水被吸入使肺泡膨胀。因此,溺水介质中的硅藻穿孔肺泡毛细血管屏障并进入肺静脉循环[9]。含有硅藻的血液会栓塞进入器官和组织,如骨髓。因此,在股骨骨髓中发现硅藻是死前溺水的迹象,这意味着一个人在水中还活着,而溺水是他死亡的原因[10]。在发现尸体的水介质中也检测到一致的硅藻,确认为一例死前溺水的病例[3,6](见图 14.1)。

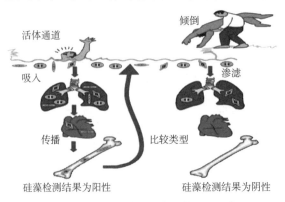

图 14.1　溺水时硅藻试验背后的机理示意图

有时考虑到医学法律的重要性,硅藻测试的有效性也受到了批评,因为在非溺水者的肺和其他周围器官中也能发现硅藻。

Hendey 推断硅藻检验的有效性为:

> 如果硅藻测试所造成的混乱和不确定性要被消除,则应从以下角度来考虑。溺水死亡的硅藻测试必须能够证明从死者身上发现的硅藻是在死者躺着的地点或死者进入水中或被认为已经进入水中的硅藻群的一部分。

根据 Ludes 和 Coste 关于死后溺水的说法,硅藻最终只会到达血液,而不会到达周围器官。因此,选择进行简单定性硅藻分析的器官不是肺,因为它有时会给出假阳性结果。错误选择硅藻样品进行硅藻试验可能会导致硅藻的吸收进入人体的假阳性试验[14-16],例如通过吸入嗜气物种和通过硅藻土[15,17]。假阳性试验的机会也出现在死后溺水的病例中,硅藻通过长时间淹没进入身体,特别是在高静水压力和分解的身体中,通过伤口和在复活过程中进入身体[18]。因此,为了避免假硅藻阳性试验,建议将左心室组织或完整股骨骨髓组织作为硅藻试验的最佳样本[13]。Guy[19],Brouardel & Vibert[20],Swann & Spafford[21],Revenstorf[22],Incze[23-24] 和他的匈牙利同事[25-26],Tamáska[27-28]和 Mueller & Gorgs[29]的先驱研究描述了硅藻和溺水检测的关系。Guy[19]在水中的老鼠身上做了实验,发现在用酸处理后,肺部的所有部分都显示出气泡。他观察到,水充满了整个空气通道,并进入气管和更大的支气管。Swann & Spaffodd[21]证明,在淡水中溺水 3 min 后,原始血容量可以被等量的水稀释。1904 年,Revenstorf[22]解释说,检测硅藻肺的溺水受害者表明死亡由于溺水,但无法检测血液中的硅藻,因为硅藻测试被认为是一个重要的工具在诊断和确认死前溺水[6]。另一方面,Incze[23]的研究表明,在正在研究的硅藻土悬浮液中,溺水兔的肺、血液和实质器官中存在硅藻。他研究了溺水的人体,发现除了肺以外的器官中还存在硅藻。Tamáska[27]进一步观察了骨髓中的硅藻,他成功地观察了三名妇女尸体的骨髓中的硅藻源于多瑙河,包括一名妇女的左心和另外两个妇女的肺和血液中存在硅藻。最后,Thomas,Van Hecke & Timperman[30-31]和 Timperman[32-33]得出结论,器官中存在硅藻表明溺水死亡。为了解决从这些问题中检测硅藻的人工错误,发展用于硅藻鉴定的 DNA 条形码技术是非常必要的,并且可能对溺水调查有用。条形码一词是由 Hebert 等[34-35]创造的,目的是提高分类单元识别的准确性,并获取隐藏的生物多样性[36-40]。条形码的重要性依赖于对包括硅藻在内的任何生物体的正确、快速和廉价的识别。条形码的想法是基于这样一个假设:即短 DNA 片段(COI、rDNA、SSU、ITS 区域、UPA)内的差异反映了物种的生物分离,因此可以用作允许物种鉴定的遗传标签[37]。因此,硅藻的 DNA 条形码分析用于溺水分析是一种新颖且优于传统的硝酸消化方法[41-42],该方法有其局限性[6],将在本章后面讨论。

随着分子生物学的发展,检测硅藻属针对基因位点像小亚基核糖体 RNA (16S rDNA),rbcL 基因座上的 RUBISCO(核酮糖 1,5 双磷酸)基因,COI(细胞色素

C 氧化酶亚基 1)和 SIT(硅转运蛋白)基因通过聚合酶链反应(PCR)提供了一个有前途的浮游生物检测组织检测溺水相关犯罪案件[2]。DNA 条形码也已成为动植物分类学中的常见做法[34]使用细胞色素 C 氧化酶 1(COI)[43-44]。但对硅藻的研究却很少[45]。Kane[46]在日本湖 Biwa 的溺水者的蓝藻中检测到属于聚球藻的微浮游生物[46]靶向 16SrRNA 基因位点。Suto 等[44]证明了即使在水样中没有微浮游生物,也可以检测到溺水导致的死亡。Suto 等[44]的工作表明,一些存在于喉咙中的特定细菌,如 *Streptococcus alivarius*(SL1), *Streptococcus sanguinis*(SN1),以及一些常见于各种水样中的 *Aeromonas hydrophila*(AH1),可以帮助使用分子技术检测溺水死亡[44]。本章旨在找出环境样本中硅藻与硅藻 DNA 的关系,作为检测死前和死后溺水的充分证明方法。

14.2 尸检取证措施

多年来,硅藻在解决溺水死亡病例中发挥了重要作用,尤其在溺水的病理体征和尸检无法确定死亡原因的情况下,硅藻测试发挥了重要的作用。因此,硅原子测试是检测死亡的黄金准则。然而,如果溺水者的硅藻数量适合检测,这个测试效果很好;例如,每 100 μL 从 10 克肺样本中提取 20 个硅藻,从身体其他器官中提取 50 个硅藻[47]。相反,Otto[48]报告说,由于持续吸入空气中的硅藻,硅藻土工业的工人的肺部含有硅藻[49]。然而,如果受害者的尸体被扔进水中,那么就不会有血液循环,因为心脏不会泵血,水介质中的硅藻由于从腐烂的皮肤内渗出硅藻等,只能传播到肺部。除了一些例外情况,还可以看到,有时从水介质或水附近中回收的尸体并不总是显示硅藻试验呈阳性。这通常见于干性溺水或因害怕掉入水中而死亡的情况,在游泳、潜水、涉水、药物、酒精等实际掉入水中之前,由于迷走神经抑制而死亡[50]。在这种情况下,硅藻不能进入身体的远端器官,如肾脏、肝脏、大脑和股骨,这些器官中没有硅藻可以被认为是死后溺水。

14.3 溺水受害者与死于其他原因的差异

溺水死亡是造成意外死亡的前三大主要原因之一。它主要涉及 1~4 岁和 15~19 岁的儿童[51]。溺水受害者与病理生理机制的并发症和威胁生命有关,这是一个非常有趣的研究课题。有许多术语与溺水和溺水相关的死亡有关,如接近溺水、二次溺水、淹没损伤、浸入没综合征[52]。溺水和接近溺水导致的死亡主要发生在年幼和健康儿童[52],由于酸中毒、低灌注、低氧血症、肺损伤、神经损伤、心血管不稳定、由于缺血、代谢异常和弥散性血管内凝血引起的肾衰竭[53-54]。众所周知,溺水死亡的诊断涉及硅藻的检测和鉴定,但不同的研究涉及检测硅藻进行分析的不同方法。Katar[55]区分了"水尸体"和"非水尸体",并制备了不同生物样本中硅藻密度的分离值(SV);即每 5 克肺 200 个硅藻,肝脏每 5 克肺 10 个硅藻,肾脏每 5 克肺 4 个硅藻,骨髓每 5 克骨髓 20 个硅藻。通过相似性指数[56]及其与物种指数(SI)[57]和优势同一性(DI)[58]的比较,确定了对器官样本和水样中硅藻物种组成的评估。这导致了许多作者[5-6,47,59-63]进行全面评估溺水现场。下面是计算物种指数(SI)和优势特征(DI)的公式:

$$SI_{1,2} = S_{1 \cap 2}/S_{1+2} \times 100 [\%]$$

$$DI_{1,2} = \sum_{i=1}^{S} q_i [\%] \tag{14.1}$$

式中,$S_{1 \cap 2}$ 为样本 1 和 2 中两个硅藻群落共有的物种数量,S_{1+2} 为样品 1 和 2 中两个硅藻群落的物种总数,q_i 为物种 I 和 S 的两个相对出现次数中的较小者(样本 1 和 2 中的所有物种)。

14.4 生物样品中硅藻的鉴定技术

与许多其他藻类群相比,硅藻具有良好的形态特征,可以通过微观分析来识别,但由于大小和脆弱性的差异,一些硅藻种类比其他种类更容易检测[64]。硅藻的详细结构描述可以通过光学显微镜看到[65],电子显微镜可以看到更详细的特征,它依赖于硅柱结构的微观特征,如对称、形状、雕塑和晶体的大小。硅藻的鉴定目前主要基于形态学分析和研究[66],因为硅藻的形态学研究分析成本较低,并且比其他生物群落(如,无脊椎动物)与环境因素的联系更好[67]。另一方面,区分密切相关的分类单元是非常困难的,并可能导致错误识别,损害硅藻指数的水质和各种其他研究的准确性[68]。自过去十年以来,科学家们正在使用不同的方法和协议从生物样和水样中分离硅藻。最常用的是酸消化法[69]、soluene-350 法[70]、酶解法[71]、膜过滤法[72-73]、干灰法[49]和胶体硅梯度法[74]。在上述六种方法中,酸消化法、soluene 法和酶解法是世界上法医学实验室硅藻提取最常用的方法。该方法具有很高的分离成功率,因为它完全从组织样本中去除有机物质,但在此过程中使用的硝酸会排放生物危害性烟雾,对进行测试的工人构成生命威胁。近年来,基于使用 DNA 序列的分子技术的替代方法已经被开发出来以避免这些问题。在分子层面方法有 FISH(荧光原位杂交)[75-76]和微阵列筛选[77],它们只能追踪已知的分类群。因此,可进行快速、高通量的筛选基于样本 DNA 的方法,如等位酶电泳[78]、DNA 指纹分析[79]、同工酶分析[80]、微卫星标记分析[81]和 DNA 元编码[82]是首选。Racz 等认为,在溺水导致死亡的情况下,即使在没有可见硅藻的情况下,用分子工具检测硅藻也将加强溺水的尸检诊断[83]。因此,鉴于这些方法的局限性,生物和水样中硅藻的分子检测比上述方法更真实、可靠、环境安全[84]。

14.4.1 水样的形态学分析

环境水样的取样通常是在发现溺水者尸体的地方附近进行的；一般来自水体的沿海地区[2]。一部分水样用于显微镜下硅藻的鉴定。根据 Biggs 和 Kilroy[85]给出的方案，首先通过准备其永久的载玻片来鉴定硅藻。该方法包括用 90% 的丙酮处理围生性的藻类样品，以去除细胞质含量。随后，样品干燥，用浓硫酸处理，用 30% 过氧化氢处理，用蒸馏水彻底清洗。干燥后，将样品安装在 Naphrax 安装介质上进行显微镜检查。利用光学显微镜（LM）和扫描电子显微镜（SEM）实现了环境样品和单个属中硅藻的形态学鉴定的准确性。光学显微镜包括使用安装介质在载玻片上永久制备切片[86]，但对于扫描电子显微镜，硅藻细胞在室温下用 3% 戊二醛溶液处理 30 min，如 Nassiri 等[87]所述的一次固定。然后将固定的细胞以 1 000 r/min 离心 10 min。进一步取出上清液，在用碳酸氢钠（0.2 mol/L）缓冲的 3% 戊二醛溶液中固定 2 h。样品按照 Beninger 等描述的方法，在 50%、70%、80%、90%、90%、95% 和 100% 的丙酮中进一步脱水 5 min，并包埋于环氧树脂中[88]。清洗后的材料安装在铝存根上，然后涂上金和钯，并进行检查[89-90]，如图 14.2 所示。

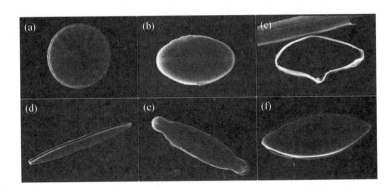

图 14.2　印度哈里亚纳水体中硅藻的 SEM 图
(a) 梅尼小环藻；(b) 丝状等带藻；(c) 假尖顶异极藻；(d) 谷皮菱形藻；(e) 羽纹藻属；(f) 谷皮菱形藻

硅藻的分类鉴定使用了几个识别键，如 ADIAC（硅藻原子自动识别和分类）[5,91-92]，美国数据库（坎伯恩和查尔斯），印度硅藻图像数据库（DIDI）[93]。使用 Omnidia 软件计算的生态指数[94]。如果从环境样本中分离的硅藻进行无菌培养，则可以进一步确保鉴定，因为这样可以减少研究中的数量有限硅藻，并为未来研究生态指标和 DNA 条形码生成相关的种质（R98）。

14.4.2 位点特异性硅藻的作用

对硅藻的定量和定性研究在文献中有充分的记录，以确定死亡原因和疑似死亡地点[7]。不同硅藻属的分布与特定的水体有关[95-96]。因此，硅藻数据库有助于了解不同水体中精确存在的硅藻类型和一些定位于特定地点的硅藻。有些硅藻只在特定季节流行,气候变化导致水体中硅藻菌群的变化,是诊断溺水死亡的有用工具。Vinayak V[96]从印度北部地区特别是哈里亚纳地区的不同水体建立了为期两年(2008~2010)的硅藻数据库以确定溺水病例的死亡,报告了56个硅藻属和112种硅藻。很少出现的硅藻是针对特定地点的并可能有助于定位溺水地点[97]。Singh等还研究了印度昌迪加尔苏克纳湖的两种特定硅藻,即弯棒杆藻和弧形汉氏藻,它们存在于所有季节(春、夏、秋和冬季)。Philipose和Karikal报道,硅藻数量在冬季最多,在季风季节最少,这也可以作为特定硅藻的溺水指标。在印度拉贾斯坦邦的加尔塔昆德发现了5种硅藻,即菱形藻、异极藻、曲壳藻、双眉藻和汉氏藻[98]。硅藻是非常适合评估河流生态完整性的生物,并对环境变化作出迅速和敏感的反应,从而对河流的生物完整性和生态系统恶化的原因提供高度信息的评估。根据群落的物种组成推断污染水平,生物指数[99-103]基于Kolkwitz和Marsson[104]以及Patrick早期监测研究[105-106]的研究主要依靠硅藻多样性作为河流健康的普遍指标。在我们早期的工作中,我们报道了[90]硅藻假尖顶异极藻在存在高水平铅和硒重金属的情况下发生硅藻壳畸形,这将作为印度库鲁克舍特拉Saraswati Dham水体的法医指标,而Pandey等[93]报道了从拉贾斯坦邦的金属(Cu和Zn)位点检测到的敏感硅藻物种尖异极藻。另一方面,在印度南部的印度半岛班加罗尔市区的浅水湿地中特别发现的两个新物种[107],它们是特定存在于印度火山口湖[108]。由此可见,对不同水体的硅藻进行定期监测,有助于研究硅藻和水体类型,这可能是确定溺水地点和季节的指标。

14.5 案例研究

14.5.1 案例1

1977~1993年,在加拿大安大略省,对771例春季、夏季、秋季和冬季的溺水病例进行了硅藻试验分析[5]。他们使用股骨髓作为硅藻提取的起始材料,这是最可靠的检测器官。他们观察到,大约205例(28%)被发现为阳性,但从家庭水源中找到的尸体很少呈阳性,由于房屋中使用的过滤和处理水而缺乏硅藻。所有阳性检测均检测到与各自的溺水介质相匹配的硅藻[13]。该研究还描述了硅藻试验发生率的月变化,显示4月、7月(约40%)和11月(约30%)硅藻试验阳性的数量较多,但冬季的频率最少。因此,季节和水中硅藻含量的百分比影响了硅藻试验的结果。

14.5.2 案例2

2013年,在印度马哈拉施特拉邦一个村庄的水井中发现三名4~9岁的未成年姐妹死亡,作者作为印度哈里亚纳邦国家法医科学实验室指定的高级科学官员,对生物和水样本进行了硅藻测试。警方怀疑这是一起强奸和谋杀案,但在进行硅藻测试后,所有三名未成年女孩的生物样本(股骨)的阳性结果使案件出现了转折。用硝酸消化法[63]从受害者的生物样本中检测出孢杆菌,并在找回尸体的井水中检测出其一致性,如图14.3所示,将病例导向死前溺水。法医科学实验室的报告进一步帮助排除了这是一起强奸案件,因为在三名小女孩的阴道载玻片、拭子和衣服中没有发现精液。法医报告,所有未成年女孩的处女膜都很完整,完全排除了强奸案,硅藻的发现有助于得出意外溺水的结论。

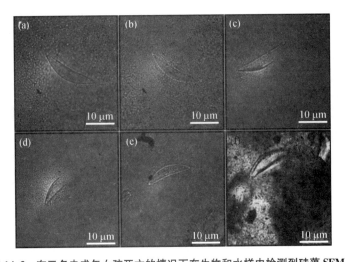

图14.3 在三名未成年女孩死亡的情况下在生物和水样中检测到硅藻SEM图
(a)和(b) 受害者-1中的近缘桥弯藻;(c)和(d) 受害者-2中的近缘桥弯藻;(e) 受害者-3中的近缘桥弯藻;
(f) 水样中检测到的仿生桥弯藻属

14.5.3 案例3

然而,在另一个案件中,作者以报告干事的身份对2009年在印度查谟和克什米尔的一条河流中发现的两名25~30岁的年轻女性死亡案件中的生物(股骨髓)和水样死亡进行了硅藻测试。这是一起涉嫌绑架、轮奸和谋杀的案件。根据宗教仪式,尸体被埋在泥土里。建议该案件重新审查,并交给中央调查机构,该机构进一步建议该案件在印度哈里亚纳邦国家法医科学实验室进行硅藻测试。两名受害者(股骨骨髓和心脏组织)的生物样本按照Pollanen的方案进行消化,结果显示存在星杆藻、双楔藻、舟形藻、破碎的针杆藻。另一方面,水样也显示出类似类型的硅藻,即双声双楔藻、舟形藻、针杆藻和盘杆藻。硅藻双声双楔藻是一种入侵淡水物种,盛于于低温恒温的水体,查谟和克什米尔是印度北部的温带地区,温度从-3~25℃不等。然而,在一致性水样中存在这些硅藻类型,证实了这是一例死后溺水的病例(见图14.4和图14.5)强奸被排除了,因为在两名溺水的女性的展品中都没有发现精液,其中一名女性的处女膜完好无损。根据这份报告,调查机构排除了这是一起强奸和谋杀案件,并根据硅藻测试的结果认为这是一起意外溺水案件。

在上述案例中,硅藻已作为解决危重案例的黄金准则,并区分了死后溺水和死前溺水的区别。此外[6,9,109]对实验动物和人类溺水病例的研究表明,小型硅藻比中心硅藻和大型硅藻具有对肺泡-毛细血管屏障的穿透能力。由于污染是溺水病例假阳性结果的原因之一,应注意各种污染来源,如在游泳时反复吞水、经胎盘通道、挖掘病例中通过棺材地板的污染[110-112]和解剖组织取样期间[113]。然而,由于法医实验室缺乏硅藻鉴定的专业知识和人工错误,硅藻检测的真实性可能会受到挑战[6]。此外,阴性硅藻测试并不表明这是一个杀人案件,因此要排除每一个可能的错误,需要开发基于PCR的引物来区分死后和死前溺水[83]。

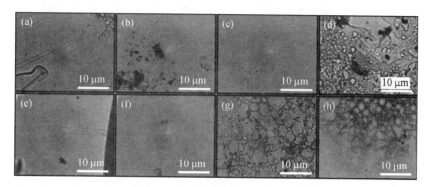

图 14.4 受害者 1 的生物组织中存在的硅藻和受害者 2 的生物组织内存在的硅藻土

(a) 美丽星杆藻和双生双楔藻;(b) 舟形藻;(c) 破碎的针杆藻;(d) 破碎的棘突藻藻细胞;(e) 和 (f) 针杆藻;(g) 和 (h) 破碎的硅藻细胞

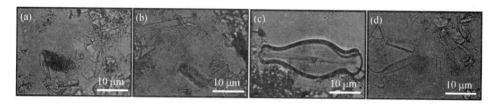

图 14.5 水样中检测到的硅藻混合菌落群

(a) 和 (b) 海滨菱形藻;(c) 双声双楔藻;(d) 舟形藻、近缘桥弯藻和棘突藻

14.6 利用分子工具鉴定组织和水样中的硅藻

从水样中获得硅藻属鉴定的真实性及其在溺水受害者组织中的一致性类型,对于建立同一溺水地点至关重要。如前所述,传统的从水和生物样品中分离和检测硅藻的分析方法有其局限性。在形态水平上的接近限制了硅藻在属和种水平上的正确鉴定。然而,法医分子生物学的最新进展证实,从溺水受害者身上检测到少量的硅藻DNA 是可能的[114]。同样,为了避免水样中其他微生物和细菌的 DNA 的交叉污染,水样中硅藻的单培养或无菌培养将有助于建立与从人体组织中提取的硅藻 DNA 相

匹配的 DNA 扩增产物[46]。从环境水样中提取无菌硅藻培养通常按照 Shishlyannikov 等采用的程序进行。Shishlyannikov 等[115]在实验室条件下(光照：黑暗16∶8小时,温度20~25℃),f/2培养基[116],轻微修改[117](见图14.6)。采用十六烷基三甲基溴化铵(CTAB)方法[118]从水样和无菌培养物中提取混合硅藻和个体硅藻物种的 DNA,利用系统发育树研究物种多样性,正确鉴定硅藻属[90]。

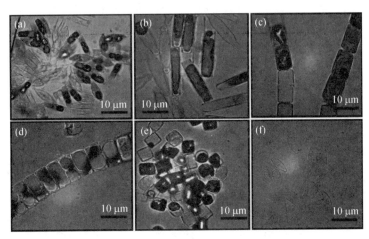

图 14.6 液体培养基和固体琼脂培养基平板中硅藻的无菌培养
(a)微小异极藻(*Gomphonema parvulum*);(b)羽纹藻(*Pinnularia saphophila*);(c)变异直链藻(*Melosira varians*);(d)丝状等带藻(*Diadesmis confervacea*);(e)梅尼小环藻(*Cycletlla meneghiniana*)

14.7 溺水受害者组织中硅藻 DNA 的分化

硅藻检测的分子方法是基于人类和植物 DNA 之间有限的同源性,植物细胞中质体基因的高拷贝数和植物界中高度保守的基因,包括参与光合作用的编码酶的基因[63]。利用分子工具鉴定人体组织中的硅藻需要扩增那些在硅藻中专门存在的基因,而不会扩增人体组织样本。主要靶向基因位点为细胞色素 C 氧化酶亚基 I、核 rDNAITS 区、核小亚基 rRNA 区、18S rDNA、16S rDNA 和 Rubisco 基因。RbcL 是一种翻译蛋白,其基因跨分类单元的序列很容易对齐,因此被认为是硅藻鉴定的最佳基因。Hamsher 等[38]调查了 rbcL 基因是鉴定硅藻不同属的最佳基因,COI-5P 是另一个区分红褐藻的优秀基因位点,因为 rbcL 比 COI-5P 更保守。由于硅藻也有一种特有的硅体蛋白相关蛋白,称为硅体蛋白蛋白[119],这些也被认为是扩增的重要靶点[13]果霉蛋白基因负责二氧化硅细胞壁有机成分的生物合成,但这些基因不参与二氧化硅的形成[120]。硅藻细胞壁的物种特异性设计可能是物种特异性蛋白结构差异的结果。不同的硅体蛋白如下：α-硅体蛋白(75 Kda)含有 α1-硅体蛋白,fruα2,fruα3,β-硅体蛋白(105 KDa),γ-硅体蛋白(140 KDa)和 δ-硅体蛋白(200 KDa)和 ε-

硅体蛋白[120]。Nils Kroger 等通过用有义引物 5′-AAT CAT GAA GTC GCC GAC CT′和反义引物 5′-TTA CCA CCA CTC CCA GAA GT-3′处理梭形筒柱藻,舟形藻,菱形藻,棱角菱形藻,三角褐指藻,隐秘小环藻和梅尼小环藻的基因组 DNA 来扩增 ε-硅体蛋白[121]。结果表明,5 种羽状硅藻均含有硅体蛋白相关的细胞壁蛋白,2 种中心硅藻与抗(硅体蛋白)血清无交叉反应。其他令人感兴趣的基因是 16S rDNA 位点,用于区分人类和植物的 DNA。由于人类(真核生物 18S rDNA)和硅藻(植物光合用 16S rDNA)之间有限的同源性[2],因此更容易区分死前和死后溺水,且误差最小。另一方面,rRNA 基因允许我们评估形态上复杂的硅藻的多样性,因为这些基因在功能和结构上高度保守[122]。

14.8 聚合酶链反应(PCR)

利用分子生物学的聚合酶链反应技术扩增从样本(水或人体组织)中提取的硅藻 DNA。PCR 反应混合物的浓度是可变的,但每 30～50 ng 浮游生物基因组 DNA 20 μL 接近 0.5 mmol/L 的正向和反向引物,10 mmol/L Tris-HCl 缓冲液(pH 8.3),50 mmol/L KCl,1.5 mmol/L MgCl$_2$,每个 dNTP 各 200 mmol/L 和 1U TaqDNA 聚合酶通常是首选[2]。PCR 的标准循环条件仍然是初始变性 95℃(2 min),变性 94℃(45 s),退火 60℃(50 s),扩展 72℃(1 min),最终扩展 72℃(10 min) 38 个循环,最后保持在 0℃ 4 min[2]。然而,不同的科学家根据硅藻的引物类型和目的基因(GOI)使用了不同的 PCR 参数,如表 14.1 所示。

表 14.1 不同基因位点 PCR 参数的不同参数方案

基因位点	预变性/时间	周期	变性/时间	退火处理/时间	伸长率/时间	最终伸长率/时间
COI-5P	94℃/4 min	35	94℃/30 s	45℃/1 min	70℃/1 min	70℃/7 min[38]
rbcL	94℃/3 min	30～40	94℃/1 min	55℃/1 min	72℃/1.5 min	72℃/5 min[39]
rbcL-3P	95℃/2 min	35	94℃/20 s	50℃/30 s	72℃/2 min	72℃/7 min[38]
LSU (D2-D3)	94℃/5 min	38	94℃/30 s	50℃/30 s	72℃/1 min	72℃/7 min[38]
UPA	94℃/2 min	35	94℃/20 s	57℃/30 s	72℃/30 s	72℃/10 min[38]

针对硅藻反应调节因子 3(DRR3)基因、16S rDNA、18S rDNA、COI、rbcL、ITS 区和硅的各种硅藻鉴定特异性引物表 14.2 列出了转运蛋白基因(SIT),以扩增不同感兴趣的基因,由于有限的相似性,这有助于将其与人类组织的 DNA 区分开来。

表14.2 用于硅藻和其他浮游植物条形码的详细清单

核苷酸序列 Nucleotide sequence (5′-3′)	PCR寡核苷酸名称 PCR oligonucleotide name	基因座 Gene locus
CYAACCAYAAAGATATTGGAAC	DiamF1	COI-5P
TCWGGGTGWCCAAARAACCA	DiamR1	COI-5P
CCAACCAYAAAGATATWGGWAC	DiamF3	COI-5P
AAACYTCWGGRTGWCCAAARA	DiamR2	COI-5P
CAACCAT/CAAAGATATA/TGGTAC	GazF2	COI-5P
GGATGACCAAAG/AAACCAAAA	GazR2	COI-5P
GAAGCWGGWGTWGGTACWGGWTG	KEint2F	COI-5P
AAACTTCWGGRTGACCAAAAA	KEintR	COI-5P
AAACTTCA/TGGG/ATGACCAAAAA	KEdtmR	COI-5P
AAGGAGAAATA/C/TAATGTCT	DPrbcL1	*rbcL*
AAG/ACAACCTTGTGTAAGTCTC	DPrbcL7	rbcL
CTCAACCATTT/CATGCG	NDrbcL5	rbcL
CTGTGTAACCCATA/TAC	NDrbcL11	rbcL
CCRTTYATGCGTTGGAGAGA	CfD	rbcL-3P
AARCAACCTTGTGTAAGTCT	DPrbcL7	rbcL-3P
AMAAGTACCRYGAGGGAAAG	T16N	LSU D2/D3
SCWCTAATCATTCGCTTTACC	T24U	LSU D2/D3
GGACAGAAAGACCCTATGAA	p23SrV_f1	UPA
TGAGTGACGGCCTTTCCACT	Diam 23Sr1	UPA
ATT CCA GCT CCA ATA GCG	D512for 18S	18S rDNA
GAC TAC GAT GGTATC TAATC	D978rev 18S	18S rDNA
CTA GTC ATA CGC TCG TCT C	DINSI	18S rDNA
GCT TGATCC TTC TGC AGG T	D1800R	18S rDNA
CTG GTT GAT CCT GCC AGTAG	Algen F	18S rDNA
GGTAAT TTA CGC GCC TGC T	Primer IR3	18S rDNA
TGTAAA ACG ACG GCC AGT	M13F (-21)	18S rDNA
CAG GAA ACA GCT ATG AC	M13R (-27)	18S rDNA
TGT AAA ACG ACG GCC AGTATT CCA GCT CCA ATA GCG	M13F-D512	18S rDNA

(续表)

核苷酸序列 Nucleotide sequence (5′-3′)	PCR 寡核苷酸名称 PCR oligonucleotide name	基因座 Gene locus
CAGGAAACAGCTATGACGACTA CGATGGTATCTAATC	M13R-D978	18S rDNA
GGAAGGTGAAGTCGTAACAAGG	ITS5_KE	18S rDNA
GCTTAAATTCGGCGGGT	LSUD_R	18S rDNA
GTC TCA AAG ATT AAG CCA TGC	18S SEQ 34 f	nuSSU
CCT TGT TAC GAC TTC ACC TTC C	ITS5-DR	V4-SSU
AGTAAGGGCGACTGAAG	LSU-DF1	D1-D2 LSU rDNA
-ACCCTATTCAGGCATAGTT	LSU-740DR	D1-D2 LSU rDNA

14.9 从一名溺水受害者的生物样本中提取硅藻 DNA

14.9.1 生物样本

随着分子工具的出现,对硅藻中特异性基因的检测为检测溺水组织中的硅藻提供了一个很有前景的工具,以确认它是一个溺水的病例。Kane 等[46,123]设计了引物来识别日本比湖的微浮游生物,并在溺水受害者组织中识别浮游生物,从而确认溺水。Nubel 设计了 16S rDNA 选择性扩增引物,用于不同蓝藻菌株的鉴定硅藻采用变性梯度凝胶电泳(DGGE)。He F 等使用 PCR 和 DGGE 协议从一个实验模型和两个人类溺水病例中的浮游生物中扩增了 16 个 S rDNA。在实验室实验由他测试浮游疾驰的存在在溺水通过放大基因 DNA 提取从肺、肾脏、肝脏、心脏血液和大脑的实验中,12 只兔子被人工淹死模拟死前溺水和十二个被击中,扔进湖水模拟死后溺水。引物(a)正向 CYA-F5′-GGGGAATYTTCCCGCAATGGG3,CYA-R 和反向(b) 5-3,CYA-R(c)5-3 的轨迹 16S rDNA 测试,以区分植物 DNA 组织的兔子在死前或死后溺水[2]。He F 等[2]实验模型表明,在一个案例(A)模拟死前溺水,十二个兔子放在笼子里,淹没在水和东湖(武汉中国)观察到水进入遥远的器官如肺、肾脏、肝脏、心脏血液和大脑和提取这些器官的 DNA 将放大 16S rDNA 基因位点在 487 个基点证实死前溺水如图 14.7 所示。他报告说,在溺水组,所有的器官(肺、肝脏、肾脏、心脏血液和大脑)分析显示积极的浮游生物的密度如肺显示 100%的浮游生物,肝脏显示 83%,肾脏显示 75%,心脏血液显示 83%和大脑显示 42%的浮游生物的存在。而淹死组,只有 2 只兔肺样本存在 16.7% 的浮游生物,而对照组样本未检测到扩增产物,如图 14.8 所示。He F 等[2]的工作表明,浮游生物 DNA 可以是非常可靠的想法来识别溺水造成的死亡,没有任何差异。

模拟凶杀案的 12 个死后样本器官没有让水穿过肺循环,因此没有显示出对植物 DNA 特异的 16S rDNA 基因位点的扩增,这些基因位点通常存在于浮游生物或硅藻中,而只有死后淹死组的肺(两只兔子/16.7%)显示扩增,因此证明为诊断溺水而采样肺可能会产生假阳性测试。

16S rDNA 的应用有助于进一步识别发现 39 岁女性尸体的死后或死前溺水在郊区。这些妇女失踪了两天后进行的尸检特征和硅藻测试结果均为阴性。发现尸体的水样也发现没有硅藻,在这种情况下检测浮游生物扩增 16S rDNA 有助于确认它是一个死前溺水的情况[2]。

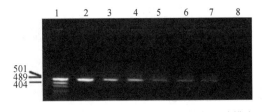 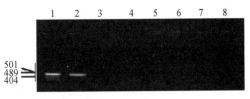

图 14.7 溺水组各组织扩增产物的琼脂糖凝胶模式
通道 1:pUC19 DNA/MspI 标记;通道 2:水样产品;通道 3~7:溺水兔子的肺、肝、肾、心血和脑中的产品;通道 8:负控制

图 14.8 对照组各组织扩增产物的琼脂糖凝胶模式
Lane 1:pUC19 DNA/MspI 标记;通道 2:水样产品;通道 3~7:来自对照组兔子的肺、肝、肾、心血和脑的产品;通道 8:负控制

14.9.2 利用硅胶梯度和酚氯仿法提取 DNA 从组织中分离浮游生物/硅藻

根据 Terazawa 和 Takatori[2,74]描述的程序,Percoll 细胞分离液是用于细胞密度梯度离心的良好参考介质,用于从生物样品中分离浮游生物物种。根据 Terazawa 和 Takatori 的说法,生物样品用高压蒸馏水和 9 mL 高酚溶液均质。溶液以 5 000 r/min 离心 60 min,丢弃上层,加入 700 μL 蒸压蒸馏水,3 000 r/min 离心 3 次,5 min。在显微镜下检查颗粒是否存在硅藻,并通过 CTAB 方法进行 DNA 提取处理[124]。另一方面,采用[125]所述的苯酚-氯仿法提取溺水受害者组织的基因组 DNA,利用硅藻特异性引物区分人类 DNA 和硅藻 DNA,有助于区分溺水相关病例。

14.10 硅藻诊断溺水的最佳条形码标记

目前许多研究者在 2011 年报道了硅藻的条形码标记。然而,到目前为止,已经系统地评估了硅藻诊断溺水病例的四种重要标志物:(1)COI:线粒体细胞色素 C 氧化酶Ⅰ基因(COI-5P)的5′端[36-37,126]。(2)质体 rbcL 基因:rbcL 大亚基(rbcL3P)的3′端。一个540 bp 的片段位于 rbcL 起始密码子下游417 bp(540 bp 的 rbcL)[38-39]。(3)核 rDNAITS 区域:大核糖体亚基的部分序列(D1-D3LSU,通常是 D1-D2 或 D2-D3)[36-37,127]。(4)核小亚基 rRNA 基因(18S):小核糖体亚基(V4SSU)的 V4 亚区[40]。

目前,DNA 条形码被设想为一种确定值得进一步探索的潜在新谱系的方法[37,128-129],而不是解决物种概念争议的灵丹妙药[128,130]。一些候选条形码标记已经在跨越与单一属的分类距离的数据集上进行了测试[37-38,40,127,131-132]到整个藻类或抗议多样性范围[133-134]。通用质体扩增子(UPA)不适合 DNA 条形码作为所有真核藻类和蓝藻的标记[134],具有较低的鉴别能力;该标记硅藻种类鉴定成功率仅为20%[38]。18S 区域在真核生物中是可变的,并在硅藻中提供了一个适当水平的信号来区分不同分类单元。根据现有的序列,本研究针对的143个碱基对区域包含足够的序列变异,可以将硅藻分解到科或属的水平。这种分辨率水平接近于高功率光学显微镜,优于色素、同位素和脂质分析[64,135]。选择了4个候选条形码标记(D1-D2 LSU[136], V4 SSU[135,137], COI-5P[36-37]和 rbcL3P[37,138])。筛选硅藻双壁藻的物种多样性[3,135],如表14.2所示。

14.10.1 细胞色素 C 氧化酶亚基 1(COI)

细胞色素 C 氧化酶亚基 1(mtCOI)的线粒体基因编码被认为是活生物体理想的 DNA 条形码。事实上,一些研究已经表明了 COI 在鳞翅目昆虫物种区分方面的成功应用[35,139],如鸟类[35]、水华植物[140]、针晶体[141-142]和其他生物体。在原生生物中,COI 是在红藻中检测出条形码[143-144]、纤毛虫[145-146]和硅藻[36]。最初,生命条形码倡议(BOLI)选择的片段是线粒体基因细胞色素 C 氧化酶 I(COI)和编码 COI 的线粒体基因5′端的片段,该片段足以识别大约动物界七个门中96%的物种[34]。一些研究表明,其他群体[147]、陆地植物[148]和真菌[149]不能很好地分离使用该基因。COI 在识别具有系统发育多样性的原生生物和具有不确定的进化关系方面的效用是混合的。选择这一标记的原因是线粒体基因组的进化速度比核基因组更快,因此物种特定信息区域的潜力更大[144]。任何编码基因的三编码特征都有助于比对[150]和它们的基因较少受到插入、缺失或其他大规模重排的影响,这些重排会在序列中引入模糊的变异[34,143]。在主要的大型藻类中,COI 是红藻植物的有用标记[143],而在一些褐藻(特别是 Alaria 属)中,基因片段不能区分明确的物种[151]。COI 标记对鞭毛藻的前景较小[152]。而其他短 DNA 序列,如 rRNA 编码基因的内部转录间隔区(ITS)[153]或线粒体细胞色素 b 基因[154]被认为更有前景。

14.10.2 核 rDNA ITS 区

Moniz & Kaczmarska[37]鉴定了核 rDNA,ITS2 转录本,该 DNA 区域从5.8S 开

始,到螺旋 III 的高度保守区域结束,并被提出作为硅藻的 DNA-条形码。Moniz 和 Kaczmarska[37]提出将 5.8S 基因与第二个内部转录间隔区(ITS-2)结合,作为硅藻的另一个潜在的条形码标记。这背后的原因包括越来越多的研究成功地使用 ITS 区域来解决密切相关的硅藻的物种问题[155-157],包括半隐物种[126],以及该地区不断扩大的参考数据集。然而即使这个标记显示足够的普遍性和良好的辨别能力时评估使用数据集包括广泛的硅藻类群[39],但是它在一些研究中被拒绝,因为大量的克隆内多样性阻碍了甚至密切相关的谱系的对齐[156-159]。COI 标记和其他对绿藻检测的条形码由于扩增不一致而无效。核糖体内部转录间隔区(ITS1 和 ITS2rDNA)是主要的真菌条形码[160-161],已被认为适用于硅藻[127],但也有一些保留。

14.10.3 核小亚基 rRNA 基因

在核 DNA 中,小核糖体亚基(SSU)具有编码基因的优势,已广泛应用于硅藻系统发育[162-166]。现有的硅藻 SSU 数据库是所有 DNA 区域中最广泛的数据库之一,它为数据库的识别提供了一个良好的初始平台。在 SSU 的 18S rDNA 标记中是最常用的标记,它是编码核糖体 rna 的基因的一部分[167]。使用 18S rDNA 没有什么优势,主要是因为它在所有真核生物中发现,在每个基因组的许多拷贝中允许遗传工作在个体水平上进行。18S rDNA 标记物高表达,允许在 RNA 水平上进行分子生态学研究。它包括一个高度保守和可变的核苷酸的镶嵌,允许在不同的分类水平上结合系统发育重建和生物群识别。18S rDNA 条形码已被证明可以有效地区分某些类群中的物种,如有孔虫[168-169]和一些硅藻[40]。然而,它们的变量还不够大,不足以解决其他几个分类单元的种间关系。

14.11 DNA 测序

对扩增后的 PCR 产物进行进一步测序,以鉴定硅藻类型,并与发现尸体的水样中发现的硅藻进行比较。这项研究不仅有助于确认死前和死后溺水,还将确定硅藻类型及其与水样中硅藻的一致性匹配,从而将同一溺水地点关联起来[2]。通过比较系统发育分析(PAUP)中的 p-t[170],支持水样和生物样本中硅藻属的分子分析。基本局部比对齐搜索工具(blast)是另一个使用 MEGA-7.0.18 比较数据库序列与参考数据库的在线分子工具[171]。一旦测序结果一致,系统发育树被设计用于鉴定共同祖先和转移单个硅藻属,以标记特定属和种的特定硅藻。Zimmermann[40]通过使用 Mega5 设计系统发育树,确定被研究菌株之间的分子关系,使用站点之间的伽马分布率,然后对树拓扑进行统计测试,进行 10 000 次的引导复制[172]用于该研究的基因座是 18S、V4 和 rbcL,最后计算 18S、V4 和 rbcL 组的单独比对的树以及连接数据集。进行各种进化距离和成对比对来计算潜在条形码的最佳正确识别总概率(PCI)。PCI 分析使用最近邻算法、全局距离、全局相似性作为分离指标和 Jukes-Cantor[173],

Kimura(2-Parameter)[174]、Jin(使用参数为 1 的伽马分布)[175-176]，Kimura(2-Parameter)以进行准确识别。

根据 Rubinoff 和 Holland[177]的说法,DNA 条形码可以被视为加速物种发现和启动新物种描述的"巨大工具"[177-178]。DNA 条形码的成本、时间和有效性使自动物种识别成为可能,这在大型取样活动中特别有用,像 Craig Venter 的全球海洋采样团队[179]。DNA 条形码使用 DNA 序列进行识别；它独立于已存在的形态学物种概念,可以与任何分类学概念相连[180]。因此,DNA 条形码的两个主要目标是将未知标本分配给物种,促进新物种的发现,促进识别,特别是在隐蔽、微观和其他形态复杂或难以接近的生物体中[147]。因此,一个成功的条形码标记[127]的必要条件是,必须可以在单次扩增中获得序列长度。它必须有超过种内距离的遗传种间距离(一个物种的代表性条形码的单系簇),并应包含保守的侧翼片段,以便于通用引物的设计。因此,设计针对硅藻特异性基因的 PCR 引物有助于加强溺水诊断的准确性。

14.12 测序技术的进步导致了数据解释技术的进步

生物样品中硅藻原子的鉴定技术 PCR 扩增环境 DNA 的核苷酸标记可以在不需要培养的情况下直接追踪环境中的原生生物。通过这种方法,可以重建季节分布和空间格局,利用序列作为代理,可以探索更多的原体多样性[181]。它最初是通过克隆文库的 sanger-测序进行的[182-184]。2013 年,一组法国研究人员使用 454 焦磷酸测序三个分子标记目标细胞核、叶绿体和线粒体基因组(SSUrDNA,rbcL 和 COI)分类特征的硅藻群落,他们使用三个样本(三倍)的模拟群落由 30 已知硅藻群属于 21 种获得的三个 DNA 参考库读取分配和准确性评估所有测试的标记表明,rbcL 基因位点在与大型 DNA 参考库结合时表现出最高的分辨能力。最后得出结论,尽管除了要求进一步优化,焦磷酸测序适合识别硅藻组合和可能发现应用领域的淡水生物监测和未来的工具来区分死亡由于溺水由于硅藻和人类 DNA 之间的同源性有限[185]。然而,DNA 条形码与下一代测序技术可以详细描述群落组成通过大量序列的组合[186-187]。因此,下一代 PCR 扩增环境 DNA 的核苷酸标记,可以追踪整个浮游原生生物群的多样性和空间格局。最近 Peng 等[188]建立了 PCR 毛细管电泳(PCR-CE)方法,从硅藻中扩增 rbcL 片段,用于检测溺水病例。他报道了 6 种硅藻的扩增,即固定化菱形藻、小环藻、舟形藻、变异直链藻、骨条藻和针杆藻与人 DNA 的阳性扩增条带为 197 bp；而蓝藻、细菌和绿藻为成功扩增出来,证实硅藻是溺水的良好检测工具。

14.13 结论

与传统的酸消化法相比,DNA 条形码法对溺水硅藻的检测更灵敏、特异性和可靠。核、线粒体和叶绿体基因组已显示出其作为条形码标记的潜力,以区分水体中存在的硅藻群落和评价水体的健康状况。然而,硅藻 DNA 序列的数据库有限,不足以覆盖整个硅藻分类。因此,为了改进硅藻数据库,我们必须定期评估条形码技术。PCR 产物的测序,以确认人体组织和硅藻细胞中 DNA 的鉴定和分化。如果对针对植物/硅藻基因位点的 PCR 引物进行靶向和扩增,则可以建立人体组织与硅藻细胞 DNA 之间有限的同源性。DNA 条形码的应用已被许多分类类群成功地证明,用于硅藻物种的鉴定[35,142,189-191]。硅藻物种识别的最佳目标基因是岩藻黄质,叶绿素"a"和"c"基因引物 SK1,正向:5'-ATGCCCGCTTTTTCCT 反向:5'-CATCCCACTCG AAGTCAA;TG 和 SK2,正向:5'-ACCCCCGGAATCCACCT 反向:5'-CCCCACTC GAAGTCAATG[114],核酮糖 1,5-二磷酸加氧酶(植物光合作用过程普遍基因)的开放阅读框;Eiler[185] 和 16S rRNA 基因[2,192]。因此,硅藻物种的 DNA 条形码不仅有助于确认死前和死后溺水,而且有助于评估不同季节和溺水死亡地点的不同物种和属。

参考文献

［1］Kumar, N.P., Rajavel, A., Natarajan, R., Jambulingam, P. (2007). DNA barcodes can distinguish species of Indian mosquitoes (Diptera: Culicidae), J. Med. Entomol., 44, 1-7.

［2］He, F. et al. (2008). A novel PCR-DGGE-based method for identifying plankton 16S rDNA for the diagnosis of drowning, Forensic Sci. Int., 176, 152-156.

［3］Auer, A. (1991). Qualitative diatom analysis as a tool to diagnose drowning, The American journal of forensic medicine and pathology, 12, 213-218.

［4］Ludes, B., Coste, M., Tracqui, A., Mangin, P. (1996b). Continuous river monitoring of the diatoms in the diagnosis of drowning, J. Forensic Sci., 41, 425-428.

［5］Cheung, C., Chiasson, D. A., Pollanen, M. S. (1997). The diagnostic value of the diatom test for drowning, I. Utility: a retrospective analysis of 771 cases of drowning in Ontario, Canada, Journal of Forensic Science, 42, 281-285.

［6］Pollanen, M.S. (1997a). The diagnostic value of the diatom test for drowning, II. Validity: analysis of diatoms in bone marrow and drowning medium, Journal of Forensic Science, 42, 286-290.

［7］Ludes, B. et al. (1999). Diatom analysis in victim's tissues as an indicator of the site of drowning, Int. J. Legal Med., 112, 163-166.

［8］Hürlimann, J. et al. (2000). Diatom detection in the diagnosis of death by drowning, Int. J. Legal Med., 114, 6-14.

［9］Lunetta, P., Penttilä, A., Hällfors, G. (1998). Scanning and transmission electron microscopical evidence of the capacity of diatoms to penetrate the alveolo-capillary barrier in drowning, Int. J. Legal Med., 111, 229-237.

［10］Pollanen, M. (1997b). The diagnostic value of the diatom test for drowning, II. Validity: analysis of diatoms in bone marrow and drowning medium, Journal of Forensic Science, 42, 286-290.

［11］Knight, B. (1992). Legal aspects of medical practice. London: Churchill Livingstone.

［12］Marshall, A. (1972). Legal aspects of medical practice, J. Clin. Pathol., 25, 1101.

［13］Pollanen, M.S. (1998a). Forensic diatomology and drowning. Elsevier Health Sciences.

［14］Timperman, J. (1972). The diagnosis of drowning. A review, Forensic science, 1, 397-409.

［15］Hendey, N. I. (1973). The diagnostic value of diatoms in cases of drowning, Medicine, Science and the Law, 13, 23-34.

［16］Yen, L. Y., Jayaprakash, P. (2007). Prevalence of diatom frustules in non-vegetarian foodstuffs and its implications in interpreting identification of diatom frustules in drowning cases, Forensic Sci. Int., 170, 1-7.

［17］Foren Peabody, A. (1977). Diatoms in forensic science, Journal of the Forensic Science Society, 17, 81-87. sic Sci. Int., 170, 1-7.

［18］Lunetta, P., Miettinen, A., Spilling, K., Sajantila, A. (2013). False-positive diatom test: A real challenge? A post-mortem study using standardized protocols, Legal Medicine, 15, 229-234.

[19] Guy, W. A. (1845). Principles of forensic medicine. Harper & brothers.
[20] Brouardel, P. (1880). Etude sur la submersion. Baillière.
[21] Swann, H., Spafford, N. (1951). Body salt and water changes during fresh and sea water drowning, Tex. Rep. Biol. Med., 9, 356-382.
[22] Revenstorf, V. (1904). Der nachweis der aspirierten ertränkungsflüssigkeit als kriterium des todes durch ertrinken, Vjschr Gerichtl Med, 27, 274-299.
[23] Incze, G. (1942). Fremdkörper im blutkreislauf ertrunkener, Zentralbl Allg Pathol Anat, 79, 176.
[24] Incze, G. (1951). Die bedeutung der phytoplankton-resorption beim ertrinkungstod, Acta mor phol. (Budapest), 7, 421.
[25] Incze, G., Tamaska, L., Gyongyosi, J. (1955a). Blood-planktons in death by drowning, Dtsch. Z. Gesamte Gerichtl. Med., 43(5-6), 517-523.
[26] Incze, G., Tamaska, L., Gyongyoysi, J. (1955b). Zur blutplanktonfrage beim tod durch ertrinken, Dtsch. Z. ges. gerichtl. Med., 43(5-6), 517-523.
[27] Tamaska, L. (1949). Diatom content of bone marrow in corpses in water, Orv. Hetil., 16, 509-511.
[28] Tamaska, L. (1961). üBer den diatomeennachweis im knochenmark der wasserleichen, Int. J. Legal Med., 51, 398-403.
[29] Mueller, B., Gorgs, D. (1949). Studien über das eindringen von corpusculären wasserbestand teilen aus den lungenalveolen in den kreislauf während des ertrinkungsvorganges, Deutsche Zeitschrift für die gesamte gerichtliche Medizin, 39, 715-725.
[30] Thomas, F., Van Hecke, W., Timperman, J. (1961). The detection of diatoms in the bone marrow as evidence of death by drowing, Journal of forensic medicine, 8, 142-144.
[31] Thomas, F., Van Hecke, W., Timperman, J. (1963). The medicolegal diagnosis of death by drowning, J. Forensic Sci., 8, 1-14.
[32] Timperman, J. (1968). Medico-legal problems in death by drowning. Its diagnosis by the diatom method. A study based on investigations carried out in Ghent over a period of 10 years, Journal of forensic medicine, 16, 45-75.
[33] Timperman, J. (1972). The diagnosis of drowning. A review, Forensic science, 1, 397-409.
[34] Hebert, P. D., Cywinska, A., Ball, S. L. (2003a). Biological identifications through DNA barcodes, Proceedings of the Royal Society of London B: Biological Sciences, 270, 313-321.
[35] Hebert, P. D., Penton, E. H., Burns, J. M., Janzen, D. H., Hallwachs, W. (2004). Ten species in one: DNA barcoding reveals cryptic species in the neotropical skipper butterfly Astraptes fulgerator, Proc. Natl. Acad. Sci. U.S.A., 101, 14812-14817.
[36] Evans, K. M., Wortley, A. H., Mann, D. G. (2007). An assessment of potential diatom "barcode" genes (cox1, rbcL, 18S and ITS rDNA) and their effectiveness in determining relationships in Sellaphora (Bacillariophyta), Protist, 158, 349-364.
[37] Moniz, M. B., Kaczmarska, I. (2009). Barcoding diatoms: Is there a good marker? Molecular Ecology Resources, 9, 65-74.
[38] Hamsher, S. E., Evans, K. M., Mann, D. G., Poulíčková, A., Saunders, G. W. (2011). Barcoding diatoms: exploring alternatives to COI-5P, Protist, 162, 405-422.
[39] MacGillivary, M. L., Kaczmarska, I. (2011). Survey of the efficacy of a short fragment of the rbcL gene as a supplemental DNA barcode for diatoms, J. Eukaryot. Microbiol., 58, 529-536.
[40] Zimmermann, J., Jahn, R., Gemeinholzer, B. (2011). Barcoding diatoms: evaluation of the V4 subregion on the 18S rRNA gene, including new primers and protocols, Organisms Diversity & Evolution, 11, 173.
[41] Kakizaki, E. et al. (2012). Detection of diverse aquatic microbes in blood and organs of drowning victims: first metagenomic approach using high-throughput 454-pyrosequencing, Forensic Sci. Int., 220, 135-146.
[42] Verma, K. (2013). Role of diatoms in the world of forensic science, Journal of Forensic Research.
[43] Suto, M. et al. (2009). PCR detection of bacterial genes provides evidence of death by drowning, Legal Medicine, 11, S354-S356.
[44] Hebert, P. D., Stoeckle, M. Y., Zemlak, T. S., Francis, C. M. (2004). Identification of birds through DNA barcodes, PLoS Biol., 2, e312.
[45] Hamilton, P. B., Lefebvre, K. E., Bull, R. D. (2015). Single cell PCR amplification of diatoms using fresh and preserved samples, Frontiers in microbiology, 6.
[46] Kane, M., Fukunaga, T., Maeda, H., Nishi, K. (1996). The detection of picoplankton 16S rDNA in

cases of drowning, Int. J. Legal Med., 108, 323-326.
[47] Ludes, B., Coste, M., Tracqui, A., Mangin, P. (1996c). Continuous river monitoring of the diatoms in the diagnosis of drowning, Journal of Forensic Science, 41, 425-428.
[48] Otto, H. (1961). uber den nachweis von diatomeen in menschlichen lungenstauben, Frankurt. Z. Path., 71, 176.
[49] Peabody, A. (1980). Diatoms and drowning—a review, Medicine, Science and the Law, 20, 254-261.
[50] Shkrum, M. J., Ramsay, D. A. (2007). Forensic pathology of trauma. Springer Science & Business Media.
[51] Organization, W. H. (1993). World Health Statistics Annual (WHO). 1992. World Health Organization.
[52] Ender, P. T., Dolan, M. J. (1997). Pneumonia associated with near-drowning, Clinical infectious diseases, 25, 896-907.
[53] Olshaker, J. (1992). Near drowning, Emerg. Med. Clin. North Am., 10, 339-350.
[54] Modell, J. H. D. (1993). New England Journal of Medicine, 328, 253-256.
[55] Kater, W. (1987). Vergleichende Diatomeenanalysen zur Differenzierung der Todesumstände bei im Wasser gefundenen Leichen.
[56] Sitthiwong, N. (2014). Detection and identification of diatoms in tissue samples of drowning victims, Month, 41.
[57] Jaccard, P. (1901). Etude Comparative de la Distribution dans une Portion des Alpes et du Jura. Bulletin de la Societe Vaudoise des Sciences Naturelle. 4.
[58] Renkonen, O. (1938). Statistisch-ökologische Untersuchungen über die terrestrische Käferwelt der finnischen Bruchmoore. Societas zoologica-botanica Fennica Vanamo.
[59] Burkhardt, W. (1938). Die feststellung des ertränkungsortes aus dem diatomeenbefund der lungen, Int. J. Legal Med., 29, 469-484.
[60] Ludes, B., Coste, M. (1996). Diatomé et médecine légale. Technique et Documentation. Paris: Lavoisier.
[61] Ludes, B., Coste, M., Coste, M. (1996). Diatomées et médecine légale: applications de la recherche des diatomées au diagnostic de la submersion vitale. Technique et doumentation Lavoisier; Editions médicales internationales.
[62] Ludes, B., Coste, M., North, N., Mangin, P. (1996a). Intérêt de l'étude de la flore des diatomées du lieu de submersion dans le diagnostic de noyade. Bulletin Français de la Pêche et de la Pisciculture, 133-137.
[63] Pollanen, M. (1998b). Laboratory procedure for the diatom test. Forensic Diatomology and Drowning. Amsterdam: Elsevier, 83-103.
[64] Passmore, A. et al. (2006). DNA as a dietary biomarker in Antarctic krill, Euphausia superba, Mar. Biotechnol., 8, 686-696.
[65] Mann, D., Droop, S. (1996). in Biogeography of freshwater algae. Springer, 19-32.
[66] Zimmermann, J., Glöckner, G., Jahn, R., Enke, N., Gemeinholzer, B. (2015). Metabarcoding vs. morphological identification to assess diatom diversity in environmental studies, Molecular ecology resources, 15, 526-542.
[67] Rimet, F., Bouchez, A. (2012). Biomonitoring river diatoms: implications of taxonomic resolution, Ecological indicators, 15, 92-99.
[68] Besse-Lototskaya, A., Verdonschot, P. F., Sinkeldam, J. A. (2006). in The Ecological Status of European Rivers: Evaluation and Intercalibration of Assessment Methods. Springer, 247-260.
[69] Pollanen, M., Cheung, C., Chiasson, D. (1997). The diagnostic value of the diatom test for drowning, I. Utility: a retrospective analysis of 771 cases of drowning in Ontario, Canada, Journal of Forensic Science, 42, 281-285.
[70] Sidari, L., Di Nunno, N., Costantinides, F., Melato, M. (1999). Diatom test with Soluene-350 to diagnose drowning in sea water, Forensic Sci. Int., 103, 61-65.
[71] Ludes, B., Quantin, S., Coste, M., Mangin, P. (1994). Application of a simple enzymatic digestion method for diatom detection in the diagnosis of drowning in putrified corpses by diatom analysis, Int. J. Legal Med., 107, 37-41.
[72] Funayama, M., Aoki, Y., Sebetan, I. M., Sagisaka, K. (1987). Detection of diatoms in blood by a combination of membrane filtering and chemical digestion, Forensic Sci. Int., 34, 175-182.
[73] Funayama, M., Aoki, Y., Sagisaka, K. (1987). Detection of diatoms in human tissue and blood by using membrane filter, Act. Crim. Japon, 53, 185-189.

[74] Terazawa, K., Takatori, T. (1980). Isolation of intact plankton from drowning lung tissue by centrifugation in a colloidal silica gradient, Forensic Sci. Int., 16, 63-66.
[75] Ishii, K., Muβmann, M., MacGregor, B. J., Amann, R. (2004). An improved fluorescence in situ hybridization protocol for the identification of bacteria and archaea in marine sediments, FEMS Microbiol. Ecol., 50, 203-213.
[76] Ahlgren, N. A., Rocap, G. (2012). Diversity and distribution of marine Synechococcus: multiple gene phylogenies for consensus classification and development of qPCR assays for sensitive measurement of clades in the ocean.
[77] Shilova, I. N. et al. (2014). A microarray for assessing transcription from pelagic marine microbial taxa, Isme J., 8.
[78] de Bruin, A., Ibelings, B. W., Van Donk, E. (2003). Molecular techniques in phytoplankton research: from allozyme electrophoresis to genomics, Hydrobiologia, 491, 47-63.
[79] Rynearson, T. A., Armbrust, E. V. (2000). DNA fingerprinting reveals extensive genetic diversity in a field population of the centric diatom Ditylum brightwellii, Limnol. Oceanogr., 45, 1329-1340.
[80] Skov, J., Lundholm, N., Pocklington, R., Rosendahl, S., Moestrup, Ø. (1997). Studies on the marine planktonic diatom Pseudo-nitzschia. 1. Isozyme variation among isolates of P. pseu dodelicatissima during a bloom in Danish coastal waters, Phycologia, 36, 374-380.
[81] Evans, K. M., Bates, S. S., Medlin, L. K., Hayes, P. K. (2004). Microsatellite marker development and genetic variation in the toxic marine diatom pseudo-nitzschia multiseries (Bacillariophyceae) 1, J. Phycol., 40, 911-920.
[82] Zimmermann, J., Glöckner, G., Jahn, R., Enke, N., Gemeinholzer, B. (2015). Metabarcoding vs. morphological identification to assess diatom diversity in environmental studies, Molecular ecology resources, 15, 526-542.
[83] Rácz, E. et al. (2016). PCR-based identification of drowning: four case reports, Int. J. Legal Med., 1-5.
[84] Singh, R., Singh, R., Thakar, M. (2006). Extraction methods of diatoms-A review, Indian Internet Journal of Forensic Medicine & Toxicology, 4.
[85] Biggs, B., Kilroy, C. (2000).
[86] Naser-Kolahzadeh, Z. P., Stavrianopoulos, J. G. (1996). Google patents.
[87] Nassiri, Y., Robert, J.-M., Rincé, Y., Ginsburger-Vogel, T. (1998). The cytoplasmic fine structure of the diatom Haslea ostrearia (Bacillariophyceae) in relation to marennine production, Phycologia, 37, 84-91.
[88] Spurr, A. R. (1969). A low-viscosity epoxy resin embedding medium for electron microscopy, J. Ultrastruct. Res., 26(1), 31-43.
[89] Massé, G. et al. (2001). A simple method for SEM examination of sectioned diatom frustules, Journal of microscopy, 204, 87-92.
[90] Gautam, S., Pandey, L. K., Vinayak, V., Arya, A. (2017). Morphological and physiological alterations in the diatom Gomphonema pseudoaugur due to heavy metal stress, Ecological indicators, 72, 67-76.
[91] Czarnecki, D. B., Blinn, D. W. (1978). Diatoms of the colorado river in grand canyon national park and vicinity: Diatoms of the Southwestern USA, II. Lubrecht & Cramer, Limited, 2.
[92] Du Buf, H., Bayer, M. M. (2002). Automatic diatom identification. World Scientific, 51.
[93] Pandey, L. K., Bergey, E. A. (2016). Exploring the status of motility, lipid bodies, deformities and size reduction in periphytic diatom community from chronically metal (Cu, Zn) polluted waterbodies as a biomonitoring tool, Sci. Total Environ., 550, 372-381.
[94] Lecointe, C., Coste, M., Prygiel, J. (1993). "Omnidia": software for taxonomy, calculation of diatom indices and inventories management, Hydrobiologia, 269, 509-513.
[95] Thakar, M. K., Singh, R. (2010). Diatomological mapping of water bodies for the diagnosis of drowning cases, Journal of forensic and legal medicine, 17, 18-25.
[96] Vinayak, V., Mishra, V., Goyal, M. (2013). Diatom fingerprinting to ascertain death in drowning cases, J Forensic Res, 4, 207.
[97] Natasha, D., Aleksej, D. (2005). Differential diagnostic elements in the determination of drowning, Rom J Leg Med, 13, 22-30.
[98] Pareek, R., Singh, R. (2011). Some fresh water diatoms of Galta kund, Jaipur, India, Journal of Soil Science and Environmental Management, 2, 110-116.

[99] Butcher, R. (1947). Studies in the ecology of rivers: VII. The algae of organically enriched waters, J. Ecol., 186-191.

[100] Fjerdinstad, E. (1950). The microflora of the river mølleaa: With special reference to the relation of the benthal algae to pollution, Gleerupska univ.-bokh.

[101] Zehlinka, M., Marvan, P. (1961). Zur prazisierung der biologischen klassifikation der Reinheit fliessender Gewasser, Arch. Hydrobiol., 57, 389-407.

[102] Lowe, R. L. (1974). Environmental requirements and pollution tolerance of freshwater diatoms.

[103] Lange-Bertalot, H. (1979). Pollution tolerance of diatoms as a criterion for water quality estimation. Beih: Nova Hedwigia, 64, 285-304.

[104] Kolkwitz, R., Marsson, M. (1908). Ökologie der pflanzlichen Saprobien. Borntraeger.

[105] Patrick, R. (1949). A proposed biological measure of stream conditions, based on a survey of the Conestoga Basin, Lancaster County, Pennsylvania, Proceedings of the Academy of Natural Sciences of Philadelphia, 101, 277-341.

[106] Patrick, R., Strawbridge, D. (1963). Variation in the structure of natural diatom communities, Am. Nat., 97, 51-57.

[107] Alakananda, B. (2012). Two new species of Nitzschia (Bacillariophyta) from shallow wetlands of Peninsular India, Phytotaxa, 54, 13-25.

[108] Alakananda, B., Karthick, B., Taylor, J. C., Hamilton, P. B. (2015). Two new species of nitzschia (bacillariophyceae) from freshwater environs of lonar crater lake, India, Phycological Res., 63, 29-36.

[109] Hrlimann, J. et al. (2001). Diatom detection in the diagnosis of death by drowning, Int. J. Legal Med., 1, 6-14.

[110] Timperman, J. (1968). Medico-legal problems in death by drowning. Its diagnosis by the diatom method. A study based on investigations carried out in Ghent over a period of 10 years, Journal of forensic medicine, 16, 45-75.

[111] Calder, I. M. (1984). An evaluation of the diatom test in deaths of professional divers, Medicine, Science and the Law, 24, 41-46.

[112] Taylor, J. (1994). Diatoms and drowning—a cautionary case note, Medicine, Science and the Law, 34, 78-79.

[113] Lunetta, P., Modell, J. H. (2005). in Forensic pathology reviews. Springer, 3-77.

[114] Abe, S., Suto, M., Nakamura, H., Gunji, H., Hiraiwa, K., Suzuki, T. et al. (2003). A novel PCR method for identifying plankton in cases of death by drowning, Med. Sci. Law, 43(1), 23-30.

[115] Shishlyannikov, S. M. et al. (2011). A procedure for establishing an axenic culture of the diatom Synedra acus subsp. radians (Kütz.) Skabibitsch. from Lake Baikal, Limnol. Oceanogr. Methods, 9, 478-484.

[116] Guillard, R. R. (1975). in Culture of marine invertebrate animals. Springer, 29-60.

[117] Vinayak, V., Gordon, R., Gautam, S., Rai, A. (2014). Discovery of a diatom that oozes oil, Advanced Science Letters, 20, 1256-1267.

[118] Doyle, J. J. (1990). Isolation of plant DNA from fresh tissue, Focus, 12, 13-15.

[119] Kröger, N., Poulsen, N. (2008). Diatoms-from cell wall biogenesis to nanotechnology, Annu. Rev. Genet., 42, 83-107.

[120] Bowler, C., Allen, A. E., Badger, J. H., Grimwood, J., Jabbari, K., Kuo, A. et al. (2008). The Phaeodactylum genome reveals the evolutionary history of diatom genomes, Nature, 456 (7219), 239-244.

[121] Kröger, N., Bergsdorf, C., Sumper, M. (1996). Frustulins: domain conservation in a protein family associated with diatom cell walls, Febs J., 239, 259-264.

[122] Nübel, U., Garcia-Pichel, F., Kühl, M., Muyzer, G. (1999). Quantifying microbial diversity: morphotypes, 16S rRNA genes, and carotenoids of oxygenic phototrophs in microbial mats, Appl. Environ. Microbiol., 65(2), 422-430.

[123] Kane, M., Maeda, H., Fukunaga, T., Nishi, K. (1997). Molecular phylogenetic relationship between strains of cyanobacterial picoplankton in Lake Biwa, Japan, J. Mar. Biotechnol., 5, 41-45.

[124] Richards, E., Reichardt, M., Rogers, S. (1994). Preparation of genomic DNA from plant tis sue. Current protocols in molecular biology, 2.3.1-2.3.7147 Gaikwad, A. B. DNA extraction: Comparison of methodologies.

[125] Moniz, M. B., Kaczmarska, I. (2010). Barcoding of diatoms: nuclear encoded ITS revisited, Protist, 161

(1), 7-34.
[126] Mann, D. G. et al. (2004). The Sellaphora pupula species complex (Bacillariophyceae): morphometric analysis, ultrastructure and mating data provide evidence for five new species, Phycologia, 43, 459-482.
[127] Trobajo, R. et al. (2013). Morphology and identity of some ecologically important small Nitzschia species, Diatom research, 28, 37-59.
[128] Mann, D.G. (1999). The species concept in diatoms, Phycologia, 38, 437-495.
[129] Mann, D.G., Sato, S., Trobajo, R., Vanormelingen, P., Souffreau, C. (2010). DNA barcoding for species identification and discovery in diatoms, Cryptogam. Algol., 31, 557-577.
[130] Pniewski, F., Friedl, T., Latała, A. (2010). Identification of diatom isolates from the Gulf of Gdańsk: testing of species identifications using morphology, 18S rDNA sequencing and DNA barcodes of strains from the Culture Collection of Baltic Algae (CCBA), Oceanol. Hydrobiol. Stud., 39, 3-20.
[131] Kaczmarska, I., Reid, C., Martin, J.L., Moniz, M.B. (2008). Morphological, biological, and molecular characteristics of the diatom pseudo-nitzschia delicatissima from the canadian maritimes this paper is one of a selection of papers published in the special issue on systematics research, Botany, 86, 763-772.
[132] Wylezich, C., Nies, G., Mylnikov, A.P., Tautz, D., Arndt, H. (2010). An evaluation of the use of the LSU rRNA D1-D5 domain for DNA-based taxonomy of eukaryotic protists, Protist, 161(3), 342-352.
[133] Hajibabaei, M., Janzen, D.H., Burns, J.M., Hallwachs, W., Hebert, P.D. (2006). DNA barcodes distinguish species of tropical Lepidoptera, Proc. Natl. Acad. Sci. U.S.A., 103(4), 968-971.
[134] Sherwood, A.R., Presting, G.G. (2007). Universal primers amplify a 23s rdna plastid marker in eukaryotic algae and cyanobacterial, J. Phycol., 43, 605-608.
[135] Urbánková, P., Veselá, J. (2013). DNA-barcoding: A case study in the diatom genus Frustulia (Bacillariophyceae, Nova Hedwigia, 142, 147-162.
[136] Veselá, J., Urbánková, P., Černá, K., Neustupa, J. (2012). Ecological variation within traditionaldiatom morphospecies: Diversity of frustulia rhomboides sensu lato (bacillariophyceae) in european freshwater habitats, Phycologia, 51, 552-561.
[137] Thüs, H. et al. (2011). Revisiting photobiont diversity in the lichen family Verrucariaceae (Ascomycota, Eur. J. Phycol., 46, 399-415.
[138] Levialdi Ghiron, J.H., Amato, A., Montresor, M., Kooistra, W.H. (2008). Plastid inheritance in the planktonic raphid pennate diatom Pseudo-nitzschia delicatissima (Bacillariophyceae), Protist, 159(1), 91-98.
[139] Kress, W.J., Wurdack, K.J., Zimmer, E.A., Weigt, L.A., Janzen, D.H. (2005). Use of DNA barcodes to identify flowering plants, Proc. Natl. Acad. Sci. U.S.A., 102(23), 8369-8374.
[140] Barrett, R.D., Hebert, P.D. (2005). Identifying spiders through DNA barcodes, Can. J. Zool., 83, 481-491.
[141] Lambert, D.M., Baker, A., Huynen, L., Haddrath, O., Hebert, P.D., Millar, C.D. et al. (2005). Is a large-scale DNA-based inventory of ancient life possible? J. Hered., 96(3), 279-284.
[142] Robba, L., Russell, S.J., Barker, G.L., Brodie, J. (2006). Assessing the use of the mitochondrial cox1 marker for use in DNA barcoding of red algae (Rhodophyta), Am. J. Bot., 93(8), 1101-1108.
[143] Saunders, G.W. (2005). Applying DNA barcoding to red macroalgae: a preliminary appraisal holds promise for future applications. Philosophical Transactions of the Royal Society of London B: Biological Sciences. 360, 1879-1888.
[144] Barth, D., Krenek, S., Fokin, S.I., Berendonk, T.U. (2006). Intraspecific genetic variation in paramecium revealed by mitochondrial cytochrome C oxidase I sequences, J. Eukaryot. Microbiol., 53 (1), 20-25.
[145] Hebert, P.D., Ratnasingham, S., de Waard, J.R. (2003b). Barcoding animal life: cytochrome c oxidase subunit 1 divergences among closely related species, Proc. Biol. Sci., 270 Suppl 1, S96-S99.
[146] Chantangsi, C., Lynn, D.H., Brandl, M.T., Cole, J.C., Hetrick, N., Ikonomi, P. et al. (2007). Barcoding ciliates: a comprehensive study of 75 isolates of the genus Tetrahymena, Int. J. Syst. Evol. Microbiol., 57(Pt 10), 2412-2423.
[147] Hollingsworth, M.L., Andra Clark, A., Forrest, L.L., Richardson, J., Pennington, R.T., Long, D.G. et al. (2009). Selecting barcoding loci for plants: evaluation of seven candidate loci with species-level sampling in three divergent groups of land plants, Mol. Ecol. Resour., 9(2), 439-457.
[148] Chase, M.W., Salamin, N., Wilkinson, M., Dunwell, J.M., Kesanakurthi, R.P., Haidar, N., et al.

[149] Cywinska, A., Hunter, F. F., Hebert, P. D. (2006). Identifying Canadian mosquito species through DNA barcodes, Med. Vet. Entomol., 20(4), 413-424.

[150] Lane, C. E., Lindstrom, S. C., Saunders, G. W. (2007). A molecular assessment of northeast pacific alaria species (Laminariales, Phaeophyceae) with reference to the utility of dna barcoding, Mol. Phylogenet. Evol., 44(2), 634-648.

[151] Zapata, M., Fraga, S., Rodríguez, F., Garrido, J. L. (2012). Pigment-based chloroplast types in dinoflagellates.

[152] Zhang, H., Lin, S. (2005). Development of a cob-18S rRNA gene real-time PCR assay for quantifying Pfiesteria shumwayae in the natural environment, Appl. Environ. Microbiol., 71(11), 7053-7063.

[153] Wayne Litaker, R et al. (2007). Recognizing dinoflagellate species using its rdna sequences, J. Phycol. 43, 344-355.

[154] Behnke, A., Friedl, T., Chepurnov, V. A., Mann, D. G. (2004). Reproductive compatibility and rdna sequence analyses in the sellaphora pupula species complex (Bacillariophyta) 1, J. Phycol., 40, 193-208.

[155] Amato, A., Kooistra, W. H., Ghiron, J. H., Mann, D. G., Pröschold, T., Montresor, M. et al. (2007). Reproductive isolation among sympatric cryptic species in marine diatoms, Protist, 158(2), 193-207.

[156] Casteleyn, G., Chepurnov, V. A., Leliaert, F., Mann, D. G., Bates, S. S., Lundholm, N. et al. (2008). Pseudo-nitzschia pungens (Bacillariophyceae): A cosmopolitan diatom species? Harmful Algae, 7(2), 241-257.

[157] Vanormelingen, P., Chepurnov, V. A., Mann, D. G., Sabbe, K., Vyverman, W. (2008). Genetic divergence and reproductive barriers among morphologically heterogeneous sympatric clones of Eunotia bilunaris sensu lato (Bacillariophyta, Protist, 159(1), 73-90.

[158] Luddington, I. A., Kaczmarska, I., Lovejoy, C. (2012). Distance and character-based evaluation of the V4 region of the 18S rRNA gene for the identification of diatoms (Bacillariophyceae), PLoS ONE, 7(9), e45664.

[159] Ozin, G, Author, A. (2003). The photonic opal—the jewel in the crown of optical information processing, Chem. Commun. (Camb.)., 2639-2643.

[160] Schoch, C. L., Seifert, K. A., Huhndorf, S., Robert, V., Spouge, J. L., Levesque, C. A. et al. (2012). Nuclear ribosomal internal transcribed spacer (ITS) region as a universal DNA barcode marker for Fungi, Proc. Natl. Acad. Sci. U. S. A., 109(16), 6241-6246.

[161] Kooistra, W. H., Medlin, L. (1996). Evolution of the diatoms (Bacillariophyta). IV. A reconstruction of their age from small subunit rRNA coding regions and the fossil record, Mol. Phylogenet. Evol., 6(3), 391-407.

[162] Medlin, L. K., Kooistra, W. H., Gersonde, R., Wellbrock, U. (1996). Evolution of the diatoms (Bacillariophyta). II. Nuclear-encoded small-subunit rRNA sequence comparisons confirm a paraphyletic origin for the centric diatoms, Mol. Biol. Evol., 13(1), 67-75.

[163] Stothard, D. R., Hay, J., Schroeder-Diedrich, J. M., Seal, D. V., Byers, T. J. (1999). Fluorescent oligonucleotide probes for clinical and environmental detection of Acanthamoeba and the T4 18S rRNA gene sequence type, J. Clin. Microbiol., 37(8), 2687-2693.

[164] Medlin, L. K., Kaczmarska, I. (2004). Evolution of the diatoms: V. Morphological and cytological support for the major clades and a taxonomic revision, Phycologia, 43, 245-270.

[165] Sarno, D., Kooistra, W. H., Medlin, L. K., Percopo, I., Zingone, A. (2005). Diversity in the genus skeletonema (bacillariophyceae). Ii. An assessment of the taxonomy of s. Costatum-like species with the description of four new species1, J. Phycol., 41, 151-176.

[166] Sorhannus, U. (2007). A nuclear-encoded small-subunit ribosomal RNA timescale for diatom evolution, Mar. Micropaleontol., 65(1-2), 1-12.

[167] de Vargas, C., Norris, R., Zaninetti, L., Gibb, S. W., Pawlowski, J. (1999). Molecular evidence of cryptic speciation in planktonic foraminifers and their relation to oceanic provinces, Proc. Natl. Acad. Sci. U.S.A., 96(6), 2864-2868.

[168] Swofford, D. L. (1993). Phylogenetic analysis using parsimony. Champaign, IL: Illinois Natural History Survey.

[169] Wylezich, C., Nies, G., Mylnikov, A. P., Tautz, D., Arndt, H. (2010). An evaluation of the use of the LSU rRNA D1-D5 domain for DNA-based taxonomy of eukaryotic protists, Protist, 161(3), 342-352.

[170] Kumar, S., Stecher, G., Tamura, K. (2016). MEGA7: Molecular Evolutionary Genetics Analysis version 7.0 for bigger datasets, Mol. Biol. Evol..

[171] Holmes, S. (2003). Bootstrapping phylogenetic trees: theory and methods, Statistical Science, 241-255.

[172] Jukes, T. H., Cantor, C. R. (1969). Evolution of protein molecules, Mammalian protein metabo lism, 3, 132.

[173] Kimura, M. (1980). A simple method for estimating evolutionary rates of base substitutions through comparative studies of nucleotide sequences, J. Mol. Evol., 16(2), 111-120.

[174] Jin, L., Nei, M. (1990). Limitations of the evolutionary parsimony method of phylogenetic analysis, Mol. Biol. Evol., 7(1), 82-102.

[175] Tamura, K. (1994). Model selection in the estimation of the number of nucleotide substitutions, Mol. Biol. Evol., 11(1), 154-157.

[176] DeSalle, R., Egan, M. G., Siddall, M. (2005). The unholy trinity: taxonomy, species delimitation and DNA barcoding. Philosophical Transactions of the Royal Society of London B: Biological Sciences, 360, 1905-1916.

[177] Desalle, R. (2006). Species discovery versus species identification in DNA barcoding efforts: response to Rubinoff, Conserv. Biol., 20(5), 1545-1547.

[178] Rusch, D. B., Halpern, A. L., Sutton, G., Heidelberg, K. B., Williamson, S., Yooseph, S. et al. (2007). The sorcerer II global ocean sampling expedition: northwest Atlantic through eastern tropical Pacific, PLoS Biol., 5(3), e77.

[179] Rach, J., DeSalle, R., Sarkar, I. N., Schierwater, B., Hadrys, H. (2008). Character-based DNA barcoding allows discrimination of genera, species and populations in Odonata, Proc. Biol. Sci., 275 (1632), 237-247.

[180] Nanjappa, D., Audic, S., Romac, S., Kooistra, W. H., Zingone, A. (2014). Assessment of species diversity and distribution of an ancient diatom lineage using a DNA metabarcoding approach, PLoS ONE, 9(8), e103810.

[181] Stoeck, T., Epstein, S. (2003). Novel eukaryotic lineages inferred from small-subunit rRNA analyses of oxygen-depleted marine environments, Appl. Environ. Microbiol., 69(5), 2657-2663.

[182] Kermarrec, L., Franc, A., Rimet, F., Chaumeil, P., Humbert, J. F., Bouchez, A. et al. (2013). Nextgeneration sequencing to inventory taxonomic diversity in eukaryotic communities: a test for freshwater diatoms, Mol. Ecol. Resour., 13(4), 607-619.

[183] McDonald, S. M., Sarno, D., Scanlan, D. J., Zingone, A. (2007). Genetic diversity of eukaryotic ultraphytoplankton in the Gulf of Naples during an annual cycle, Aquatic microbial ecology, 50, 75-89.

[184] Potvin, M., Lovejoy, C. (2009). PCR-Based diversity estimates of artificial and environmental 18S rRNA gene libraries, J. Eukaryot. Microbiol., 56(2), 174-181.

[185] Eiler, A., Drakare, S., Bertilsson, S., Pernthaler, J., Peura, S., Rofner, C. et al. (2013). Unveiling distribution patterns of freshwater phytoplankton by a next generation sequencing based approach, PLoS ONE, 8(1), e53516.

[186] Bittner, L., Gobet, A., Audic, S., Romac, S., Egge, E. S., Santini, S. et al. (2013). Diversity patterns of uncultured haptophytes unravelled by pyrosequencing in naples bay, Mol. Ecol., 22(1), 87-101.

[187] Shokralla, S., Spall, J. L., Gibson, J. F., Hajibabaei, M. (2012). Next-generation sequencing technologies for environmental DNA research, Mol. Ecol., 21(8), 1794-1805.

[188] Peng, F., Xu, Q., Liu, C., Liu, X., Xiao, C., Zhu, X. et al. (2018). Drowning Diagnosis by Detecting Diatoms rbc L Genes with PCR-Capillary Electrophoresis, nanosci. nanotechnol. lett., 10(1), 138-144.

[189] Hajibabaei, M. et al. (2005). Critical factors for assembling a high volume of DNA barcodes. Philosophical Transactions of the Royal Society of London B: Biological Sciences, 360, 1959-1967.

[190] Ward, R. D., Zemlak, T. S., Innes, B. H., Last, P. R., Hebert, P. D. (2005). DNA barcoding Australia's fish species. Philosophical Transactions of the Royal Society of London B: Biological Sciences, 360, 1847-1857.

[191] Pollanen, M. S. (1998c). Diatoms and homicide, Forensic Sci. Int., 91(1), 29-34.

[192] Kakizaki, E. et al. (2012). Detection of diverse aquatic microbes in blood and organs of drowning victims: first metagenomic approach using high-throughput 454-pyrosequencing, Forensic Sci. Int., 220, 135-146.

第15章
使用中的硅藻土：
性质、修改、商业应用和未来趋势

Mohamed M Ghobara，Asmaa Mohamed

15.1 硅藻土的性质

　　硅藻土是一种沉积岩，属于一类生化（生物成因）沉积岩。这些岩石主要由化石遗迹组成，主要分为两大类：碳酸盐和非碳酸盐生物岩沉积物[1-2]。非碳酸盐生物岩成因岩根据岩石中丰富的分类单元化石的主要化学成分，可细分为硅质岩和钙质岩，也可以由以下方式命名：例如，放射性石是由放射性石遗骸组成的[3]，而硅藻土是由硅藻遗骸组成的。硅藻岩或硅藻土（DE）是一种柔软的、脆弱的多孔的轻质硅质沉积岩，主要是通过硅藻化石遗骸的积累而形成的[4]。硅藻是光合单细胞微生物，属于硅藻门[5]。像其他微生物一样，它们用无定形水合二氧化硅在原生质体周围形成细胞壁[6]，称为硅藻壳（包括两个瓣膜和带，有时有棘和其他突出的二氧化硅结构）。硅藻壳表现出多种形式，可视为特殊的自然艺术，具有多种形态，从针状、盘状、鼓形和棒状，通常具有三维复杂的超微结构，包括功能性微纳米结构[5,7]。除了通常碎裂的硅藻壳遗骸之外，通常硅藻土还含有少量的其他微生物群化石，如硅鞭藻和球石藻，以及微型动物化石，如放射虫的骨骼和海绵的骨针[8-9]。

　　硅藻岩被称为硅藻土（DE），特别是当它被压碎成细粉末时。它有世界范围内常见的名字，包括：硅藻土、化石贝壳粉或一个不正确的名字（infusorial earth）。这个名字在生物学上是不正确的，因为Ehrenberg认为它是由异养生物的骨骼 *Infusoria* 组成的，即微小的水生单细胞原生动物，如纤毛虫[10-12]。此外，一种从利比亚的黎波里市提取的硅藻土，以该城市命名为"Tripolite"[13]。今天，这个名字指的是另一种由非常细的非硅藻粉末组成的材料[14]。此外，由于其独特的属性，硅藻土应用程序数量的增加，给它提供了大量令人困惑的商品名，其大部分商号取决于应用程序的性质。例如在美国，用于甘蔗精炼的硅藻土被称为"Filter-Cel""Filter-Pac""Dicalite"或其他名称。此外，在混凝土行业中，DE被称为"Non-Pard""Armstrong Brick""Diatex""Feather-Stone""Super-Brick"等[11]。

　　硅藻土的主要化学成分是水合无定形二氧化硅（$SiO_2 \cdot nH_2O$），此外还有次要成分（小于0.1%）如氧化铝（Al_2O_3）、铁氧化物（Fe_2O_3）、方解石（CaO）、氧化镁（MgO）

以及除有机物含量外[10]的其他微量(大于0.1%)无机成分[15]。硅藻有一系列颜色，包括白色、黄灰色、浅灰色、深灰色和褐灰色。棕色硅藻土的颜色是由于其中嵌入的有机残留物[16]。

15.1.1 硅藻土的形成

硅藻土的形成是一个漫长的过程，有四个不同的步骤。

(1) 集中在活硅藻细胞在环境条件下产生硅质晶体的能力。硅藻通过优雅的生物矿化机制来合成这些小晶体。活细胞把硅酸从环境分离，然后把它变成硅纳米球，再安排在一起形成不同的部分在高尔基体的帮助下形成其他蛋白质的模板纳米结构矩阵的过程[17-19]。即使在活细胞死亡、石化以及许多清洗程序之后，一些不溶性和被捕获的有机化合物仍然嵌在硅藻壳中[20]。

(2) 硅藻获得足够高的细胞浓度以达到水华状态的能力。即活细胞的过度生长，这需要获得营养物质(富营养化)和其他环境需求[21]。在任何水生生态系统中，特定活微生物的生长速度的增加除了与自然捕食者和毒素的浓度有关外，还与营养物质的可用性(特别是氮、磷，如果是硅藻，还与硅酸盐有关)以及它们最佳的环境的理化条件的存在有关，例如合适的温度、溶解氧、盐度和pH。只有合适的条件才会导致硅藻水华[21]。

(3) 沉积死硅藻遗骸，完整的细胞或其碎片。到达底部的硅质碎片(底栖生物)形成了一层非常微妙的沉积物，其中通常有其他成分。如果在海洋中一次出现一个硅藻壳或一个瓣膜，其沉积速度通常太慢[22]。然而，有一些机制可以减少硅藻停留时间的存在。其主要机制是捕食放牧的浮游动物，然后形成粪便颗粒，其中包括果茎碎片。粪便颗粒加速了硅藻遗骸的沉积速度[23]。另一种机制没有被食草动物摄取活细胞，这是由于物理化学参数的变化，例如季节性变化，特别是在温带、极地和沿海地区。这一机制可能包括水华的絮凝和原生质体的快速解体[22]。

在硅藻水华的情况下，在活细胞死亡后，它们开始与其他碎片和细菌等凝固，形成更大的凝固颗粒，下沉速度更快[24]。当浮游植物细胞的粘性增加时，这种絮凝过程。它取决于藻类细胞的有效大小，藻类的生理和形态粘性，以及周围环境的其他物理化学参数和细菌。当水华进入下降阶段时，与水华细胞相关的细菌数量迅速增加，导致浮游植物细胞的粘性增加，并可能涉及分泌黏液[24]。

(4) 最后发生在较长的时间跨度内，通过硅藻硅藻壳积累一层变得更厚。这些沉积层可转化为硅藻土，并可能涉及成岩作用过程[25]。

15.1.2 硅藻对溶解的抵抗力(它们能保存数百万年的原因)

硅藻壳和其他生物二氧化硅能够抵抗其水介质中的溶解[26]，这可能与以下两种可能的原因有关：

第一个原因是在硅藻壳内嵌入不溶性有机基质，在其外部可能存在有机外壳，内外部的这种不同组分被认为与抵抗溶解的生物机制相关[27-28]。即使在死亡后,硅质碎片也表现出抗溶解能力,这可能是因为有机物质的残留物嵌在二氧化硅内和/或覆盖在二氧化硅上。

第二个原因可能是硅藻周围环境中存在一些无机阳离子,包括:镁、铁和铝,它们可能产生一层不溶性硅酸盐,从而抑制二氧化硅的溶解。在这一过程中最重要的元素是铝,它降低了硅的溶解速度,降低了其溶解度。人们可以在硅藻土样品中发现大量的铝含量可以超过1%[10,15]。

许多结果支持这两个原因的有效性,即是最可能的防止溶解的保护机制。通过 EDTA 处理,可以观察到溶解速率的加快,它可以从硅藻壳表面去除有机物质和无机阳离子[26]。此外,已经证明了用热处理[29]、蛋白质变性剂、有机溶剂或碱性 pH 处理[30]特别是酸处理后的样品[26]会影响硅藻壳二氧化硅的溶解速率。因此,在工业加工过程中,部分或完全去除有机质或无机阳离子可以显著加速二氧化硅的溶解速率。

因此在某些情况下,活细胞死亡后,硅藻壳在适当的条件下部分或完全溶解(见图 15.1)。图 15.1 显示了热过氧化氢处理后 *Nitzschia palea*(新鲜培养)硅藻壳的溶解(未发表数据)。我们可以看到瓣膜内光晕的孔隙阻塞消失了[见图 15.1(a)]在溶解过程中,瓣膜出现了严重的损伤[见图 15.1(b)]。然而,在某些情况下,硅藻化石的良好保存可能会持续数千年甚至更长时间(见图 15.2)。

图 15.1　不同溶解程度的 *Nitzschia palea* 清洁阀门 SEM 图

(a)出现了两种程度的溶解,一种是溶解性孔隙闭塞,另一种是肋骨和条纹断裂;(b)溶解过程的进展是明显的,也反映了阀门的二氧化硅纳米球性质

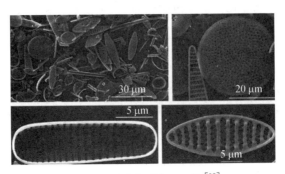

图 15.2　硅藻化石的 SEM 图[32]

在自然界中,保存质量是可变的,取决于在沉积、石化和成岩作用过程中的许多因素[31];环境物理化学因素(如 pH、盐度、水深和温度)、生物因素(如放牧和细菌)、沉降因素(如沉降率、沉积物积累速率和聚集率、洋流、硅藻浓度和沉积物纹理),以及与硅藻硅藻分类单元(硅藻壳部分的形状和结构、机械强度)[31]。一般来说健壮硅藻生活在 pH、温度、盐度、放牧和细菌低、下沉速率高、沉积物积累速率和细沉积物的硅藻浓度的深水中,很可能有更好的保存。此外,聚集的硅藻细胞可能相互保护不溶解,降低溶解速率[33]。

综上所述,硅藻颗粒的溶解并不依赖于保存时间,而是依赖于上述影响保存质量的因素。因此,新鲜的硅藻壳可能溶解并显示形态损伤,而其他硅藻在数百万年内保存完好,反之亦然。

15.2 发现和古代应用的历史

包括埃及、希腊和罗马人在内的古代文明曾利用硅藻土作为一种轻质岩石,在磨料和轻质建筑材料等应用上与石灰岩非常相似[34-35]。此外,几个世纪以来,农民们可能一直用它来控制昆虫和寄生虫。

虽然我们确信古代文明并不知道硅藻土的真实性质,但对他们对硅藻土性质的正确认识仍有推测。例如,古埃及人在迪麦(Faiyum,Egypt)使用硅藻土建造房屋,可能是为了降低房屋内部的温度[35]。此外,古埃及曾使用过富含硅藻原子的泥砖,但他们对其性质的理解尚未完全建立[36]。

一些最令人印象深刻的例子使用硅藻土建筑块位于土耳其,包括 30 m 直径(见图 15.3),独特的圣索菲亚大教堂在伊斯坦布尔建成[34],建于罗马帝国时代晚期。

图 15.3　土耳其伊斯坦布尔圣索菲亚圆顶的内部视图(穹顶由硅藻土块和径向对称艺术品组成)[37]

1836 年,Peter Kasten 是第一个发现硅藻土的人,当时他正在位于德国北部 Lüneburg Heath 的 Haußelberg 山钻探一口井。在此之前,它可能在大多数时候被归类为一种特殊的类型石灰石。与此同时，Bailey 于 1839 年在纽约西点附近发现了它，现在世界上最大的硅藻土生产商[38]。

直到第一次世界大战，世界上许多硅藻土的生产仅限于德国北部。那里的矿床厚度可达 28 米，而且都来自淡水。直到 1994 年，硅藻土的开采在不同的地点进行，包括 Neuohe、Hammerstorf、Wiechel、Hetendorf 和 Steinbeck。美国第一个硅藻土的商业生产于 1884 年在马里兰州开始[39-40]。

硅藻土帮助 Alfred Nobel 成功地开发了炸药，随后导致其改进了采矿、加工和大规模生产[38,41]。此后，它被用于啤酒和水过滤，被认为是当今硅藻土的主要应用之一。从第二次世界大战开始，它也被美国陆军和美国海军广泛用作过滤介质，用于去除锅炉和发动机水中的石油，以及船舶的低光反射涂料[42]。

在 20 世纪 20 年代，煅烧、通量煅烧、分级和上浆技术得到发展，从而导致硅藻土应用的增长[38]，如吸收剂、填料和绝缘体，并继续用于过滤目的（啤酒、食用油和葡萄酒），导致其商业价值的增加。

15.3 硅藻土的发生和分布

最古老的可识别的硅藻化石已经在侏罗纪晚期（大约 1.5 亿年前）的化石记录中被观察到，并且可能更古老[35]，这导致了对最古老的硅藻祖先的推测[43]。也许即使存在更古老的硅化石，可能保存完好也可能已经溶解。

迄今为止，第一批保存完好的硅藻化石是海洋中心化石，可以追溯到白垩纪。后来，可以观察到海洋的凹陷形态，其次是淡水形态，出现在俄罗斯的古新世和北美的始新世晚期。今天，大多数硅藻土的商业产品属于始新世和中新世时期[5,43-44]以及第四纪遗存。具有商业价值的硅藻土沉积物包括美国、德国、中国、丹麦、日本、法国、法国、西班牙、秘鲁、阿尔及利亚、阿根廷、澳大利亚、墨西哥、智利和哥斯达黎加[38]。

硅藻土可以通过风化作用限制其重新分布，通过风转移，从而显著影响地球上的许多地方。例如，大量的硅藻土通过空气从撒哈拉沙漠转移到亚马逊森林，从而丰富了森林[45]。撒哈拉沙漠是世界上第三大沙漠，但在过去，它是世界上最大的湖泊[46]，从而形成了大量的全新世沉积物。

Bodélé 洼地，一个被称为"palaeolake Megachad"的老湖，是撒哈拉沙漠中有大量轻型河床的地方之一。这个湖现在已经完全干燥，湖床沉积物暴露在空气中[47]。研究发现，这些轻质沉积物均为硅藻土层，主要由淡水类群组成，如 *Aulacosira* spp.[47]。最近，美国宇航局的一项分析证实了撒哈拉沙漠，特别是 Bodélé 洼地，在全球尘埃生产中的重要作用。然而，撒哈拉沙漠中的这个地方相对较小（只有 200 千米），但仍然可以被认为是世界上最大的单一尘埃来源，可能是所有撒哈拉尘埃产生的一半以上的原因[48]。因此，硅藻是在跨大西洋旅行中通过风传输的气溶胶的主要组成部分之一[49]。

15.4 硅藻土的开采和加工

虽然硅藻土已经被使用了很长一段时间,但它的大规模生产是在过去的两个世纪才开始的[35,38]。在商业生产中,硅藻土沉积层主要有两种工艺:开采和加工。

通常,硅藻土存在于地质地层中,在其他层间或地层之间,通过沉积循环嵌入在其他层之间。通过地层学科学解释,硅藻土层可能与其他层交替出现,这可以在意大利西西里岛卡波达索组的硅藻土-粘土交替中看到[50],或美国科罗拉多州弗洛里桑特地层中的灰粘土沉积层与硅藻土间隔交替[24]。因此,为了专门提取硅藻土层和去除其他层,需要进行采矿过程。

地表开采成为提取硅藻的常用工具,这意味着去除不重要的层(覆盖层),以到达埋藏的纯硅藻土地层。一旦硅藻土层暴露,通常用强大的刮刀切割并储存。有时,地下开采也有助于提取硅藻土。大规模采矿过程依赖于大型设备,包括推土机、刮铲运输车、动力铲和卡车(见图 15.4)。之后,硅藻土被送往加工厂进行进一步加工和制备成最终形式(IDPA 2017)。

图 15.4 EP 一个矿场的硅藻土矿开采[51]

硅藻土的加工目的是生产一种具有理想的物理和化学性质的商业产品,使该产品适合市场。处理步骤和处理方法取决于应用程序。一般来说,使用不同设计的破碎机,将矿山或工厂内的轻质硅藻土大块变成小碎片。研磨后,使用特殊搅拌机进行除团聚,获得天然硅藻土造粒粉(EPA 1995)。下一步是去除这种粉末中的水分,这可能达到 60% 以上。闪光灯和旋转式干燥器都用于将粉末水分降低到大约 15%。这种无需进一步加工的干粉被认为是天然级硅藻土,是市售的硅藻土等级之一。

总的来说,商业硅藻土主要有三类:自然等级、煅烧和通量煅烧。方法和处理步骤的数量取决于自然等级的组成部分和应用程序类型。在减少粘土和云母含量后,在空气或氧气的存在下,通过加热到 700～1 100℃,以去除有机污染物、剩余的水分,并引起相变,从而产生煅烧品位。通量煅烧版本是最高的硅藻土等级,是通过在 1 200℃的高温下煅烧过程中添加通量剂而产生的。最后一种工艺的产物是一种具有高孔隙度的白色粉末。可能需要进一步的处理来获得超纯硅藻土,如酸洗,以制备用于制药工业中等敏感应用的硅藻土。

15.5 硅藻土的特征

利用 X 射线荧光(XRF)分析,将硅藻土的化学成分分为三类:主要、次级和次要成分[15]。这种分类反映了硅藻土的化学成分,无论其在地质地层中的年龄如何。主要成分始终是二氧化硅(SiO_2),其次是 Al_2O_3、CaO 和 Fe_2O_3 作为次要成分,而次要成分可能包括 Na_2O、SO_3、V_2O_5、TiO_2、MnO_2 和其他成分。在大多数情况下,二氧化硅含量在 50%～90%之间变化[52-53];在沉积、石化和成岩过程中,根据其性质和周围环境,次要成分的比例不同。此外,在硅藻土矿石中还可能存在重金属杂质和稀有元素,并可以利用元素分析技术进行检测。在这些元素中有 Sb、As、Ba、Ni、Sr、Zn、Pb、La 和 Ce。此外,根据硅藻土矿的比例可以预测某一硅藻土矿的经济价值,从而确定开采和加工获得硅含量高于 80%的最终产品的可能成本。硅藻土矿石中硅氧化硅含量越多,其经济价值越高。

未经处理的天然硅藻土,即没有煅烧或其他化学处理,主要由非晶态二氧化硅(类似于蛋白石-a)组成,在 X 射线分析中显示出一个非常宽的峰[52-54]。其他的晶体相通常以次要组分的形式出现。这些相可能包括石英、方解石、高岭石、长石、白云石、方长石和云母[52-57]。据记录,蛋白石-a 相煅烧后转化为方石,而云母、高岭石和方解石可以通过酸处理(盐酸)降低或通过煅烧后酸处理完全去除[53,55]。

作为非晶态二氧化硅意味着在硅藻土表面有大量的硅烷醇和硅氧烷桥[6]。拉曼光谱和 FTIR 光谱能够研究这些官能团。该分析证实了二氧化硅和有机碳特征谱带的存在,此外还有一个在 3 600 cm^{-1} 附近具有宽峰的大谱带,这可能与二氧化硅表面上的吸附水有关[55-56]。不同的作者报道了 493～607 cm^{-1} 附近的两个主要峰,这可能分别与 O_3SiOH 四面体振动模式和 $(SiO)_3$ 环呼吸模式有关。与 Si—O—Si 振动相关的 IR 波段在 1 100 cm^{-1} 附近表现为非常强且宽的波段,通常分配给 Si—O—Si 不对称伸缩振动的 TO 和 LO 模式。799 cm^{-1} 附近的 IR 波段可以归属于 Si—O—Si 对称伸缩振动,Si—O—Al 的峰出现在 533 cm^{-1} 处,并且可以显示在煅烧后降低[55,58]。

15.6 硅藻壳改性

原硅藻土(DE)在某些应用中可以直接使用而不需要任何处理,而大多数应用则需要进一步的处理和修改。通常,需要通过适当的清洗程序去除有机材料和其他污染物以准备为某些应用和进一步的表面修饰。Wang[59]做了一个彻底的回顾,说明了制备栽培硅藻和化石硅藻壳的不同清洗程序。

研究了几种用金属、聚合物、聚合物/金属、碳基材料和磁性纳米颗粒修饰硅藻表面的物理和化学方法[60-61]。二氧化硅壳与目标材料(即二氧化钛、金、聚合物或碳)或将贵金属或半导体纳米颗粒固定在活化的二氧化硅表面已被用于不同的目的[62]。在纳米颗粒固定化的情况下,在功能化过程之前激活硅表面的硅醇基团,增加了产率。二氧化硅壳的硅表面像任何非晶硅一样是化学修饰的,被认为是硅烷功能化的优良材料[63]。清洗后的硅藻壳表面的功能化可以将硅藻壳粘附在底物上,也可以固定具有共价键的有机分子、聚合物、无机纳米颗粒,特别是用于催化和药物传递应用[60-62,64]。

根据应用程序,人们可以从可用的不同种类的硅烷(类似于烷烃的硅氢化物)中选择,它们已在别处列出[63]。硅烷的表面改性可以改变许多性质,包括疏水性、吸收、附着力、释放、亲水性、介电常数和电导率。例如,对于以硅藻硅藻壳作为药物载体的给药系统,表面应该用硅烷修饰,以增强其亲水性。另一方面,对于基于硅藻土的超疏水表面时,应使用合适的疏水硅烷。硅烷对硅表面改性机理如图15.5所示。

除了修改表面化学,化学工程师广泛研究不同修改硅藻壳的方法通过改变壳本身的基本化学成分,例如,通过代谢插入如插入锗、钛氧化物[65],或镍[66]。在培养过程中,或通过气体/硅置换反应,保留了硅藻壳特征的三维结构[67-68],或通过Diab和Mokari[69]和Ragni等[61]中回顾的其他方法。

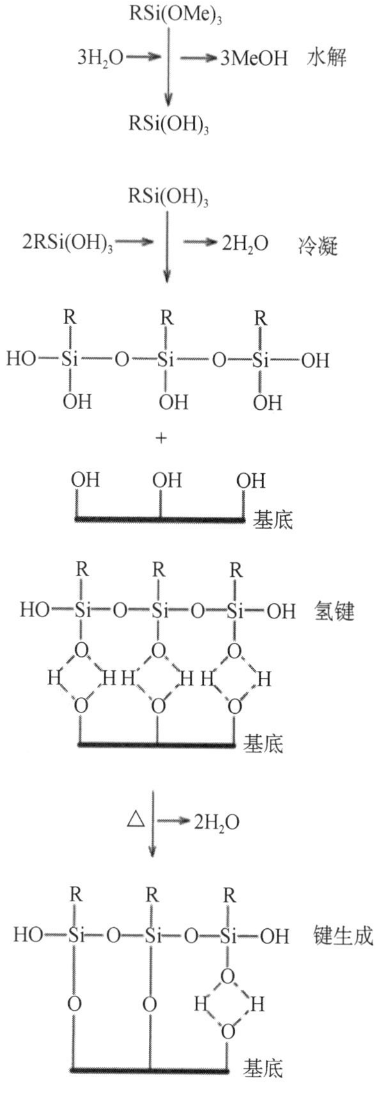

图15.5 硅烷对二氧化硅表面改性的机理[70]

15.7 硅藻土改性

如前所述,硅藻土长期作为建筑材料、保温材料和磨料使用。在过去的两个世纪中,硅藻土的真实性质已经被揭示出来,导致了一个持续的大规模生产随着时间的推移而增加。这些材料的开采和加工在20世纪初显示出了一个飞跃,这与新应用的发明有关,特别是在水、啤酒和油过滤方面。除了其他预期的趋势外,这里还回顾了市场上已经有的一些重要的应用程序。

15.7.1 基于硅藻土的过滤技术

过滤和分离过程是一种深层的物理技术,用于从流体混合物中分离出一定大小的颗粒。过滤有非常广泛的应用,包括水、空气、油和啤酒过滤器,用于生物技术应用、发动机、水处理、电子、食品和制药工业[71-72]。

硅藻土(DE)由于其物理性质,自第二次世界大战以来被广泛用于液体过滤和净化[42]。基于硅藻土介质的过滤有两种:表面过滤和深层过滤。表面(筛)过滤,其中过滤介质的孔径小于不良的污染物的尺寸,防止这些颗粒通过过滤器,并将其保留在表面图[见图 15.6(a)]。在深层过滤的浆液中,即需要过滤的溶液,渗透过滤介质到一个点,不希望的颗粒停留在通道或过滤图[见图 15.6(b)]中的曲折空隙内。与表面过滤相比,深度过滤过程能够更有效地收集不良颗粒,并具有更大的容量。图 15.6 是这两种技术之间的示意图比较。一般来说,使用硅藻土介质的过滤可以包括应用于进入侧的压力泵,或应用于流出侧的真空泵[73]。

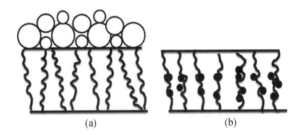

图 15.6 表面过滤和深度过滤之间的差异的示意图

(a) 颗粒积聚在表面的表面过滤;(b) 深度过滤,可以用具有不同梯度孔径的层数表示,以保留和吸附过滤介质结构内的污染物

使用硅藻土的一种常见的过滤技术是预涂层过滤,它通过在过滤元件或隔膜的表面建立一层硅藻土来实现。预涂覆技术的主要目的是防止过滤介质的堵塞,并通过反洗过程提供简单的清洗。然而,在此过程中必须使用超纯的硅藻土以提高过滤的性能[74]。预滤波可以用修正的达西定律来描述,该定律指出速度通过这些层的流量与通过这些层的压力梯度成正比[75],根据以下公式:

$$\frac{k}{\mu}\frac{dP}{dL} = \frac{dV}{A\,dt} \qquad (15.1)$$

其中,k:与粘度无关的修正渗透系数;μ:动态或绝对粘度;$\frac{dP}{dL}$:跨床层的压力梯度;A:该床的横截面面积;$\frac{dV}{dt}$:单位时间内通过床的滤液量。

Sentmanat[76]研究了这一过程并得出结论,随着预涂层的增加,净化水所需的时间减少了。在网络市场上,全球有数百家 DE 过滤供应商。美国约 60％的硅藻土生产用于过滤[34]。它用于啤酒、油、酒、饮用水和游泳池过滤,也可用于加热石油脱盐[77]。

15.7.1.1 水过滤

最近,大部分自然水资源(包括河流、湖泊和地下水)受到人为的强烈污染的威胁,通过不断释放有害物质,包括化学、工业和放射性废物,可能影响人类和环境[78-79]。因此,水的过滤和净化是目前特别要考虑的问题。水过滤的最终目标是生产出满足饮用、药理、医疗和其他工业应用要求的水。这些技术可能有助于保护人类健康免受严重的水生污染。水的净化可以通过不同的物理方法(如煮沸、过滤和沉淀)以及生物或化学过程(如氧化、絮凝、氯化和使用紫外线等电磁辐射)来实现。

天然的和改性的硅藻土可以有效地去除饮用水中的各种污染物,包括重金属,如 Pb^{2+}、Cu^{2+}、Cd^{2+}、Zn^{2+}、Cr^{+3}[80-84],有机染料[85-87],以及有害微生物,和一些通过深度过滤技术传播的病毒[88]。此外,硅藻土已成功用于去除石油泄漏和石油产品,如苯、甲苯、乙苯、二甲基和甲基叔丁基醚[89-90],因此,如果开发出合适的大规模过滤技术,如 Fu 等所报道的,这可能有助于水生生物从石油泄漏事故中生存下来。

硅藻土已被用于经济型的水过滤装置的设计。例如,Baumann 和 LaFrenz[91]开发了一个基于硅藻土的过滤工厂,用于水净化,成本、水质和最佳过滤率都可以接受。过滤成本取决于水中不良杂质的浓度。后来,本研究扩展到三个城市,表明最大城市的生产成本相对较低,因为成本与水的产量不是线性正比。

此外,对于大型水站直接大规模过滤,天然硅藻土可与合成聚合物如聚氯化铝(PAC)、聚硫酸铁(PFS)、聚丙烯酰胺结合启动凝絮凝步骤,悬浮颗粒如粘土、藻类、病毒、有机或无机污染物在表面或底部絮凝,便于表面或底部收集去除[92]。

15.7.1.2 啤酒过滤

硅藻土长期被用作啤酒过滤的助过滤剂。啤酒过滤过程采用深度过滤技术,即杂质被困在过滤器内或吸附在过滤介质上,以生产高度纯化的啤酒。它成为啤酒行业的一项基本技术,从啤酒中以消除任何浊度和去除酵母和其他微生物[93]。此外,硅藻土还降低了啤酒的铁含量,增加了其保质期,而不影响其颜色和风味。然而,当过滤器辅助器被阻塞时,必须更换硅藻土层,这是一个昂贵的过程,并留下可能威胁环境的副产品。因此,寻求膜取代硅藻土的过滤,但膜污染是一个大问题,首先形成小块,将杂质吸附到过滤介质中阻塞膜孔,如图 15.7 所示,导致过滤能力下降(过滤啤酒通量)下降,进而增加过滤过程的成本[94]。过滤性的确定是预

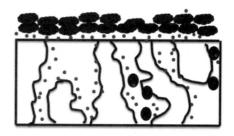

图 15.7 膜污染现象示意图

测过滤过程生产力的一个重要因素,它取决于跨膜的压力梯度、啤酒浊度、过滤率和作为过滤辅助剂的硅藻土的量等因素。

15.7.1.3 硅藻基分离技术的最新趋势

硅藻土作为分离介质并非只用于水、啤酒或油过滤,但也可以提取或分离RNA[96]、DNA[97-100]、蛋白质[101]或制药[102]。例如,Buyel等[101]发现,与只使用纤维素纤维的膜相比,使用硅藻土深度过滤器可以将烟叶提取物的浊度降低约500倍。同时他们还建议使用硅藻土过滤器来分离植物衍生的生物制药蛋白。

15.7.1.4 重复使用的硅藻土过滤介质

在过滤过程中使用硅藻土后变成废物需要替换,留下我们使用的硅藻土,并且它被证明是可重复使用的。例如动物饲料中的填充料[103]、沥青混合物或建筑砖[95]。此外,从啤酒行业处理的硅藻土可以用作农业肥料(见图15.8),因为它含有特定数量的吸附有机和无机材料以及酵母,可以提高栽培产品。因此被认为是环保和具有成本效益的[95]。

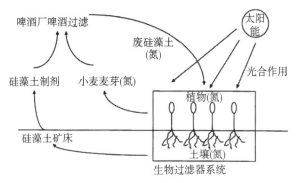

图 15.8　生物滤池处理系统示意图(其中啤酒行业的废 DE 用作肥料)[95]

15.7.2　保温用硅藻土

在阿拉伯湾、非洲和其他炎热地区,夏季气温升高,高达50℃,有时甚至更高。另一方面,在加拿大、北欧、俄罗斯和其他靠近地球两极的地方的冬季,温度可能会降到-40℃以下。当前的冷却和加热系统的高功率要求,以及它们的效率限制,促使我们寻找更合适的解决方案,提供更低的功耗和更高的空调效率。这些解决方案之一是在建筑设计中加入隔热器砖或涂料,从而减少能源损失。带有工程气隙的构件可以被建立并被广泛使用。

此外,将聚苯乙烯珠和木炭等材料混合到砖中,可以获得更好的保温性能[105]。

硅藻土可能是制造具有非常有效的隔热性能的建筑砖的一个很好的选择。硅藻石绝缘砖已经在市场上上市。然而,使用这种砌块的主要缺点之一是它们的吸水性。

因此,本章作者建议在将硅藻土与其他构件混合之前对其进行修改以获得,获得疏水性,但保留部分亲水性可能有助于控制室内湿度[106-107]。

此外,硅藻土已经以绝缘层的形式用于火电厂和冶金厂炉的保温和隔热,以及热传输[108]。例如,它可以提高高温炉的保温效率(见图15.9)[104]。

硅藻土的使用也可以扩展到制造管道或其他表面的涂层层,以实现有效的保温效果[109]。Ye,Chen,Chang,Mo和Wang[110]通过与另外两种保温填料平行添加硅藻土,可以降低水基丙烯酸多保温涂层的导热系数。此外,Martinez[111]制定了一种特殊的涂层,其中包括硅藻土也可以抵抗火灾和腐蚀。

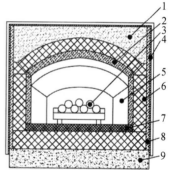

1—屋面保温;2—耐火纤维材料屋顶;
3—加热材料;4—炉体外部;5—工作腔;
6—泡沫硅藻土的高温隔热;7—致密耐
火砖基层;8—低温绝缘;9—炉座。

图15.9 炉的建议隔热示意图[104]

15.7.3 硅藻土基建筑材料

尽管,硅藻土的性质直到17世纪第一个光学显微镜发明[112]才被揭示,随后在18世纪进行了物种鉴定和描述[113],使用硅藻土作为建筑材料的一种成分可能与在古埃及一样古老。古埃及人使用石灰和石膏灰泥。即石灰砖或石膏、沙子和一些水的混合物,以促进混合物的反应形成硅酸盐钙,已持续数千年。据推测,这种砂浆,特别是用于金字塔建筑块之间的抹灰的类型,包括来自埃及开罗西南部法尤姆的全新世湖泊硅藻土沉积物的硅藻化石[114-115]。此外,一些用于土耳其石灰砂浆的石灰岩含有大量的硅藻化石,如用于捆绑一些历史悠久的奥斯曼浴场的砂浆,以及历史悠久的Kamanlı Mosque, Izmir[116]。有人声称,使用富含硅藻的石灰可能有助于古代文明降低砂浆煅烧所需的温度,这是获得水力特性的必要过程[117]。

除此之外,硅藻土被用作历史建筑的轻型建筑块,如土耳其伊斯坦布尔的HagiaSophia圆顶(见图15.3),以减轻重量并使其持续1 500多年[118]。

然而,因为现代建筑的力学性能要求较高,高于纯天然硅藻土的机械强度,因此不适合像古代文明直接使用天然硅藻土制作建筑砌块。但是,如果与其他部件混合,就可以制作出合适的建筑砖。Sisman,Karaman和Oztoprak[119]测试了在混凝土制造中不同比例的原始硅藻土,特别是对农业建筑。对硅藻土轻质构建块进行了优化,以达到足够的强度和55%的低导热率。最有希望的硅藻土百分比在10%~50%。此外,在高温(1 000℃)煅烧过程中获得的粗硅藻土团聚料,非常适合具有合适力学性能的轻质块[120]。在工业中,原料硅藻土通常与粘土混合,并在高温(1 000℃)下烧制。最近,人们发现硅藻土基建筑材料和一些硅藻土复合材料可以控制室内湿度,而室内湿度受到煅烧硅藻土的孔隙度和材料的表面自由能的影响[106-107]。此外,还制作了隔热和防火建筑块[120-121]。

研究了硅藻土在混凝土生产中对水泥的部分替代[122-124]。混合原料和煅烧水泥可增加其稳定性,降低其密度,增加隔热性能,并有助于混凝土的水下硬化[125-126]。

15.7.4 硅藻土作为一种杀虫剂

硅藻土作为一种无化学物质的无毒杀虫剂已长期成功应用,特别是用于粮食储存。它在干燥的条件下是一种非常有效的杀虫剂。在其他地方对硅藻土作为杀虫剂使用的所有方面进行了回顾[127-128]。另一方面,硅藻土在潮湿的条件下[129]或对一些有害的昆虫如 Bradysia spp.[130]并不有效。这些研究和专利,包括以硅藻土为基础的新杀虫剂配方,仍然是一个积极发展的领域。

15.7.5 硅藻土作为土壤改良剂

最近,硅被认为是植物生长和发展的重要元素[131-133]。在土壤中添加硅可能是通过其他地方描述的机制来缓解植物胁迫的一个重要工具[131]。与普通土壤相比,硅藻土具有大量的植物水溶性硅含量[131]。因此,在土壤中添加硅藻土可以增强其功能,包括改善土壤的物理结构,即使在沙质土壤中,也能保持水分[134-135],帮助受污染地区的重金属和烃的修复,以及植物的生产力和质量,提高生长、伸长和机械性能[131,136-139],以及增加水稻的叶数、分蘖数和圆锥花序数,提高植物耐受重金属、干旱或盐度等不同胁迫的能力[140-142]。Abdalla[140]是关于在栽培过程中在土壤中添加硅藻土的潜在好处的合适参考。

在市场上,有很多公司为土壤改良提供硅藻土。硅藻土的粒径从非常细的粉末到较大的砖不等。使用超细粉与土壤或生长介质混合,可保持高性能,而大砖只适用于保持土壤中的水分。对于农业大规模的应用,可以使用合适的搅拌机和调整 pH 制备硅藻土粉浆,然后在耕作前向土壤中添加所需量的硅藻土浆,类似于 Abdalla 进行的小规模实验。犁制改善了硅藻土浆料与土壤的均匀混合。

最近,通过对尿素颗粒的两层涂层合成了一种控释肥料,其内层由多糖组成,外层由硅藻土组成(见图 15.10)[143]。从作者的角度来看,所得到的趋势是极好的,因为它控制了肥料的释放速率,从而对植物更有效,同时也降低了排水水中的尿素量。此外,它还可以为土壤提供足够量的硅藻土,作为一种修正作用,有助于保持水分。

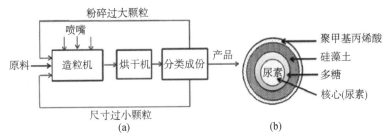

图 15.10 新型肥料的制造程序示意图(a)和多层肥料(包括 DE)作为主要层示意图(b)[143]

此外,硅藻土和其他土壤成分的混合物作为水培培养的新基质非常成功[144-145]。使用硅藻土作为水培培养基,而不是只使用沙子,可使根系更长,更密集,从而提高生产力,并帮助水培研究人员容易从生长培养基中分离根系进行进一步研究[144]。

15.7.6 硅藻土作为填料

原料和改良的硅藻土已被用作造纸工业、橡胶工业,特别是轮胎[146]、沥青、陶瓷工业、制药[147]和喂养家禽[148],油漆[149],塑料行业,例如聚丙烯管[150]和牙科材料[151]。然而,这里只能讨论有限数量的例子。改性硅藻土作为造纸工业的活性填料,可以增强纸张物理性能[152]阻燃和消烟[153-155]以及获得轻质印刷纸[156]。

此外在适当的修改后,硅藻土在橡胶工业中作为半增强填料,以增加橡胶的弹性模量[157]。已发现有三种橡胶类型适用于硅藻土的添加;氟橡胶、EPDM 橡胶和丙烯酸酯橡胶[158]。此外,有机中和的硅藻土可以被制备出来,以成功地用于硅橡胶配方[159]。

此外,硅藻土还被认为是不同类型沥青中活性填料的候选材料(即沥青是一种混合物,包括沥青、骨料和惰性或活性矿物填料),用于改善沥青的抗剥离、抗老化[160-161]以及一些力学性能,包括抗变形[162-164]。例如,玄武岩纤维和硅藻土的复合材料可用于提高热拌沥青的疲劳寿命和抗拉强度[165]。有趣的是,防冻沥青复合材料可以以硅藻土为原料获得[166]。其中一些复合材料可能会增加对低温裂纹的抗性,如改性砂沥青[162]。

硅藻土纳米银复合材料也可以合成具有潜在用途的抗菌填料[167-168]。然而,可能必须考虑一种更环保的方法来制造环保的抗菌硅藻土填料,例如,二氧化钛/硅藻土复合材料[169]。我们建议在游泳池和浴室的陶瓷和油漆中使用二氧化钛/硅藻土复合材料作为抗菌填充物。

最近,硅藻土被成功地用作强化填料,与聚(乳酸)和咖啡粉提取物结合,用于制造食品包装的增强环保薄膜[170]。

15.7.7 硅藻质土作为硅砂材料

在工业和国内应用中,天然和合成磨料都用于切割、抛光、研磨料、研磨料和钻孔表面。磨料是指通过广泛的摩擦来切割、光滑或完成其他较软的材料的材料。如今,市场上出售的金属、玻璃、陶瓷和高质量抛光的塑料,很大程度上取决于用于去除表面不希望的粗糙度的磨料类型。

天然硅藻土作为一种研磨料本身可能不够好,因为它具有中等的研磨力,由于其相对较低的硬度值,只表现出轻微的刮痕[171]。然而,煅烧会导致更硬的结晶形式,可以通过添加其他成分与细粉结合或反应,获得更高硬度较高的更硬的颗粒[172]。煅烧硅藻土已被用作牙膏的一种成分,以提高抛光能力,以去除牙齿上的咖啡和茶污渍[173]。此外,它还可以用于金属的表面清洁,特别是不锈钢,以及用于制造目镜玻璃镜片的热塑性聚合物材料[174]。此外,它已被用作洗碗液中的一种成分,以增强清洗过程[171]。

15.7.8 硅藻土作为动物和人类的食物添加剂

用粘土喂养动物已经实行了几百年了。最近,研究人员重新发现了粘土矿物作为饲料添加剂的重要性,以促进生长、控制胃肠道疾病和感染,并利用对动物和人类的健康益处[175]。硅藻土还被用作农场动物的饲料添加剂,以促进健康、控制肠道寄生虫感染和其他相关疾病[176-177]。此外,有人认为硅藻土增加了 Wistar 大鼠的免疫活性[178]。有很多硅藻土供应商生产硅藻土食品级,可用于喂养动物,可能适用于人类饮食。

15.7.9 硅藻土和纳米技术

硅原子纳米技术关注的是在该技术中使用硅藻的纳米多孔颗粒的优点,无论是通过直接使用,特别是在适当的修饰后,还是通过模拟[127]。一般来说,所有的应用程序,包括硅藻颗粒和利用孔隙率,光学和机械性能除了微和纳米流体,可以被认为是硅藻纳米技术和密切相关这个纳米孔隙及其颗粒内的模式。

虽然有一些工作涉及新制备的硅藻化石在纳米技术中的应用,但也有许多关于硅藻土化石的文章。使用化石碎片有时不适合用于纳米技术。例如,在许多情况下,我们无法得到一个完整的中心硅藻瓣膜,包括某些应用所需的所有纳米细节,特别是光子学,但我们可以很容易地从新鲜培养的样品中获得完整的瓣膜[179]。与新鲜栽培的硅藻壳相比,硅藻土的主要优点是成本低,市场上的可用性高。

1. 太阳能收集系统中的硅藻土

硅藻晶体独特的光学特性已被揭示[180-182],并发现它们与纳米多孔结构有关,具有许多可用于太阳能收集系统的潜在功能。光伏发电和集中太阳能热能是收集太阳能并将其转化为电力等其他有用能源的两种主要技术。硅藻可以被合并到这两种体系中。例如,它们已被用于染料敏化太阳能电池[183-184]。虽然新鲜培养的硅藻硅藻壳对这类太阳能电池显示出良好的效果[183,185],但是硅藻土支架相对简单和便宜,与标准电池相比,效率仍提高了 30%[186]。此外,一些具有稳定低温相变的硅藻土复合材料可以被认为是潜在的热能存储材料[187-192]。该技术可用于多种应用,包括集中式太阳能热电厂或建筑系统[191,193]。

2. 硅原子质土基超疏水表面

如果表面与水的接触角大于 150°,则会发生超疏水性。起初,超疏水性被发现是一种有趣的自然现象,出现在植物、昆虫和动物的几个例子中[194]。然后,这种现象出现了模仿。这可以被证明是微米和纳米尺度的粗糙度和表面化学修饰的联合效应的结果[195]。它的广泛应用于防污、防水、自清洗和防冻涂层的制造。

硅藻土已被用于制造超疏水材料[196]。两种氟化的[196]和非氟化硅烷已被用于给予硅藻土的疏水性。此外,还发现硅烷修饰的硅藻土晶体的形态与其疏水性之间存在相关性[197]。此外,研究还发现所选硅烷的烷基链长度和碳含量与硅藻土的疏水性有关[195,198]。可以得出结论,使用链长超过 12 个碳原子的硅烷可以获得超疏水性[198]。

3. 硅原子土复合材料作为催化剂

催化对于工业和生物技术至关重要,因为大多数化学反应都需要催化剂[199],这就是为什么需要不断开发和增强催化剂的原因。加工后的 DE 由于其纳米孔隙率具有巨大的表面积[200]。在用金属氧化物或金属纳米颗粒进行适当的改性后,使其能够用作催化剂。一些硅藻土复合材料具有较高的催化活性,这已被广泛研究[201-211]。

4. 硅藻基超级电容器

改性硅藻土,特别是二氧化锰改性硅藻土,已被用于制备新型超级电容器的复合材料[208,212-215],制备电极的性能得到了显著提高。

5. 基于硅藻土的药物和生物医学应用

传统治疗药物的局限性不断迫使科学界寻找合适的具有高溶解度、毒性最小、特异性传递机制、控释和高治疗指数的替代药物系统[216]。他们发现了一个潜在的答案,纳米技术,和不同类型的有机和无机纳米颗粒研究药物传递目的包括胶束[217]、脂质体[218]、碳纳米管[219]、石墨烯[220]和介孔二氧化硅[221]。新的药物系统使治疗具有靶向性、高效、无痛,在到达目标之前通过血液循环时得到保护,并有助于传递水不溶分子的分子如大多数抗癌药物[222]。近年来,介孔二氧化硅因其独特的高表面积、生物相容性和高载药能力等化学和物理性质,在给药系统中表现出了良好的效果。

由于与合成介孔二氧化硅的使用有关缺点,包括高价格和毒性材料和高能量参与合成,硅藻土被认为是一种具有相当大的表面积、通过功能化过程控制其表面化学的的良好能力的替代品,以及优秀的生物相容性[222-223]。Gordon, Losic, Tiffany, Nagy 和 Sterrenburg 是第一个设想使用硅藻硅藻壳作为药物传递载体,甚至讨论了更复杂的系统,如游泳微型机器人(见图 15.11)。在那之后,Losic 团队开始了实验来证明这个概念[223-225]。这是随后是越来越多的出版物和专利[226-227]。

硅藻土微粒可能不是完美的植入药物传递系统,所以硅藻土纳米颗粒已经被制造出来[228]并作为治疗药物的纳米载体进行了研究[229]。另一方面,它们可以直接作为微粒作为口服药物使用。此外,硅藻土(食品级)也可以作为片剂配方的填充物[147],这已获得美国 EPA 的批准。

此外,还提出将纯化的硅藻土复合材料用于硬组织再生应用[230-232]。与添加普通二氧化硅相比,具有更高的性能增强[232]。首次制备了新型壳聚糖硅藻土复合膜[233],结果包括改善了细胞的表面积、膨胀性能、表面粗糙度和蛋白质吸附能力,这些都是成骨细胞粘附和增殖的重要因素。

6. 基于硅藻土的芯片实验室

近年来,硅藻土已成功地用于制造超灵敏、无标签的芯片实验室器件,用于从各种复杂样品的不同生物分子和危险化学物[234-236]。

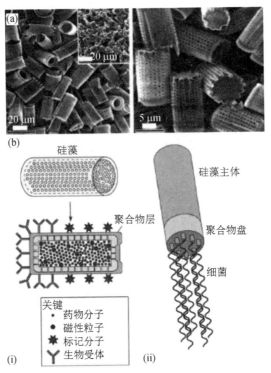

图 15.11 直链藻化石 SEM 图(a)和功能化化石上的药物负荷以及与截头体相连的建议的运动细菌并可能赋予其运动能力的示意图(b)

15.7.10 非工业应用

硅藻化石有时被发现是沉积层的一小部分,有时部分溶解。这使得沉积物的商业价值较低;然而,它对其他奇怪的非工业应用仍然有价值。这些化石可以作为很多事情的指标,包括对并发生态系统的探索和不同的古代水生生态系统和湿地的古环境变化[237]。

在考古学中,除了了解古生态学和某一文明周围的古环境重建外,硅藻分析可能有助于了解这种文明对环境的影响,特别是生活在水生资源附近的文化[238-239]。此外,它可能有助于确定粘土的来源,粘土已被广泛使用[240],因此作者推荐它作为不同文化之间贸易联系的可能指标。此外,它还可以揭示许多关于古代文明的秘密,因为它是他们建筑的一个组成部分[239]。

在天然气和石油勘探中,硅藻分析有时成为岩石年代测定和相关岩石的关键程序,特别是如果这些岩石因钙质微化石或湖泊沉积物保存不佳。硅藻层可能更有助于以更高的分辨率低估该盆地的古环境事件,生物地层学是评估盆地碳氢化合物潜力和研究储层分布的重要任务[241]。

15.8 硅藻土的制造及未来的发展前景

在自然界中,当所有条件都变得合适时,硅藻可以产生大量水华[242-244]。因此,硅藻土的形成在自然界中是一个连续的过程。虽然硅藻土有大量的硅藻土天然沉积物,但它们仍然像化石燃料一样不可再生,因为形成有用的硅藻土需要很长的时间和特定的条件。如果我们幸运的话,未来可以开发出一种模拟、优化和加速硅藻土形成机制的新方法。硅藻水华的维持和下降机制是形成硅藻沉积层的必要条件,可以在自然界中发现,需要进行更多的研究[244]。硅藻的培养早已成功,然而大规模生产仍有很多局限性。大量种植硅藻对水产养殖和生物柴油的生产都至关重要。最近,硅藻已经开始大规模生产[245],这也可以通过去除污染物,同时吸收大量的 CO_2 来帮助废水处理。Wang & Seibert[245] 的工作是思考硅藻商业生产进展的一个很好的开始,然而,他们对硅藻生物产品而不是硅氧化硅感兴趣。从这开始,我们可以通过废水处理实现三个目标,获得有用的副产品,并获得人工池塘底部的"化石"沉淀。在中国,每年每公顷干重超过 120 吨[245]。海洋硅藻也可以通过人工富集在开阔海洋中培养。寻找优化硅藻大规模生产的新方法是获得人工硅藻土的第一步。

15.9 结论

硅藻土是一种非常有前途的材料,有着丰富的历史和辉煌的未来,从古代文明的建筑材料和诺贝尔炸药延伸到光电子和药物传递系统。其开采和加工已被开发出来,以满足市场需求。全球有数百家硅藻土供应商。尽管已经有大量的旧的和最近的应用,但许多新的趋势已经被广泛探索,目前正在出现在研究文章、专利,以及潜在的市场上。除了其特殊的物理性质外,硅藻土还以其复杂的结构和孔隙度、图案和间距的规模成功地吸引了材料工程师的注意,从而导致大量的纳米技术应用可能随着时间的推移而增加。硅藻土可以通过模仿其形成步骤来再生,特别是所需的水华条件和良好的硅藻壳保存。所有的事实都支持这样一种观点,即硅藻和它们的化石是非常独特的,并且充满了可能隐藏在它们精细的超微结构中的新机会。

参考文献

[1] Boggs, S.J. (2012). Origin, classification, and occurrence of sedimentary rocks Petrology of Sedimentary Rocks. UK: Cambridge University Press.

[2] Nichols, G. (2009) Sedimentology and stratigraphy. UK: Wiley-Blackwell.

[3] Radiolarite (2017). In Wikipedia, the Free Encyclopedia, https://en.wikipedia.org/w/index.php?title=Radiolarite&oldid=787658767.

[4] Harwood, D. M. (2010). Diatomite. In The Diatoms Applications for the Environmental and Earth Sciences 2nd. pp. 570-574. J.P. Smol and E.F. Stoermer (eds.). UK.

[5] Round, F. E., Crawford, R. M., Mann, D. G. (1990). the Diatoms: Biology and Morphology of the Genera. UK: Cambridge University Press.

[6] Gordon, R., Drum, R. W. (1994). The chemical basis of diatom morphogenesis, Int. Rev. Cytol., 150 (243-372), 544-542.

[7] De Stefano, M., De Stefano, L. (2005). Nanostructures in diatom frustules: functional morphology of valvocopulae in Cocconeidacean monoraphid taxa, J. Nanosci. Nanotech., 5(1), 15-24.

[8] Hüneke, H., Mulder, T. (eds).(2011). Deep-Sea Sediments. 1st. 63. New York: Elsevier.

[9] Lyle, M. (2014). Deep-Sea Sediments. In Encyclopedia of Marine Geosciences. pp.1-21. J. Harff, M. Meschede, S. Petersen and J. Thiede (eds.). Dordrecht: Springer.

[10] Bakr, H. E. (2010). Diatomite: Its characterization, modifications and applications, Asian Journal of Materials Science, 2(3), 121-136.

[11] Calvert, R. (1930). Diatomaceous earth, J. Chem. Educ., 7(12), 2829.

[12] Ehrenberg, C. G. (2018). Wikipedia, the free encyclopedia, https://en.wikipedia.org/wiki/Christian_Gottfried_Ehrenberg.

[13] Dolley, T. P. (2002). Diatomite. In: 2002 Minerals Yearbook, 24.1-24.6, https://minerals.usgs.gov/minerals/pubs/commodity/diatomite/diatmyb02.pdf.

[14] Pohl, W. L. (2011). Economic Geology: Principles and Practice. Somerset: Wiley-Blackwell.

[15] El-dernawi, A. M., Rious, M. J., Al-Samarrai, K. I. (2014). Chemical, physical and mineralogical characterization of Al-Hishah diatomite at subkhat Ghuzayil area, Libya, International Journal of Research in Applied, 2(4), 165-174.

[16] Ivanov, S. É., Belyakov, A. V. (2008). Diatomite and its applications, Glass and Ceramics, 65(1-2), 48-51.

[17] Gröger, P., Poulsen, N., Klemm, J., Kröger, N., Schlierf, M. (2016). Establishing super-resolution imaging for proteins in diatom biosilica, Sci. Rep., 6(1).

[18] Kotzsch, A., Pawolski, D., Milentyev, A., Shevchenko, A., Scheffel, A., Poulsen, N., et al. (2015). Biochemical Composition and Assembly of Biosilica-associated Insoluble Organic Matrices from the DiatomThalassiosira pseudonana, Journal of Biological Chemistry, 291(10), 4982-4997.

[19] Kröger, N. (2007). Prescribing diatom morphology: toward genetic engineering of biological nanomaterials, Current Opinion in Chemical Biology, 11(6), 662-669.

[20] Ingalls, A. E., Whitehead, K., Bridoux, M. C. (2010). Tinted windows: The presence of the UV absorbing compounds called mycosporine-like amino acids embedded in the frustules of marine diatoms, Geochim. Cosmochim. Acta, 74(1), 104-115.

[21] Shaw, G. R., Moore, D. P., Garnett, C. (2003). Eutrophication and algal blooms. In Encyclopedia of Life Support Systems (EOLSS. A. Sabljic (ed). Oxford: Developed under the Auspices of the UNESCO, Eolss Publishers.

[22] Alldredge, A. L., Gotschalk, C. C. (1989). Direct observations of the mass flocculation of diatom blooms: characteristics, settling velocities and formation of diatom aggregates, Deep Sea Research Part A. Oceanographic Research Papers, 36(2), 159-171.

[23] Liu, H., Wu, C. (2016). Effect of the silica content of diatom prey on the production, decompositionand sinking of fecal pellets of the copepod [Calanus sinicus, Biogeosciences, 13(16), 4767-4775.

[24] O'Brien, N. R., Meyer, H. W., Reilly, K., Ross, A. M., Maguire, S. (2002). Microbial taphonomic processes in the fossilization of insects and plants in the late Eocene Florissant Formation, Colorado, Rocky Mountain Geology, 37(1), 1-11.

[25] Wallace, A. R., Frank, D. G., Founie, A. (2006). Freshwater Diatomite Deposits in the Western United States. U.S. Geological Survey Fact Sheet 2006-3044, https://pubs.usgs.gov/fs/2006/3044/.

[26] Lewin, J. C. (1961). The dissolution of silica from diatom walls, Geochim. Cosmochim. Acta, 21(3-4), 182-198.

[27] Kotzsch, A., Pawolski, D., Milentyev, A., Shevchenko, A., Scheffel, A., Poulsen, N., et al. (2015). Biochemical Composition and Assembly of Biosilica-associated Insoluble Organic Matrices from the DiatomThalassiosira pseudonana, Journal of Biological Chemistry, 291(10), 4982-4997.

[28] Tesson, B., Hildebrand, M. (2013). Characterization and Localization of Insoluble Organic Matrices Associated with Diatom Cell Walls: Insight into Their Roles during Cell Wall Formation, PLoS ONE, 8(4).

[29] Kamatani, A. (1982). Dissolution rates of silica from diatoms decomposing at various tempera tures, Mar. Biol., 68(1), 91-96.

[30] Barker, P., Fontes, J.-C., Gasse, F. (1994). Experimental dissolution of diatom silica in con centrated salt solutions and implications for paleoenvironmental reconstruction, Limnol. Oceanogr., 39(1), 99-110.
[31] Flower, R., Ryves, D.B. (2009). Diatom preservation: differen tial preservation of sedimentary diatoms in two saline lakes, Acta Botanica Croatica, 68(2), 381-399.
[32] Crespin, J., Yam, R., Crosta, X., Massé, G., Schmidt, S., Campagne, P., et al. (2014). Holocene glacial discharge fluctuations and recent instability in East Antarctica, Earth and Planetary Science Letters, 394, 38-47.
[33] Passow, U., Engel, A., Ploug, H. (2003). The role of aggregation for the dissolution of diatom frustules, FEMS Microbiol. Ecol., 46(3), 247-255.
[34] Dolley, T.P., Moyle, P.R. (2003). History and overview of the U.S. diatomite mining industry, with emphasis on the Western United States. In Contributions to industrial minerals research: U.S.J.D. Geological Survey Bulletin, P.R. Moyle and K.R. Long (eds.). 2209. E1-E8.
[35] Flower, R.J. (2014). DIATOM METHODS | Diatomites: Their Formation, Distribution, and Uses Reference Module in Earth Systems and Environmental Sciences: Encyclopedia of Quaternary cience. pp. 501-506.
[36] Flower, R.J. (2006). Diatoms in ancient building materials: application of diatom analysis to Egyptian mud bricks, Nova Hedwigia, 245-263.
[37] Jenkins, D. (2011). St. Sophia's-once a church, then a mosque, now a museum, http://www.pbase.com/mrtoad/image/136205040/original.
[38] Dolley, T.P. (2000). Diatomite. In: 2000 Minerals Yearbook, 25.1-25.6, https://minerals.usgs.gov/minerals/pubs/commodity/diatomite/250400.pdf.
[39] Dolley, T.P. (2002). Diatomite. In: 2002 Minerals Yearbook, 24.1-24.6, https://minerals.usgs.gov/minerals/pubs/commodity/diatomite/diatomyb03.pdf.
[40] Founie, A. (2005). Diatomite. In: 2005 Minerals Yearbook, 23.1-23.6, https://minerals.usgs.gov/minerals/pubs/commodity/diatomite/diatomyb05.pdf.
[41] Crangle, R.D. (2015). Diatomite. In: 2015 Minerals Yearbook, 22.1-22.5, https://minerals.usgs.gov/minerals/pubs/commodity/diatomite/myb1-2015-diato.pdf.
[42] LaFrenz, R.L. (1961). Design of municipal diatomite filters for iron removal. Retrospective Theses and Dissertations. 2463, http://lib.dr.iastate.edu/rtd/2463.
[43] Speer, B.R. (1997). Diatoms: Fossil Record, http://www.ucmp.berkeley.edu/chromista/diatoms/diatomfr.html.
[44] Olney, M. (2002). Diatoms, http://www.ucl.ac.uk/GeolSci/micropal/diatom.html.
[45] Koren, I., Kaufman, Y.J., Washington, R., Todd, M.C., Rudich, Y., Martins, J.V., et al. (2006). The Bodélé depression: a single spot in the Sahara that provides most of the mineral dust to the Amazon forest, Environmental Research Letters, 1(1), 014005.
[46] Drake, N., Bristow, C. (2006). Shorelines in the Sahara: geomorphological evidence for an enhanced monsoon from palaeolake Megachad, The Holocene, 16(6), 901-911.
[47] Bristow, C.S., Drake, N., Armitage, S. (2009). Deflation in the dustiest place on earth: The bodélé depression chad, Geomorphology, 105(1-2), 50-58.
[48] Washington, R., Todd, M.C., Lizcano, G., Tegen, I., Flamant, C., Koren, I., et al. (2006). Links between topography, wind, deflation, lakes and dust: The case of the Bodélé Depression, Chad, Geophysical Research Letters, 33(9).
[49] Todd, M.C., Washington, R., Martins, J.V., Dubovik, O., Lizcano, G., Mbainayel, S., et al. (2007). Mineral dust emission from the Bodélé Depression, northern Chad, during BoDEx 2005, J. Geophys. Res., 112(D6).
[50] Suc, J., Violanti, D., Londeix, L., Poumot, C., Robert, C., Clauzon, G., et al. (1995). Evolution of the Messinian Mediterranean environments: the Tripoli Formation at Capodarso (Sicily, Italy, Rev. Palaeobot. Palynol., 87(1), 51-79.
[51] Brown, J. (2017). Ep engineered clays buys bleaching clay, activated clay catalysts from basf, http://blog.epminerals.com/ep-engineered-clays-buys-bleaching-clay-activated-clay-catalysts-from-basf.
[52] Fragoulis, D., Stamatakis, M.G., Chaniotakis, E., Columbus, G. (2004). Characterization of lightweight aggregates produced with clayey diatomite rocks originating from Greece, Materials Characterization, 53(2-4), 307-316.

[53] Ibrahim, S. S., Selim, A. Q. (2011). Evaluation of Egyptian diatomite for filter aid applications, Physicochemical Problems of Mineral Processing, 47, 113-122.

[54] Yuan, P., He, H., Wu, D., Wang, D., Chen, L. (2004). Characterization of diatoma-ceous silica by Raman spectroscopy, Spectrochimica Acta. Part A Molecular and Biomolecular Spectroscopy, 60(12), 2941-2945.

[55] Bakr, H., Burkitbaev, M. (2009). Elaboration and characterization of natural diatomite in akty ubinsk/kazakhstan, The Open Mineralogy Journal, 3(1), 12-16.

[56] Hadjar, H., Hamdi, B., Jaber, M., Brendlé, J., Kessaïssia, Z., Balard, H., et al. (2008). Elaboration and characterisation of new mesoporous materials from diatomite and charcoal, Microporous and Mesoporous Materials, 107(3), 219-226.

[57] Hassan, M. S., Ibrahim, I. A., Ismael, I. S. (1999). Diatomaceous deposits of Fayium, Egypt; characterization and evaluation for industrial application, Chin. J. Geochem., 18(3), 233-241.

[58] Nayab, S., Farrukh, A., Oluz, Z., Tuncel, E., Tariq, S. R., Rahman, H., et al. (2014). Design and Fabrication of Branched Polyamine Functionalized Mesoporous Silica: An Efficient Absorbent for Water Remediation, ACS Applied Materials & Interfaces, 6(6)), 4408-4417.

[59] Wang, Y., Cai, J., Jiang, Y., Jiang, X., Zhang, D. (2012). Preparation of biosilica structures from frustules of diatoms and their applications: current state and perspectives, Appl. Microbiol. Biotechnol., 97(2), 453-460.

[60] Leonardo, S., Prieto-Simón, B., Camp s, M. (2015). Past, present and future of diatoms in biosensing, Trends in Analytical Chemistry, 79, 276-285.

[61] Ragni, R., Cicco, S. R., Vona, D., Farinola, G. M. (2017). Multiple Routes to Smart Nanostructured Materials from Diatom Microalgae: A Chemical Perspective, Adv. Mater. Weinheim., 30(19), 1704289.

[62] Jantschke, A., Herrmann, A., Lesnyak, V., Eychmüller, A., Brunner, E. (2011). Decoration of Diatom Biosilica with Noble Metal and Semiconductor Nanoparticles, Chemistry-An Asian Journal, 7(1), 85-90.

[63] Dolatabadi, J. E. N., de la Guardia, M, Guardia, M. D. (2011). Applications of diatoms and silica nanotechnology in biosensing, drug and gene delivery, and formation of complex metal nanostructures, TrAC Trends in Analytical Chemistry, 30(9), 1538-1548.

[64] Arkles, B. (2011). Hydrophobicity, hydrophilicity and silane surface modification. Gelest, Inc, https://www.pcimag.com/ext/resources/PCI/Home/Files/PDFs/Virtual_Supplier_Brochures/Gelest_Additives.pdf.

[65] Jeffryes, C. S. (2009). Biological insertion of nanostructured germanium and titanium oxides into diatom biosilica [Ph. D. Thesis]. USA: Oregon State University.

[66] Townley, H. E., Woon, K. L., Payne, F. P., White-Cooper, H., Parker, A. R. (2007). Modification of the physical and optical properties of the frustule of the diatomCoscinodiscus wailesiibynickel sulfate, Nanotechnology, 18(29), 295101.

[67] Sandhage, K. H., Dickerson, M. B., Huseman, P. M., Zalar, F. M., Carroll, M. C., Rondon, M. R., et al. (2008). A Novel Hybrid Route to Chemically-Tailored, Three-Dimensional Oxide Nanostructures: The Basic (Bioclastic and Shape-Preserving Inorganic Conversion) Process. Ceramic Engineering and Science Proceedings 26th Annual Conference on Composites, Advanced Ceramics, Materials, and Structures: B: Ceramic Engineering and Science Proceedings, 23(4), 653-664.

[68] Shian, S., Cai, Y., Weatherspoon, M. R., Allan, S. M., Sandhage, K. H. (2006). Three-Dimensional Assemblies of Zirconia Nanocrystals Via Shape-Preserving Reactive Conversion of Diatom Microshells, Journal of the American Ceramic Society, 89(2), 694-698.

[69] Diab, M., Mokari, T. (2018). Bioinspired Hierarchical Porous Structures for Engineering Advanced Functional Inorganic Materials, Adv. Mater., 30(41), 1706349.

[70] Kregiel, D. (2014). Advances in biofilm control for food and beverage industry using organosilane technology: A review, Food Control, 40, 32-40.

[71] 3Kuiper, S. (2000). Development and application of microsieves Ph. D. Dissertation. p. 133. Enschede, the Netherlands: University of Twente.

[72] Nogue, M. G. (2005). Inorganic and polymeric microsieves: strategies to reduce fouling Ph. D. Dissertation. p. 203. Enschede, the Netherlands: University of Twente.

[73] Bhardwaj, V., Mirliss, M. J. (2001). Diatomaceous earth filtration for drinking water. Morgantown: NDWC.

[74] Sulpizio, T. E. (1999). Advances in Filter Aid and Precoat Filtration Technology. Presentation at the

American filtration and Separations Society Annual Technical Conference. Boston, MA.
[75] Bridges, H. R. (1970). Design requirements of precoat filters for water filtration, Retrospective Theses and Dissertations, 4287.
[76] Sentmanat, J. M. (2011). Clarifying Liquid Filtration: A practical guide to liquid filtration, Chemical Engineering, 38-47.
[77] Elhaddad, E., Abdel-Raouf, M. E. S. (2013). New model to eliminate salts from Sarir crude oil: a case study, International Journal of Engineering Research and Science & Technology, 2(4), 1111-1116.
[78] Ghobara, M., Salem, Z. E. (2017). Spatiotemporal Fluctuations in Phytoplankton Communities and Their Potential Indications for the Pollution Status of the Irrigation and Drainage Water in the Middle Nile Delta Area, Egypt The Handbook of Environmental Chemistry. Berlin: Springer.
[79] Salem, Z., Ghobara, M., Nahrawy, A. A. (2017). Spatio-temporal evaluation of the surface water quality in the middle Nile Delta using Palmer's algal pollution index, Egyptian Journal of Basic and Applied Sciences, 4(3), 219-226.
[80] Khraisheh, M., Aldegs, Y., Mcminn, W. (2004). Remediation of wastewater containing heavy metals using raw and modified diatomite, Chemical Engineering Journal, 99(2), 177-184.
[81] Du, Y., Fan, H., Wang, L., Wang, J., Wu, J., Dai, H. (2013). α-Fe_2o_3 nanowires deposited diatomite: highly efficient absorbents for the removal of arsenic, J. Mater. Chem. A, 1(26), 7729.
[82] Caliskan, N., Kul, A. R., Alkan, S., Sogut, E. G., Alacabey, I. (2011). Adsorption of Zinc(II) on diatomite and manganese-oxide-modified diatomite: A kinetic and equilibrium study, Journal of Hazardous Materials, 193, 27-36.
[83] Dobor, J., Perényi, K., Varga, I., Varga, M. (2015). A new carbon-diatomite earth composite adsorbent for removal of heavy metals from aqueous solutions and a novel application idea, Microporous and Mesoporous Materials, 217, 63-70.
[84] Du, Y., Wang, L., Wang, J., Zheng, G., Wu, J., Dai, H. (2015). Flower-, wire-, and sheet-like Mno_2-deposited diatomites: Highly efficient absorbents for the removal of Cr(VI), Journal of Environmental Sciences, 29, 71-81.
[85] Khraisheh, M., Al-Ghouti, M., Allen, S., Ahmad, M. (2005). Effect of OH and silanol groups in the removal of dyes from aqueous solution using diatomite, Water Res., 39(5), 922-932.
[86] Li, S., Li, D., Su, F., Ren, Y., Qin, G. (2014). Uniform surface modification of diatomaceous earth with amorphous manganese oxide and its adsorption characteristics for lead ions, Applied Surface Science, 317, 724-729.
[87] Mohamed, A., Ghobara, M. M., Abdelmaksoud, M. K., Mohamed, G. G. (2019). A novel and highly efficient photocatalytic degradation of malachite green dye via surface modified polyacrylonitrile nanofibers/biogenic silica composite nanofibers, Separation and Purification Technology 210, 935-942.
[88] Michen, B., Meder, F., Rust, A., Fritsch, J., Aneziris, C., Graule, T. (2012). Virus Removal in Ceramic Depth Filters Based on Diatomaceous Earth, Environ. Sci. Technol., 46(2), 1170-1177.
[89] Aivalioti, M., Papoulias, P., Kousaiti, A., Gidarakos, E. (2012). Adsorption of BTEX, MTBE and TAME on natural and modified diatomite, Journal of Hazardous Materials, 207-208, 117-127.
[90] Aivalioti, M., Vamvasakis, I., Gidarakos, E. (2010). BTEX and MTBE adsorption onto raw and thermally modified diatomite, Journal of Hazardous Materials, 178(1-3), 136-143.
[91] Baumann, E. R., LaFrenz, R. L. (1963). Optimum economical design for municipal diatomite filter plants, Journal-American Water Works Association, 55(1), 48-58.
[92] Zhao, S., Huang, G., Fu, H., Wang, Y. (2014). Enhanced Coagulation/Flocculation by Combining Diatomite with Synthetic Polymers for Oily Wastewater Treatment, Separation Science and Technology, 49(7), 999-1007.
[93] Zarnkow, M., McGreger, C., McGreger, N. (2016). German Beer. In Traditional Foods. Integrating Food Science and Engineering Knowledge Into the Food Chain. 10. K. Kristbergsson and J. Oliveira (eds.). Boston, MA: Springer.
[94] Skantz, M. (2004). Crossflow microfiltration of beer. Department of Chemical Engineering, Lund Institute of Technology.
[95] Johnson, M. (1997). Management of Spent Diatomaceous Earth from the Brewing Industry. [Thesis] the University of Western Australia.
[96] Horn, N., Marquet, M., Meek, J., Budahazi, G. (1996). Process for reducing RNA concentration in a

mixture of biological material using diatomaceous earth. US 5576196A.
[97] Hilbrig, F., Freitag, R., Schumacher, I. (2008). Method for treating biomass for producing cell lysate containing plasmid DNA. US7378238B2.
[98] Little, M. C. (1991). Process for the purification of DNA on diatomaceous earth. US 5075430.
[99] Machesky, L. (1996). Plasmid Preparations with Diatomaceous Earth. In Methods in Molecular Biology^{TM}, 58. Humana Press. D. N. A. Basic, R. N. A. Protocols and A. J. Harwood (eds.).
[100] Tanaka, J., Ikeda, S. (2002). Rapid and Efficient DNA Extraction Method from Various Plant Species Using Diatomaceous Earth and a Spin Filter, Breed. Sci., 52(2), 151-155.
[101] Buyel, J. F., Gruchow, H. M., Fischer, R. (2015). Depth filters containing diatomite achieve more efficient particle retention than filters solely containing cellulose fibers, Front. Plant Sci., 6.
[102] Tan, S. C., Yiap, B. C. (2009). DNA, RNA, and Protein Extraction: The Past and The Present, Journal of Biomedicine and Biotechnology, 1-10.
[103] Johal, S., Sachtleben, S., Edmonds, M., Dennis, R. (2016). Animal feed compositions including spent filter media containing diatomaceous earth. US 20160295884 A1.
[104] Kashcheev, I. D., Popov, A. G., Ivanov, S. E. (2009). Improving the thermal insulation of hightemperature furnaces by the use of diatomite, Refractories and Industrial Ceramics, 50(2), 98-100.
[105] Al-Jabri, K. S., Hago, A. W., Al-Nuaimi, A. S., Al-Saidy, A. H. (2005). Concrete blocks for thermal insulation in hot climate, Cement and Concrete Research, 35(8), 1472-1479.
[106] Hu, Z., Zheng, S., Jia, M., Dong, X., Sun, Z. (2017). Preparation and characterization of novel diatomite/ground calcium carbonate composite humidity control material, Advanced Powder Technology, 28(5), 1372-1381.
[107] Zheng, J., Shi, J., Ma, Q., Dai, X., Chen, Z. (2017). Experimental study on humidity control performance of diatomite-based building materials, Applied Thermal Engineering, 114, 450-456.
[108] Saran, G. (2012). Investigation of refractory thermal insulation material. 357515. 7th International Forum on Strategic Technology (IFOST) Russia, IEEE: Tomsk.
[109] Wenhua, Z., Faai, Z. (2007). Study on the Effect of Material and Film Thickness of Coatings on Reflection and Heat-Insulation Performance. Paint & Coatings Industry. Retrieved from, http://en.cnki.com.cn/Article_en/CJFDTOTAL-TLGY200-706008.htm.
[110] Ye, X., Chen, D., Chang, M., Mo, Y., Wang, Q. (2017). preparation of a Novel Water-based Acrylic Multi-Thermal Insulation Coating, Materials Science, 23(2).
[111] Martinez, M. F. (2014). Fire-resistant, insulating, ecological and corrosion-inhibiting coating. US20170327699A1.
[112] Van Helden, A., Dupré, S., van Gent, R. (2010). The Origins of the Telescope. Amsterdam University Press.
[113] Ussing, A. P., Gordon, R., Ector, L., Buczko', K., Desnitskiy, A. G., Van Landingham, S. L. (2005). the colonial diatom "Bacillaria paradoxa": chaotic gliding motility, Lindenmeyer model of colonial morphogenesis, and bibliography, with translation of O. F. Muoller (1783) "about a peculiar being in the beach-water". In: Diatom monographs Witkowski, A.. (ed). p.139 Koenigstein, Germany: Koeltz.
[114] Barsoum, M. W., Ganguly, A., Hug, G. (2006). Microstructural evidence of reconstituted limestone blocks in the great pyramids of egypt, J. American Ceramic Society, 89(12), 3788-3796.
[115] Philokyprou, M. (2012). The Earliest Use of Lime and Gypsum Mortars in Cyprus, Historic Mortars, 25-35.
[116] Teomete, E. (2004). Finite element modeling of historical masonry structures; case study: Urla Kamanlı Mosque, Izmir [MS thesis]. Izmir, Turkey: Institute of Technology.
[117] Böke, H., Çizer, Özlem., İpekoğlu, B., Uğurlu, E., Şerifaki, K., Toprak, G. (2008). Characteristics of lime produced from limestone containing diatoms, Construction and Building Materials, 22 (5), 866-874.
[118] Engh, K. R., Staff, Ub. (2014). Diatomite. In Kirk-Othmer Encyclopedia of Chemical Technology. J. I. Kroschwitz (ed). Wiely. 4. pp.1-11.
[119] Sisman, C., Karaman, S., Oztoprak, B. (2015). Usage Possibilities of Diatomite in the Concrete Production for Agricultural Buildings, Journal of Basic & Applied Sciences, 11, 31-38.
[120] Posi, P., Lertnimoolchai, S., Sata, V., Chindaprasirt, P. (2013). Pressed lightweight concrete containing calcined diatomite aggregate, Construction and Building Materials, 47, 896-901.

[121] Turner, T. A. (1995). Fire-resistant building component, U.S. 5391245 A.
[122] Ergün, A. (2011). Effects of the usage of diatomite and waste marble powder as partial replacement of cement on the mechanical properties of concrete, Construction and Building Materials, 25(2), 806-812.
[123] Janotka, I., Krajci, L., Uhlík, P., Bačuvčik, M. (2014). Natural and calcined clayey diatomite as cement replacement materials: microstructure and pore structure study, International Journal of Research in Engineering and Technology, 03(25), 20-26.
[124] Letelier, V., Tarela, E., Muñoz, P., Moriconi, G. (2016). Assessment of the mechanical properties of a concrete made by reusing both: Brewery spent diatomite and recycled aggregates, Construction and Building Materials, 114, 492-498.
[125] Degirmenci, N., Yilmaz, A. (2009). Use of diatomite as partial replacement for Portland cement in cement mortars, Construction and Building Materials, 23(1), 284-288.
[126] Yılmaz, B., Ediz, N. (2008). The use of raw and calcined diatomite in cement production, Cement and Concrete Composites, 30(3), 202-211.
[127] Korunic, Z. (1998). Review Diatomaceous earths, a group of natural insecticides, Journal of Stored Products Research, 34(2-3), 87-97.
[128] Losic, D., Korunic, Z. (2017). CHAPTER 10: Diatomaceous Earth, A Natural Insecticide for Stored Grain Protection: Recent Progress and Perspectives Diatom Nanotechnology: Progress and Emerging Applications. pp. 219-247.
[129] Korunić, Z., Rozman, V., Liška, A., Lucić, P. (2016). A review of natural insecticides based on diatomaceous earths, Poljoprivreda, 22(1), 10-18.
[130] Cloyd, R. A., Dickinson, A. (2005). Effects of Growing Media Containing Diatomaceous Earth on the Fungus Gnat Bradysia sp. nr. coprophila (Lintner) (Diptera: Sciaridae, HortScience, 6, 1806-1809.
[131] Abdalla, M. (2010). Sustainable effects of diatomite on the growth criteria and phytochemical contents of Vicia faba plants, ABJNA, 1(5), 1076-1089.
[132] Chen, J., Caldwell, R. D., Robinson, C. A., Steinkamp, R. (2000). Lets put the Si back into the soil, Part I. Plant Nutrition, 4, 44-46.
[133] Liang, Y., Sun, W., Zhu, Y., Christie, P. (2007). Mechanisms of silicon—mediated alleviation of abiotic stresses in higher plants: A review, Environ. Pollut., 147, 422-428.
[134] Aksakal, E. L., Angin, I., Oztas, T. (2013). Effects of diatomite on soil consistency limits and soil compactibility, CATENA, 101, 157-163.
[135] Angin, I., Kose, M., Aslantas, R. (2011). Effect of diatomite on growth of strawberry, Pakistan Journal of Botany, 43(1), 573-577.
[136] Crooks, R., Prentice, P. (2011). The benefits of silicon fertilizer for sustainably increasing crop productivity. The 5th International Conference on. Beijing, China: Silicon in Agriculture.
[137] El-Sherif, F., El Zaina, D., Yap, Y. K. (2018). Diatomite improves productivity and quality of moringa oleifera grown in greenhouse, Electronic Journal of Biology, 14(1), 1-6.
[138] Ogbaji, P. O., Shahrajabian, M. H., Xue, X. (2013). Changes in germination and primarily growth three cultivars of tomato under diatomite and soil materials in auto-irrigation system, Int. J. Biol., 5(3), 80.
[139] Sandhya, K., Prakash, N. B., Meunier, J. D. (2018). Diatomaceous earth as source of silicon on the growth and yield of rice in contrasted soils of Southern India, Journal of Soil Science and Plant Nutrition, (Ahead).
[140] Abdalla, M. (2011a). Impact of diatomite nutrition on two Trifolium alexandrinum cultivars differing in salinity tolerance, International Journal of Plant Physiology and Biochemistry, 3(13), 233-246.
[141] Abdalla, M. (2011b). Beneficial effects of diatomite on the growth, the biochemical contents and polymorphic DNA in Lupinus albus plants grown under water stress, ABJNA, 2(2), 207-220.
[142] Matichenkov, V. V., Kosobrukhov, A. A. (2004). Si effect on the plant resistance to salt toxicity ISCO 2004-13th International Soil Conservation Organisation Conference. p. 626. Brisbane.
[143] Mukerabigwi, J. F., Wang, Q., Ma, X., Liu, M., Lei, S., Wei, H., et al. (2015). Urea fertilizer coated with biodegradable polymers and diatomite for slow release and water retention, Journal of Coatings Technology and Research, 12(6), 1085-1094.
[144] Bugbee, B. (2004). Nutrient management in recirculating hydroponic culture, Acta Hortic., 648(648), 99-112.
[145] Dykyjová, D., Véber, K., Pribáň, K. (1971). Productivity and root/shoot ratio of reedswamp species

growing in outdoor hydroponic cultures, Folia Geobot. Phytotax., 6(3), 233-254.
[146] Giannini, L., Nahmias, N. M. (2007). Tire and crosslinkable elastomeric composition comprising diatomite particles. WO2009080091A1.
[147] Mikulásik, E., Albrecht, O. (2013). Use of diatomaceous earth in the pharmaceutical industry, US, 20130149380, A1.
[148] Liu, J. D., Fowler, J. (2016). Effect of inert fillers with changing energy-protein ratio on growth performance and energy digestibility in broilers, The Journal of Applied Poultry Research.
[149] Gysau, D. (2006). Fillers for paints: fundamentals and applications Vincentz Network GmbH & Co KG. Germany: Hannover.
[150] Zhao, Y., Du, M., Zhang, K. X., Gao, L. (2018). Effect of Modified Diatomite on Crystallinity and Mechanical Properties of Polypropylene, Materials Science Forum, 913, 551-557.
[151] Lu, X., Xia, Y., Liu, M., Qian, Y., Zhou, X., Gu, N., et al. (2012). Improved performance of diatomite-based dental nanocomposite ceramics using layer-by-layer assembly, International Journal of Nanomedicine, 7, 2153-2164.
[152] Shang, W., Qian, X., Yu, D. (2018). Preparation of a modified diatomite filler via polyethyleneimine impregnation and its application in papermaking, Journal of Applied Polymer Science, 135(20), 46275.
[153] Sha, L. Z., Chen, K. F. (2014). Preparation and characterization of ammonium polyphosphate/diatomite composite fillers and assessment of their flame-retardant effects on paper, Bioresources, 2014(9), 3104-3116.
[154] Sha, L., Chen, K. (2016). Surface modification of ammonium polyphosphate-diatomaceous earth composite filler and its application in flame-retardant paper, Journal of Thermal Analysis and Calorimetry, 123(1), 339-347.
[155] Zhao, H., Sha, L. (2017). Effect of surface modification of ammonium polyphosphate—diatomite composite filler on the flame retardancy and smoke suppression of cellulose paper, Journal of biosources and bioproducts, 2(1).
[156] Jin-bao, L., Hui-juan, X., An-dong, W. (2008). Pilot Study of Diatomite as Filler of Light Printing Paper. Bulletin of the Chinese Ceramic Society, http://en.cnki.com.cn/Article_en/CJFDTOTAL-GSYT200804039.htm.
[157] Lamastra, F., Mori, S., Cherubini, V., Scarselli, M., Nanni, F. (2017). A new green methodology for surface modification of diatomite filler in elastomers, Materials Chemistry and Physics, 194, 253-260.
[158] Wu, W., Chen, Z. (2017). Modified-diatomite reinforced rubbers, Materials Letters, 209, 159-162.
[159] Greene, M., Danvers, N., Hu, Q. (2009). Organo-neutralized diatomaceous earth, methods of preparation, and uses thereof. WO2009137394A1.
[160] Cheng, Y., Tao, J., Jiao, Y., Guo, Q., Li, C. (2015). Influence of Diatomite and Mineral Powder on Thermal Oxidative Ageing Properties of Asphalt, Advances in Materials Science and Engineering, 2015(1), 1-10.
[161] Das, A. K., Singh, D. (2017). Investigation of rutting, fracture and thermal cracking behavior of asphalt mastic containing basalt and hydrated lime fillers, Construction and Building Materials, 141, 442-452.
[162] Cheng, Y., Zhu, C., Tao, J., Jiao, Y., Yu, D., Xiao, B. (2018). Effects of Diatomite-Limestone Powder Ratio on Mechanical and Anti-Deformation Properties of Sustainable Sand Asphalt Composite, Sustainability, 10(3), 808.
[163] Shukry, N. A., Hassan, N. A., Abdullah, M. E., Hainin, M. R., Yusoff, N. I., Mahmud, M. Z., et al. (2018). Influence of diatomite filler on rheological properties of porous asphalt mastic, International Journal of Pavement Engineering, 1-9.
[164] Yang, C., Xie, J., Zhou, X., Liu, Q., Pang, L. (2018). Performance Evaluation and Improving Mechanisms of Diatomite-Modified Asphalt Mixture, Materials, 11(5), 686.
[165] Davar, A., Tanzadeh, J., Fadaee, O. (2017). Experimental evaluation of the basalt fibers and diatomite powder compound on enhanced fatigue life and tensile strength of hot mix asphalt at low temperatures, Construction and Building Materials, 153, 238-246.
[166] Wei, H., Li, Z., Jiao, Y. (2017). Effects of Diatomite and SBS on Freeze-Thaw Resistance of Crumb Rubber Modified Asphalt Mixture, Advances in Materials Science and Engineering, 1-14.
[167] Hanh, T. T., Thu, N. T., Quoc, L. A., Hien, N. Q. (2017). Synthesis and characterization of silver/diatomite nanocomposite by electron beam irradiation, Radiation Physics and Chemistry, 139, 141-146.

[168] Xia, Y., Jiang, X., Zhang, J., Lin, M., Tang, X., Zhang, J., et al. (2017a). Synthesis and characterization of antimicrobial nanosilver/diatomite nanocomposites and its water treatment application, Applied Surface Science, 396, 1760-1764.

[169] Fernández, M. A., Bellotti, N. (2017). Silica-based bioactive solids obtained from modified diatomaceous earth to be used as antimicrobial filler material, Materials Letters, 194, 130-134. Fischer, C., Adam, M., Mueller, A. C., Sperling, E., Wustmann, M., van Pée, K.-H, Pée, K. V., et al.

[170] Cacciotti, I., Mori, S., Cherubini, V., Nanni, F. (2018). Eco-sustainable systems based on poly(lactic acid), diatomite and coffee grounds extract for food packaging, International Journal of Biological Macromolecules, 112, 567-575.

[171] Chirash, W., Crosier, H. E., Proulx, C. R. (1988). Light duty liquid dishwashing composition containing abrasive. US 4772425 A.

[172] Olson, H. M. (1923). Abrasive composition. US 1444479 A.

[173] Yeh, K., Synodis, J. (1986). Stain removal toothpaste. US 4612191 A. Gordon, C. (1971). Metal cleaner. US 3619962 A.

[174] Liu, Y. (2010). Diatomaceous Earth-Containing Slurry Composition And Method For Polishing Organic Polymer-Based Ophthalmic Substrates Using The Same. US 20100330884 A1.

[175] Slamova, R., Trckova, M., Vondruskova, H., Zraly, Z., Pavlik, I. (2011). Clay minerals in animal nutrition, Applied Clay Science, 51(4), 395-398.

[176] Limpitlaw, U. (2006). Palliative and curative Earth materials [master's thesis. USA: University of Northern Colorado.

[177] Limpitlaw, U. G. (2010). Ingestion of Earth materials for health by humans and animals, International Geology Review, 52(7-8), 726-744.

[178] Pristiazhniuk, I. E., Gorchakov, V. N. (1997). Changes in the lymph nodes of rats under the Remediation of wastewater containing heavy metals.

[179] Shady, A. A., Zalat, A., Al-Ashkar, E., Ghobara, M. (2018). The porosity of some Egyptian Diatom frustules and their potential in Nanotechnology, Nanoscience &. Nanotechnology-Asia, , 08.

[180] De Tommasi, E., Gielis, J., Rogato, A. (2017). Diatom Frustule Morphogenesis and Function: a Multidisciplinary Survey, Marine Genomics, 35, 1-18.

[181] De Tommasi, E. (2016). Light Manipulation by Single Cells: The Case of Diatoms Journal of Spectroscopy, 2016, 2490128.

[182] Ghobara, M. M., Vinayak, V., Smith, D. R., Schoefs, B., Gebeshuber, I. C., Gordon, R. (2015). Diatom frustules as photo-regulators of diatom photobiology. In National symposium onHorizons of light in molecules, materials and daily life, December 18-19, 2015, Departmentof Chemistry (School of Chemical Sciences &. Technology), Dr. H. S. Gour Central University, Sagar, MP 470003, India. Edited by R. N Yadava. Department of Chemistry (School of Chemical Sciences &. Technology), Dr. H. S. Gour Central University, Sagar, MP India.

[183] Jeffryes, C., Campbell, J., Li, H. Y., Jiao, J., Rorrer, G. (2011). The potential of diatom nanobiotechnology for applications in solar cells, batteries, and electroluminescent devices, Energy &. Environmental Science, 4(10), 3930-3941.

[184] Jeffryes, C. S. (2009). Biological insertion of nanostructured germanium and titanium oxides into diatom biosilica [Ph. D. Thesis]. USA: Oregon State University.

[185] Gautam, S., Kashyap, M., Gupta, S., Kumar, V., Schoefs, B., Gordon, R., et al. (2016). Metabolic engineering of TiO_2 nanoparticles in Nitzschia palea to form diatom nanotubes: an ingredient for solar cells to produce electricity and biofuel, RSC Adv., 6(99), 97276-97284.

[186] Toster, J., Iyer, K. S., Xiang, W. C., Rosei, F., Spiccia, L., Raston, C. L. (2013). Diatom frustules as light traps enhance DSSC efficiency, Nanoscale, 5(3), 873-876.

[187] Karaman, S., Karaipekli, A., Sarı, A., Biçer, A. (2011). Polyethylene glycol (PEG)/diatomite composite as a novel form-stable phase change material for thermal energy storage, Solar Energy Materials and Solar Cells, 95(7), 1647-1653.

[188] Rao, Z., Zhang, G., Xu, T., Hong, K. (2018). Experimental study on a novel form-stable phase change materials based on diatomite for solar energy storage, Solar Energy Materials and Solar Cells, 182, 52-60.

[189] Sarı, A., Bicer, A., Al-Sulaiman, F., Karaipekli, A., Tyagi, V. (2018). Diatomite/CNTs/PEG composite PCMs with shape-stabilized and improved thermal conductivity: Preparation and thermal energy

storage properties, Energy and Buildings, 164, 166-175.

[190] Sun, Y., Wang, R., Liu, X., Dai, E., Li, B., Fang, S., et al. (2018). Synthesis and Performances of Phase Change Microcapsules with a Polymer/Diatomite Hybrid Shell for Thermal Energy Storage, Polymers, 10(6), 601.

[191] Wen, R., Zhang, X., Huang, Z., Fang, M., Liu, Y., Wu, X., et al. (2018). Preparation and thermal properties of fatty acid/diatomite form-stable composite phase change material for thermal energy storage, Solar Energy Materials and Solar Cells, 178, 273-279.

[192] Xu, G., Leng, G., Yang, C., Qin, Y., Wu, Y., Chen, H., et al. (2017). Sodium nitrate—Diatomitecomposite materials for thermal energy storage, Solar Energy, 146, 494-502.

[193] Xu, B., Li, P., Chan, C. (2015). Application of phase change materials for thermal energy storage in concentrated solar thermal power plants: A review to recent developments, Applied Energy, 160, 286-307.

[194] Darmanin, T., Guittard, F. (2015). Superhydrophobic and superoleophobic properties in nature, Materials Today, 18(5), 273-285.

[195] Perera, H.J., Mortazavian, H., Blum, F.D. (2017). Surface Properties of Silane-Treated Diatomaceous Earth Coatings: Effect of Alkyl Chain Length, Langmuir, 33(11), 2799-2809.

[196] Polizos, G., Winter, K., Lance, M.J., Meyer, H.M., Armstrong, B.L., Schaeffer, D.A., et al. (2014). Scalable superhydrophobic coatings based on fluorinated diatomaceous earth: Abrasion resistance versus particle geometry, Applied Surface Science, 292, 563-569.

[197] Sedai, B.R., Alavi, S.H., Harimkar, S.P., Mccollum, M., Donoghue, J.F., Blum, F.D. (2017). Particle morphology dependent superhydrophobicity in treated diatomaceous earth/polystyrene coatings, Applied Surface Science, 416, 947-956.

[198] Perera, H.J., Blum, F.D. (2018). Alkyl chain modified diatomaceous earth superhydrophobic coatings. 2018 Advances in Science and Engineering Technology International Conferences (ASET).

[199] Fischer, C., Adam, M., Mueller, A.C., Sperling, E., Wustmann, M., van Pée, K.-H., Pée, K.V., et al. (2016). Gold nanoparticle-decorated diatom biosilica: A favorable catalyst for the oxidation of d-Glucose, ACS Omega, 1(6), 1253-1261.

[200] Vrieling, E.G., Beelen, T.P., Sun, Q., Hazelaar, S., Santen, R.A., Gieskes, W.W. (2004). Ultrasmall, small, and wide angle X-ray scattering analysis of diatom biosilica: interspecific differences in fractal properties, Journal of Materials Chemistry, 14(13), 1970.

[201] Bahramian, B., Ardejani, F.D., Mirkhani, V., Badii, K. (2008). Diatomite-supported manganese Schiff base: An efficient catalyst for oxidation of hydrocarbons, Applied Catalysis A: General, 345(1), 97-103.

[202] Cai, L., Gong, J., Liu, J., Zhang, H., Song, W., Ji, L. (2018). Facile Preparation of Nano-Bi_2MoO_6/Diatomite Composite for Enhancing Photocatalytic Performance under Visible Light Irradiation, Materials, 11(2), 267.

[203] Jabbour, K., Hassan, N.E., Davidson, A., Massiani, P., Casale, S. (2015). Characterizations and performances of Ni/diatomite catalysts for dry reforming of methane, Chemical Engineering Journal, 264, 351-358.

[204] Padmanabhan, S.K., Pal, S., Haq, E.U., Licciulli, A. (2014). Nanocrystalline TiO_2—diatomite composite catalysts: Effect of crystallization on the photocatalytic degradation of rhodamine B, Applied Catalysis. A General, 485, 157-162.

[205] Shen, M., Fu, L., Tang, J., Liu, M., Song, Y., Tian, F., et al. (2018). Microwave hydrothermalassisted preparation of novel spinel-$NiFe_2O_4$/natural mineral composites as microwave catalysts for degradation of aquatic organic pollutants, J. Hazard. Mater., 350, 1-9.

[206] Tang, W., Qiu, K., Zhang, P., Yuan, X. (2016). Synthesis and photocatalytic activity of ytterbiumdoped titania/diatomite composite photocatalysts, Applied Surface Science, 362, 545-550.

[207] Wang, J., Seibert, M. (2017). Prospects for commercial production of diatoms, Biotechnology for Biofuels, 10(1).

[208] Zhang, Y.X., Huang, M., Li, F., Wang, X.L., Wen, Z.Q. (2014). One-pot synthesis of hierarchical MnO_2-modified diatomites for electrochemical capacitor electrodes, Journal of Power Sources, 246, 449-456.

[209] Zhu, Q., Zhang, Y., Zhou, F., Lv, F., Ye, Z., Fan, F., et al. (2011). Preparation and characterization of Cu_2O—ZnO immobilized on diatomite for photocatalytic treatment of red water produced from

manufacturing of TNT, Chemical Engineering Journal, 171(1), 61-68.

[210] Zhu, P., Chen, Y., Duan, M., Liu, M., Zou, P. (2018). Structure and properties of Ag_3PO_4/diatomite photocatalysts for the degradation of organic dyes under visible light irradiation, Powder Technology, 336, 230-239.

[211] Mohamed, A., Ghobara, M. M., Abdelmaksoud, M. K., Mohamed, G. G. (2019). A novel and highly efficient photocatalytic degradation of malachite green dye via surface modified polyacrylonitrile nanofibers/biogenic silica composite nanofibers, Separation and Purification Technology 210, 935-942.

[212] Guo, X. L., Kuang, M., Li, F., Liu, X. Y., Zhang, Y. X., Dong, F., et al. (2016). Engineering of three dimensional (3-D) diatom@TiO_2@MnO_2 composites with enhanced supercapacitor performance, Electrochimica Acta, 190, 159-167.

[213] Jiang, D. B., Zhang, B. Y., Zheng, T. X., Zhang, Y. X., Xu, X. (2018). One-pot synthesis of η-Fe_2O_3 nanospheres/diatomite composites for electrochemical capacitor electrodes, Materials Letters, 215, 23-26.

[214] Le, Q. J., Wang, T., Tran, D. N., Dong, F., Zhang, Y. X., Losic, D. (2017). Morphology-controlled MnO_2 modified silicon diatoms for high-performance asymmetric supercapacitors, Journal of Materials Chemistry. A, 5(22), 10856-10865.

[215] Zhang, Y., Guo, W. W., Zheng, T. X., Zhang, Y. X., Fan, X. (2018). Engineering hierarchical Diatom@CuO@MnO_2 hybrid for high performance supercapacitor, Applied Surface Science, 427, 1158-1165.

[216] Farokhzad, O. C., Langer, R. (2009). Impact of nanotechnology on drug delivery, ACS Nano, 3(1), 16-20.

[217] Wu, W., Jiang, X. (2016). Polymeric Micelles for Drug Delivery, in Biomedical Nanomaterials. In Wiley-VCH Verlag GmbH & Co. KGaA. Y. Zhao and Y. Shen (eds.). Weinheim, Germany.

[218] Allen, T. M., Cullis, P. R. (2013). Liposomal drug delivery systems: From concept to clinical applications, Advanced Drug Delivery Reviews, 65(1), 36-48.

[219] Bianco, A., Kostarelos, K., Prato, M. (2005). Applications of carbon nanotubes in drug delivery, Current Opinion in Chemical Biology, 9(6), 674-679.

[220] Iannazzo, D., Pistone, A., Salamò, M., Galvagno, S., Romeo, R., Giofré, S. V., et al. (2017). Graphene quantum dots for cancer targeted drug delivery, International Journal of Pharmaceutics, 518(1-2), 185-192.

[221] Slowing, I., Viveroescoto, J., Wu, C., Lin, V. (2008). Mesoporous silica nanoparticles as controlled release drug delivery and gene transfection carriers, Adv. Drug Deliv. Rev., 60(11), 1278-1288.

[222] Terracciano, M., Stefano, L. D., Santos, H. A., Martucci, N. M., Tino, A., Ruggiero, I. (2016). In Chapter 9: Silica-Based Nanovectors: From Mother Nature to Biomedical Applications. 191. N. Thajuddin and D. Dhanasekaran (eds.). In: Algae-Organisms for Imminent Biotechnology.

[223] Aw, M. S., Bariana, M., Yu, Y., Addai-Mensah, J., Losic, D. (2013). Surface-functionalized diatom microcapsules for drug delivery of water-insoluble drugs, J. Biomater. Appl., 28(2), 163-174.

[224] Aw, M. S., Simovic, S., Addai-Mensah, J., Losic, D. (2011). Silica microcapsules from diatoms as new carrier for delivery of therapeutics, Nanomedicine, 6(7), 1159-1173.

[225] Losic, D., Yu, Y., Aw, M. S., Simovic, S., Thierry, B., Addai-Mensah, J. (2010). Surface functionalisation of diatoms with dopamine modified iron-oxide nanoparticles: toward magnetically guided drug microcarriers with biologically derived morphologies, Chemical Communications, 46(34), 6323.

[226] Lodriche, S., Soltani, S., Mirzazadeh, R. (2013). Silicon nanocarrier for delivery of drug, pesticides and herbicides, and for waste water treatment, US, 20130225412, A1.

[227] Uthappa, U., Brahmkhatri, V., Sriram, G., Jung, H., Yu, J., Kurkuri, N., et al. (2018). Nature engineered diatom biosilica as drug delivery systems, Journal of Controlled Release, 281, 70-83.

[228] Ruggiero, I., Terracciano, M., Martucci, N. M., Stefano, L. D., Migliaccio, N., Tatè, R., et al. (2014). Diatomite silica nanoparticles for drug delivery, Nanoscale Research Letters, 9(1), 329.

[229] Managò, S., Migliaccio, N., Terracciano, M., Napolitano, M., Martucci, N. M., Stefano, L. D., et al. (2017). Internalization kinetics and cytoplasmic localization of functionalized diatomite nanoparticles in cancer cells by Raman imaging, Journal of Biophotonics, 11(4).

[230] Hertz, A., FitzGerald, V., Pignotti, E., Knowles, J. C., Sen, T., Bruce, I. J. (2012). Preparation and characterisation of porous silica and silica/titania monoliths for potential use in bone replacement, Microporous and Mesoporous Materials, 156, 51-61.

[231] Le, T. D., Bonani, W., Speranza, G., Sglavo, V., Ceccato, R., Maniglio, D., et al. (2016). Processing and characterization of diatom nanoparticles and microparticles as potential source of silicon for bone tissue engineering, Materials Science and Engineering: C, 59, 471-479.

[232] Lopez-Alvarez, M., Solla, E. L., González, P., Serra, J., Leon, B., Marques, A. P., et al. (2009). Silicon-hydroxyapatite bioactive coatings (Si—HA) from diatomaceous earth and silica. Study of adhesion and proliferation of osteoblast-like cells, Journal of Materials Science. Materials in Medicine, 20(5), 1131-1136.

[233] Tamburaci, S., Tihminlioglu, F. (2017a). Diatomite reinforced chitosan composite membrane as potential scaffold for guided bone regeneration, Materials Science and Engineering: C, 80, 222-231.

[234] Kong, X., Chong, X., Squire, K., Wang, A. X. (2018a). Microfluidic diatomite analytical devices for illicit drug sensing with ppb-Level sensitivity, Sensors and Actuators. B Chemical, 259, 587-595.

[235] Kong, X., Squire, K., Wang, A. X. (2018b). Microfluidic Diatomite Analytical Devices for UltraSensitive Detection of Hazardous Chemicals, Conference on Lasers and Electro-Optics.

[236] Squire, K., Kong, X., LeDuff, P., Rorrer, G. L., Wang, A. X. (2018). Photonic crystal enhanced fluorescence immunoassay on diatom biosilica, Journal of Biophotonics, e201800009.

[237] Smol, J. P., Stoermer, E. F. (2010). the diatoms: applications for the environmental and earth sciences. 2nd ed. Cambridge, UK: Cambridge University Press.

[238] Brugam, R. B., Munoz, S. E. (2018). A 1 600-year record of human impacts on a floodplain lake in the Mississippi River Valley, J. Paleolimnol., 60(3), 445-460.

[239] Juggins, S., Cameron, N. G. (2010). Diatoms and archaeology. In the diatoms: applications for the environmental and earth sciences. 2nd ed., J. P. Smol and E. F. Stoermer. Cambridge, UK: Cambridge University Press.

[240] Riederer, J. (2004). Thin Section Microscopy Applied to the Study of Archaeological Ceramics, Hyperfine Interactions, 154(1-4), 143-158.

[241] Krebs, W. N., Gladenkov, A. Y., Jones, G. D. (2010). Diatoms in oil and gas exploration. In the diatoms: applications for the environmental and earth sciences (2nd J. P. Smol and E. F. Stoermer (eds.). Cambridge, UK: Cambridge University Press.

[242] Brzezinski, M. A., Nelson, D. M., Franck, V. M., Sigmon, D. E. (2001). Silicon dynamics within an intense open-ocean diatom bloom in the Pacific sector of the Southern Ocean, Deep Sea Research Part II: Topical Studies in Oceanography, 48(19-20), 3997-4018.

[243] Villareal, T. A., Brown, C. G., Brzezinski, M. A., Krause, J. W., Wilson, C. (2012). Summer Diatom Blooms in the North Pacific Subtropical Gyre: 2008-2009, PLoS ONE, 7(4), e33109.

[244] Onitsuka, G., Shikata, T., Kitatsuji, S., Abe, K., Yamamoto, T., Ochiai, H., et al. (2016). Factors influencing maintenance and decline of a diatom bloom in the Yatsushiro Sea, Japan, Journal of Oceanography, 72(4), 617-627.

[245] Wang, J., Seibert, M. (2017). Prospects for commercial production of diatoms, Biotechnology for Biofuels, 10(1).

第 16 章
生物医学用硅

Shaheer Maher，Moom Sin Aw，Dusan Losic

16.1 简介

在过去的20年里,纳米技术及其应用领域取得了显著进展,对包括药物输送、组织工程、生物传感和生物成像在内的众多生物医学研究领域产生了重大影响。在药物输送领域,随着生物技术和药物化学的迅猛发展,许多新药得以研发。然而,这些新药以及大部分现有药物都面临诸多限制,例如不理想的溶解度(过低或过高)、稳定性差以及高毒性。因此,对先进药物输送系统(DDS)的需求迅速增长。这类系统的主要目标是将药物运送至特定部位(即靶向药物输送),减少潜在副作用并改善药物的物理化学性质。同时,用于此类递送系统的药物载体应具有生物相容性、可降解性、低成本、易于制造,并能在较长时间内维持最佳药物浓度(即在治疗窗口内),以减少给药频率,降低患者的不适感[1-2]。利用工程化纳米粒子的纳米级DDS为更有效的多样化治疗递送提供了新的机会,解决了传统药物输送方式中的一些关键问题。其优势在于能够携带活性治疗分子并将其集中于特定部位,而目前使用的传统药物输送系统则会将药物分布到全身各处[3]。在此背景下,各种合成纳米材料被开发和探索为药物纳米载体,以创建复杂的药物输送载体,将治疗分子特异性地传送至病变部位,这一点是传统系统无法实现的[4-5]。为开发微纳米载体,各种材料和技术得到了广泛研究,包括聚合物纳米粒子、脂质体、胶束、碳纳米管、树枝状大分子、磁性纳米粒子、二氧化硅纳米粒子和量子点[1,3,6-9]。此外,基于聚合物、金属和无机纳米粒子或其复合材料的几种有机、无机和混合DDS已在临床上被证明可用于药物输送应用[10-13]。

考虑到不同的材料,基于硅/二氧化硅材料的药物载体因其独特的物理、化学和光学特性,已被公认为药物输送应用中聚合物材料的有前途的替代品。特别是通过溶胶-凝胶法或电化学蚀刻法获得的合成二氧化硅,因其可调节的孔径、化学稳定性、高比表面积、高载药容量、安全性以及可调节的药物释放特性,受到了广泛关注[14-15]。然而,这些材料也存在一些缺点,如合成成本高且耗时、需要精细且不友好的实验条件、使用有毒材料以及产生大量废料[16-17]。

为了克服这些合成二氧化硅载体的局限性,并避免昂贵且对环境不友好的制造工艺,可以通过较少的加工和低成本从天然或生物资源中获得多孔二氧化硅。大自然通过数百万年的进化,发展出了最优雅的基因驱动方法,以创建三维多孔结构的纳米结构生物硅材料,这些结构来自一种被称为硅藻的单细胞光合藻类[18]。硅藻作为天然生物硅的来源,可以通过培养轻松生产,但最重要的来源是其化石形式,称为硅藻土(DE)矿物。这种硅藻土矿物是通过硅藻在海洋或湖泊底部沉积数百万年后形成的,被广泛开采并以低成本(约 200 美元/吨)用于许多行业,包括食品工业、农业和天然药物。通过可扩展、可重复和具有成本效益的工艺,可以从这种廉价且丰富的天然材料中获得多孔二氧化硅纳米颗粒(SiO_2 NPs)。

本章将介绍硅藻生物硅材料在生物医学领域中的应用,包括药物传递。本章首先描述硅藻生物硅微/纳米颗粒的功能化过程,这是改善其物理化学性质和特性、实现可调节和响应性药物释放的第一步。然后介绍从硅藻中制备可生物降解的多孔硅(pSi)的最新进展,这为开发新型可生物降解的天然微纳米药物载体提供了有前景的机会。最后讨论硅藻在组织工程和止血控制等其他生物医学应用中的潜在用途。

16.2 硅藻:用于治疗药物输送的天然二氧化硅微胶囊

16.2.1 结构

硅藻是占据阳光充足水域的单细胞真核藻类。这些微生物是微观硅纳米制造工厂,它们通过两个重叠的壳瓣由腰带状的结构连接在一起,封闭在一个称为硅质壳的硅壳中。硅质壳是以几十纳米的精度分层制造的。壳瓣由堆叠的六边形腔室构成,这些腔室由硅质板隔开。硅质板上有均匀分布的孔洞,这些孔洞的直径从外到内逐渐减小或增大,为硅藻提供机械保护、分子和胶体筛选以及光收集功能。图 16.1 显示了硅藻硅质壳三维结构的示意图。这幅图展示了典型的硅藻种类,*Coscinodiscus* sp. 的结构。外表面板具有最小的孔径,直径为 40 nm,被称为外筛板。分隔两个腔室层的板有中等直径的孔,约为 120 nm,被称为内筛板。外筛板和内筛板也被称为筛板,与它们的假定功能有关。最内层靠近筛板表面的孔直径为 1 100 nm,被称为门孔。

目前已鉴定出超过 100 000 种不同的硅藻物种,每种硅藻都有独特的硅质壳形状,并装饰有特定的纳米级特征,如孔洞、脊状物和棘刺,这些特征具有特定的功能和性质。它们的空心盒状结构具有多种尺寸,范围从 500 nm 到 50 μm[16-17]。它们的形状和尺寸从圆形到三角形不等(见图 16.2)。物种之间在形状、大小、几何结构、孔结构的组织和密度方面存在的巨大差异,代表了材料结构的细微变化,以优化在复杂生物环境中的运动和相互作用。硅藻壳具有许多独特的特性,如高可用性、低成本、优异的生物相容性以及具有微米和纳米级孔隙的高表面积。硅藻硅质壳因其独特的3D 多孔空心盒状结构,适合在水环境中运动,激发了新型纳米结构材料的设计和生产。它们的空心多孔微胶囊结构使其成为开发各种医疗疗法(包括诊疗一体化和微型机器人技术)中的纳米/微米药物载体的理想选择[19-21]。

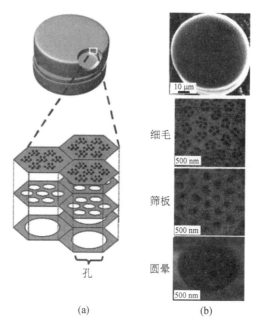

图 16.1 具有硅质壁横截面轮廓的中心硅藻硅质壳(通常由三层重叠的多孔层组成)示意图(a)和圆筛藻以及这些层上的孔隙结构 SEM 图(b)[19]

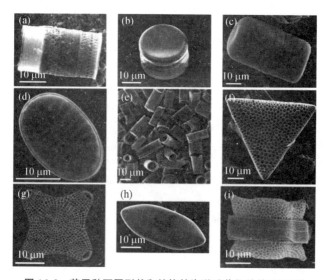

图 16.2 若干种不同形状和结构的海洋硅藻物种的 SEM 图

16.2.2 硅藻的表面修饰

原矿硅藻土和由矿业供应的煅烧硅藻土粉末含有一些需要净化的杂质。原始硅藻土通常通过酸处理或氧化处理进行纯化。这些清洗后的硅藻多孔硅结构的特性可以轻松地进行调整,从而为开发用于生物医学应用的新型生物工程材料铺平了道路。关于活性生物分子固定到改性后的硅藻表面,广泛采用了两种方法:一种是涉及物理吸附和其他弱相互作用的非共价相互作用,另一种是涉及强化学吸附相互作用和共价结合的共价固定。非共价结合(如静电作用)的主要问题在于,键合强度取决于溶液条件(如 pH 值或离子强度的变化),这些层的稳定性并不高。从稳定性和可重复性方面考虑,生物分子与硅藻表面的共价结合更为优越,预计将用于实际应用[22]。因此,将硅藻用于生物医学应用尤其是药物输送之前的第一步是表面功能化。

过去 10 年,硅藻结构的表面改性取得了显著进展,采用了多种不同的方法,主要基于为合成硅颗粒开发的策略,这些策略包括应用有机单分子层、聚合物、蛋白质以及金属和无机氧化物层的涂层[23]。最常见的方法基于使用硅藻表面可用的活性硅羟基(SiOH)基团,这些基团可以很容易地与许多活性物质(如—NH_2、—COOH、—SH 和—CHO)功能化,提供了稳固的偶联点,用于固定生物或化学部分,如药物、酶、蛋白质、抗体、适配体、DNA、传感探针等[24-25]。值得一提的是,硅烷化通过形成 Si—O—Si 共价键,广泛应用于在硅藻表面稳定附着不同活性部分[22,26]。图 16.3 展示了通过 Si—O—Si 共价键形成自组装层,使用有机硅烷对硅藻微胶囊进行典型表面功能化的示意图[27]。

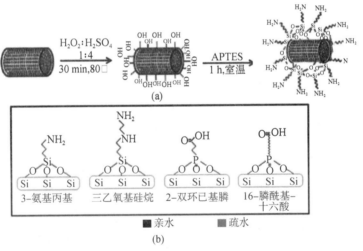

图 16.3　使用(3-氨丙基)三乙氧基硅烷(APTES)对硅藻进行表面功能化的示意图(a)和使用疏水性和亲水性有机硅烷和膦酸的硅藻功能化示意图(b)

16.2.3　硅藻原子作为药物载体的应用

药物输送(DD)的主要挑战之一是将治疗剂以有效浓度递送到人体特定区域(即病变组织),并且对健康组织的副作用最小[28-29]。此外,许多用于传统递送系统(DDS)的药物分子,尤其是在癌症治疗中,存在诸如在生理环境中溶解度低、细胞摄取不足等不良理化性质,这限制了它们的生物分布和治疗效果。为克服这些缺点,药物分子可以加载到特定的纳米载体上,这些载体可以保护药物免受降解/快速清除,增强它们的理化性质,并增加它们的细胞摄取。这将有助于改善药物分布,增强在靶向组织中的治疗效果,并尽量减少不良反应[30-32]。

如上所述,多孔硅(宏孔、中孔或微孔)由于其独特的理化性质,已被广泛研究作为药物载体[33-35]。然而,这些材料的优点受到昂贵且耗时的制备程序的挑战,这些程序涉及使用可能在最终产品中残留的有毒溶剂[36-37]。为了解决这些问题,引入了使用硅藻或其化石形式作为更安全、成本效益更高的生物载体用于药物输送的概念[38]。

Morse 在 1999 年首次提出利用硅藻的生物结构来构建新材料的概念,他认为硅藻生物生成的硅质具有遗传控制的纳米尺度架构精度,超出了当时人类工程的能力[39]。随后,Rosi 等在 2004 年的实验中首次展示了有目的地利用硅藻作为递送系统的可能性,他们成功控制了从 DNA 功能化的硅藻表面释放金纳米颗粒的过程[40]。

硅藻微胶囊口服作为药物载体的应用首次由 Losic 等提出,成为最常见的药物输送途径[41-42]。研究结果证实,硅藻在递送水不溶性药物吲哚美辛方面有效,药物负载量约为 22%①,并且在 14 天内实现了持续释放。值得注意的是,观察到硅藻微胶囊中药物释放的两个阶段:第一阶段是由于表面吸附的药物脱附导致的 6 小时内的爆发性释放;第二阶段是缓慢且延长的释放,在两周内以零级动力学方式释放,归因于从硅藻孔内部的药物释放。在相同的背景下,Zhang 等评估了负载在硅藻上的泼尼松和美沙拉嗪的口服递送。结果证实了两种药物的持续释放。同时,体外毒性评估显示硅藻壳在浓度高达 1 000 μg/mL 时的毒性可以忽略不计。此外,当使用 Caco-2/HT-29 共培养的单层细胞测试时,发现两种药物的渗透性有所增强(见图 16.4)[43]。

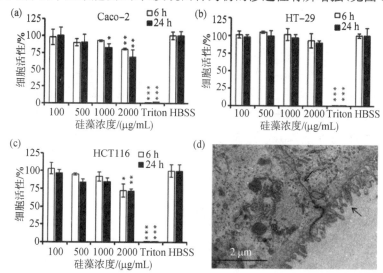

图 16.4 使用 CellTiter Glo® 发光分析法对硅藻进行细胞毒性评估

---

① 指质量分数。

为了采用表面修饰作为控制药物释放的策略,研究了表面功能化对硅藻微胶囊在水不溶性药物吲哚美辛的药物负载和释放中的影响。在这项研究中,测试了两种不同的改性方法:使用有机硅烷3-氨丙基三乙氧基硅烷(APTES)和 N-(3-(三甲氧基硅基)丙基)乙二胺(AEAPTMS),以及磷酸(2-羧乙基磷酸和16-磷酸基十六烷酸),以赋予微胶囊亲水和疏水性质(见图16.3b)[26]。结果显示,适当的硅藻表面功能化能够调节药物负载量(15%~24%)和药物释放(6~15天),从缓慢释放到超过两周的持续释放。值得注意的是,亲水功能化增加了药物负载量并延长了药物释放时间。相反,疏水改性导致了较低的药物负载量和更快的药物释放。类似地,Bariana等[44]使用了甲氧基-聚乙二醇-硅烷(mPEG-硅烷)、7-十八烷基三氯硅烷(OTS)、3-(缩水甘油氧基丙基)三甲氧基硅烷(GPTMS)和两种磷酸[即2-羧乙基磷酸(2 CEPA)和16-磷酸基十六烷酸(16 PHA)],以控制水不溶性药物(吲哚美辛)和水溶性药物(庆大霉素)的药物负载和释放特性。通过对硅藻微粒表面不同官能团的修饰,成功确认了这些功能基团的不同界面性质导致了药物释放时间(13~26天)和负载能力(14%~24%)的差异。结果表明,由于极性羧基、胺基或水解的环氧基团的存在,亲水表面有利于延长吲哚美辛的释放,而由有机烃类修饰的疏水硅藻表面则为庆大霉素提供了延长的持续释放特性[44-45]。

为了探索形成靶向药物输送系统的可能性,Losic等通过用多巴胺修饰的氧化铁纳米颗粒对硅藻进行表面功能化,制造了磁性引导的药物微载体。他们证明了从这种磁化硅藻中获得持续2周的水不溶性药物(吲哚美辛)释放的可能性[22]。

硅藻土也被用于控制其药物载荷的释放。Kumeria等首次通过3-氨丙基三乙氧基硅烷表面功能化,将二维材料氧化石墨烯(GO)附着到硅藻壳上,制备了一种纳米杂化硅藻材料,如图16.5所示。结果显示,这种方法可以显著改变非甾体抗炎药物吲哚美辛在不同pH条件下的释放行为[46]。药物释放行为受加载药物与氧化石墨烯在不同pH缓冲液中化学相互作用变化的影响。在另一种控制硅藻表面药物释放的尝试中,Voelcker等制备了一种依赖于聚合物作为物理致动器来控制药物释放的递送系统,而不是化学相互作用。他们采用表面引发的原子转移自由基聚合(ATRP)技术,将寡聚乙二醇甲基丙烯酸酯[O(EG)MA]的热响应共聚物接枝到硅藻微胶囊表面,以展示所得复合材料在热响应药物输送中的应用。抗菌剂左氧氟沙星从共聚物修饰的微胶囊中的药物释放在临界溶解温度(LCST)以下和以上的释放动力学表现出强烈的温度依赖性,如图16.6所示[47]。

除了化学功能化硅藻二氧化硅表面以实现稳定且高药物负载之外,许多研究强调了聚合物掺入的好处,如增强生物相容性、减少聚集、增强细胞内化和改善水相稳定性。Terracciano等用聚乙烯醇(PEG)和细胞穿透肽(CPP)对硅藻纳米颗粒(DNP)进行涂层,以递送抗癌药物索拉非尼[48]。DNP通过机械粉碎和超声处理微米级硅藻壳获得。之后,使用3-氨丙基三乙氧基硅烷(APT)进行表面功能化,并通过1-乙基-3-[3-二甲氨丙基]氨基碳酸酯/N-羟基琥珀酰亚胺(EDC/NHS)在硅藻表面的

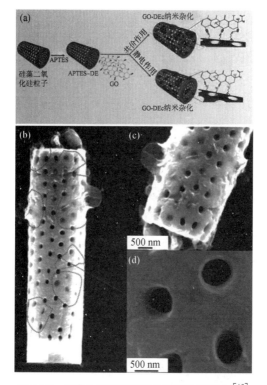

图 16.5 纳米杂化硅藻材料的制备及 SEM 图[46]

(a) 硅藻壳表面共价功能化的示意图；(b) 通过静电附着制备的氧化石墨烯-硅藻纳米杂化材料的 SEM 图，显示完整的硅藻结构并附着的 GO 部分覆盖孔洞；(c) 由较大 GO 片覆盖的硅藻孔洞的高倍 SEM 图；(d) 覆盖有较小 GO 纳米片的硅藻二氧化硅的高倍 SEM 图

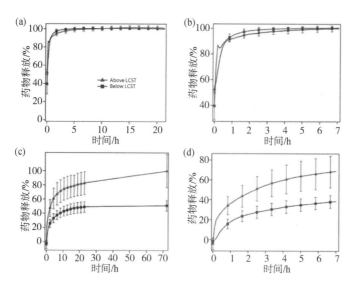

图 16.6 左氧氟沙星在[O(EG)MA]共聚物临界溶液温度以下和以上的硅藻微囊中的药物释放曲线[47]

(a) 总释放量；(b) 未改性微胶囊的前 7 小时释放量；(c) 总释放量；(d) MA 改性微囊的前 7 个小时释放量

APT 氨基(—NH₂)与 PEG 的羧基(—COOH)之间形成酰胺键。最终,CPP 被附着到 PEG 修饰的硅藻纳米颗粒表面,如图 16.7 所示。所制备的纳米颗粒在水溶液中稳定且与红细胞和乳腺癌细胞(MCF-7 和 MDA-MB-231)具有生物相容性。同时,CPP 的掺入显著提高了细胞摄取量。此外,与未涂层 DNP 相比,PEG 涂层使抗癌药物的负载量增加了两倍。

图 16.7　DNP 功能化示意图

(Ⅰ) DNPs-APT 通过 EDC/NHS 的聚乙二醇化,在室温下搅拌过夜;(Ⅱ) DNPs-APT-PEG 通过 EDC/NHS 的 CPP 肽生物偶联,在室温下搅拌过夜;双重生物功能化基于 EDC/NHS 化学促进的纳米颗粒表面与生物分子之间的共价结合

纳米技术的应用为多种治疗药物的递送提供了新的机会,尤其是在癌症治疗方面。基于纳米技术的载体具有靶向递送难溶于水的抗癌药物的能力,并通过被动或主动靶向在肿瘤部位积累[49-50]。Javalkote 等制备了磁性硅藻壳,并装载了姜黄素,然后研究了其化疗活性。研究表明,制备的姜黄素负载微壳可以作为潜在的载体用于高效的化疗[51]。在类似的研究中,Todd 及其合作伙伴证明了与氧化铁纳米颗粒(IONP)联合使用的硅藻作为磁性递送载体的体内应用。当施加磁场时,观察到粒子在肿瘤部位的积累显著增加(比对照组高 6.4 倍)[52]。

小干扰 RNA(siRNA)的递送已被确立为癌症治疗的有效策略。siRNA 可以用于选择性地抑制癌细胞中缺陷基因的表达(基因沉默)。这些缺陷基因主要负责癌细胞的扩散或不受控制的生长,因此基因沉默可能导致癌细胞死亡。然而,siRNA 在到达其作用部位(即癌细胞)之前容易被核酸酶降解。因此,需要将其装载在能够保护它们直到到达目标的载体上。为此,开发了多种纳米载体,如脂质体、聚合物纳米粒子等[53-54]。首次尝试使用硅藻递送 siRNA 是将聚 d-精氨酸肽/siRNA 复合物共价连接到 APTES 表面修饰的硅藻上[55]。在这项研究中,未加载药物的二氧化硅硅藻纳米颗粒在与人表皮癌细胞(H1355)孵育 72 h 后表现出极低的毒性。同时,还展示了有效的 siRNA 递送至细胞质并实现有效的基因沉默活性,如图 16.8 所示。

在一种有趣的方法中,一些研究涉及调控硅藻的基因表达(如基因沉默/过表达)以生成具有独特特性的硅藻种类[56-57]。Delalat 等开发了基因工程改造的硅藻生物硅体,其表面展示了蛋白 G 的 IgG 结合域(GB1-生物硅体),用于递送水不溶性细胞毒性剂[36]。我们使用了两种化疗药物,喜树碱(CPT)及其更强效的衍生物 7-乙基-10-羟基喜树碱(SN38)。它们首先被制备为含有 CPT 的 DOTAP 脂质体和含有

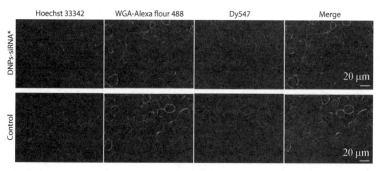

图 16.8 用 siRNA* 修饰的硅藻土纳米载体处理的细胞(第一行)和作为对照的未处理细胞(第二行)的共聚焦显微镜图

SN38 的 CTAB 胶束。为了特异性靶向癌细胞,IgG 抗体[例如,针对 p75 神经营养因子受体(p75NTR)的抗体(抗 p75NTR)]被连接到 GB1-生物硅体上,形成了抗 p75NTR-GB1 生物硅体。

这些化疗药物通过静电作用被加载到抗 p75NTR-GB1 生物硅体上。结果显示,这些基因工程的生物硅纳米颗粒能够特异性地靶向神经母细胞瘤和 B 型淋巴瘤细胞。当将 SN38 载药的抗 p75NTR-GB1 生物硅纳米颗粒注射到携带 SH-SY5Y 神经母细胞瘤的 BALB/c 裸鼠体内时,肿瘤的生长显著减少,如图 16.9 所示。

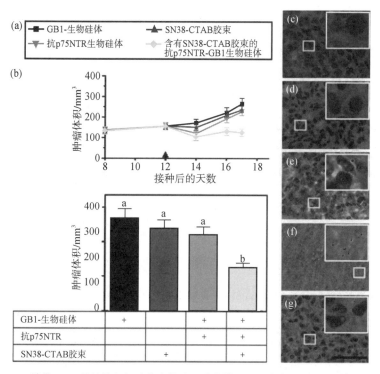

图 16.9 携带 SN38 的抗体标记硅藻生物硅可减少神经母细胞瘤 SH-SY5Y 肿瘤的生长,并对 SH-SY5 Y 异种移植物进行组织化学分析

16.2.4 硅藻作为药物输送应用的可生物降解载体的来源

所有上述研究均表明,硅藻壳能够实现广泛治疗剂的持续和控制释放。因此,使用硅藻作为药物载体的药物输送可能成为一种通用方法。然而,由于硅藻壳的构建单元由二氧化硅纳米粒子组成,它们可能存在一个主要缺点,即在生物体液中降解性差[58-63]。这可能导致二氧化硅颗粒的积累,尤其是在重复给药的情况下,可能会在血液供应有限的器官中(如玻璃体)积聚。此外,大量系统性给药的靶向药物输送系统在到达目标之前会被吞噬系统清除,这增加了意外急性或慢性毒性的可能性[64]。

在一项研究中,Borak 等研究了二氧化硅颗粒在大鼠体内的生物分布。结果显示,二氧化硅颗粒主要被困在肺部的气囊和肾小球中[65]。因此,将二氧化硅硅藻转化为可生物降解的形式(即硅复制品)将具有巨大的优势。这一假设基于以下事实:硅是人体内常见的微量元素,其降解产物硅酸(H_4SiO_4)是人体主要吸收的形式,并且在许多组织中天然存在。此外,先前的研究证实,向人体内注射硅酸后,它将通过肾脏有效排出体外[66-67]。

为了克服二氧化硅的局限性,合成的多孔硅(pSi)似乎是作为药物载体的一种有吸引力的替代方案。然而,值得注意的是,pSi 在其制造过程中存在许多限制,这些限制与其基于硅片的电化学蚀刻工艺有关。首先,硅片的蚀刻是一种耗时的合成过程,因为它涉及多个步骤。其次,由于该工艺的高成本(包括硅片、化学品、相关设备,如蚀刻单元、电源、安全设备等),制造工艺的可扩展性具有挑战性。最后,使用高毒性化学品,如氢氟酸(HF),在处理过程中需要采取特殊的安全预防措施[68]。此外,在 pSi 合成过程中使用 HF 可能最终导致 pSi 微粒中残留毒性,从而限制该系统在实际生物医学应用中的适用性[69-70]。

在 Koynov 等的研究中,他们证明了在使用常见的蚀刻技术制造 pSi 颗粒过程中,会形成残留的有毒硅四氟化物(SiF_4)和氟硅酸(H_2SiF_6)[见式(16.1)和式(16.2)]。这些有毒残留物通常会被困在 pSi 的介孔结构中。在这项研究中,使用 AY-27 大鼠膀胱癌细胞探讨了 pSi 的毒性。结果表明,应用 pSi 纳米颗粒后,细胞的存活率显著下降[71]。Liu 等也得出了类似的结果,当他们研究 pSi 对结直肠癌细胞(HT-29)的影响时,也发现了相似的细胞毒性[72]:

$$Si + 4HF \longrightarrow SiF_4 + 4H^+ \quad (16.1)$$

$$SiF_4 + 2HF \longleftrightarrow H_2SiF_6 (在 H_2O 的存在下) \quad (16.2)$$

此外,pSi 的光子晶体特性已被证明与角度有关,因为 pSi 颗粒的光谱和颜色不仅取决于所载/释放的药物量,还取决赖于光的入射/观察角度。尽管最近的研究已证明 pSi 光子颗粒可以实现体内自我报告的眼内药物输送,但由于眼球玻璃体的流动性和 pSi 的角度依赖性光子特性,这种能力受到了限制。这可能导致通过 pSi 光子晶体的自我报告功能(如反射光谱或干涉颜色)对药物释放量的误判[73-74]。然而,这一限制可以通过开发具有非角度依赖性光学特性的多孔硅结构来克服。

基于上述论述，迫切需要具备先进特性的 pSi，以克服通过电化学蚀刻硅生产的合成 pSi 的固有局限性。新型 pSi 应通过工业上可扩展的、具有成本竞争力和时间效率的制造工艺生产，且不使用有害化学品。此外，这些 pSi 材料应完全可降解且无细胞毒性，以适用于生物医学应用，同时具有非角度依赖性的光致发光特性。最近，在一项前沿研究中，Bao 等通过应用镁热还原反应，将硅藻的二氧化硅壳复制为纯硅壳，用于气体传感器[75-76]。这种基于镁催化剂的还原过程不仅使二氧化硅硅藻壳完全转换为硅藻壳复制体成为可能，而且保留了硅藻壳的结构和形态。这项研究为从这些天然存在的硅藻壳中获取硅颗粒开辟了道路。Losic 课题组首次开发出一种新型自我报告药物纳米载体，该载体基于天然存在的硅藻二氧化硅，具有 3D 生物形态的可降解和发光多孔硅[20,77]。在他们的研究中，探讨了所制造硅颗粒的生物相容性、自我报告能力和药物释放特性。研究者成功实现了将天然硅藻二氧化硅壳转换为用于生物医学应用的具有典型 3D 多孔结构的硅藻壳复制体，这是首次实现的技术，如图 16.10(a)~(c)所示。与原始硅藻相比，硅微粒表现出显著更高的比表面积。这是由于在 SiO_2 转化为 Si 的还原过程中，二氧化硅壳的硅纳米颗粒构建模块体积减少所致。此外，硅复制体表现出高结晶性和改善的降解率。此外，硅颗粒表现出 pH 依赖性降解行为，在酸性 pH 介质中降解受到抑制，而在生理 pH=7.4 条件下降解得到增

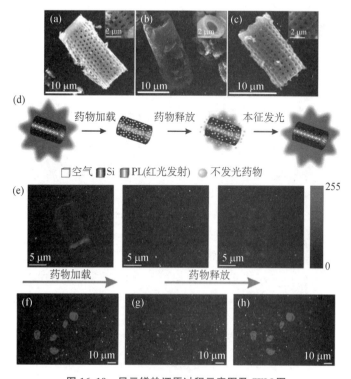

图 16.10 显示镁热还原过程示意图及 SEM 图

强。用于预防增殖性玻璃体视网膜病变(PVR)的硅藻壳载药硅颗粒的体外释放研究显示，药物可持续释放长达30天。此外，硅复制体还可以作为自我报告载体，通过简单的视觉分析监测发光和非发光药物的释放，如图16.10(d)~(h)所示[20]。

研究表明，除了成像和自我报告应用之外，注入体内的多孔硅纳米颗粒会通过增强的渗透和滞留效应(EPR效应)积聚在肿瘤部位，而不是在正常组织或器官中，这一过程与实体肿瘤中存在的渗漏血管和淋巴引流不良有关[49-50]。然而，值得注意的是，使用载药合成pSi纳米载体的理论基础是减少与各种治疗相关的副作用和毒性，使药物在注射后能够持续释放。例如，抗癌药物阿霉素(DOX)的应用受到其狭窄的治疗指数及其副作用的限制，包括骨髓抑制和心脏毒性[78-79]。先前的研究证实，通过在肿瘤部位局部释放DOX，同时减少其在健康组织中的无差别释放，可以显著降低这些不良反应。这可以通过将DOX负载在能够增强其摄取并延长其在肿瘤细胞中释放的药物载体上实现[80-81]。

在这种情况下，我们实验室提出了一个新的概念，即从矿石到癌症治疗的硅藻土应用，其中由硅藻衍生的天然且可降解的硅纳米颗粒(SiNPs)用于阿霉素(DOX)的持续输送[77]。研究结果表明，通过镁热还原、研磨和超声波粉碎的组合，成功生产了由硅藻土微结构衍生的纳米级多孔硅颗粒。将阿霉素载入SiNPs的体外研究显示出长达30天的持续释放，证实了它们作为局部癌症治疗药物输送载体的潜力。药物释放表现出pH依赖性，在酸性环境(pH=5.5)下释放的药物比在生理环境(pH=7.4)下更多。研究发现，释放机制依赖于药物分子从纳米孔硅藻结构中的扩散以及硅结构的降解，这通过降解与释放之间的强相关性得到了证明。此外，这项研究首次证明了这种SiNPs是生物相容的。此外，体外细胞培养研究表明，与相应浓度的游离DOX相比，DOX-SiNPs的细胞毒性作用增强。这些结果证实，使用可扩展的转化工艺可以从廉价且易得的天然资源(硅藻土)中获得硅藻基载体，为设计适用于广泛生物医学应用的先进微/纳米载体提供了一个有前途的替代方案。

最近，我们团队利用微流控技术制备了多功能微球，将硅藻硅纳米颗粒(SiNPs)和磁性纳米颗粒结合起来，用于靶向结直肠癌治疗，如图16.11所示[82]。这些微球通过微流控工艺组装，包含载药的SiNPs和磁性纳米线，并被封装在pH敏感的聚合物基质中。SiNPs和磁性纳米线均载有两种抗癌药物，即5-氟尿嘧啶(5FU)或姜黄素(CUR)。结果显示，这些颗粒能够在所需pH条件下选择性打开，并在外部磁场的作用下靶向定位。在体外使用结肠癌细胞SW480进行测试时，观察到5FU和CUR之间强大的协同抗癌活性，显示了这些混合物在靶向和局部癌症治疗中的巨大潜力。

16.2.5 用于其他生物医学应用的硅藻

1. 组织工程

硅在骨形成和维持中发挥了重要作用，它可以改善成骨细胞的功能并诱导矿化。骨变形和长骨异常经常与二氧化硅/硅缺乏有关[83]。硅藻土是硅藻骨架的自然沉积

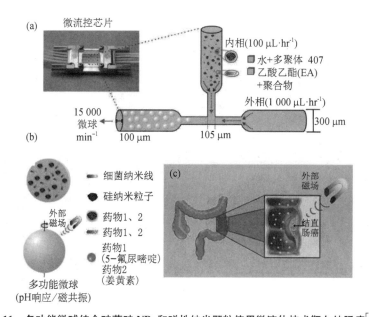

图 16.11 多功能微球结合硅藻硅 NPs 和磁性纳米颗粒使用微流体技术靶向结肠癌[82]

(a) 用于单分散液滴形成的具有 T 形连接几何形状的微流控系统示意图；(b) 含有抗癌药物(5-氟尿嘧啶或姜黄素)负载的硅纳米颗粒(SiNP)和细菌磁性纳米线(BacNWs)；(c) 结直肠癌肠道联合治疗的实际应用

物,一种廉价且丰富的生物硅原材料,可用于再生医学应用。

利用 $Thalassosira weissflogii$ 的硅藻壳来增强小鼠 L 成纤维细胞和人骨肉瘤细胞系 MG63 的粘附和增殖[84]。在细胞生长之前,两种试剂被附着在硅藻表面。第一种是抗生素环丙沙星,通常用于治疗与骨骼或牙科植入物相关的细菌感染；第二种是抗氧化剂环硝基氧化物 2,6,6-四甲基哌啶-N-氧化物(TEMPO),用于清除活性氧物种以避免炎症。通过 APTES(3-氨丙基三乙氧基硅烷)和物理吸附分别实现了 TEMPO 和环丙沙星的附着。细胞在涂有纯硅藻壳(F)或功能化硅藻壳(FT)的玻璃基底上培养,并与未涂层的玻璃基底作为对照进行比较。为了评估细胞活力和发育,使用了 MTT 检测和扫描电子显微镜。结果显示,与对照组相比,在纯硅藻壳或功能化硅藻壳上培养 48 小时和 7 天后,两个细胞类型的细胞活力均显著提高,如图 16.12 所示。

2. 出血控制

未控制的出血可能导致低血容量性休克,如果不迅速通过体液替代和止血处理,可能导致死亡。目前的止血剂如 QuikClot 沸石和 HemCon 壳聚糖绷带存在各种局限性。例如,QuikClot 沸石可能由于高温产生(温度可达 95℃)导致组织烧伤。硅藻被引入作为一种无细胞毒性、无免疫原性且廉价的止血剂,能够克服常用止血剂的缺点[85]。硅藻提供了高血浆吸收性而没有任何热量产生(即与沸石相反)。研究中,两种类型的纯化硅藻骨架,即商购的硅藻土和实验室培养的硅藻,用不同浓度的壳聚糖(0.5%、1%、3% 和 5%)进行涂层处理,然后进行了体外溶血和体内凝血活性测试。结果显示,与未涂层硅藻相比,壳聚糖涂层导致的溶血活性可以忽略不计。另一方

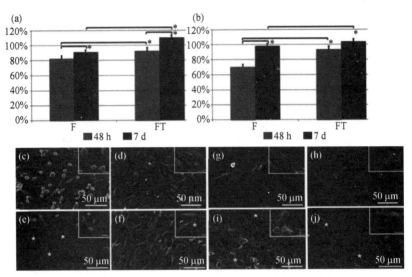

图 16.12　培养 48 小时和 7 天后,将 L 成纤维细胞和 MG63 类成骨细胞接种到 F 涂层和 FT 涂层玻片上进行 MTT 分析的结果

面,当用壳聚糖涂层硅藻测试其在大鼠尾巴截肢后的止血能力时,获得了最短的凝血时间(98.34 s±26.54 s),相比之下,纱布的凝血时间为 510.26 s±63.22 s,商业 QuikClot 沸石的凝血时间为 133.66 s±21.84 s。此外,观察大鼠尾巴也显示,与纱布和 QuikClot 沸石相比,壳聚糖涂层硅藻没有导致大的血栓,如图 16.13 所示。

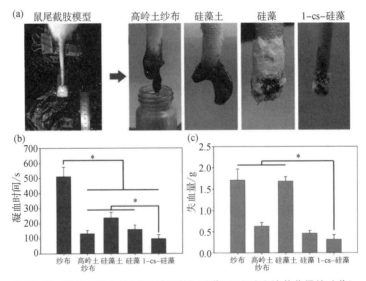

图 16.13　来自硅藻土(即商业购买的)、硅藻(即实验室培养获得的硅藻)、壳聚糖(1%)涂层硅藻(1-CS-硅藻)、纱布和商业 Quikclot 沸石中鼠尾截肢模型的数据[85]

(a) 与伤口接触的止血效果照片;(b) 凝血时间;(c) 失血数据表示平均值±SD($n=6$)

16.3 结论

硅藻凭借其在自然环境下形成的独特三维硅结构,为开发与合成多孔硅材料性能极为相似的新一代生物材料提供了蓝图。硅藻骨架具有独特的分层多孔结构、高比表面积、可定制的表面修饰、高生物相容性、化学稳定性以及特定的光学和光致发光特性,因此被广泛应用于低成本的生物医学支架。本章中描述了硅藻作为整体骨架或还原为微/纳米颗粒的药物输送应用的最新研究成果,重点介绍了硅藻载体的制备、表面修饰、药物装载及其将治疗分子运输至细胞内的能力。研究显示,将硅藻硅转化为可降解的微纳米载体是一种有前景和启发性的策略,可以进一步探索硅藻材料作为低成本且有价值的替代品,用于制备新一代自然生成的药物输送系统。

然而,要将硅藻作为药物输送载体用于临床应用,仍有相当多的工作要做。这包括在动物和临床前模型中,通过体内条件下的研究来评估硅藻药物输送系统的有效性和生物相容性。尽管硅藻土已经获得美国环保署(EPA)、美国农业部(USDA)和美国食品药品监督管理局(FDA)的批准,可作为家畜饲料中的抗结块剂和化学杀虫剂的成分,但硅藻材料在药物领域的应用尚未获得监管批准。只有通过广泛的生物学特征研究,硅藻材料才能在监管途径上获得批准。

参考文献

[1] Natarajan, J. V., Nugraha, C., Ng, X. W., Venkatraman, S. (2014). Sustained-release from nanocarriers: a review, Journal of Controlled Release, 193, 122-138.
[2] Wang, Y., Santos, A., Evdokiou, A., Losic, D. (2015b). An overview of nanotoxicity and nano medicine research: principles, progress and implications for cancer therapy, J. Mater. Chem. B, 3(36), 7153-7172.
[3] Yan, L., Chen, X. (2014). Nanomaterials for Drug Delivery. In Nanocrystalline Materials. Second Edition. (ed.). S Tjong. Oxford: Elsevier. pp. 221-268.
[4] LaVan, D. A., McGuire, T., Langer, R. (2003). Small-scale systems for in vivo drug delivery, Nat. Biotechnol., 21(10), 1184-1191.
[5] Torchilin, V. P. (2005). Recent advances with liposomes as pharmaceutical carriers, Nat. Rev. Drug Discov., 4(2), 145-160.
[6] Farokhzad, O., Langer, R. (2006). Nanomedicine: Developing smarter therapeutic and diagnostic modalities, Advanced Drug Delivery Reviews, 58(14), 1456-1459.
[7] Nokhodchi, A., Tailor, A. (2004). In situ cross-linking of sodium alginate with calcium and aluminum ions to sustain the release of theophylline from polymeric matrices, Il Farmaco, 59(12), 999-1004.
[8] Wenzel, J., Balaji, K. S. S., Koushik, K., Navarre, C., Duran, S. H., Rahe, C. H. (2002). Pluronic® F127 gel formulations of Deslorelin and GnRH reduce drug degradation and sustain drug release and effect in cattle, Journal of Controlled Release, 85(1-3), 51-59.
[9] de Jesus, M. B., Zuhorn, I. S. (2015). Solid lipid nanoparticles as nucleic acid delivery system: Properties and molecular mechanisms, Journal of Controlled Release, 201, 1-13.
[10] M. Bimbo, L., Peltonen, L., Hirvonen, J., A. Santos, H. (2012). Toxicological profile of therapeutic nanodelivery systems, CDM, 13(8), 1068-1086.
[11] Herranz-Blanco, B., Liu, D., Mäkilä, E., Shahbazi, M.-A., Ginestar, E., Zhang, H., et al. (2015). Drug Delivery: On-Chip Self-Assembly of a Smart Hybrid Nanocomposite for Antitumoral Applications (Adv. Funct. Mater. 10/2015, Adv. Funct. Mater., 25(10), 1612-1612.
[12] Russo, L., Colangelo, F., Cioffi, R., Rea, I., Stefano, L. D. (2011). A Mechanochemical Approach to

Porous Silicon Nanoparticles Fabrication, Materials, 4(6), 1023-1033.
[13] Sailor, M. J., Park, J.-H. (2012). Hybrid nanoparticles for detection and treatment of cancer, Adv. Mater., 24(28), 3779-3802.
[14] Muhammad, F., Guo, M., Qi, W., Sun, F., Wang, A., Guo, Y., et al. (2011). pH-Triggered controlled drug release from mesoporous silica nanoparticles via intracelluar dissolution of ZnO nanolids, J. Am. Chem. Soc., 133(23), 8778-8781.
[15] Zhao, Y., Trewyn, B. G., Slowing, I. I., Lin, V. S.-Y. (2009). Mesoporous silica nanoparticle-based double drug delivery system for glucose-responsive controlled release of insulin and cyclic AMP, J. Am. Chem. Soc., 131(24), 8398-8400.
[16] Losic, D., Mitchell, J. G., Voelcker, N. H. (2009). Diatomaceous Lessons in Nanotechnology and Advanced Materials, Adv. Mater., 21(29), 2947-2958.
[17] Sumper, M., Brunner, E. (2006). Learning from Diatoms: Nature's Tools for the Production of Nanostructured Silica, Adv. Funct. Mater., 16(1), 17-26.
[18] Round, F. E., Crawford, R. M., Mann, D. G. (1990). Diatoms: Biology and Morphology of the Genera. Cambridge University Press.
[19] Losic, D., Pillar, R. J., Dilger, T., Mitchell, J. G., Voelcker, N. H. (2007). Atomic force microscopy (AFM) characterisation of the porous silica nanostructure of two centric diatoms, J. Porous Mater., 14(1), 61-69.
[20] Maher, S., Alsawat, M., Kumeria, T., Fathalla, D., Fetih, G., Santos, A., et al. (2015). Luminescent Silicon Diatom Replicas: Self-Reporting and Degradable Drug Carriers with Biologically Derived Shape for Sustained Delivery of Therapeutics, Adv. Funct. Mater., 25(32), 5107-5116.
[21] Rea, I., Terracciano, M., De Stefano, L. (2017). Synthetic vs Natural: Diatoms Bioderived Porous Materials for the Next Generation of Healthcare Nanodevices, Adv. Healthc. Mater., 6(3), 1601125.
[22] Losic, D., Yu, Y., Aw, M. S., Simovic, S., Thierry, B., Addai-Mensah, J. (2010). Surface functionalisation of diatoms with dopamine modified iron-oxide nanoparticles: toward magnetically guided drug microcarriers with biologically derived morphologies, Chem. Commun., 46(34), 6323-6325.
[23] Howarter, J. A., Youngblood, J. P. (2006). Optimization of Silica Silanization by 3-Aminopropyltriethoxysilane, Langmuir, 22(26), 11142-11147.
[24] De Stefano, L., Oliviero, G., Amato, J., Borbone, N., Piccialli, G., Mayol, L., et al. (2013). Aminosilane functionalizations of mesoporous oxidized silicon for oligonucleotide synthesis and detection, Journal of The Royal Society Interface, 10(83), 20130160.
[25] Terracciano, M., Rea, I., Politi, J., De Stefano, L. (2013). Optical characterization of aminosilanemodified silicon dioxide surface for biosensing, 8.
[26] Aw, M. S., Bariana, M., Yu, Y., Addai-Mensah, J., Losic, D. (2013). Surface-functionalized diatom microcapsules for drug delivery of water-insoluble drugs, J. Biomater. Appl., 28(2), 163-174.
[27] Rea, I., Terracciano, M., Chandrasekaran, S., Voelcker, N. H., Dardano, P., Martucci, N. M., et al. (2016). Bioengineered Silicon Diatoms: Adding Photonic Features to a Nanostructured Semiconductive Material for Biomolecular Sensing, Nanoscale Res. Lett., 11(1), 405.
[28] Ferrari, M. (2005). Cancer nanotechnology: opportunities and challenges, Nat. Rev. Cancer, 5(3), 161-171.
[29] Wagner, V., Dullaart, A., Bock, A.-K., Zweck, A. (2006). The emerging nanomedicine landscape, Nat. Biotechnol., 24(10), 1211-1217.
[30] El-Aneed, A. (2004). An overview of current delivery systems in cancer gene therapy, Journal of Controlled Release, 94(1), 1-14.
[31] Erhardt, P. W., Proudfoot, J. R. (2007). 1.02-Drug Discovery: Historical Perspective, Current Status, and Outlook A2-Taylor, John B. In Comprehensive Medicinal Chemistry II. (eds.). D.J Triggle. Oxford: Elsevier. pp. 29-96.
[32] van der Meel, R., Vehmeijer, L. J. C., Kok, R. J., Storm, G., van Gaal, E. V. B. (2013). Ligand-targeted particulate nanomedicines undergoing clinical evaluation: Current status, Advanced Drug Delivery Reviews, 65(10), 1284-1298.
[33] Ezzati Nazhad Dolatabadi, J., Omidi, Y., Losic, D. (2011). Carbon Nanotubes as an Advanced Drug and Gene Delivery Nanosystem, CNANO, 7(18), 297-314.
[34] Shahbazi, M.-A., Herranz, B., Santos, H. A. (2012). Nanostructured porous Si-based nanoparticles for

targeted drug delivery, Biomatter, 2(4), 296-312.
- [35] Tran, P. A., Zhang, L., Webster, T. J. (2009). Carbon nanofibers and carbon nanotubes in regenerative medicine, Advanced Drug Delivery Reviews, 61(12), 1097-1114.
- [36] Delalat, B., Sheppard, V. C., Rasi Ghaemi, S., Rao, S., Prestidge, C. A., McPhee, G., et al. (2015). Targeted drug delivery using genetically engineered diatom biosilica, Nat. Commun., 6(1).
- [37] Losic, D., Mitchell, J. G., Voelcker, N. H. (2009). Diatomaceous Lessons in Nanotechnology and Advanced Materials, Adv. Mater., 21(29), 2947-2958.
- [38] Anderson, M. W., Holmes, S. M., Hanif, N., Cundy, C. S. (2000). Hierarchical pore structures through diatom zeolitization, Angewandte Chemie, 112(15), 2819-2822.
- [39] Morse, D. E. (1999). Silicon biotechnology: harnessing biological silica production to construct new materials, Trends in Biotechnology, 17(6), 230-232.
- [40] Rosi, N. L., Thaxton, C. S., Mirkin, C. A. (2004). Control of Nanoparticle Assembly by Using DNA-Modified Diatom Templates, Angew. Chem. Int. Ed., 43(41), 5500-5503.
- [41] Aw, M. S., Simovic, S., Addai-Mensah, J., Losic, D. (2011). Silica microcapsules from diatoms as new carrier for delivery of therapeutics, Nanomedicine, 6(7), 1159-1173.
- [42] Aw, M. S., Simovic, S., Yu, Y., Addai-Mensah, J., Losic, D. (2012). Porous silica microshells from diatoms as biocarrier for drug delivery applications, Powder Technology, 223, 52-58.
- [43] Zhang, H., Shahbazi, M.-A., Mäkilä, E. M., da Silva, T. H., Reis, R. L., Salonen, J. J., et al. (2013). Diatom silica microparticles for sustained release and permeation enhancement following oral delivery of prednisone and mesalamine, Biomaterials, 34(36), 9210-9219.
- [44] Bariana, M., Aw, M. S., Kurkuri, M., Losic, D. (2013). Tuning drug loading and release properties of diatom silica microparticles by surface modifications, International Journal of Pharmaceutics, 443(1-2), 230-241.
- [45] Bariana, M., Aw, M. S., Losic, D. (2013). Tailoring morphological and interfacial properties of diatom silica microparticles for drug delivery applications, Advanced Powder Technology, 24(4), 757-763.
- [46] Kumeria, T., Bariana, M., Altalhi, T., Kurkuri, M., Gibson, C. T., Yang, W., et al. (2013). Graphene oxide decorated diatom silica particles as new nano-hybrids: towards smart natural drug microcarriers, J. Mater. Chem. B, 1(45), 6302-6311.
- [47] Vasani, R. B., Losic, D., Cavallaro, A., Voelcker, N. H. (2015). Fabrication of stimulus-responsive diatom biosilica microcapsules for antibiotic drug delivery, J. Mater. Chem. B, 3(21), 4325-4329.
- [48] Terracciano, M., Shahbazi, M.-A., Correia, A., Rea, I., Lamberti, A., De Stefano, L., et al. (2015). Surface bioengineering of diatomite based nanovectors for efficient intracellular uptake and drug delivery, Nanoscale, 7(47), 20063-20074.
- [49] Maeda, H., Wu, J., Sawa, T., Matsumura, Y., Hori, K. (2000). Tumor vascular permeability and the EPR effect in macromolecular therapeutics: a review, Journal of Controlled Release, 65(1-2), 271-284.
- [50] Maeda, H. (2001). The enhanced permeability and retention (EPR) effect in tumor vasculature: the key role of tumor-selective macromolecular drug targeting, Advances in Enzyme Regulation, 41(1), 189-207.
- [51] Javalkote, V. S., Pandey, A. P., Puranik, P. R., Deshmukh, P. K. (2015). Magnetically responsive siliceous frustules for efficient chemotherapy, Materials Science and Engineering: C, 50, 107-116.
- [52] Todd, T., Zhen, Z., Tang, W., Chen, H., Wang, G., Chuang, Y.-J., et al. (2014). Iron oxide nanoparticle encapsulated diatoms for magnetic delivery of small molecules to tumors, Nanoscale, 6(4), 2073-2076.
- [53] Gomes, M. J., Kennedy, P. J., Martins, S., Sarmento, B. (2017). Delivery of siRNA silencing P-gp in peptide-functionalized nanoparticles causes efflux modulation at the blood—brain barrier, Nanomedicine, 12(12), 1385-1399.
- [54] Yu-Wai-Man, C., Tagalakis, A. D., Manunta, M. D., Hart, S. L., Khaw, P. T. (2016). Receptortargeted liposome-peptide-siRNA nanoparticles represent an efficient delivery system for MRTF silencing in conjunctival fibrosis, 6, 21881.
- [55] Rea, I., Martucci, N. M., De Stefano, L., Ruggiero, I., Terracciano, M., Dardano, P., et al. (2014). Diatomite biosilica nanocarriers for siRNA transport inside cancer cells, Biochimica et Biophysica Acta (BBA)-General Subjects, 1840(12), 3393-3403.
- [56] De Riso, V., Raniello, R., Maumus, F., Rogato, A., Bowler, C., Falciatore, A. (2009). Gene silencing in the marine diatom Phaeodactylum tricornutum, Nucleic Acids Research, 37(14), e96-e96.

[57] Lavaud, J., Materna, A. C., Sturm, S., Vugrinec, S., Kroth, P. G. (2012). Silencing of the Violaxanthin De-Epoxidase Gene in the Diatom Phaeodactylum tricornutum Reduces Diatoxanthin Synthesis and Non-Photochemical Quenching, PLoS ONE, 7(5), e36806.

[58] Cauda, V., Schlossbauer, A., Bein, T. (2010). Bio-degradation study of colloidal mesoporous silica nanoparticles: Effect of surface functionalization with organo-silanes and poly(ethylene glycol, Microporous and Mesoporous Materials, 132(1-2), 60-71.

[59] Hao, N., Liu, H., Li, L., Chen, D., Li, L., Tang, F. (2012). In Vitro Degradation Behavior of Silica Nanoparticles Under Physiological Conditions, j. nanosci. nanotechnol., 12(8), 6346-6354.

[60] Lauwers, A. M., Heinen, W. (1974). Bio-degradation and utilization of silica and quartz, Arch. Microbiol., 95(1), 67-78.

[61] Martinez, J. O., Chiappini, C., Ziemys, A., Faust, A. M., Kojic, M., Liu, X., et al. (2013). Engineering multi-stage nanovectors for controlled degradation and tunable release kinetics, Biomaterials, 34(33), 8469-8477.

[62] Neethirajan, S., Gordon, R., Wang, L. (2009). Potential of silica bodies (phytoliths) for nanotechnology, Trends in Biotechnology, 27(8), 461-467.

[63] Vrieling, E. G., Sun, Q., Beelen, T. P. M., Hazelaar, S., Gieskes, W. W. C., Van Santen, R. A., et al. (2005). Controlled silica synthesis inspired by diatom silicon biomineralization, J. Nanosci. Nanotech., 5(1), 68-78.

[64] Park, J.-H., Gu, L., von Maltzahn, G., Ruoslahti, E., Bhatia, S. N., Sailor, M. J. (2009). Biodegradable luminescent porous silicon nanoparticles for in vivo applications, Nature Mater., 8(4), 331-336.

[65] Borak, B., Biernat, P., Prescha, A., Baszczuk, A., Pluta, J. (2012). In vivo study on the biodistribution of silica particles in the bodies of rats, Advances in Clinical and Experimental Medicine, 21(1), 13-18.

[66] Anderson, S. H. C., Elliott, H., Wallis, D. J., Canham, L. T., Powell, J. J. (2003). Dissolution of different forms of partially porous silicon wafers under simulated physiological conditions, phys. stat. sol., 197(2), 331-335.

[67] Popplewell, J. F., King, S. J., Day, J. P., Ackrill, P., Fifield, L. K., Cresswell, R. G., et al. (1998). Kinetics of uptake and elimination of silicic acid by a human subject: A novel application of 32Si and accelerator mass spectrometry, Journal of Inorganic Biochemistry, 69(3), 177-180.

[68] Guo, M., Zou, X., Ren, H., Muhammad, F., Huang, C., Qiu, S., et al. (2011). Fabrication of high surface area mesoporous silicon via magnesiothermic reduction for drug delivery.

[69] Microporous and Mesoporous Materials, 142(1), 194-201.

[70] Vallet-Regí, M., Ruiz-González, L., Izquierdo-Barba, I., González-Calbet, J. M. (2006). Revisiting silica based ordered mesoporous materials: medical applications, J. Mater. Chem., 16(1), 26-31.

[71] Vallet-Regí, M., Izquierdo-Barba, I., Rámila, A., Pérez-Pariente, J., Babonneau, F., GonzálezCalbet, J. M. (2005). Phosphorous-doped MCM-41 as bioactive material, Solid State Sciences, 7(2), 233-237.

[72] Koynov, S., Pereira, R. N., Crnolatac, I., Kovalev, D., Huygens, A., Chirvony, V., et al. (2011). Purification of Nano-Porous Silicon for Biomedical Applications, Adv. Eng. Mater., 13(6), B225-B233.

[73] Liu, D., Zhang, H., Herranz-Blanco, B., Mäkilä, E., Lehto, V.-P., Salonen, J., et al. (2014). Microfluidic Assembly of Monodisperse Multistage pH-Responsive Polymer/Porous Silicon Composites for Precisely Controlled Multi-Drug Delivery, Small, 10(10), 2029-2038.

[74] Diener, J., Shen, Y. R., Kovalev, D. I., Polisski, G., Koch, F. (1998). Two-photon-excited photoluminescence from porous silicon, Phys. Rev. B, 58(19), 12629-12632. Dorvee, J. R., Derfus, A. M., Bhatia, S. N., Sailor, M. J. (2004). Manipulation of liquid droplets using amphiphilic, magnetic one-dimensional photonic crystal chaperones, Nature Mater., 3(12), 896-899.

[75] Bao, Z., Weatherspoon, M. R., Shian, S., Cai, Y., Graham, P. D., Allan, S. M., et al. (2007). Chemical reduction of three-dimensional silica micro-assemblies into microporous silicon replicas, Nature, 446(7132), 172-175.

[76] Bao, Z., Ernst, E. M., Yoo, S., Sandhage, K. H. (2009). Syntheses of Porous Self-Supporting MetalNanoparticle Assemblies with 3D Morphologies Inherited from Biosilica Templates (Diatom Frustules, Adv. Mater., 21(4), 474-478.

[77] Maher, S., Kumeria, T., Wang, Y., Kaur, G., Fathalla, D., Fetih, G., et al. (2016). From The Mine to Cancer Therapy: Natural and Biodegradable Theranostic Silicon Nanocarriers from Diatoms for Sustained

Delivery of Chemotherapeutics, Adv. Healthcare Mater., 5(20), 2667-2678.

[78] Shalviri, A., Raval, G., Prasad, P., Chan, C., Liu, Q., Heerklotz, H., et al. (2012). pH-Dependent doxorubicin release from terpolymer of starch, polymethacrylic acid and polysorbate 80 Diatom Silica for Biomedical Applications Journal of Pharmaceutics and Biopharmaceutics, 82(3), 587-597.

[79] Thorn, C. F., Oshiro, C., Marsh, S., Hernandez-Boussard, T., McLeod, H., Klein, T. E., et al. (2011). Doxorubicin pathways: pharmacodynamics and adverse effects, Pharmacogenet. Genomics, 21(7), 440-446.

[80] Osminkina, L. A., Tamarov, K. P., Sviridov, A. P., Galkin, R. A., Gongalsky, M. B., Solovyev, V. V., et al. (2012). Photoluminescent biocompatible silicon nanoparticles for cancer theranostic applications, J. Biophoton., 5(7), 529-535.

[81] Wang, C.-F., Sarparanta, M. P., Mäkilä, E. M., Hyvönen, M. L. K., Laakkonen, P. M., Salonen, J. J., et al. (2015a). Multifunctional porous silicon nanoparticles for cancer theranostics, Biomaterials, 48(0), 108-118.

[82] Maher, S., Santos, A., Kumeria, T., Kaur, G., Lambert, M., Forward, P., et al. (2017). Multifunctional microspherical magnetic and pH responsive carriers for combination anticancer therapy engineered by droplet-based microfluidics, J. Mater. Chem. B, 5(22), 4097-4109.

[83] Le, T. D. H., Bonani, W., Speranza, G., Sglavo, V., Ceccato, R., Maniglio, D., et al. (2016). Processing and characterization of diatom nanoparticles and microparticles as potential source of silicon for bone tissue engineering, Materials Science and Engineering: C, 59, 471-479.

[84] Cicco, S. R., Vona, D., De Giglio, E., Cometa, S., Mattioli-Belmonte, M., Palumbo, F., et al. (2015). Chemically Modified Diatoms Biosilica for Bone Cell Growth with Combined Drug-Delivery and Antioxidant Properties, ChemPlusChem, 80(7), 1104-1112.

[85] Feng, C., Li, J., Wu, G. S., Mu, Y. Z., Kong, M., Jiang, C. Q., et al. (2016). Chitosan-Coated Diatom Silica as Hemostatic Agent for Hemorrhage Control, ACS Appl. Mater. Interfaces, 8(50), 34234-34243.

第 17 章
Diafuel™（硅藻生物燃料）与电动汽车的基本比较：使印度能源独立的高潜力可再生能源

Vandana Vinayak，Khashti Ballabh Joshi，Priyangshu Manab Sarma

17.1 简介

印度与碳排放相关产生的空气污染日益严重，导致现任政府支持销售电动汽车[1-3]。我们认为，这一决定可能为时过早，会适得其反。它还抑制了生物燃料作为汽油替代品的研究[4-8]。在替代生物燃料来源中，藻类有丰富的油含量，在藻类分类中，硅藻是很有前途的候选者[9]。因此，我们将比较硅藻生物燃料的前景和汽车供电的前景，假设最终将以某种方式发生完全更换。

从硅藻中提取的生物燃料在这里被命名为 Diafuel™[10]（见图 17.1）。硅藻属于硅藻纲[11]，自侏罗纪时代以来一直存在[12]，占地球原油的 30%[13]。它们还提供了我们呼吸的 20% 的 O_2，在光合作用过程中，它们在隔离 25% 的大气溶解性 CO_2 方面发挥了主要作用[14]。如果能从活硅藻中提取石油，那么在不久的将来，活硅藻可能会成为纳米生物工厂[15]。与农业植物不同，硅藻生长和繁殖迅速，每公顷土地产生 10 倍以上的油脂，理论估计是大豆的 200 倍[16-18]。Ramachandra 等[19]提出，在不损伤细胞的情况下从硅藻中榨油不仅可以解决印度的能源危机，而且还可以解决全球的能源危机。

图 17.1 Diafuel™ 的商标形象
这张图片描绘了一个具有代表性的硅藻舟形藻 (*Fistulifera saprophila*)，它具有榨取油的潜力，因为获得的油来自硅藻，因此取名为 Diafuel™ 的生物燃料

硅藻几乎可以在具有阳光、水、CO_2 和少量营养物质的任何地方生长[20-22]。大多数硅藻的最关键的条件通常是温度在 18～25℃ 之间。Diafuel™[23]也几乎没有硫，因此减少了二氧化硫的排放[24-29]。

17.2 关于温室气体排放(GHG)与 CO_2 和温度关系的探讨

《巴黎协定》试图将全球变暖温度减缓到 2℃ 以下,并努力在本世纪末将气温上升限制在 1.5℃(见图 17.2)[30-31]。该协议于 2016 年 11 月生效,当时提交了 160 份国家意向捐款(INDC),代表联合国气候变化框架公约 195 个缔约方中的 187 份。然而,美国最近退出了计划,表示将停止向绿色气候基金捐款。自 2013 年以来,发达国家已向该基金承诺 100 亿美元,帮助低收入国家应对所谓的气候变化影响[32]。

这些相互冲突的政策是由于正在进行的关于人为产生 CO_2 是否导致全球变暖[33-34],全球变冷[34],或者是最重要的人为气候因素[35-36]的辩论导致的。不同的工作人员进行了三项不同的研究,绘制了温差与 CO_2 浓度的关系图,范围从基本成比例到完全不成比例(见图 17.2)。

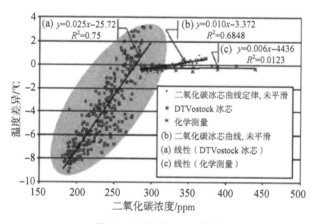

图 17.2 温差与 CO_2 浓度

另一方面,对地质年代大气 CO_2 的评估表明,大约 5.5 亿年前古时代的 CO_2 水平远高于现在。根据 Berner 和 Kothavala[37] 的说法,与现在相比,最高的 26 倍 CO_2 水平发生在冰河时期,或者根据 Royer 等[38] 的说法,比现在高 20 倍,这进一步支持了 CO_2 不是驱动地球表面温度的结论。最近二十年的"变暖间歇期"已经在精神上被逆转,各种解释表明长期变暖趋势正在发生[39-40]。

Vostok 冰芯数据记录了 42 万年前的环境条件,如图 17.3 所示,表明通过 4 个冰期-间冰期旋回,CO_2 和温度之间存在重复的模式相关性。它清楚地描述了在每一次冰川间冰期过渡中,CO_2 的上升滞后于 400 到 1 000 年的温度变化,从而表明温度和 CO_2 之间的关系与假设的完全相反。除 CO_2 外,还有其他因素被提出影响全球温度,如太阳活动和云量或海洋环流的变化[41-42]。

在所有三个冰期-间冰期过渡过程中,CO_2 浓度的上升滞后于温度变化 400~

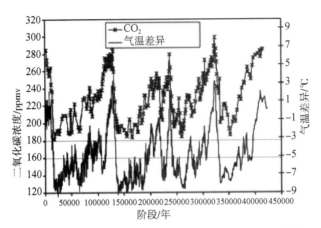

图 17.3　Vostok 冰芯数据通过四个冰期-间冰期旋回在过去 42 万年中 CO_2 浓度和温差之间的相关性

1 000 年,因此,这表明温度和 CO_2 之间的关系似乎与传统气候模型研究中假设的完全相反。很明显,在自然过程中,温度首先上升,然后大气 CO_2 值增加[43]。

Lashof 等[44-45]计算了甲烷、一氧化碳、氧化亚氮和氯氟烃的全球变暖潜力,估计在 90 年代 80% 的全球变暖是由 CO_2 造成的,而 80 年代是 57%。Solomon 等[46]预测,如果大气 CO_2 浓度从 385/百万体积(ppmv)增加到 450~600 ppmv 的峰值,就会出现不可逆转的旱季,伴随着降雨减少和海平面上升。然而,彼得·考克斯发现,在"一切如常"的场景下,陆地生物圈充当碳汇;可能到 2050 年变成来源[47]。他报告说:"到 2100 年,海洋每年吸收 5 Gt C 的速率被陆地碳源平衡,在我们完全耦合的模拟中,大气 CO_2 浓度比非耦合的碳模型高 250 ppmv,导致全球平均变暖 5.5 K,而没有碳循环反馈时为 4 K。"

与此同时,尽管辩论仍在进行,但印度等国家需要就如何最好地满足其人口日益增长的能源需求做出合理的决定。由于印度代表最近决定强调电动汽车,以减少碳排放,我们的目标是将其与硅藻生物燃料的前景进行比较。

17.3　2015 年巴黎协议的结果

联合国气候科学小组警告说,我们必须在 2070 年实现净零 CO_2 排放,以避免危险的全球变暖[48],因为可以排放到大气中而不损害大气的 CO_2 总量被认为是有限的[49-50]。因此,从地球物理的角度来看,我们需要实现总 CO_2 的零排放[51-52]。巴黎协议设定了控制 CO_2 排放和温度升高的目标,如图 17.4 所示。

政府间气候变化专门委员会(IPCC)预计,到 2030 年,每年 CO_2 排放量将达到 550 亿公吨。无政策基准情景设定为到 2030 年每年减少约 9 Gt CO_2-eq,与当前政策情景估计相比,每年减少约 4(2~8)Gt CO_2,仅仅为当前政策减少量的一半值。问题仍然是全球气温升高能否被限制在 2℃ 以下,政策制定者认为这一可能性不超过 66%。如图 17.4(b)所示,一些情景显示 2030 年后 CO_2 排放量将快速下降,到 2060 年和 2080 年,全球 CO_2 排放量将接近于零[53]。

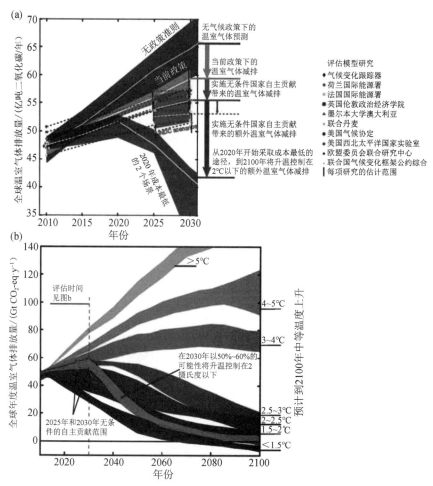

图 17.4 巴黎协定设定的控制 CO_2 排放和温度升高的目标[59]

这里值得一提的是,2013 年印度的煤炭产量为 3.4 亿吨煤当量(Mtce)[54],计划到 2020 年增加到 15 亿吨[55]。继中国、美国和欧盟之后,印度是世界上第四大 CO_2 排放国[56],CO_2 排放总量为 25.97 亿吨($MtCO_2$)。然而在 2015 年,与卡塔尔(39.74 吨)、美国(16.07 吨)和中国(7.73 吨)相比,印度的人均排放量相对较低,为 1.9 吨/人[57-58]。

17.4 印度的能源需求

图 17.5(a)显示了过去几十年印度的能源需求是如何增长的。印度的 CO_2 人均

排放量从 1990 年的 0.71 公吨增加到 2016 年的 1.73 公吨[60]。人均用电量几乎增加了 3 倍,从 1990 年的 273 千瓦时增加到 2016 年的 806 千瓦时。如图 17.5(b)所示,煤炭发电量从 1971 年的 49% 增加到 2014 年的 75%。如图 17.5(c)所示,印度国内生产总值(GDP)以每年 7% 的速度增长,因此,到 2040 年,印度需要近 800 千兆瓦的额外电力容量。如图 17.5(d)所示,新政策情景下印度能源系统的变化预示着不同能源部门的一些显著变化,该图预测:

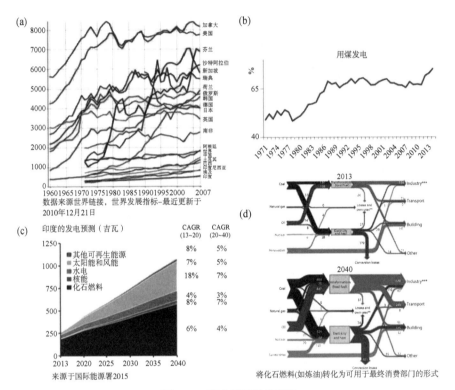

图 17.5 印度能源需求数据

煤炭在该组合中保持着中心地位,在一次能源中的总体份额从 2013 年的 44% 上升到 2040 年的 49%(与全球趋势相反,煤炭下降了 4 个百分点至 25%)[61],石油和天然气的份额小幅上升。然而,最大的一些变化是对非化石燃料的使用。一方面,主要用于烹饪的固体生物质的比例已经从 2013 年的一次能源的近四分之一下降到 2040 年的 11%;但是,另一方面,以太阳能和风能为主导的现代可再生技术的部署有了强劲的增长[62]。印度表示,"气候变化是一个重大的全球挑战,但它不是我们创造的,而是由化石燃料推动的工业时代的繁荣和进步所导致的全球变暖的结果"[63-64]。因此,为了实现这一承诺,印度的煤炭政策将被可再生资源取代,到 2022 年提高可再生能源产能 175 亿瓦(GW)的目标[65]。然而,这并不是一个巨大的成就,因为到 2040 年,印度对化石燃料的需求将上升,将成为世界上最大的石油消费国和进口国;近 980 万桶(mmb/d)的石油产品(乙烷、液化石油气、石脑油、汽车汽油、煤油、柴油、燃油和其他产品)和 700 万桶的原油进口。

需要注意的是,如果印度人的人均能源使用量与加拿大人和美国人一样多,他们的人均能源使用量将增加约 16 倍(见图 17.5)。印度越来越多的煤炭使用导致了全球温室气体排放(GHG)。因此,我们必须看看是否能够利用低净碳排放的来源来逆转这种能源需求。不同的化石燃料为产生相同的能量释放不同水平的 CO_2。例如,燃烧煤炭产生的 CO_2 是天然气中同等能量的两倍[66]。主要的 CO_2 排放是人为的,来自化石燃料燃烧和水泥制造[67-68]。

燃煤电厂和燃料运营车辆排放的 CO_2 是空气污染的主要问题。到 2050 年,印度的人口预计将增加到 15.93 亿,到 2050 年将超过中国(13.92 亿)[69]。这导致了车辆数量的增加,达到近 5 亿辆,主要使用汽油/柴油。印度政府决定到 2021 年将原油在印度能源生产份额中的所占比例从目前的 12% 降到 7%,取而代之的是天然气(到 2021 年从目前的 13% 增到 16%)、核能(到 2021 年从 2.5% 增到 5%)和可再生能源(从 1.5% 增到 3%)[70]。如果印度不转向替代能源资源来运行其车辆,预测不断增加的原油进口成本将导致印度经济赤字[54]。此外,道路运输造成的 CO_2 排放份额大于火车、飞机或船只。

因此,印度政府决定到 2030 年实现乘用车电气化,使用煤炭以外的可再生资源,以避免温室气体排放。预计该国将在未来十年增加电力消耗[71]。Abhyankar 等[72]通过比较在印度电动汽车的三种仿真模型,产生了一份报告来支持在 2030 年打造电动汽车时代:(1)插电式电动汽车基础设施模型(PEVI);(2)PLEXOS 基于行业模型与最低成本投资,经济调度和经济性;(3)基于环境。不过,这份提交良好的报告预计,到 2030 年,印度将独立于所有天然或产生的运输燃料来源。假设原油价格约为每桶 40 美元,电池电动汽车(BEVs)每年可以减少 70 亿美元的石油进口。到 2050 年,每年可减少 510 亿美元[49]。

到 2030 年,电动汽车的市场份额预计将达到新车销量的 40% 至 50%[73]。混合动力汽车(HEV)或电池电动汽车(BEV)的概念非常好,在印度应该为所有车提供动力。因此,到 2030 年,不仅所有的电价都会降低,而且所有的运输都只能使用电动汽车;这意味着不再需要原油。我们将研究这一目标是否可以实现,或者印度愿意尝试廉价和可再生的 Diafuel™,这将对印度经济更好,Diafuel™ 碳足迹接近零,仅需要太阳能作为获取这种生物燃料的主要能源投入。

在汽车电气化的道路上,法国也计划到 2040 年停止所有汽油和柴油汽车,到 2050 年实现全国碳中和,并将停止使用煤炭发电。法国并不是唯一一个禁止燃油动力汽车的国家,荷兰和挪威已经表示,他们将在 2025 年前停止使用汽油和柴油汽车,而德国和印度也将在 2030 年前实现类似的计划[74]。电动汽车的车轮 CO_2 排放范围从 0 gm/km(用于驱动可再生能源电力需求的车辆)到 70 gm/km[用于由电网供电的插电式混合动力汽车(PHEV)]不等,250 gm/km 适用于煤电驱动的电池电动汽车和排放 163 gm/km CO_2 气体的当前油车[75]。因此,电力来源对于评估电动汽车是否真的减少 CO_2 排放非常重要。

17.5 批评家谈论电动汽车进入市场

根据 NITI(印度国家转型机构)的报告[76],到 2030 年,电动汽车、公共汽车和地铁可以帮助印度节省 600 亿美元。如果印度转向公共交通、电动汽车和拼车等更绿色的交通解决方案,到 2030 年,印度将节省高达 390 亿卢比(约 600 亿美元)。报告称,到 2030 年,印度可以通过转向共享电力交通,节省 64% 的能源。随后汽油和柴油消耗的下降将达到 1.56 亿吨石油当量(MTOE)或 1.8 万亿瓦特时的能源——足以为该国 17.963 亿家庭提供电力。

然而,电动汽车仍然需要电力来为使用寿命有限的电池充电。美国退出了巴黎协议,尽管许多美国公司反对这一决定。法国总统马克龙强调,"如果美国在气候变化问题上变得孤立,甚至可能失去明天的创新者"[77]。尽管印度政府认为,通过在不久的将来采用电动汽车,CO_2 排放量在 2030 年将下降 37%。据英国广播新闻报道,一些即将上市的电动汽车品牌已经为推出电动汽车设定了一定的目标:特斯拉已经设定了 2018 年每年生产 50 万辆电动汽车的目标,2020 年每年生产 100 万辆[78]。从 2019 年开始,沃尔沃汽车将专注于电动汽车,届时将只关注电动汽车,并慢慢逐步淘汰纯内燃机(ICE)汽车,在 2019 年至 2021 年期间,沃尔沃将推出 5 款全电动汽车车辆[79],雷诺和高通公司展示了动态无线电动汽车充电[80],到 2030 年,德国所有的新车都必须是电动汽车。

另一方面,《在线编年报》严重反对特斯拉电动马达,导致了更多的行业,如风力涡轮机和太阳能光伏电池的发展。《编年史》的报告提到,无论是他们,还是碳封存装置等许多其他反碳发明,在任何商业意义上都不是"真实的"。全球可再生能源产业浪费了数万亿美元建设经济上不合理的基础设施,这是商业史上最大的财富损失[81]。

记者引用了梅赛德斯-奔驰印度总经理和首席执行官罗兰·福尔杰的话[82]:"政府能投资数千亿美元建立充电站和相关基础设施吗?如果没有,那么谁来买单呢?当然不是私营部门。如果政府最终能够筹集到资金,那么在实现减少污染的关键目标方面的努力值得吗?"事实上,福尔杰认为,此举"在额外的电力需求方面会更适得其反,因为你仍在建设和支持热电厂。"是的,以目前的煤基发电模式,随着电力需求的急剧增加,这将造成更多的污染,或者我们有资金来升级我们所有的旧热电厂吗?或者我们能完全摆脱污染严重的燃煤电厂吗?如果是,这样的计划所需要的成本是多少?

尽管有这些重要的报告,但电动汽车将在 2030 年左右在世界上进行普及,这份报告是建议我们要预见电动汽车可能陷入的混乱时代,因此政府应该看看其研究开发团队寻找类似于无需大量能量、仅需低排放就能自主生产的替代燃料。事实上,硅藻负责固定 25% 的 CO_2[14]，而通过酯交换反应获得的 Diafuel™ 的碳足迹几乎为零。如果 Diafuel™ 在比其他热带国家太阳能充足的印度取得成功，那么印度就有黄金机会出口 Diafuel™，从而增加其经济。

17.6　电动汽车与 Diafuel™ 大型汽车的比较

17.6.1　电动汽车

电动汽车无疑被视为可持续的交通选择，但目前尚不清楚如何以低成本和低 CO_2 排放向它们提供电力。不同的混合动力电动汽车具有不同的成本和 CO_2 排放量，到 2030 年，它们的份额将从 3% 增加到 30%，其中 75% 的插电式混合动力电动汽车将占主导地位[83-84]。电动汽车不断增长的需求肯定会带来电力供应增加的新问题，例如：充电站和剩余锂离子电池系统。通过利用化石燃料满足需求间接导致 CO_2 的排放，来满足来自可再生能源的额外电力供应具有挑战性[85]。有学者提出了一个使用热电联产作为可再生能源的电力需求模型框架[86]。使用热电联产装置的优势在于其高效率（80%～90%），因为电动汽车和热电联产技术之间可以快速协同使用电力和热量。然而，模型研究表明，基于低碳电力的电网电力实现了 60～190 t CO_2 排放量下降。基于热电联产的电动汽车的 CO_2 排放量比汽油车低 40%，只有燃料电池联合充电系统（FC-CCS）和联合充电系统-热电联产（CCS-CHP）系统更贵，成本是电网参考成本的 3.6～4.7 倍，减排成本为 165 t CO_2。然而，热电联产和电动汽车的协同作用并没有发挥作用，减排成本也没有降低，如下图 17.6 所示。此外，汽油车的 CO_2 排放量与任何热电联产工厂或低碳排放（LCE 混合）电网的排放量一样高。图

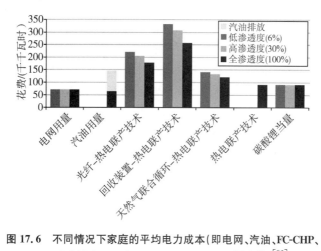

图 17.6　不同情况下家庭的平均电力成本（即电网、汽油、FC-CHP、FC-CCS-CHP、NGCC-CHP、μ-CHP、LCE 混合物）[86]

在不同的情况下，汽油车排放物是在平均行驶距离为 38 千米/天的情况下排放的

17.6 显示了不同情况下家庭的平均电费,即:电网、汽油、FC-CHP、FC-CCS-CHP、天然气联合循环-热电联产(NGCC-CHP)、μ-CHP、LCE-mix[87]。这表明,在不同的情况下,平均驾驶距离为 38 千米/天时会排放汽油。

在电动汽车中,插电式混合动力电动汽车(PHEV)是交通和电力领域的新技术,具有 4 kWh 的电池存储系统,在所有电动模式下至少可行驶 10 英里[①][88]。它们可以使用化石燃料、电力或两者的组合,从而降低 GHG 排放、车辆到电网技术(V2G)以及对进口石油的依赖[89-91]。此外,数据表明,以每千瓦时 10 美分的电力成本,为每加仑汽油添加相当于 70 美分的 PHEV 燃料,行驶 40 英里可将油耗降低至原值的 2/3[92]。丰田普锐斯是 PHEV 最好的例子,它早期是一种混合动力电动汽车(HEV),使用替代套件转换成 PHEV,现在作为混合动力电动汽车和 PHEV 生产[93]。

1. 印度电动汽车的现状

现在市场上有许多电动汽车,包括雪佛兰 Volt(2010 年)、Cooper Mini E(当前)、Fisker Karma(2010 年)、日产 LEAF(2010 年)和特斯拉 Roadster(2011 年)、欧宝 Ampera(2012 年)、Wheego Whip LiFe、Mia electric、沃尔沃 C30 电动汽车、福特福克斯电动汽车(2011 年)、三菱 iMiEV(2011 年)、宝马 ActiveE、Coda、雷诺 Fluence Z.E、特斯拉 Model S、本田飞度电动汽车、丰田 RAV4 电动汽车、雷诺 Zoe、weE50 ScioniQ EV 和 Kandi EV(2012~2013)、华晨宝马 Zinoro1E、奇瑞 eQ、吉利-Kandi Panda EV、众泰智斗 E20、起亚 Soul EV、大众 e-Golf、奔驰 B 级电动驱动和启辰 30(2014 年)、雪佛兰 Bolt EV 概念车、BYDe5 和特斯拉 Model X(2015 年)、特斯拉 Model 3、现代 Ioniq Electric(2016 年)、欧宝 Ampera-e(2017 年);自行车:geo by(2009 年中国)、pede go(2010 年美国),中国于 1997 年首次推出电动自行车,名称为 Pedelec(Raghavan 2016)。

印度于 2015 年 4 月启动了加快混合动力和电动汽车的采用和制造计划,通过各种补贴支持电动汽车市场的增长。印度的第一辆电动汽车列瓦于 2001 年由马恒达和马恒达私人有限公司推出,即使在 15 年后,情况也没有改变,甚至也更好的电动汽车推出[94]。列瓦是印度推出的第一款电动汽车,其大部分汽车在欧洲市场销售。除列瓦外,在印度流行的电动汽车包括丰田推出的普锐斯混合动力车、丰田凯美瑞混合动力车、宝马 i8、Mahindra Scorpio 微型混合动力车。这是在印度商业销售的仅有的六辆 EV/HEV 汽车[83,87]。塔塔汽车正在创造历史,从位于古吉拉特邦的 Sanand 工厂推出首款 Tigor 电动汽车,并在 2017 年 9 月获得了能效服务有限公司对其 10 000 辆电动汽车的投标。这些汽车将于 2018 年在印度上路[95]。目前,印度的电动汽车依靠铅酸电池运行,因为暂时缺乏锂离子电池来提高效率[96]。锂离子电池的主要组件从中国台湾和日本进口,关税为 4%,而锂离子电池进口关税为 18%,然而,电动汽车整体关税为 14%,功率控制器从美国进口,这使得普通人的汽车非常昂贵。虽然政府

---

① 1 英里 = 1.609 344 千米。

已将电池驱动车辆中三种进口部件的关税降至10%，但对于此类车辆的降价来说，激励措施似乎还不够。印度道路上没有足够的混合动力电动汽车充电站，也没有足够的电动汽车基础设施[97]。目前，印度有近56 000个加油站，而电动汽车只有206个充电站。印度那格浦尔成为第一个拥有由印度石油公司在其加油站设立有充电站的城市[98]。目前，印度对电动汽车的需求非常少，只有1%的两轮车市场是电动两轮车，电动汽车市场仅占整个汽车市场的2%~3%。电动汽车行业正在寻求中央政府的补贴，德里政府已经宣布提供29.5%的补贴。然而，印度目前的经济状况对于像快速充电基础设施这样的技术来说无利可图。此外，还有固定资产投资和运营成本的额外成本，电池成本、续驶里程、充电技术的持续发展；充电器成本推动快速充电基础设施业务的经济性。在2025年之前，快速充电器似乎很难在印度实现经济可行性[99]。

随着电动汽车的蓬勃发展，电动摩托车开始出现在印度的道路上。继Okinawa Ridge之后，冲绳在印度推出了第二款电动摩托车Okinawa Praise[100]。前者拥有1 000瓦的电动马达，功率为2 500瓦(3.35马力)，时速为75千米，成为印度最快的电动摩托车。该公司声称，冲绳普雷兹可以跑175~200千米，一次充电电池每千米约10派提，完全没有竞争对手，还可以选择锂电池[101]。

2. 电动汽车对全球和印度网络及其能源的预测影响

随着这些电动汽车在市场上的出现，2016年将达到约150万辆，到2030年超过5 000万辆，其中25%的电动汽车为PHEV[102]。Judd和Overbye[103]研究了PHEV的经济可行性，这不仅代表了PHEV减少温室气体排放，而且内燃机的PHEV发电量为61%，内燃机为12.6%。Li等研究了光伏发电作为给PHEV充电的电源[104]，作者计算了一个为PHEV-40充电的光伏面板的尺寸（一次充满电可以行驶40英里）。在太阳辐射最多的一年，电池板的尺寸为20平方米，在12月（太阳辐射较少），电池板的尺寸为78平方米。作者最终得出结论，利用太阳能辐射可以减少对电动汽车发电的污染，但这种电池板的成本太贵了。尽管维纳汽车等已经研究了汽车到电网(V2G)模型为电动汽车充电。他们提出了两种智能公园模型，每一种模型都由四辆车组成，通过升压变压器(208 V/22.9 kV)通过输电线路相互连接[105]。对电力传输/充电时间进行监控和标准化，以尽量减少每天电网交易的收入。然而，我们观察到，在V2G交易过程中，存在许多对车辆有害的电网故障。

关于PHEV，美国需要73%的能源来驱动它们，这将进一步为电网增加9 100亿千瓦时的额外负荷，迫使电网昼夜运行，从而鼓励采用新技术开发廉价的替代能源[106]。

17.6.2　Diafuel™

Diafuel™是最好的替代来源之一，硅藻大量存在于水体中，利用实验室芯片技术，我们可以从活硅藻中获取石油，准备用于国内或可经酯化用于工业目的[17]。除了如此多的全球工业化来开发电动汽车之外，我们还不需要修改或开发新的汽车来驾驶 Diafuel™ 驱动的汽车。不需要额外的电池和充电站。生物柴油,特别是 Diafuel™ 具有比汽油或柴油更好的性能。生物柴油含有 10% O_2 这有助于它完全燃烧,并且比柴油有更高的闪点,使它成为一种更安全的燃料。微藻状硅藻的生物柴油含有 O_2,从而减少燃烧过程中 CO 和未燃烧的碳氢化合物的数量。它们不同于电动汽车,不需要对现有的汽车进行任何发动机改装,因此可以自主生产和使用。Diafuel™ 将是硅藻生物柴油,不含硫氧化物,减少一氧化碳以及未燃烧的碳氢化合物的排放,因此有害污染物的排放可以忽略不计[107]。Diafuel™ 的基本化学成分富含脂肪酸、醇类、酮类、酯类和醛类,如图 17.7(a) 中所示,Diafuel™ 的气相色谱(GC)色谱图通过利用 Abubakar 和 Mutie 的方法[108]提取的 Pinnularia saprophila。12 种硅藻菌株的无菌培养物来自从印度哈里亚纳邦不同水域采集的水样,标记为 V1 至 V12(见表 17.1),其 FTIR 研究显示了各种相似性,表现为 3 000~3 600 cm^{-1} 范围内的一些常见峰模式,这些峰型用于—O—H 吸收(数据未显示)。所有的硅藻物种,在 3 700~3 100 cm^{-1} 波段中,在 3 400 cm^{-1} 附近都表现出吸收下降。这种更宽的下降反映了羟基 O—H 拉伸官能团的存在,这反映了由于—NH 和—OH 基团而存在的氢键。2 850 cm^{-1} 和 2 925 cm^{-1} 附近的吸收下降证实了 CH_2 和甲基 C—H 不对称和对称拉伸的脂肪族官能团[109]。此外,在 1 650 cm^{-1} 附近的吸收下降可能表明存在对于伯酰胺官能团具有>C=O 伸展的蛋白质。

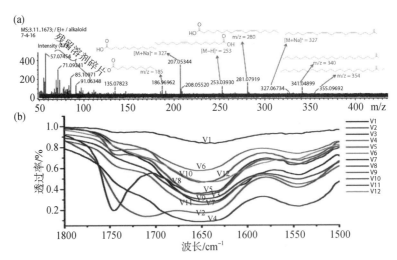

图 17.7 从 Pinnularia saprophila 藻硅中获得的脂质的 GC-MS 图谱及其所含成分分析(a)和各种硅藻物种(如表 17.1 所示)的 FTIR 光谱图谱(b)

表 17.1 不同硅藻物种中的常见官能团及其相应的 FTIR 振动频率

官能团	波长范围/cm^{-1}	振动类型
脂肪族	2 925	C—H 强不对称伸缩振动
—CH_2—CH_3	2 850	C—强对称伸缩振动

(续表)

官能团	波长范围/cm^{-1}	振动类型
蛋白质 —NH—C—O	1 650 1 545	>C=O N—H 弯曲
羧酸	~3 400 1 735	O—H >C=O 脂肪酸
羟基	~3 400 ~1 540	O—H O—H 弯曲
二氧化硅	1 060,1 072,1 086	Si—O—Si
多糖	1 000~1 260	C—O

在 1 000~1 300 cm^{-1} 范围内,宽峰反映了一个以上官能团的存在,而锐峰显示了单一的主要官能团。V1 到 V7 在 1 000~1 250 cm^{-1} 的较宽下降反映了具有 Si—O—Si 结构的多糖,特别是二氧化硅(1 250~1 100 cm^{-1},1 072 cm^{-1})和多糖(1 086 cm^{-1})。在这个区域中反映的其它官能团是醇的 C—O 键。如图 17.6 所示,除了上述的常见吸收带,一些物种还在波数范围 1 800~1 500 cm^{-1} 显示了其他主要吸收带。V2 和 V8 分别在 1 710 cm^{-1} 和 1 745 cm^{-1} 附近有吸收下降。这些谱带反映了酮>C=O(1 700~1 725 cm^{-1})的存在,这可能来自酮和醛,而在 1 735~1 750 cm^{-1} 时>C=O 可能分别是由于 V2 和 V8 中的酯类。表 17.1 总结了 V1 到 V12 的 FTIR 光谱的不同主带区域中所有硅藻物种中各种官能团的存在。

1. Diafuel™ 工业生产

硅藻微藻的培养可以在开放或封闭系统中进行;然而,我们建议将系统封闭,以避免污染。Diafuel™ 的商业化并没有迈出巨大的一步,但有一点是需要谨慎的,即我们使用的主要原油含有硅藻油是 24-去甲胆甾烷碳氢化合物的一种成分[110]。

为了培养所需的硅藻,人们可以准备好对附近或房屋内的环境开放的浅水盆地。现在的光生物反应器是一些透明设备覆盖的系统,如塑料袋、试管平板等。易于维护和高表面积体积比使垂直管成为最受欢迎的生物反应器系统[112],但资本成本相对更昂贵。这种类型的生物反应器优于开放式池塘系统的其他优点是,它们对野生型藻类菌株的污染具有高抗性,可以提供人造光,因此可以在白天和夜晚都起作用,并且脂质生产率可以根据需要通过使用不同的光源来调节。另一方面,这些需要由寿命不超过 5 年的酸性电池组成的高能量备份[26]。虽然硅藻太阳能电池板(DSP)的概念是在 2009 年提出的,但设计和原型是在微观层面上设计的[113],一旦其工业原型准备就绪;这可能是能源行业的福音。在该系统中,与开放系统相比,物理化学条件的控制要容易得多,这导致高脂质生产,因此产生更多的能量。

Diafuel™ 生产的能源自给自足的想法是将硅藻放在室内的营养液中,并将其暴露在室外天气条件下享受阳光。住在房子里的居民会呼出 CO_2,这些 CO_2 会被储存在室内蓄水池中的硅藻吸收,从而通过它们不断产生的 O_2 使居住在房子里的人更加健康。当需要时,使用电机泵对抗重力(见图 17.8),将蓄水池中的硅藻送至屋顶蓄水池。屋顶上是一个储层,具有一组微管道通道,其共振频率约为 350 mHz,足以撞击硅藻壁,使油从具有高 750 μN 机械强度的硅藻中渗出。这个想法是"我们用硅藻太阳能电池板代替屋顶上的二氧化硅太阳能电池板,硅藻太阳能板由海藻状硅藻组成,无需进口,一家印度家庭只需在屋顶安装一块硅藻太阳能面板,就能满足自己的石油需求"。

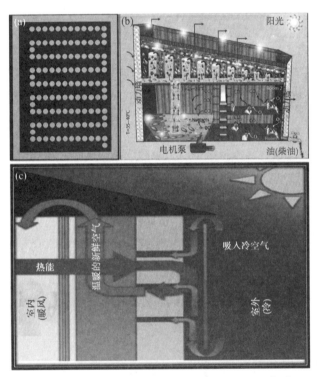

图 17.8 描述了优化硅藻太阳能电池板的原理模型[111]

(a) 显示了纳米共振微腔中的一系列球体/圆柱体,硅藻在其中移动、碰撞,并因其壁上的应变而在压力下释放油;(b) 屋顶水箱的三维视图,水箱有一个圆柱形管道阵列,从房屋内部的一个主要大管道中抽取介质中的硅藻;(c) 外墙动态传热

2. 设计一个能源自给自足的印度住宅生产 Diafuel™

因此,我们建议建造硅藻生物燃料太阳能电池板,其中硅藻置于一个封闭的平板容器中,外部覆盖着透明的玻璃板,只允许阳光进入。在透明玻璃板旁边是太阳能透明绝缘板[114],这将防止屋顶水库的温度损失。介质中的硅藻将被保存在屋顶和建筑物的侧壁上,在那里它们将暴露在与我们人类生活在室内相同的温度下。房子的围护结构将是动态围护结构的一个变体,也为居住者提供更健康的生活。例如,动态围护结构将被设计在印度住宅的侧墙和屋顶上,以证明这种太阳能电池板在凉爽的夜

晚增加能源需求的可行性[111]。当室外温度高于室内时,不会影响硅藻的生长或活力,因为室内设置的动态壁保持了优化的温度,如图 17.8(c)所示。在印度环境中,营养罐中的硅藻将被保存在室内,室温为 25～30℃,人类居住的温度见图 17.8(b)。它们会被宏观管道泵送到屋顶顶部,这将导致硅藻介质和硅藻进入到一个具有锯齿状排列孔的扁平容器,如图 17.8(a)所示。

因此,假设介质中的硅藻在到达屋顶时被室内电机泵送,它们从微管阵列中掉下来,由于机械压力,应力作用在在硅藻柱的壁上。750 μN 以上的应力会使硅藻中的油流出,就像在 *Terpsinoë musica*[7] 介绍的情况一样。根据印度气候,室外房屋的温度保持在 30～40℃之间。由于硅藻是在室温和屋顶水库之间循环利用的,温度保持在一些世界性菌株 *Nitzschia palea* 和 *Diadesmis confervacea* 能够承受的平均温度。也可以保持硅藻生长在 20～25℃的理想温度,完整的封闭式的屋顶油罐可以转移在地下室,带有一排中空的、大端穿孔的微柱,可以从油罐底部直到屋顶,收集所有阳光,通过这些管道流动,从而为硅藻生长所需的封闭系统提供光线。当一个家庭的居民对石油的需求较低时;硅藻不会从室内地面储层泵到屋顶储层进行光合作用和生产脂质。因此,它们仍然沉积在室内底层储层的底部。硅藻可以在漫射光中生活 3～4 个月,漫射光可能是正常的管光或阳光,当它们保存在室内时,在低温和低要求或无额外要求的情况下不需要阳光直射,除了从房屋居民释放的 CO_2。

硅藻将被选择性地培育,以生产只需要很少,或者最好不需要进一步处理供人类使用的石油。最终,这种硅藻太阳能电池板可能会成为每个印度房屋屋顶的一部分。然而,在寒冷的气候条件下,高效住宅创新(EHI)的理念奏效了,建造房屋的静态隔热暖风墙被转换为从室外恶劣天气条件下抽出的动态冷空气[111]如图 17.8(b)所示。其想法是为了保持硅藻生长的理想室内室温,保持硅藻生长的最佳状态。

3. 印度房屋硅藻板的工作原型

图 17.8(a)显示了 Vinayak 等[113]构建的纳米摩擦装置,该装置可在 2 mm 硅片上使用锆钛酸铅(PZT)圆盘,在不杀死硅藻的情况下,从硅藻中释放油。图 17.8(b)显示了房屋在屋顶上有微阵列通道和在室内的主水库的三维图像。硅藻的主要储层被重力泵到屋顶,然后到微通道流向重力。硅藻的下落会产生机械压力,导致油从硅藻中渗出。原型类型清楚地显示硅藻介质和硅藻被从室内(温度 25～30℃)泵到屋顶顶部的一个宏观管道阵列的微型管道,从硅藻介质包含在主水库保持室内辐射的管道。当硅藻和介质向重力下降时,由于机械压力,细胞承受压力,以渗出油。屋顶覆盖着一层透明的玻璃板,让阳光进入面板,让硅藻从微通道中掉落,以进行光合作用。根据印度次大陆的纬度,硅藻保持在 32°的角度。印度夏季的外部温度达到 30～40℃,室内硅藻到屋顶的持续循环保持了温度,硅藻在到达屋顶水库时获得足够的光照,而不会因为室外天气条件而散热。连续循环对硅藻产生足够的压力,使油渗出。外膜/玻璃片层允许光线进入屋顶储层,其中有一系列微型管道,硅藻被拉出来。玻璃片旁边的层是保温层,防止温度损失,从而保持温度,内部疏水透膜只允许脂质

分子通过,如图 17.8(b)所示。比水轻的油在顶部收集,通过疏水渗透板吸收,从沿侧壁连接的通风管道下降,到达房屋地面,可用于车辆泵入等。

我们提出的此类房屋的概念设计包含以下尺寸,可能会相应变化,如图 17.8 所示:(1)硅藻太阳能电池板的倾斜度在 32 度;(2)蓄水池的最小容量为 10 桶,可根据要求进行调整;(3)房屋尺寸为 100×100 平方英尺;(4)顶部蓄水池可以使用,也可以不使用,整个屋顶顶部取决于要求;(5)释放出 CO_2 被硅藻消耗。

原型的第二部分将是极端气候条件下的动态墙流动,而不是这个概念证明的一部分。例如在北半球和南半球,温度低至-60℃,内部房间比外部天气条件更温暖。图 17.8(c)显示了动态墙,在寒冷的天气中,通过从室外吸入空气,帮助保持室内的温度,而在炎热条件下,室内温暖在第二阶段也会进行同样的变化。由于硅藻是在室内生活的,所以它们会吸收室内居民释放的 CO_2,从而使生活环境更新鲜,富含氧气。

虽然有很多技术可以不牺牲藻类的情况下从藻类中提取油脂,但自然渗出是最便宜的,正如 Vinayak 等[115]在 *Diadesmis confervacea* 中发现,但并不适用于所有的硅藻菌株。因此,我们需要开发一种新的技术,以成本效益和更便宜的方式的炼油。当硅藻细胞被加工到不会杀死它们的极限时,每 10 桶介质的预期油产量约为 1 kg 原油。

4. Diafuel™ 的优势

来自硅藻的 Diafuel™(硅藻生物燃料)的主要优势是直接生产燃料油/脂质,以取代太阳能电池板中的石油燃料,这比其他能源需求方式具有优势。由于汽油产品是可互换的,所以没有必要用电力发动机来代替燃油发动机。因此,它们产生的废物更少,也更经济。由于脂质能源生产脱离电网,电网短缺导致限电[22]。使用 Diafuel™,客户可以摆脱使用寿命短、昂贵、笨重、污染严重的铅电池来储存太阳能。使用 Diafuel™ 的另一个优点是,储存 Diafuel™ 不需要新技术,与储存汽油和柴油一样。汽油的能量密度是电池的 100 倍[25]。硅藻易于培养,它们在 5~24 h[116]内快速分裂近两次。从硅藻中收集油的过程包括在屋顶下以人体温度进行局部生产,使用透明隔热材料[117],在固定阶段收集硅藻用于 Diafuel™[17]。由于我们将脂质生产与低足迹技术相结合,该系统在很少或没有能量的情况下运行,不会与粮食作物竞争。作者提出了 Diafuel™ 生产原型硅藻太阳能电池板的优势[7,17],其具有使用硅藻电池板和高效住宅技术,具有模块化、低维护房屋。Diafuel™ 可能适用于运输、供暖和发电。

硅藻太阳能电池板的唯一前提条件是能够在培养室外培养硅藻,从而让硅藻渗出油,而不需要从细胞中提取油。在印度发现了这样一种硅藻,称为 *Diadesmis confervacea*,它连续分泌油 30 天以上[115]。通过改变生化途径、胁迫氮、磷、二氧化硅或基因工程,自然或实验室选择富含油的菌株[118]可以帮助增加用于硅藻太阳能板中使用的脂质。其次,硅藻生物燃料太阳能电池板的生产成本应该适中,并允许在当地屋顶持续生产生物燃料;在这种情况下,我们建议使用在印度气候下生长最佳的硅藻,如 *Diadesmis confervacea*。

17.7 电动汽车的发电来源

尽管截至 2017 年 7 月 31 日,工业中储存的装机容量为 50 289 兆瓦,但印度的气候条件已经影响了未来 10 年的电力需求[119]。到 2022 年,减少 CO_2 排放,生产 175 亿瓦的可再生能源,这是相当雄心勃勃的计划,从目前的情况来看似乎是困难的。由于煤炭不是人为制造的,对 10 年的规划不应忽视未来 100 年的未来。虽然印度盛产软煤,没有任何污染物,但要在后代中得到软煤,需要三件事,即:水、阳光、植物、树木[21]。印度将因其能源依赖性而依赖煤炭,因为其在印度的储量接近 58.9%,仅印度就生产 944 861 兆瓦的能源,在满足自己的需求后,与世界其他地区共享 50%[29]。毫无疑问,印度是第三大总发电生产国。根据国际能源署(IEA)的数据,到 2040 年,印度的能源价格将增长 25%[54]。到 2022 年,印度将产生 175 万千瓦的可再生能源。能源需求将进一步增加,使太阳能发电的化石燃料的需求增加一倍。现在,由于需求增加,化石燃料的消费也在增加。决策者是否强调了生产化石燃料的因素? 值得批评的是,非耗竭型水库,特别是煤炭和水电发电是天然的,一旦消耗也需要补充,但如何补充呢? 印度的政策制定者如果未能实现 175 兆瓦的能源需求,将耗尽大部分煤炭和其他能源储备,无需补充和对 GHE 的任何重大控制(温室效应)[20]。当然,如果我们依赖电动汽车,未来 20 年我们将不会出现赤字[33]。大型水力水储量接近 33.5%,小型水力水储量高达 1.5%。印度的水文储量主要依赖于季风,而季风则依赖于许多地理环境和气候因素。私人拥有的汽车从 2002 年的 2 900 万辆增加到 2013 年的 1.6 亿辆,预计到 2030 年将增加超过 5 亿辆[120]。印度政府已经做出了承诺,并签署了条约,但在一个 2.4 亿人口的国家,那里仍然没有电力来运行这种基于电力的车辆[121]。电力成本本身并不稳定,它们随着季节变化,不同类型的消费者也不同。如果只有 2 000 辆汽车充电,充电从充电站进行,但当不到 3.5 亿汽车日夜从任何国家、山或平原的任何地方进行充电时,首先需要建立一个备用电源,并考虑到太阳能光伏发电和近 1% 的核电的有限容量。在印度,除了改用煤炭发电,别无选择,其产生的温室气体将远远超过汽油、柴油或压缩天然气[122]。因此,到 2030 年,我们需要多达 25 万个可充电站,而且在缺电的小城镇也是如此[120]。同样,在停电期间,这些电站将由低压发电机运行,这需要柴油在大气中添加温室气体。在印度,2.4 亿人没有电,两级目标意味着印度将不得不通过燃煤发电厂的排放。印度约 60% 的电力来自煤炭,但印度确实很难转向可再生能源,它必须在几年内关闭其燃煤电厂[29]。

17.7.1 具有零碳排放的资源

可再生能源(RE)是发电和其他能源用途的唯一清洁替代品。可再生能源发电是经济的,特别是在扩展传统燃料电网不经济的偏远地区。近年来,印度的可再生资源产量从 2002 年的 2% 提高到了目前的 13%。印度主要的可再生能源资源来自风能(59%)、生物质能(23%)、小型水力发电项目(11%)和太阳能(6%)[33]。印度消耗的约 65% 的电力来自火力发电厂,22% 来自水力发电厂,3% 来自核电站,其余 10% 来自太阳能、风能、生物质能等其他替代来源。印度 53.7% 的商业能源需求是通过该国巨大的煤炭储量来满足的[123]。到 2030 年转向电动汽车发电的替代措施,我们需要不仅经济的资源,由于碳足迹为零,因此风能、太阳能和核能仍然是唯一的选择。

1. 核电

因为核能是零碳排放的替代可再生能源。因此,印度也一直专注于核能,以应对印度的能源短缺问题。该计划是从目前的 20 个反应堆基础上再增加 30 个反应堆[44]。然而,核反应堆的成本太高,高达 850 亿美元,因此如果没有外国的财政支持,它似乎无法发挥作用[124]。核能虽然对零碳足迹是有益的,但它会产生其他危险的废物[125]。印度最大的上游公共部门是石油天然气公司(ONGC),它约占该国石油总产量的 61.5%。然而,随着能源需求的增加和化石燃料的限制,印度已经转向核能以满足目标。目前,印度有三个核反应堆,并计划再建造 18 个核反应堆[44]。

2. 太阳能促进印度更快地采用和制造电动汽车和混合动力汽车

印度是一个仍然缺电的国家,他们已经下了很大的决心,想要依靠太阳能生产电动汽车[112]。政府希望将其太阳能装机容量从 2017 年 5 月的 5 700 万千瓦增加到 2022 年的 17 500 万千瓦。印度政府希望,到 2027 年,印度的可再生能源能够满足其 40% 的能源需求[47]。太阳能电价大幅下降,使太阳能成为一种有效的替代方案,最新的出价是每千瓦时 2.44 卢比,这几乎是 2007 年要求的价格的五分之一[116]。太阳能光伏发电很难实现为 2030 年设定的减少温室气体排放和控制气候变化的目标[126]。甚至政府也采取了这一措施,因为太阳能发电厂的安装成本很低,而且没有风险。印度将通过部署所有现代可再生能源领先的太阳能来发挥主导作用,使印度在 2040 年成为第二大太阳能市场[58]。据《印度时报》2017 年 12 月 22 日报道,印度投资了 450 亿卢比,为位于中央邦里瓦古尔的世界上最大的大型太阳能发电厂奠基[78]。据报道,该工厂生产 750 兆瓦的电力,所生产的 24% 的绿色能源将用于首都的地铁运营,其余将用于该国贫困家庭供电。另一方面,印度新能源和可再生资源部已就印度太阳能改造提出了 SRISTI 可持续屋顶实施计划,以激励印度屋顶太阳能项目的安装[80]。印度政府的努力令人难以置信,但太阳能光伏(SPV)电厂的成本比 2010 年高出一半[127]。由于太阳能电池板中使用的二氧化硅格栅的高成本,家庭屋顶太阳能电池板仍然无法接近。人们认为 2030 年将被称为电动汽车时代;因此,计划到 2022 年建立 10 万兆瓦的太阳能发电能力,但起步缓慢[79]。2015 年仅新增 2 133 兆瓦,2016 年略低于 4 000 兆瓦,在建 10 000 兆瓦到 2018 年投入使用,未来 4 年仍需要 84 000 兆瓦。为了实现这一目标,政府将太阳能目标到 2030 年提高到 25 万兆瓦[126]。SPV 电厂的红色转移点是 3 个 SPV 电厂,前两个电厂每年只能运行 1 700 小时,这意味着每天不到 5 小时。光伏电站安装的 25 万兆瓦发电将不会超过

6万兆瓦,因此到2030年只能满足该国所需额外电力的10%(2016年世界经济论坛)。为了实现这些目标,印度将需要对可再生资源进行更多的私人投资,因此"到2020财政年度,印度政府将推动超过8 000亿瓦的太阳能项目和30 000亿瓦的风能项目",据《经济时报纪事报》[82]报道,2017年11月26日,印度新能源和可再生能源部秘书阿南德·库马尔在新德里说。印度将需要吸引更多的私人投资来实现这些目标。即使在达到这些目标之后,印度的最高赤字水平也将会膨胀,直到该国实施改革,允许输配电运营商收回成本并产生利润,从而维持和改善自己的网络。印度的电力行业还有其他重大问题,包括由于缺乏燃料供应或融资而滞留的资产。

3. 风力

风能是另一种零排放的资源,将用于电动汽车的发电,但在没有资源的地方,它仍然很昂贵。肖特和德霍尔姆的工作[128]展示了一种用于发电的风力部署系统(WinDS)模型,特别是用于美国车辆。WinDS 模型是一种检验美国风能发电效率的计算机模型。据估计,到2050年发电208 GW。在 PHEV-20 项目中,WinDS 项目预计将达到到 235 GW,而 PHEV-60WinDS 项目则声称将产生 443 GW 的电力。PHEV-60 的风力发电比基础外壳安装增加了近110%,风力提供了美国16%的电力(10 082 千瓦时中的1 554 千瓦时),图 17.9(a)和(b)显示了 PHEV-60 中所有发电机的容量和发电量。

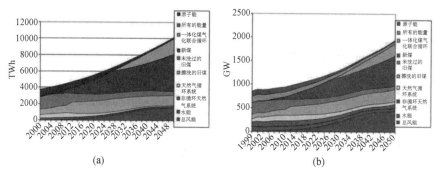

图 17.9　WinDSHEV-60 的扩容(a)和 WinDSPHEV-60 的发电(b)[128]

所有这些微型混合电动力器都是 V2G,充电效率接近85%。PHEV 的大部分充电时间是晚上和夜间,还有时间在早上7点到下午1点和6点到晚上10点。尽管印度在年度风力发电市场中排名全球第三,每年增长2.1 GW 的安装[127],但这种类型的模型可以有效地使用。根据国际能源署的项目,印度到2020年将需要327 GW 的发电能力,预计到2020年每年生产81万亿吨能源,到2030年生产174万亿吨,2020年节省4 800万吨CO_2,2030年节省10.55亿吨。到2030年,印度的风力发电预计将减少9.1亿美元,到2020年将从每年37亿美元减少到24亿美元[129]。印度从2009年至2030年的累计风能预测如图17.10所示。

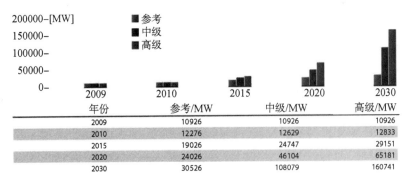

图 17.10 2009 年至 2030 年印度的累计风电容量

4. 风能和太阳能的障碍

在印度,电力主要来自燃煤火力发电,这将持续到 2050 年及以后,导致健康问题和温室气体排放[130]。除了未来几年印度拥有丰富的太阳能和风能外,2009—2022 年的全球市场可能会增长,如图 17.11 所示。它在大规模用电利用中面临着各种障碍。有些是专门针对技术、政策、地点或地区的[127],其中描述了需要对太阳能光伏和风力发电厂的研发进行改进,以提高这些发电厂的安装和能源生产成本。虽然印度计划在 2017 年实现太阳能并网,在 2022 年实现风能并网,但仍然有必要采取最好的政策措施,太阳能发电厂生产硅和晶片,以最便宜的价格采用新技术生产,这在目前仍属罕见[93]。

松下印度公司能源系统主管 Hindu 在接受采访时,表示由于电网容量低,该国可再生能源发电总量的 15%～20% 被浪费了。因此,对于绿色能源走廊的建设,需要一个 35 亿美元的可再生能源输电项目来抵消这些损失。[131]

图 17.11 显示了 2014—2022 年,SPV 和风力发电的复合年增长率(CAGR)预计将随着成本的下降而增加,但是这对印度来说是幻想,因为印度没有制造硅片和足够的稀有金属储层的公司。

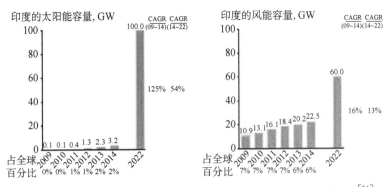

图 17.11 太阳能光伏成本随时间的推移显示,SPV 成本下降了 50% 以上[94]

17.8 电动汽车与汽油驱动汽车的 CO_2 排放量

Huo 等[132]在 2008 年和预计在 2030 年对中国的 CO_2、SO_2 和氮氧化物排放量进行了比较。与美国不同的是，2008—2009 年用电量保持在 0.5% 左右，预计到 2035 年将达到 0.8%(EIA)[133]。预计 2010—2035 年，中国的能源消耗为 5%[39]，中国的石油和天然气资源有限，主要依靠煤炭来满足电力需求。虽然中国正计划扩大其风能和太阳能发电资源，但中国电动汽车对环境的影响仍然是一个需要评估的关键参数。国际能源机构(IEA)预计，到 2030 年，煤炭发电量为 78%～81%[134]，到 2030 年，国内煤炭市场的份额将下降到 65%[133]。中国有六个以其所在地区命名的省际电网：东北、华北、华中、华东、西北和华南。图 17.12 显示了汽油混合动力电动汽车(HEV)和内燃机电动汽车(ICEV)的 CO_2 排放量，这些汽车由包括煤电网在内的 6 个电网供电。可以看出，与基于汽油的 ICEV 电动汽车相比，由煤电网供电的电动汽车会增加 7.3% 的 CO_2 排放量[132]。然而，使用 35% 非化石能源电网的华南地区排放的 CO_2 水平几乎与汽油混合动力汽车排放的 CO_2 水平相同，比 ICEV 汽油混合动力汽车排放的 CO_2 水平低 30%，因此表明混合动力汽车并不适合所有地区，混合动力汽车可能更适合 CO_2 排放量最少的华北和华东地区。另一方面，中国的西北、中部和南部地区以水力和风力资源的形式拥有良好的非化石能源。如果电动车 50% 的电能来自这些非化石能源发电厂，那么与使用石油燃料的混合动力汽车相比，在不使用汽油的情况下，中国的电动车可以减少 18% 的 CO_2 排放。但是，如果使用目前的电网（即煤炭）充电，与汽油车相比，电动汽车可能会增加 3～10 倍的二氧化硫排放量和两倍的氮氧化物排放量。电动车产生的二氧化硫是 ICEV 的 9～18 倍，因此在中国控制二氧化硫排放面临更大的挑战（见图 17.10）。

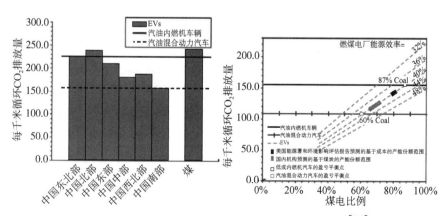

图 17.12　中国电动汽车燃料循环 CO_2 排放量与发电结构[132]

虽然汽油汽车尾气对全国二氧化硫排放总量的贡献很小（2006 年为 0.2%），但如果被电动汽车取代,其贡献将上升到2%～4%(电动汽车的二氧化硫排放量是汽车运行阶段 ICEV 排放量的9～18倍)。电动汽车将对中国控制二氧化硫排放总量的目标提出新的挑战。在未来,即使有更先进的燃烧技术和100%烟气脱硫(FGD)渗透,电动汽车的二氧化硫排放仍将是 ICEV 排放的1.3～5倍,混合动力汽车排放的3～7倍。即使增加100%的洗煤,这在实践中是不可行的,也不可能将电动汽车的二氧化硫排放降低到中国大多数地区的 ICEV 和混合动力汽车的水平。电动汽车肯定可以取代石油汽车,因为进口石油的成本高和空气污染造成的这些石油汽车的燃烧。但燃煤电厂的高污染将不利于中国电动汽车的发展。印度和中国是人口最多的国家,CO_2 排放量很高,如果这些国家不关注他们的政策,电动汽车可能会使它们的经济完全破产。补充说明,由于两种燃煤发电厂如果用于发电都会增加汞排放,因此电动汽车每行驶1千米会产生0.01毫克的汞排放[135]。中国的汽车数量预计在20年内将超过美国。如果非化石工厂不能匹配相应的电力需求运行电动汽车,中国不得不切换到燃煤电厂使电动汽车失败,因为它是当前的高排放发电厂会使电动汽车比 HEV 便宜和环保。中国市场的 HEV 售价近20 000～25 000美元,比传统的 ICEV 价格(8 000～10 000美元)贵得多,但电动汽车的价格(>30 000美元)甚至更高。因此,作为一个国家,我们需要一种碳排放少、廉价资源、所有人都现成的替代来源。2016年中国的一项调查显示,由于基础设施的改善,2011—2016年,BHEV 和 PHEV 的销量增长了两倍。因此如果电动汽车必须在印度取得成功,印度要发展充电站基础设施,有自己的太阳能和风力发电厂基础设施以及自己的锂离子(电动汽车电池)和钕储备(用于风力发电厂的齿轮箱)。

17.9 耗尽地球金属来运行电动汽车与 Diafuel™ 丰富资源的比较

传统上,电动汽车有铅酸电池,因为方便且经济。然而,自20世纪90年代以来,电池已经发展,现在使用锂离子组合的电池由于其效率更高、充电时间短、寿命更长、功率输出好以及对电池处理的环境影响更少,因此要好得多[95]。由于铅酸电池是不可回收的,而且会向我们的环境收集大量的酸废物。目前常用的四种电池:铅酸、镍镉(NiCd)、镍氢化物(NiMH)和锂离子。锂离子电池相对于其他离子电池具有更高的比能量。悲剧是,印度的锂离子储备主要位于玻利维亚和智利[136]。这永远无法降低电动汽车中使用的锂离子电池的成本。此外,子午线国际研究公司声称,没有足够的经济可回收的锂来制造全球电动汽车经济所需的电动汽车锂离子电池[96]。回收利用可能会改变这种情况,但这取决于行业是否会让锂离子电池的可回收性价格下降[98]。同样,用于2 000多万辆燃料电池汽车的长期使用取决于可回收性。大多数用于风力涡轮机的钢和混凝土是可回收的,因此没有任何损耗的威胁,但用于风力涡轮机的大多数可能面临灭绝的威胁元素是其变速箱中通常使用的钕[136]。然而,制造商正在转向建造无齿轮涡轮机,以克服这一限制。另一方面,光伏电池需要非晶态或晶体态硅、碲化镉、或硒化铟铜和硫化物。碲和铟的有限供应可能会降低某些类型的薄膜太阳能电池的前景，尽管不是全部。大规模生产可能会受到电池所需银的限制，但找到降低银含量的方法可以克服这一障碍。回收旧电池的零件也可以改善材料的困难。如图 17.13 所示，三种成分可能对制造数百万辆电动汽车构成挑战：用于电动机的稀土金属、用于锂离子电池的锂和用于燃料电池的铂[136]。

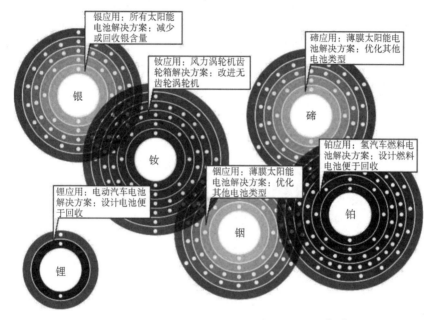

图 17.13　电动汽车成功所需的可能材料及其短缺[136]

电动汽车的高成本与昂贵的锂离子电池有关，因此使电动汽车价格昂贵。一个千瓦时的电力足以行驶大约 6 千米，所以一个 200 千米的"全油箱"范围需要大约 35 千瓦时的电池。目前全球锂离子电池的价格约为 250 美元/千瓦时，不包括进口关税在内，电池成本为 57 万卢比。即使有 8 年的使用寿命和 12% 的利率，仅凭每千米节省的电池成本就意味着人们每年必须行驶超过 2.5 万千米。最大的问题是，如果锂离子电池灭绝，电动汽车将成为改变全球经济下滑的游戏规则改变者。除此之外，印度政府已经发誓在 2030 年成为电动汽车时代，但似乎没有后备计划。如果太阳能、风力发电厂不能满足电力需求，我们是否计划从煤炭中获取电力，让情况变得更糟。是时候重新考虑并寻找其他生物燃料来源了，这些燃料丰富，不会死亡，廉价生产，碳足迹为零。

17.9.1　Diafuel™ 能成为答案吗

除煤炭外，印度没有能源资源来满足电动汽车项目平稳运行所需的电力需求。电动汽车不仅仅限于电力需求，而且可回收环保的锂离子电池也不是前面解释的印度产品。自2014年6月以来,原油价格下跌了70%,每桶159升,最近一次价格为1956卢比,每升为12.3卢比[99]。Diafuel™的质量是可以直接使用或转酯化;硅藻富含脂肪酸,主要是C14,C16,C16:1,C16:2,7,10和C22:5n³,生物燃料产率为60~40 μg/mL,电池数量约为1189 kg/10桶/209 m²/年[138]。很少有菌株,如单独比其他的多14.6%的油,每毫升产生5%的油含有$2.5×10^6$细胞/毫升[115]。图17.14显示了2010年7月1日至2015年12月1日,中国深圳的硅藻产量,在直径0.5 mm、高1 m的有机硅气缸中生长,以测试硅藻油的长期产品产量[137]。该实验包括硅藻的收获,包括半连续收获(每天进行一次)、选择富含油的菌株、控制入侵菌株和控制以硅藻为食的水生动物。水热液化[101]硅藻干重的1/3可转化为生物原油,即36 000升生物原油/公顷,生产300天。图17.14中的数据显示干硅藻平均日产量为0.44吨。300有效日年平均产量为132吨干硅藻/公顷。

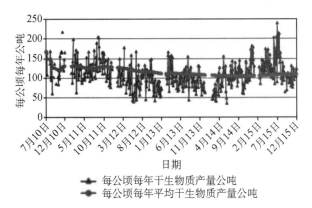

图17.14 2010年7月至2015年12月1日,中国深圳开放式光生物反应器的硅藻产率[137]

17.9.2 从硅藻中获取 Diafuel™

相比之下,Diafuel™需要丰富的基本营养物质的阳光,而CO_2几乎存在于几乎每个地理位置和合适的温度来生长。如果全球数百万经济体将用于开发电动汽车充电资源,而其增长仅将5%用于开发PBRs以收获Diafuel™,那么印度很快就会实现能源独立。

图17.15代表了一个封闭的系统,其中只需要一些基本的营养物质,油可以从微藻中开采出来,而在传统的收获/提取过程中,下游过程需要大量的能量输入来产生高价值的代谢物。收集和粉碎微生物细胞以从藻类生物量中提取脂质是生产藻类生物燃料的标准的、昂贵的、需要能源的、复杂的程序。我们希望选择(或基因工程)硅藻来增加脂质产量,并开发一种技术,在不影响硅藻生长的情况下从硅藻中提取油,允许在保持细胞存活的同时持续挤取生物燃料。我们已经发现了一种这样的印度物种,被称为 *Diadesmis confervaceae*,它有时会自动渗出油,恢复,然后产生更多的

Diafuel™。我们打算寻找、选择和/或基因工程更多的硅藻物种,特别是在中等的嗜热性,能够在温暖的印度条件下产生石油的菌株。Diafuel™ 太阳能电池板新技术的开发将允许当地经济的生产生物燃料,相当于当今太阳能电池板的发电量。由于硅藻利用大气中的 CO_2 来制造石油,硅藻生物燃料太阳能电池板不会增加大气中的 CO_2。

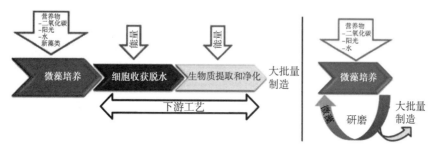

图 17.15　将自榨取作为当前工艺的替代方案的方案[7]

17.10　目前状况

Diafuel™ 的创新将是能源行业的一个福音,这将使印度变成能源充足的国家,每个人都可以在自己的家里轻松收获自己所需要的石油。通过调整含硅藻的储层的大小和相应的数量,可以调整个体的能源需求。

Diafuel™ 是富含脂肪酸的第三代生物燃料,在早期消化过程中转化为浓缩脂质[139]。由于世界原油因消费量超过产量而减少,许多国家正面临危机。仅美国就从其他国家进口了三分之二的石油[140]。印度和中国等发展中国家的能源需求增加,因此从外国市场进口更多的原油越来越多[141]。石油的消耗不仅会导致有限的化石燃料的消耗,而且还会增加温室气体的排放[142]。目前的生物燃料生产来源,如大豆油、玉米油和废植物油,都有其局限性。这些资源不仅需要大量的土地面积,而且它们产生的石油产量远远低于硅藻的产量(见表 17.2)。在藻类中,硅藻本身就能产生 200 倍的油[17]并通过从大气中吸收 CO_2,以脂质的形式生产碳水化合物来减少温室气体的影响[18,143]。常见的酯交换反应可以描述,在催化剂的存在下,形成硅藻原料,即母油。表 17.2 显示了(L/ha)L 和 (Mha)所需面积以及满足美国所有运输燃料需求 50% 的比较。

表 17.2　三代生物燃料的来源比较

庄稼	产油量(L/ha) L	区域需要(M ha)
玉米	172	1 540

(续表)

庄稼	产油量(L/ha) L	区域需要(M ha)
大豆	446	594
油菜籽	1 190	223
麻风树	1 892	140
椰子	2 689	99
油棕	5 950	45
微藻(硅藻)	136 900	2

目前使用的生物柴油的粘度是石油柴油的 10～17 倍。生产生物柴油有四种主要方法：直接使用和混合、微乳剂、热裂解(热解)和最常用的方法是酯交换反应[144]。母油的酯交换反应可以达到接近石油柴油[145]的粘度。生物原油还可用于生产几种增值产品，如生物汽油和生物乙醇。

因此在未来，硅藻的生物炼油厂可能集成几种不同的转换技术生产生物燃料包括生物柴油、绿色柴油和绿色汽油(由催化水处理和催化开裂的植物油)，分别是航空燃料(商业和军事)、乙醇和甲烷，以及有价值的副产品，包括油、蛋白质和碳水化合物。

Diafuel^{TM} 太阳能电池板有潜力使印度独立于进口石油。在像印度这样的发展中国家，不断上涨的油价和产品消费正在增加。随着消费品需求的增加，我们需要通过廉价和资源丰富的方法和措施来开发我们自己的原油来源。硅藻太阳能电池板的创新理念[17]在所有低或高天气条件的国家使用，这可能有利于衡量能源危机，作为影响国家经济的重要参数。我们可以预见，随着研发工作的进行，我们必须解决以下问题：①让硅藻释放它们的油，而我们只需要很少或不需要能量输入；②为硅藻提供足够的气体交换，使它们可以从空气中吸收 CO_2 并释放 O_2；③选择一种对硅藻需要的波长透明，透气，但不透水的材料；④将硅藻保持在室内中等的温度，使用被动透明绝缘；⑤重新设计透明隔热材料，以优化其集光能力，同时实现足够的强度和重量轻的屋顶；⑥确定 Diafuel^{TM} 太阳能电池板的面积，以满足一个典型的印度家庭的能源需求；⑦将 Diafuel^{TM} 太阳能电池板集成到住宅建筑计划中，为新的建设和改造项目提供节能且廉价的服务。

17.10.1 EV 与 Diafuel^{TM} 的数据分析与比较

一个拥有大约 8×10^{10} 个细胞/升培养基的活硅藻培养物将产生 1 189 千克油/年/209 平方米[146]。然而，美国能源记录显示，42 加仑的原油生产 12 加仑的生物燃料和 19 加仑的汽油(美国能源信息管理局，2018)，所以我们的粗略估计显示，1 加仑的原油将产生 0.28 加仑的生物燃料/加仑和 0.45 加仑的汽油/加仑。这种产品对像印度这样的国家来说是一种繁荣,那里的夏天足够强度的自然阳光。在未来,印度有可能成为一个石油出口国,因为许多国家都没有印度那样丰富的自然阳光。表 17.3 显示了电力与 Diafuel™ 驱动车辆的比较。

表 17.3 两种有效能源 diafue™ 的比较(Vinayak,2018 年)和交通、建筑和工业用电

号数	进口	柴油(生物燃料)	电力
1	物理状态	液体(R109)	通过导体的电子流动速率[338]
2	对于即时能量解决方案	最佳选择(远程传输控制 2011)	好选择
3	资源	完全可再生[57]	可或部分可再生[5]
4	成本	非常便宜[41]	昂贵
5	过剩的产能	需要低过剩容量	需要高过剩容量
6	原材料来源	独立	依赖
7	可扩展性	非常容易,不需要任何其他选项	由于依赖于有限的供应,这种可扩展性是困难的;因此选择锌电池,这也是不可再生的[138]
8	消费者的选择	一旦商业屋顶燃料"印度房屋"被安装,这将是第一次投资,每个人都可以拥有自己的油箱,然而,现有的汽车油箱和其他发动机将几乎零成本,并将在国内生产[3,5,19]	大多数人将能够在家里给汽车充电,因此,家庭充电需要升级电网系统,然而,除了安装新的汽车发动机外,安装充电站的成本很高,不是本地生产的[4]
9	加油与充电时间	燃料"将需要五分钟填满一辆车"	电动汽车和其他汽车一样需要几个小时 如果要在公共充电站充电,请办理[17,24,253]
10	加油与充电时间	能不能超过 150 英里/小时	限制车速 40~60 英里/小时[6,19,44]
11	计费管理	消费者自己自己掌握	消费者无法管理,因此由于愤怒/管理不善,无法获得进一步的好处
12	农场主的观点	更合适的	不适当的
13	地理条件	不受任何地理条件的影响,如果阳光较少,LED 灯可以在几乎零能量输入的情况下使用	高度依赖于地理条件,因此并不适合所有人,这就像在车辆退役之前需要更换电池一样[76]
14	维护问题	密封件和软管可能会受到较高百分比的混合物的影响,但可以很容易地改善	高度依赖于地理条件,因此并不适合所有人,这就像在车辆退役之前需要更换电池一样[76]

(续表)

号数	进口	柴油(生物燃料)	电力
15	能源安全影响	这种燃料是国内生产的,可再生的,在其整个生命周期中可以减少高达95%的石油使用量[14]	电力是通过煤炭产生的,其他来源仍然可以显示出依赖性,因此需要首先建立一个备用来源,这增加了能源安全风险[210]
16	零碳排放量	更合适的选择(R99)	太合适的选择
17	温室气体(GHG)和二氧化碳排放	一旦我们用Diafuel™取代目前的汽油,二氧化碳排放量就不会进一步增加,温室气体和二氧化碳排放量也会降低	煤炭电网驱动的电动汽车可能会增加eco。电动汽车并非适用于每个地区,因此需要混合动力汽车以减少CO排放,或电动汽车增加更多的温室气体和CO[120]
18	二氧化硫排放	没有	是的
19	可用资源的直接影响	没有影响	使用四种常见类型的电池:铅酸电池、镍镉电池(NiCd)、镍氢电池(NiMH)和锂离子电池(Li-ion)。在电动汽车。锂离子电池由于其相对于其他类型的电池具有更高的比能量,因此印度没有锂离子储备,这反过来又不得不面对高成本车辆的问题,并可能耗尽地球金属[97]
20	回收利用	简单	回收部分是非常困难的,高度依赖于工业头脑[102,110]
21	行驶距离	可长距离行驶	由于电池续航里程有限,不能用于所有车辆行驶里程,因此在电池电量耗尽后,需要再次使用插电式混合动力[124]

缩写词列表

CHP—热电联产

CO_2—二氧化碳

EV—电动车辆

HEV—混合动力电动汽车

IPCC—政府间气候变化专门委员会

mmbd—百万桶/天

DSP—硅藻太阳能电池板

MTOE—百万吨石油当量

PHEV—插入混合动力汽车

CINDC—有条件的意向国家确定的贡献

Diafuel™—硅藻生物燃料商标

GHG—全球温室气体

INDC—预期的由国家决定的贡献

LCE—低碳排放

Mtce—百万吨煤当量

$MtCO_2$—百万吨CO_2

NITI—改造印度的国家机构

SPV—太阳能光伏

UINDC—无条件预期国家确定的贡献　　WinDS—风力部署系统

17.11　结论

因此相比高价的电动汽车由于高科技基础设施，限制或依赖锂离子电池，持续损耗的地球金属，制造自己的太阳能和风力发电厂基础设施，Diafuel[TM][147]可以是这样的替代生物能源。由硅藻合成的战略和技术合成燃料具有一些环境优势，如减少或可忽略一氧化碳(CO)、碳氢化合物(HC)、颗粒物(PM)的排放，以及其对绿色环境的生物降解性。因此，Diafuel[TM]作为替代能源，如果为像印度这样接受主要太阳能的国家工作，可以很容易地解决与能源相关的问题。由于印度发展得非常快，因此可以为每一个印度家庭开发硅藻太阳能电池板(DSP)。无论是穷人还是富人，DSP都可以为个人家庭生产生物柴油。因此，以太阳、CO_2、H_2、H_2O、少量营养物为代价获得的生物柴油导致油的连续生产，所述油可回收用于家用以及运输目的。因此，Diafuel[TM]将成为领先的可再生资源，有望成为下一代的补充燃料。

参考文献

[1] J, M. (2017). Biofuels vs evs: The union of concerned scientists responds, Biofuel Digest, 1-4.
[2] Pillai, R.K., Ahuja, A. (2015). Electric Vehicles: A Sustainable Solution to Air Pollution in Delhi [ISGF ♯2015/00012 Version 1.0]. New Delhi, India: India Smart Grid Forum.
[3] Vidhi, R., Shrivastava, P. (2018). A review of electric vehicle lifecycle emissions and policy recommendations to increase EV penetration in India. Energies, 11.
[4] Guliyev, I.S., Feizulayev, A.A., Huseynov, D.A. (2001). Isotope geochemistry of oils from fields and mud volcanoes in the South Caspian Basin, Azerbaijan, Petroleum Geoscience, 7(2), 201-209.
[5] Fortman, J.L., Chhabra, S., Mukhopadhyay, A., Chou, H., Lee, T.S., Steen, E. et al. (2008). Biofuel alternatives to ethanol: pumping the microbial well, Trends in Biotechnology, 26(7), 375-381.
[6] Gordon, R., Losic, D., Tiffany, M.A., Nagy, S.S., Sterrenburg, F.A.S. (2009). The glass menagerie: Diatoms for novel applications in nanotechnology, Trends in Biotechnology, 27(2), 116-127.
[7] Vinayak, V., Manoylov, K.M., Gateau, H., Blanckaert, V., Pencréac'h, G., Hérault, J. (2015). Diatom milking: a review and new approaches [with Correction], Marine Drugs, 13(5,12), 2629-2665.
[8] Gautam, S., Kashyap, M., Gupta, S., Kumar, V., Schoefs, B., Gordon, R. et al. (2016). Metabolic engineering of TiO_2 nanoparticles in Nitzschia palea to form diatom nanotubes: an ingredient for solar cells to produce electricity and biofuel, RSC Adv., 6(99).
[9] Aoyagi, K., Omokawa, M. (1992). Neogene diatoms as the important source of petroleum in Japan, Journal of Petroleum Science and Engineering, 7(3-4), 247-262.
[10] Camdana Vinayak., Richard Gordon., Khashti Ballabh Joshi and Benoit Schoefs. "Diafuel." (2018). Trademark application no 3778882. Trade Marks Journal No: 1846, 23/04/2018 Class 4, http://www.ipindia.nic.in/journal-tm.htm GRANTED.
[11] Chisti, Y. (2007). Biodiesel from microalgae, Biotechnology Advances, 25(3), 294-306.
[12] Kooistra, W.H.C.F., Medlin, L.K. (1996). Evolution of the diatoms (Bacillariophyta), IV. A. reconstruction. of. their. age. from. small. subunit. rRNA. coding. regions. and. the. fossil. record. Molecular Phylogenetics &. Evolution, 6(3), 391-407.
[13] Krebs, W.N. (1999). Diatoms in oil and gas exploration The Diatoms. Applications for the Environmental and Earth Sciences. E.F. Stoermer and J.P. Smol. 402-412. Cambridge: University Press, Cambridge.
[14] Gordon, R., Losic, D., Tiffany, M.A., Nagy, S.S., Sterrenburg, F.A.S. (2009). The glass menagerie: Diatoms for novel applications in nanotechnology, Trends in Biotechnology, 27(2), 116-127.
[15] Vinayak, V., Manoylov, K.M., Gateau, H., Blanckaert, V., Pencréac'h, G., Hérault, J. (2015).

Diatom milking: a review and new approaches [with Correction], Marine Drugs, 13(5,12), 2629-2665.
[16] Nouni, M. R., Mullick, S. C., Kandpal, T. C. (2009). Providing electricity access to remote areas in India: Niche areas for decentralized electricity supply, Renewable Energy, 34(2), 430-434.
[17] Ramachandra, T. V., Mahapatra, D. M., B, K., Gordon, R. (2009). Milking diatoms for sustainable energy: Biochemical engineering versus gasoline-secreting diatom solar panels, Ind. Eng. Chem. Res., 48 (19), 8769-8788. Complex Materials II special issue, October.
[18] Sheehan, J., Dunahay, T., Benemann, J., Roessler, P. (1998). A look back at the u. S. Department of energy's aquatic species program: Biodiesel from algae Close-Out Report [NREL/TP-580-24190. Golden, Colorado: National Renewable Energy Laboratory.
[19] Ramachandra, T. V., Mahapatra, D. M., B, K., Gordon, R. (2009). Milking diatoms for sustainable energy: Biochemical engineering versus gasoline-secreting diatom solar panels, Ind. Eng. Chem. Res., 48 (19), 8769-8788. Complex Materials II special issue, October.
[20] Edenhofer, O., Pichs-Madruga, R., Sokona, Y., Matshoss, P., Seyboth, K. (2012). Renewable energy sources and climate change mitigation: Summary for policymakers and technical summary Intergovernmental Panel on Climate Chan.
[21] Reddy, P. J. (2013). Clean Coal Technologies for Power Generation. CRC Press.
[22] Ramachandra, T. V., Hegde, G. (2015). Energy Trajectory in India: Challenges and Opportunities for Innovation, RED, 12(1-2), 1-24.
[23] Vandana Vinayak., Richard Gordon., Khashti Ballabh Joshi and Benoit Schoefs. "Diafuel." (2018). Trademark application no 3778882. Trade Marks Journal No: 1846, 23/04/2018 Class 4, http://www.ipindia.nic.in/journal-tm.htm GRANTED.
[24] Rampen, S. W., Schouten, S., Abbas, B., Elda Panoto, F., Muyzer, G., Campbell, C. N., et al. (2007).
[25] Fischer, M., Werber, M., Schwartz, P. V. (2009). Batteries: Higher energy density than gasoline? Energy Policy, 37(7), 2639-2641.
[26] Chen, C.-Y., Yeh, K.-L., Aisyah, R., Lee, D.-J., Chang, J.-S. (2011). Cultivation, photobioreactor design and harvesting of microalgae for biodiesel production: A critical review, Bioresource Technology, 102(1), 71-81.
[27] Liu, X., Sheng, J., Curtiss, R. (2011). Fatty acid production in genetically modified cyanobacteria, Proc. Natl. Acad. Sci. U.S.A., 108(17), 6899-6904.
[28] Ramachandra, T. V., Hegde, G. (2015). Energy Trajectory in India: Challenges and Opportunities for Innovation, RED, 12(1-2), 1-24.
[29] Wikipedia. (2017). Electricity sector in India, https://en.wikipedia.org/wiki/Electricity_sector_in_India.
[30] Rogelj, J., den Elzen, M., Höhne, N., Fransen, T., Fekete, H., Winkler, H., et al. (2016). Paris agreement climate proposals need a boost to keep warming well below 2℃, Nature, 534(7609), 631-639.
[31] Wikipedia (2018a). Paris agreement list of countries by carbon dioxide emissions, https://en.wikipedia.org/wiki/Paris_Agre emen tWik iped iahttps://en.wikipedia.org/wiki/List_of_countries_by_carbon_dioxide_emissions.
[32] Bordoff, J. (2017). Withdrawing from the paris climate agreement hurts the US, Nat. Energy, 2 (9), 17145.
[33] Hinojosa, F. C., Okonjo-Iweala, N., Polman, P., Stern, N. (2014). The new climate economy, The Global Commission on the Economy and Climate.
[34] Sorokhtin, O. G., Chilingar, G. V., Khilyuk, L. F. (2007). Global warming and global cooling: Evolution of climate on earth, Global Warming and Global Cooling: Evolution of Climate on Earth, 5, 1-313.
[35] Caillon, N., Severinghaus, J. P., Jouzel, J., Barnola, J.-M., Kang, J., And Lipenkov, V. Y. (2003). CO_2 and Antarctic temperature changes across Termination III, Science, 299(5613), 1728-1731.
[36] Fischer, H., Wahlen, M., Smith, J., Mastroianni, D., Deck, B. (1999). Ice core records of atmospheric CO_2 around the last three glacial terminations, Science, 283(5408), 1712-1714.
[37] Berner, R. A., Kothavala, Z. (2001). GEOCARB III: A revised model of atmospheric CO_2 over Phanerozoic time, American Journal of Science, 301(2), 182-204.
[38] Royer, D. L., Berner, R. A., Montañez, I. P., Tabor, N. J., Beerling, D. J. (2004). CO_2 as a primary driver of Phanerozoic climate, GSA Today, 14(3), 4-10.
[39] Medhaug, I., Stolpe, M. B., Fischer, E. M., Knutti, R. (2017). Reconciling controversies about the "global warming hiatus", Nature, 545(7652), 41-47.

[40] Wei, M., Qiao, F. (2017). Attribution analysis for the failure of CMIP5 climate models to simulate the recent global warming hiatus, Sci. China Earth Sci., 60(2), 397-408.
[41] Holba, A. G., Tegelaar, E. W., Huizinga, B. J., Moldowan, J. M., Singletary, M. S., McCaffrey, M. A. et al. (1998). 24-norcholestanes as age-sensitive molecular fossils, Geol, 26(9), 783-786.
[42] Pielke, R. A., Mahmood, R., McAlpine, C. (2016). Land's complex role in climate change, Physics Today, 69(11), 40-46.
[43] Florides, G. A., Christodoulides, P. (2009). Global warming and carbon dioxide through sciences, Environment International, 35(2), 390-401.
[44] GeographyAndYou.com (2017). Energy consumption and energy security a rising concern Integrated Energy Policy: Report of an Expert Committee, Government of India.
[45] Lashof, D. A., Ahuja, D. R. (1990). Relative contributions of greenhouse gas emissions to global warming, Nature, 344(6266), 529-531.
[46] Solomon, S., Plattner, G. K., Knutti, R., Friedlingstein, P. (2009). Irreversible climate change due to carbon dioxide emissions, Proc. Natl. Acad. Sci. U. S. A., 106(6), 1704-1709.
[47] Upadhaya, A. (2017). India to step beyond renewable goal with Chinascale tenders, https://www.bloomberg.com/news/articles/2017-11-24/india-to-step-beyond-renewable-goal-with-china-scale-tenders.
[48] Rogelj, J., Schaeffer, M., Meinshausen, M., Knutti, R., Alcamo, J., Riahi, K., et al. (2015). Zero emission targets as long-term global goals for climate protection, Environ. Res. Lett., 10(10), 105007.
[49] Abhyankar, N., Gopal, A. R., Sheppard, C., Park, W. Y., Phadke, A. (2017). All Electric Passenger Vehicle Sales in India by 2030: Value proposition to Electric Utilities, Government, and Vehicle Owners. Ernest Orlando. Berkeley, California, USA: Lawrence Berkeley National Laboratory.
[50] Stocker, T. F., Qin, D., Plattner, G.-K., Tignor, M. M. B., Allen, S. K., Boschung, J. (2014). Climate change 2013: The physical science basis, Contribution of Working group I to the Fifth assessment Report of IPCC the Intergovernmental Panel on Climate Change. Cambridge University Press, Cambridge.
[51] Knutti, R., Rogelj, J. (2015). The legacy of our CO_2 emissions: a clash of scientific facts, politics and ethics, Climatic Change, 133(3), 361-373.
[52] Matthews, H. D., Caldeira, K. (2008). Stabilizing climate requires near-zero emissions, Geophys. Res. Lett., 35(4).
[53] Rogelj, J., Schaeffer, M., Meinshausen, M., Knutti, R., Alcamo, J., Riahi, K., et al. (2015). Zero emission targets as long-term global goals for climate protection, Environ. Res. Lett., 10(10), 105007.
[54] Van der Hoeven, M. (2011). CO_2 emissions from fuel combustion: highlights. Paris, France: International Energy Agency.
[55] Kearney, D. (2015). India's coal industry in flux as government sets ambitious coal production targets, https://www.eia.gov/todayinenergy/detail.php?id=22652.
[56] Wikipedia (2018a). Paris agreement list of countries by carbon dioxide emissions, https://en.wikipedia.org/wiki/Paris_Agreemen tWik iped iahttps://en.wikipedia.org/wiki/List_of_countries_by_carbon_dioxide_emissions.
[57] EDGAR. (2017). CO_2 time series 1990-2015 per capita for world countries, http://edgar.jrc.ec.europa.eu/overview.php?v=CO_2 ts_pc1990-2015.
[58] Ebinger, C. K. (2016). India's energy and climate policy: Can india meet the challenge of industrialization and climate change? [Policy Brief 16-01]. Washington, DC, USA: Brookings Institution.
[59] Rogelj, J., den Elzen, M., Höhne, N., Fransen, T., Fekete, H., Winkler, H., et al. (2016). Paris agreement climate proposals need a boost to keep warming well below 2℃, Nature, 534(7609), 631-639.
[60] Van der Hoeven, M. (2011). CO_2 emissions from fuel combustion: highlights. Paris, France: International Energy Agency.
[61] Agarwal, P., Ahluwalia, S., Alberti, M., Becerril, C., Benaissa, A., Bharadwaj, R. (2015). International energy agency World energy outlook. 191.
[62] Van der Hoeven, M. (2011). CO_2 emissions from fuel combustion: highlights. Paris, France: International Energy Agency.
[63] Chauhan, C. (2015). Climate change not of India's making, rich nations to blame: Modi, http://www.hindustantimes.com/india/climate-change-not-of-india-s-making-rich-nations-toblame-modi/story-eIBz 5hUG R1CF NNic zGVCWP.html.
[64] Ebinger, C. K. (2016). India's energy and climate policy: Can india meet the challenge of industrialization

and climate change? [Policy Brief 16-01]. Washington, DC, USA: Brookings Institution.
[65] Van der Hoeven, M. (2011). CO_2 emissions from fuel combustion: highlights. Paris, France: International Energy Agency.
[66] Salovaara, J. (2011). CO_2 Emissions Reduction [BA Thesis].
[67] Andres, R. J., Marland, G., Fung, I., Matthews, E. (1996). A $1°\times1°$ distribution of carbon dioxide emissions from fossil fuel consumption and cement manufacture, 1950-1990, Global Biogeochem. Cycles, 10(3), 419-429.
[68] Andrew, R. M. (2018). Global CO_2 emissions from cement production, Earth Syst. Sci. Data. Discuss, 10, 195-217.
[69] Hubacek, K., Guan, D., Barua, A. (2007). Changing lifestyles and consumption patterns in developing countries: A scenario analysis for China and India, Futures, 39(9), 1084-1096.
[70] Deorukhkar, S., Garcia-Herrero, A., And Xia, L. (2014). India Economic Watch, Economic Analysis. India is becoming key for world energy demand: What are the main opportunities and challenges? BBVA Research.
[71] Buckley, T., And Sharda, J. (2015). India's electricity-sector transformation. Cleveland, Ohio, USA: Institute for Energy Economics and Financial Analysis.
[72] Abhyankar, N., Gopal, A. R., Sheppard, C., Park, W. Y., Phadke, A. (2017). All Electric Passenger Vehicle Sales in India by 2030: Value proposition to Electric Utilities, Government, and Vehicle Owners. Ernest Orlando. Berkeley, California, USA: Lawrence Berkeley National Laboratory.
[73] Hassan, M. H., Kalam, M. A. (2013). An overview of biofuel as a renewable energy source: development and challenges, Procedia Engineering, 56, 39-53.
[74] Pressman, M. (2017). Electric Car Incentives In Norway, UK, France, Germany, Netherlands, & Belgium, https://cleantechnica.com/2017/09/02/electric-car-incentives-norway-uk-france-germany-netherlands-belgium/.
[75] Campanari, S., Manzolini, G., De la Iglesia, F. G. (2009). Energy analysis of electric vehicles using batteries or fuel cells through well-to-wheel driving cycle simulations, Journal of Power Sources, 186(2), 464-477.
[76] Khullar, S. (2015). Report on India's Renewable Electricity Roadmap 2030: Toward Accelerated Renewable Electricity Deployment. National Institution for Transforming India, NITI Aayog. New Delhi: Government of India.
[77] Butler, D. (2017). Climate scientists flock to france's call, Nature, 547(7663), 269.
[78] Times of India (2017). MP CM lays foundation for "world's largest" solar power plant, https://m.timesofindia.com/city/bhopal/mp-cm-lays-foundation-for-worlds-largest-solar-powerplant/amp_articleshow/62211204.cms.
[79] Ramesh, M. (2018). Rooftop solar is still out in the cold. 21.1.2018, http://www.thehindu.com/business/Industry/rooftop-solar-is-still-out-in-the-cold/article22486534.ece.
[80] Hindu (2017). Centre moots home rooftop solar grants, http://www.thehindu.com/business/Industry/centre-moots-home-rooftop-solar-grants/article22174311.ece.
[81] Solomon, L. (2017). How tesla's elon musk became the master of fake business, http://business.financialpost.com/opinion/lawrence-solomon-how-teslas-elon-musk-became-the-masterof-fake-business.
[82] Bureau, E. T. (2017). Govt to float tender for adding 20 GW of solar energy capacity., https://economictimes.indiatimes.com/industry/energy/power/india-can-achieve-200-gw-renewable-energy-by-2022-r-k-singh/articleshow/61786415.cms.
[83] Salvi, R., Bbiar, S. B. (2011). Electric vehicle india, /Phttp://www.tekes.fi/fi/gatewayTARGS_0_201_403_994_2095_43/; http://3B/tekesali2%3B7087/publishedcontent/publish/programmes/eve/documents/finpro_electric_mobility_in_india_2013.Pdf
[84] Tanaka, N. (2011). Technology roadmap: Electric and plug-in hybrid electric vehicles, International Energy Agency, Tech. Rep.
[85] Campanari, S., Manzolini, G., De la Iglesia, F. G. (2009). Energy analysis of electric vehicles using batteries or fuel cells through well-to-wheel driving cycle simulations, Journal of Power Sources, 186(2), 464-477.
[86] Brouwer, A. S., Kuramochi, T., van den Broek, M., Faaij, A. (2013). Fulfilling the electricity demand of electric vehicles in the long term future: An evaluation of centralized and decentralized power supply systems, Applied Energy, 107, 33-51.

[87] Garg, A. (2018). All the electric-hybrid cars you can buy in india—mahindra, toyota and more, http://www.news18.com/news/auto/all-the-electric-hybrid-cars-you-can-buy-in-india-tatatiago-tigor-ev-mahindra-toyota-camry-1665293.html.

[88] Green, R. C., Wang, L., Alam, M. (2011). The impact of plug-in hybrid electric vehicles on distribution networks: A review and outlook, Renewable and Sustainable Energy Revie.

[89] Dickerman, L., Harrison, J. (2010). A new car, a new grid, IEEE Power and Energy Mag., 8(2), 55-61.

[90] Duvall, M. (2002). Comparing the benefits and impacts of hybrid electric vehicle options for compact sedan and sport utility vehicles. Palo Alto, California, USA: Electric Power Research Institute.

[91] Li, X., Lopes, L. A. C., Williamson, S. S. (2009). "On the suitability of plug-in hybrid electric vehicle (PHEV) charging infrastructures based on wind and solar energy." In Power & Energy Society General Meeting. PES'09, Ieee Xplore, 1-8.

[92] Dickerman, L., Harrison, J. (2010). A new car, a new grid, IEEE Power and Energy Mag., 8(2), 55-61.

[93] Shukla, S., Patel, N., Kharul, R., Pillai, G. M., Sawyer, S. (2012). India wind energy outlook: 2012. Brussels, Belgium: Global Wind Energy Council.

[94] Sawin, J. L., Seyboth, K. M., Sverrisson, F. (2017). Renewables 2017 global status report. Paris, France: REN21 Secretariat. Nam.

[95] Becker, T. A., Sidhu, I., Tenderich, B. (2009). Electric vehicles in the united states: A new model with forecasts to 2030 center for entrepreneurship & technology, university of california. Berkeley, California, USA: Berkeley.

[96] Tahil, W. (2007). The trouble with lithium; implications of future PHEV production for lithium demand, http://meridian-int-res.com/Projects/Lithium_Problem_2.pdf.

[97] Salvi, R., Nambiar, S. B. (2011). Electric vehicle india, /Phttp://www.tekes.fi/fi/gatewayTARGS_0_201_403_994_2095_43/; http://3B/tekesali2%3B7087/publishedcontent/publish/programmes/eve/documents/finpro_electric_mobility_in_india_2013.pdf.

[98] Nykvist, B., Nilsson, M. (2015). Rapidly falling costs of battery packs for electric vehicles, Nature Clim. Change, 5(4), 329-332.

[99] BankBazaar.com (2018). Petrol price in india today, https://www.bankbazaar.com/fuel/petrolprice-india.html.

[100] Toor, S. S., Rosendahl, L., Rudolf, A. (2011). Hydrothermal liquefaction of biomass: A review of subcritical water technologies, Energy, 36(5), 2328-2342.

[101] Biller, P., Ross, A. B. (2011). Potential yields and properties of oil from the hydrothermal liquefaction of microalgae with different biochemical content, Bioresource Technology, 102(1), 215-225.

[102] NYISO. (2009). Alternate route: Electrifying the transportation sector. Potential impacts of plug-in hybrid electric vehicles on new york state's electricity system, New York Independent System Operator, New York, NY, USA.

[103] Judd, S. L., Overbye, T. J. (2008). An evaluation of PHEV contributions to power system disturbances and economics 40th North American Power Symposium. 1-8. Calgary, AB, Canada: NAPS'08. IEEE.

[104] Li, X., Lopes, L. A. C., Williamson, S. S. (2009). "On the suitability of plug-in hybrid electric vehicle (PHEV) charging infrastructures based on wind and solar energy." In Power & Energy Society General Meeting. PES'09, Ieee Xplore, 1-8.

[105] Venayagamoorthy, G. K., Mitra, P., Corzine, K., Huston, C. (2009). Real-time modeling of distributed plug-in vehicles for V2G transactions Energy Conversion Congress and Exposition. pp.3937-3941. IEEE, San Jose, CA, USA: ECCE 2009.

[106] Green, R. C., Wang, L., Alam, M. (2011). The impact of plug-in hybrid electric vehicles on distribution networks: A review and outlook, Renewable and Sustainable Energy Reviews, 15(1), 544-553.

[107] Ma, F. R., Hanna, M. A. (1999). Biodiesel production: a review, Bioresour. Technol., 70(1), 1-15.

[108] Abubakar, L., And Mutie, A. (2012). Characterization of algae oil (oilgae) and its potential as biofuel in Kenya, Journal of Applied Phytotechnology in Environmental Sanitation, 1(4), 147-153. Agarwal, P., Ahluwalia, S., Alberti, M., Becerril, C., Benaissa, A., Bharadwaj, R. (2015).

[109] Swann, G. E. A., Patwardhan, S. V. (2011). Application of fourier transform infrared spectroscopy (ftir) for assessing biogenic silica sample purity in geochemical analyses and palaeoenvironmental research,

Clim. Past, 7(1), 65-74.
[110] Rampen, S. W., Schouten, S., Abbas, B., Elda Panoto, F., Muyzer, G., Campbell, C. N., et al. (2007). On the origin of 24-norcholestanes and their use as age-diagnostic biomarkers, Geol, 35(5), 419-422.
[111] Gillies, T., And Poulin, B. (2015). The efficient house innovation: Healthful, efficient and sustainable housing for northern and southern climates. In Building Today—Saving Tomorrow: Sustainability In Construction And Deconstruction Conference Proceeding. 44-57 M Panko and L Kestle (eds.) Auckland, New Zealand: Unitec Institute of Technology.
[112] Singh, R. K., Sundria, S. (2017, 25.1.2017). Living in the dark: 240 million indians have no electricity. 50 million rural homes without power despite idle generation, https://www.bloomberg.com/news/features/2017-01-24/living-in-the-dark-240-million-indians-have-no-electricity.
[113] Vinayak, V., Kumar, V., Kashyap, M., Joshi, K. B., Gordon, R., Schoefs, B. (2017). Fabrication of resonating microfluidic chamber for biofuel production in diatoms (Resonating device for biofuel production). In 3rd International Conference on Emerging Electronics (ICEE). Mumbai, IndiaS Ganguly and D Saha. IIT BombayIEEE Electron Devices Society.
[114] Kaushika, N. D., Sumathy, K. (2003). Solar transparent insulation materials: a review, Renewable and Sustainable Energy Reviews, 7(4), 317-351.
[115] Vinayak, V., Gordon, R., Gautam, S., Rai, A. (2014). Discovery of a diatom that oozes oil, adv. sci. lett., 20(7), 1256-1267.
[116] Garg, V. (2017). India's Energy Transition: Mapping subsidies to fossil fuels and clean energy in India. GSI Report. Winnipeg, Manitoba, Canada: International Institute for Sustainable Development.
[117] Vinayak, V. (2016). TITLE [local, under roof production at human temperatures, using transparent insulation] Conference. Editor. Place: Publisher.
[118] Trentacoste, E. M., Shrestha, R. P., Smith, S. R., Gle, C., Hartmann, A. C., Hildebrand, M., et al. (2013). Metabolic engineering of lipid catabolism increases microalgal lipid accumulation without compromising growth, Proc. Natl. Acad. Sci. U.S.A., 110(49), 19748-19753.
[119] Wikipedia. (2018c). Energy policy of India, https://en.wikipedia.org/wiki/Energy_policy_of_India.
[120] Jha, S. P. (2017). India Has a Better Option Than Electric Cars, https://thewire.in/144021/india-better-option-electric-cars/.
[121] Singh, R. K., Sundria, S. (2017, 25.1.2017). Living in the dark: 240 million indians have no electricity. 50 million rural homes without power despite idle generation, https://www.bloomberg.com/news/features/2017-01-24/living-in-the-dark-240-million-indians-have-no-electricity.
[122] McGlade, J. (2016). The Emissions Gap Report 2016: A UNEP Synthesis Report. Nairobi, Kenya: United Nations Environment Programme.
[123] Commission, E. P. (2011). Energy BOOK? pp. 342-393.
[124] Times, E. (2017). Units 5, 6 at Kudankulam nuclear power plant to cost Rs 50 000 crore, https://economictimes.indiatimes.com/industry/energy/power/units-5-6-at-kudankulam-nuclearpower-plant-to-cost-rs-50000-crore/articleshow/58959079.cms.
[125] Nouni, M. R., Mullick, S. C., Kandpal, T. C. (2009). Providing electricity access to remote areas in India: Niche areas for decentralized electricity supply, Renewable Energy, 34(2), 430-434.
[126] Tuli, V., Khera, A. (2014). India: Towards energy independence 2030. McKinsey & Company. U.S. Administration, E. I. (2018). Refining crude oil, https://www.eia.gov/energyexplained/index.cfm?page=oil_refining#tab3.
[127] Khare, V., Nema, S., Baredar, P. (2013). Status of solar wind renewable energy in India, Renewable and Sustainable Energy Reviews, 27, 1-10.
[128] Short, W., Denholm, P. And . And (2006). A preliminary assessment of plug-in hybrid electric vehicles on wind energy markets [Technical Report NREL/TP-620-39729] National Renewable Energy Laboratory. Colorado, USA: Golden.
[129] van der Hoeven, M. (2013). World energy outlook [slides]. Tokyo, Japan: International Energy Agency.
[130] Ghosh, S., And Sengupta, P. P. (2011). Energy management in the perspective of global environmental crisis: An evidence from India. In International Conference on Management and Service Science. 1-5. MASS (ed). China: IEEE.
[131] Raghavan, S. (2016). India wastes 15%-20% of its renewable energy due to lack of storage: Panasonic

Energy head, http://www.thehindu.com/business/Industry/India-wastes-15-20-of-its-renewable-energy-due-to-lack-of-storage-Panasonic-Energy-head/article14638013.ece.

[132] Huo, H., Zhang, Q., Wang, M. Q., Streets, D. G., He, K. (2010). Environmental implication of electric vehicles in China, Environ. Sci. Technol., 44(13), 4856-4861.

[133] Conti, J. J. (2007). Annual energy outlook 2007, with projections to 2030 [DOE/EIA-0383]. energy information administration, office of integrated analysis and forecasting, U, S. Department of Energy, Washington, DC, USA.

[134] Birol, F. (2007). World Energy Outlook 2007: China and India Insights. Paris, France: International Energy Agency.

[135] Streets, D., Hao, J., Wu, Y., Jiang, J., Chan, M., Tian, H., et al. (2005). Anthropogenic mercury emissions in China, Atmospheric Environment, 39(40), 7789-7806.

[136] Jacobson, M. Z., Delucchi, M. A. (2009). A path to sustainable energy by 2030, Sci. Am., 301(5), 58-65.

[137] Wang, J.-K., Seibert, M. (2017). Prospects for commercial production of diatoms, Biotechnol. Biofuels, 10(1), #16.

[138] Graham, J. M., Graham, L. E., Zulkifly, S. B., Pfleger, B. F., Hoover, S. W., Yoshitani, J. (2012). Freshwater diatoms as a source of lipids for biofuels, J. Ind. Microbiol. Biotechnol., 39(3), 419-428.

[139] Levitan, O., Dinamarca, J., Hochman, G., Falkowski, P. G. (2014). Diatoms: a fossil fuel of the future, Trends in Biotechnology, 32(3), 117-124.

[140] Darzins, A., Pienkos, P., Edye, L. (2010). Current status and potential for algal biofuels production: A report to iea bioenegy task 39 Report T39-T2, A report to IEA Bioenergy Task, 39.

[141] Khan, S. A., Rashmi, H., Hussain, M. Z., Prasad, M. Z., S. Banerjee, Banerjee, U. C., (2009). Prospects of biodiesel production from microalgae in India, Renewable and Sustainable Energy Reviews, 13(9), 2361-2372.

[142] Dincer, I. (2000). Renewable energy and sustainable development: a crucial review, Renewable and Sustainable Energy Reviews, 4(2), 157-175.

[143] Hu, Q., Sommerfeld, M., Jarvis, E., Ghirardi, M., Posewitz, M., Seibert, M. et al. (2008). Microalgal triacylglycerols as feedstocks for biofuel production: perspectives and advances, Plant J., 54(4), 621-639.

[144] Ma, F. R., Hanna, M. A. (1999). Biodiesel production: a review, Bioresour. Technol., 70(1), 1-15.

[145] Samantray, S., Aakanksha, S. G., Ramachandra, T. V. (2010). Prospects of diatoms as third generation biofuel Lake 2010: Wetlands, Biodiversity and Climate Change. PLACE: PUBLISHER.

[146] Walker, D. A. (2009). Biofuels, facts, fantasy, and feasibility, J. Appl. Phycol., 21(5), 509-517.

[147] Vandana Vinayak., Richard Gordon., Khashti Ballabh Joshi and Benoit Schoefs. "Diafuel." (2018). Trademark application no 3778882. Trade Marks Journal No: 1846, 23/04/2018 Class 4, http://www.ipindia.nic.in/journal-tm.htmGRANTED.

第 18 章

泡沫农业：硅藻生物燃料和下一次绿色革命的可扩展缩影

Richard Gordon，Clifford R Merz，Shawn Gurke，Benoît Schoefs

18.1 简介

藻类生物燃料尚未实现与化石燃料和其他绿色能源竞争的资本[2-6]，尽管越来越多的证据表明它们是合理的替代品[7-11]。与几年前的预期相反[12-14]，化石燃料的价格目前很低（2019 年），并可能持续很长一段时间[15]。尽管政策企业家、政治家和环保人士强调电动汽车[15-19]，全球原油产量和消费正在增长，预计从目前每天 1 亿桶/天增长（见图 18.1），价格为每桶 60 美元[20]，因此每年总值为 2 万亿美元。相比之下，2016 年全球生物燃料市场为 1 680 亿美元，与原油相比为 8.4%，生物柴油价格为每桶 110 至 120 美元[21]。

图 18.1　全球石油和其他液体燃料[1]

1 桶＝42 加仑或 158.987 3 升原油

已经发现用于生长藻类物质的两种主要技术不能与化石燃料竞争，尤其是在目前的低成本下：露天跑道反应器[22-23]（见图 18.2）和封闭式光生物反应器[24-26]（见图 18.3）。光生物反应器的财政可行性大大低于开放式跑道反应器。在基本情况下，包括财务成本在内的脂类生产平均总成本，对于开放式跑道和光生物反应器分别为 12.73 美元/加仑（3.36 美元/升）和 31.61 美元/加仑（8.35 美元/升）[27]。

图 18.2 美国新墨西哥州 Qualitas Health 跑道藻类生产农场[28]

图 18.3 葡萄牙的藻类光生物反应器藻类农场[29]

为了抵消生产藻类生物燃料的成本,人们已经提出其他目标,如废水处理[30-42], CO_2 隔离[43-48],或生产高价值的产品(HVPs)(见表 18.1)[49-56],以及政府补贴[57]和激励措施[58]。低价值的生物燃料生产本身在经济上是不可行的[59-61]。相反,所谓的综合方法,即生物炼制服务于多个过程,一些有利可图的,已被提出作为必要的,以抵消藻类生物燃料的高成本[62-68]。事实上,一些多年来一直有生产生物燃料的坚定目标的公司,在试图生产替代产品后,现在正在软化向生物燃料过渡的立场,这已经是一种财务生存战略[69]。

因此,大宗商品中最需要的生物燃油成了事后诸葛亮,特别是当 HVPs 的值比原油高 7 个数量级(1 700 万倍)时(见表 18.1 和图 18.4)。等待"石油峰值"及其对化石燃料价格高企的预期后果已被证明是徒劳的,因为石油峰值的概念是一个不断移动的目标[70-73]。如果石油需求得以实现,预计它将从 2024 年至 2030 年开始下降[74],如果它实现也可能有助于将原油价格保持在较低水平[75]。2030 年的目标是每加仑 3 美元(每升 0.79 美元)的藻类生物燃料[69,76],可能不够低,特别是考虑到预计每桶石油 40 美元作为其新的低盈亏平衡点[77]。生物燃料的经济可行性仍然难以捉摸[78],部分原因是进口石油价格的波动决定初创公司的命运,因为当生物燃料成本超过化石燃料油成本时,投资者就会退出[79-80]。

表 18.1 可通过泡沫养殖生产的藻类高价值产品示例[83-85]

高价值产品(HVP)	价格/千克
原油(供比较)	$0.15~$1.11
硅藻土,食品级,用于比较	$10
人类食物中的蛋白质[467]	$22
葡萄糖胺[40]	$35~$106
B-藻红蛋白[41]	$50

(续表)

高价值产品(HVP)	价格/千克
欧米茄-3 油(50%～75%纯)[76,78,82,86]	$60～$185
贝塔胡萝卜素[91,100,102]	$300～$3 000
虾青素[103,105,122,125]	$2 000～$14 000
欧米茄-3 油(99%纯)[132]	$2 154
C-藻青蛋白[152]	$500～$100 000
藻胆蛋白[154]	$15 000～$69 000
人类抗体[31,51,53]	$2 500 000
用于太阳能电池板、电池和其他在体内或分离后功能化的纳米技术的纯化硅藻壳(结晶体)[112,126,128,134,153]	$2 000

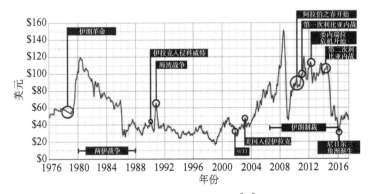

图 18.4　石油价格变化图[92]

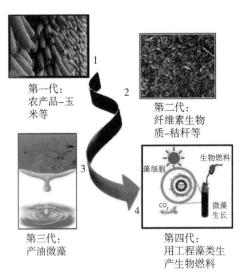

图 18.5　生物燃料研发的四代人[81]

生物燃料的历史被分为"四代"[81-82](见图 18.5)。第一代和第二代的原料主要来自陆生植物和动物的废物流,这使人们对生物燃料生产和粮食生产之间的土地和营养物质的竞争感到担忧。相比之下,第三代和第四代则以藻类原料为基础,从而减少了传统粮食作物基生物燃料中固有的对食物和燃料的担忧。这种 4 部分的生物燃料分类没有明确包括"藻类榨取"(生物相容性提取),它在不杀死藻类的情况下去除油,使它们可以重复使用[86-90]。榨取可能属于第三代和/或第四代,这取决于如何实现,或者可能应该被标记为第五代。传统上,藻类生物量被生长,细胞被破坏和杀死,以提取脂质[91],这些脂质被加工成生物燃料。我们不是通过将奶牛磨碎并提取乳液来获取乳液的[89],那么,为什么我们要这样对待藻类呢? 其结果是,将藻类油从其他细胞成分中分离出来并进行脱水的成本巨大[78]。就像奶牛产奶一样(见图18.6),我们可以预计在未来几十年,"乳液"藻细胞的脂质生产成本将大幅下降。

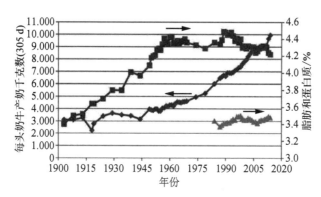

图 18.6　丹麦全国奶牛群体的产奶量、脂肪和蛋白质含量的历史变化

18.1.1　泡沫农业概念

我们想提供一种替代滚道和光生物反应器的方法:大量相对较小的"气泡",形成孤立的、模块化的容器,用于生长藻类。我们设想用充满水的气泡中的藻泡覆盖土地,作为一种廉价的生产生物燃料的替代方式。取代每个气泡内的空气的将是一个水溶液中的藻类培养基。例如,在气泡中培养的硅藻可能包括不同物种的混合物[96],可能是与特定的共生细菌[97]或蓝藻细菌[98]组成的联盟[99]。气泡的大小将覆盖特定地区的生长季节,有足够的质量使气泡包裹起来压住风,并紧密包装覆盖大部分地面。但与开放的跑道池相比,每个气泡都很小,其内容彼此保护,不受蒸发、污染、真菌[100]和食草动物[101-102],气泡内表面是无菌的[103-104]。模块化使得气泡农场可扩展到任何规模的农场,包括那些包括部分/平方千米的小麦和其他主要作物的农场,提供专门的装载设备[105-106],铺设、收获、回收、再利用和回收气泡膜。泡沫农业实例化了我们的生物燃料生产硅藻太阳能电池板的概念[89],但规模很大。

在农业中使用大面积塑料已经变得很普遍(2015年使用量为1 100万吨[107],用干草包的塑料包装(见图18.7)[108-110],棉包(见图18.8)和塑料覆盖,在土地上铺设塑料薄膜,防止有害植物生长并与选择的

图 18.7　一个正在工作的捆包装器(形成一长排塑料包装的干草捆)[93-94]

作物竞争[111-117]（见图 18.9），所以我们提出的大片被泡沫包裹的农田的概念并不奇怪。

图 18.8　塑料包装的棉花包[95]

图 18.9　用于覆盖的塑料膜（可以控制杂草、害虫和土壤温度和保持水分）[117]

使用气泡包装进行藻类培养的想法来自于对作物植物和藻类之间的根本区别的思考，植物提供物理支持和内部隔室，生长在我们收获的每一种作物的部分。谷物生长在草鞘内的草茎的末端。水果生长在节水和抗病原体的皮肤或果皮中。螺肉通常在壳内形成。根状作物有上皮细胞作为屏障。棉纤维、豆类和豌豆会留在豆荚里，直到完全形成。但藻类没有这些保护，所以我们需要提供他们需要的水环境的物理支持和膜保护他们免受病原体，捕食者，竞争对手和失水，同时提供的气体交换叶子：白天 CO_2 进入，晚上排出 O_2，反之亦然。气泡膜中的气泡可以被认为是藻类中光合叶绿体的叶状支撑，即人造叶子[118-120]，尽管里面有正常的光合作用。类似于它的食品保护功能[121]，气泡膜将延长藻类生长的时间，保护藻类免受腐败和浪费，这是开放的跑道无法实现的，而且成本低于使用封闭的光生物反应器。

在气泡包装中生长微生物的想法始于当时 13 岁的 Matthew Huber[122-124]，并已被用于实验室规模的培养和材料存储[103-104,125-131]。其他已经设想了大面积用于光催化制氢的充水塑料袋[132]或用于港口城市近海藻类废水处理的漂浮塑料袋类似的漂浮袋光生物反应器，也不是孤立的微观世界[133-135]。多孔塑料袋正在测试在太空种植藻类[136-137]。CO_2 吸收一个含有需要低光照的异囊蓝藻的壁板已用于室内使用。在较小的规模上，水被封装在 PDMS（聚二甲基硅氧烷）中[138-142]，一种典型的透气性塑料。然而，藻类的气泡养殖的概念似乎是新颖的，它结合了封装、气体交换、保水和覆盖在生物燃料农场地面上的大片气泡包裹的物理完整性。

藻类油的生产现在涉及到许多步骤[见图 18.10(a)～(e)]。通过选择性育种[143-144]或基因工程，有可能将藻类石油生产的生物化学转向类似汽油或柴油的石油，不需要很少或不需要进一步处理，并可能诱导自我榨取[145-146]。然后，实际上，所有的生物炼制步骤[147]都将由藻类自己来完成[见图 18.10(f)]。最终目标将是实现高辛烷值汽油的获取[89,148-149]和柴油燃料,允许继续使用内燃机[150],与最近的预测相反[61]。

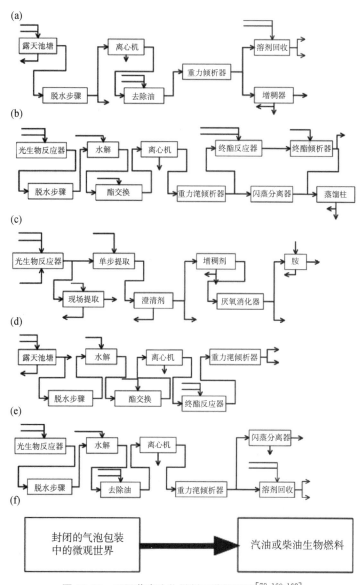

图 18.10 不同藻类生物燃料工艺流程图[78,160-162]

(a) 产生三酰甘油酯(TAG)的开放式池塘系统;(b) 产生游离脂肪酸甲酯(FAME)的太阳能光生物反应器;(c) 发光二极管(LED)发光光生物反应器;(f) 泡沫农业,终极,一步到位的愿景;目前,藻类生物量将被加工成(a)~(e)中的任何一种,脱水和离心通常都很昂贵

18.1.2 可能通过无人机进行气泡注射、取样、收获和密封

"用带针或移液管尖端的注射器注入样品,然后用指甲硬化剂密封孔,很容易填

充气泡包裹的气泡"[103],这显然是可以实现自动化的。例如,农业无人机[151-154]可以通过光学或取样评估藻类[155]和石油[156]的生长,并且只收获那些准备好的。更大的无人机[157],就像那些用于喷洒的无人机,可以携带水,注入水,密封孔,使气泡干燥,对内容物进行取样,或收集每个气泡顶部的石油。容量、能量使用和飞行距离之间的平衡必须通过计算机进行编程。有了无人机,气泡膜可能会在田野上停留超过一个生长季节。尽管会越冬,但较少的处理可能会延长它的寿命。无人机还克服了设备相对于气泡的精确放置的问题,因为无人机软件可以定位每个气泡。密封和避免撕裂可以用具有粘性表面的贴片来实现,该贴片粘合到上部塑料片的外表面上。已经存在用空气[106,158]或液体食品[159]填充预制空气泡的设备,并且可以作为用水填充或清空气泡的设备的模型。湍流可用于去除生物膜以进行收获。

18.1.3 方法

我们的方法是概述所预见的问题和解决方案,并提出调查它们所需的研究成果。但我们可以把它简化为三个概念。

(1) 在密封的气泡微型群落中种植藻类的气泡包装农场提供了一种即时、廉价的用于生物燃料提取的藻类生物量种植方法,作为工业炼油厂的原料(见图18.11)。

(2) 考虑到气泡包装的使用范围和年度更换情况,评估可重复使用或可回收/可降解的气泡包装塑料是必要的。

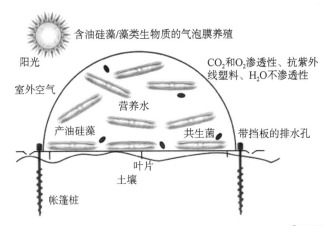

图18.11 一种用于生物燃料提取的藻类生物量种植方法示意图[102,167-171]

(3) 从长远来看,可以选择或设计在气泡中直接产生生物燃料的藻类,以便直接收获和立即使用,而无需进一步处理(见图18.12)。石油应该上升到每个气泡的顶部,在那里它可以定期或在一个生长季节后被清除。例如,对于"自我榨取"硅藻,对榨取的干预[90]将是不必要的。

第一个目标可以在短时间内实现,只需要CO_2和透氧塑料气泡,合理的不透水,

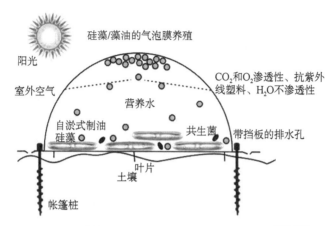

图 18.12 藻类在气泡中直接产生生物燃料的示意图[171-172]

这样水就可以持续整个生长周期或季节。第二个和第三个目标要求更高,需要许多研究途径,预计还需要几年时间。接下来的大部分目标都是针对第三个目标,即最困难的目标。

18.2 机械性能

农田可能是塑料制品的恶劣环境。地面很粗糙。气泡膜的任何相对运动都可能引起磨损。在温带气候下,它会受到加热、冷却,在季节开始或结束时受到寒冷的温度。它的内部是湿的,外面是干燥的,可能出现弯曲(这就是为什么很少有水族馆是由塑料制成的)。它必须承受冰雹,让雨水流失,不能被泥巴覆盖,泥巴会阻挡阳光。它必须抑制或保护木质或杂草杂草植物的生长,这可能导致穿刺,并被土壤细菌消化[163]底部和顶部的塑料薄膜需要不同的机械性能和其他性能。底部的薄膜需要更抗刺,而顶层必须抵抗冲击,如冰雹或鸟类。气泡之间的凹槽可能成为针晶体[164-166]和其他生物的栖息地。

气泡弯曲超过滚道的一个优点是土地不需要非常平坦,因为每个气泡都保留了自己的水。然而,会有一个最大的坡度,这将取决于水的质量和底部塑料和下面的地面之间的摩擦系数,气泡膜会滑出位置。可能需要从底层突出的叶片或肋骨,或带刺的尖刺(如帐篷钉)来抓住土壤,以防止滑动(见图 18.11 和图 18.12)。如果设计得当,这种支撑和锚点也可以增加地下低洼作物的空气空间。即使是很小的气隙也可以缓解土壤真菌生长的问题[117]。

18.2.1 最佳气泡尺寸

在真空模板制造过程中,气泡尺寸的最佳选择由真空模板控制,将取决于许多因素:

(1) 较大的气泡将对温度波动产生较慢的、阻尼的响应,减少过热或冻结的机会。

(2) 较大的气泡的表面体积比较小,从而使水的损失更慢。然而,它们的散热速度也会更慢。

(3) 更大的气泡需要更强的塑料,因为它们进一步投射到边界层进行空气流动,而且因为塑料片会受到水的重量和风切变的拉伸。

(4) 每单位面积的总质量(如大部分是封装的水)必须足够高,以防止气泡膜拍打、提起和吹走,或在风中爆裂[173],这可能需要现场试验和/或风洞试验,以允许阵风[174-177]。

(5) 在较小的气泡中,气体交换更好,在这些气泡中,对流对整个表面的气体交换并不重要[178],然而,与同等体积的半球相比,大气泡会倾向于变平变宽,气体交换的表面积增加。大的气泡可能会产生热梯度,从而驱动其内部的对流。

(6) 生物量的能力随着气泡大小的增大而增加。

(7) 水对紫外线的过滤随着气泡厚度的增加而增加[179],这将保护较低的塑料片和生长在它上面的藻类。

(8) 用于填充或收获气泡内容物的农场设备不应该有紧密的水平定位公差,这表明较大的气泡会更好。

(9) 较大的气泡需要更少的填充/排空操作,尽管并行处理允许补偿较小的气泡所需的时间。

(10) 较大的气泡聚焦的光较少。

一些目前可用的气泡包装产品及其尺寸见表 18.2。

表 18.2　泡沫纸和相关空气枕商业产品的尺寸参数[185]

产品名称	泡尺寸	泡面积/m^2	泡体积/cm^3	每平方米水的质量/kg	表面积/体积/cm^{-1}
小泡膜	3/8″(0.952 cm)直径×3/16″(0.476 cm)高度	15 500	0.34	5.2	8.4
中泡膜	1″(2.54 cm)直径×5/16″(0.794 cm)高度	2 200	4.0	8.8	4.1
大泡膜	1.25″(3.175 cm)直径×1/2″(1.27 cm)高度	1 400	10	14	2.8
空气枕	10 cm×20 cm×~5 cm 厚度	50	1 000	50	0.7
空气枕	20 cm×20 cm×~5 cm 厚度	25	2 000	50	0.6
可能	1 m×1 m×10 cm 厚度	1	100 000	100	0.24

18.3 光学性质

气泡膜的顶层对于支持光合作用的波长应该是透明的,如果红外(IR)会在特定的区域加热,也应该是红外(IR)波长的反射。当然,红外反射器应该传输可见光[180]。防反射涂层会增加进入气泡的阳光量,并减少被气泡包裹的土壤的反照率,这很重要,因为在任何给定的时间,气泡包裹都有可能覆盖相当一部分的土地。它应该外部涂层紫外线不透明,以保护塑料,将暴露在充分的阳光下,以延长其寿命[181]。打捆塑料防紫外线,使用一年。而温室用塑料用 50 μm 涂层保护 10 年[182-183]。基本上,上层应该反射或吸收水的可见光透明窗口外的部分或所有波长[179],如作为一个带通滤波器[184]。

无柄气泡的形状使它成为一个平凸透镜,因此光线将在一定程度上聚焦在气泡内,水滴越大,它就越平坦,焦距也就越长[186-187]。形状取决于上部塑料片的弹性张力(而不是水的张力),但在气泡的体积和气泡内部光分布的平坦程度之间存在权衡。这为气泡大小提供了一个优化参数。此外,表面塑料可以纹理像在菲涅耳透镜用于一些混合电力太阳能电池板[188-190]。生活在弱光下的植物有光聚焦表面结构[191]。如果运动的羽状(针晶体)硅藻在气泡中生长,它们会沿着内表面移动,自我优化照明[192-194],并随着太阳方向的变化继续这样做。典型速度为 10 $\mu m/s$[195],例如,一种硅藻可以在一小时内移动到 3.6 cm 的距离。因此,在设定最佳气泡大小时,也需要考虑硅藻速度和焦点轴运动之间的平衡。此外,光梯度下的运动行为随硅藻种类的不同而变化[196]。

底部的塑料片在阳光下可能比气泡中的较高水平更温暖,特别是考虑到其藻类垫或下面的土壤/表面吸收光。这将在气泡内建立对流[42,197-198],这可以阻止任何浮游藻类沉降,可能会增加生物量。此外,底层可以反射支持光合作用的波长,以最大限度地利用阳光,使其第二次通过气泡内的介质,例如,铝化聚酯薄膜[199]。

在晴朗的天空中,夏季正午的阳光强度约为 3 000 μmol 光子 $m^{-2} \cdot s^{-1}$,而最佳光合作用已经达到约 300 μmol 光子 $m^{-2} \cdot s^{-1}$[65]。气泡顶部的光致变色层[200]可以用来阻挡过量的阳光,硅藻还有可以聚焦机制和叶绿体迁移(核光),保护它们免受过量光[201]。

18.4 表面性质

塑料顶层的外表面最好是自清洁的,可能是超疏水的,这是通过复制龙舌兰叶表面获得的特性[202]。底层的外表面可以是粗糙的或鳍状的,以增加滑动摩擦,从而减少由于风和斜坡引起的气泡包裹的运动。它必须对土壤细菌的生物降解具有抗性[203]。

内底表面可以具有疏水性,允许硅藻的最佳粘附性[204-206]和生物膜的形成[207]。

有图案的内表面可能有利于某些物种[208],并增加产量[209],粗糙的表面也一样[210]。内表面可以是亲水的[211],抑制硅藻粘附,防止流出的油粘在其上。

18.4.1 气体交换特性

泡沫养殖需要高 CO_2 渗透性的塑料[212-213]。同样,我们需要高 O_2 渗透性,因为气体不透水的光生物反应器中会导致 O_2 积累[214-215],有时甚至室外培养[216]也会抑制光合作用。但是气泡养殖也需要低水渗透性,所以水停留在气泡膜的每个气泡中,最好是在整个生长季节,或直到收获。在本节中,我们将考虑许多材料,其中一些材料为实现这种渗透性能的组合提供了前景。因此,我们寻求的是一种对 O_2 和 CO_2(而不是水)具有选择性的膜,但是一般来说,渗透性和选择性之间存在权衡关系,高渗透性材料的选择性较低,反之亦然[217]。

从文献中可以收集到相同塑料的 CO_2、O_2 和 H_2O 的绝对扩散系数(见表18.3)。选择这些扩散系数是为了一种比较运输的统一方法,尽管偶尔会使用其他措施。由于水分子小于氧分子,一般认为其扩散系数应该较大[218]。因此,寻找一种对 CO_2 和具有高度渗透性的不透水塑料可能是有问题的,需要进行权衡。因此,我们将讨论一些特殊的情况,而不是那么常见的膜替代品,一些是受生物学的启发。

表 18.3 各种塑料中二氧化碳、氧气和水的扩散系数[136,227-235]

材料	CO_2	O_2	H_2O	比例
全氟磺酸基聚合物[121,122]		$0.24 \sim 6 \times 10^{-6}$	$2 \sim 4 \times 10^{-12}$	这里提出 3 000 000
尼龙 6[45,47,48,51,52]	22×10^{-9}	$3 \sim 40 \times 10^{-10}$ 10.6×10^{-9}	$3.4 \times 10^{-10} \sim 0.8 \times 10^{-7}$	6.5
聚三氟氯乙烯[63]	7.08×10^{-8}			
聚二甲基硅氧烷+30%沸石[56]		1.32×10^{-5}		
二甲基硅氧烷[61,63,65,69]	$2.2 \sim 11.3 \times 10^{-5}$	$0.054 \sim 3.4 \times 10^{-5}$	2×10^{-5}	1.7
二甲基硅氧烷交联[62]	7.08×10^{-10}			
聚对苯二甲酸乙二醇酯[71-76]	$2.16 \times 10^{-9} \sim 5.3 \times 10^{-7}$	8.31×10^{-20}	$4 \times 10^{-13} \sim 2.7 \times 10^{-9}$	1 300 000
聚羟基丁酸酯[77,79]	$3.2 \sim 43 \times 10^{-8}$	4×10^{-10}	$1.1 \sim 7 000 \times 10^{-11}$	39 000
聚-3-羟基丁酸-3-羟基戊酸共聚物[81]	3×10^{-9}	1.2×10^{-7}		
聚乳酸[83,85]	0.35×10^{-4}	10^{-4}	10^{-4}	1

第 18 章 泡沫农业:硅藻生物燃料和下一次绿色革命的可扩展缩影

(续表)

材料	CO_2	O_2	H_2O	比例
聚甲基丙烯酸甲酯[69]	1.04×10^{-6}			
聚酰亚胺-芳香族聚酯[91,92]	$0.32 \sim 1.62 \times 10^{-8}$	$1.98 \sim 2.38 \times 10^{-8}$		
聚乳酸 4013-D[93]		4.33×10^{-10}		
聚酰胺-66[94]	$1.8 \sim 2.8 \times 10^{-9}$	$1.6 \sim 2.0 \times 10^{-9}$		
聚乙烯[95-101]	$0.045 \sim 0.51 \times 10^{-6}$	10^{-6}	$0.109 \sim 8.20 \times 10^{-6}$	9.2
聚乙烯-石油烯 NA[102]		2.17×10^{-8}		
低密度聚乙烯[103]	$3.9 \sim 12.5 \times 10^{-7}$	$3.2 \sim 7.4 \times 10^{-7}$		
聚酰亚胺[104]			2×10^{-9}	
聚异戊二烯[105]	$0.057 \sim 1.5 \times 10^{-5}$	$7.51 \sim 9.21 \times 10^{-7}$		
聚烯烃[106]	$0.06 \sim 1.48 \times 10^{-6}$			1 343
聚苯乙烯[107]	1.79×10^{-7}			
聚氨酯[108]	$1.9 \sim 5.7 \times 10^{-6}$			
聚氯乙烯[109]	0.22×10^{-8}	$0.4 \sim 2 \times 10^{-8}$	6.885×10^{-8}	0.03
pp=聚丙烯[110]	$0.31 \sim 0.7 \times 10^{-7}$	$1.45 \sim 2 \times 10^{-7}$	$4.90 \sim 196 \times 10^{-9}$	41
有机硅[111-114]	0.196×10^{-4}	$0.34 \sim 0.452 \times 10^{-4}$	1.14×10^{-4}	0.40
硅橡胶[115]	2×10^{-7}			
二氧化硅-硅杂化物[119]	未测量	低	5.4×10^{-17}	
特氟龙=聚四氟乙烯[120]	$0.027 \sim 2.9 \times 10^{-6}$	0.17×10^{-6}	0.243×10^{-6}	12
特氟龙-氟塑料 4[121]		1.83×10^{-8}		
角质层[122]		$1.9 \sim 3.1 \times 10^{-5}$	$3.3 \sim 4.2 \times 10^{-10}$	94 000

(续表)

材料	CO_2	O_2	H_2O	比例
植物角质层[123]	$0.23 \sim 1.1 \times 10^{-8}$	$2.5 \sim 2700 \times 10^{-8}$	$15.7 \sim 496 \times 10^{-8}$	170
空气[124]	0.16×10^0	0.2×10^0	0.219×10^0	0.9
水[116]	$1.92 \sim 2.233 \times 10^{-5}$	1.97×10^{-5}	2.299×10^{-5}	0.97
气体囊泡[117]	自由渗透	1.06×10^{-5}		
气囊/SiO_2-硅杂化复合材料[118]	1.06×10^{-5}	1.06×10^{-5}	5.4×10^{-17}	196 000 000 000
经验法则:相对于氮气的扩散系数[119]	1	1.7	5	0.34

聚二甲基硅氧烷(PDMS)已被用于控制微流控光生物反应器中的氧水平[219](见表18.3)。然而,长期的保水可能是一个重大的问题。通过添加8%的胶原蛋白,PDMS对水的渗透性可以减半[220]。在相反方向的过程中,增加PDMS的水渗透性[221],可能会为如何继续进行提供线索。

抗蒸腾剂长期以来一直被用来保留水果中的水分,并帮助作物度过干旱时期。普遍的共识是,它们同时减少了水和CO_2的蒸腾作用[222]。然而,也有一些相反的证据[223],所以他们考虑作为涂层或气泡膜的成分可能是值得的。天然植物角质层对水的渗透性可以相差四个数量级[224-226],尽管一份报告提供的扩散系数范围为32倍。这是角质层厚度的部分功能[236]。通过多糖萃取可以降低吸水率[237]。其主要成分聚酯角质蛋白已被合成。一些叶角质层[238],不包括气孔[239],对水的渗透性比对CO_2要强得多。角质层对O_2的渗透性从种子的低[240]到叶片角质层突变体的高[241]。植物根系明显调节其角质层对当地环境O_2的渗透性[225]。由于塑料现在是由角质层化合物制成的[242],有可能使用它们来获得低水渗透性和高氧和CO_2渗透性。

氧电极包括一种透氧性聚四氟乙烯膜[243],这是一种疏水的材料。在有水的情况下,聚四氟乙烯膜的CO_2渗透性下降了10%,假设这是由于疏水微孔区域的大量水簇阻碍了CO_2的扩散[244]。透明特氟隆是可以利用的[245-246]。

Janus塑料薄膜,对具有不同孔隙度和纳米结构的功能化逐卷印刷塑料[247-250]可能提供有趣的作用,例如一侧疏水而另一侧亲水。由于这些是两相结构(空气和塑料),通过薄膜的路径的弯曲度对于确定不同分子的有效扩散系数是很重要的。

在寻找不透水、透气的膜时,我们可以考虑三种生物学情况:
(1) 细菌和真核生物共同的单极脂质双层膜。
(2) 古生菌的双极脂质单层膜。
(3) 蓝藻细菌和古生菌的气泡。

然而,脂质体膜对水的渗透性与塑料聚合体膜大致相同[251],因此我们将不再进一步考虑生物脂质膜,案例 1 和 2。但案例 3 有一些潜力。

从蓝藻微囊藻中分离出的气体囊泡已被完整地分离出来,并用于创建一个 1 毫米厚的层,按体积计算为 35% 的气体囊泡。在一个蛋白酶处允许 100% 的 O_2 通过该层扩散[252]。表 18.3 的最后一行显示了创建一种潜在的不透水、透气性混合塑料/蛋白质材料的方法:在二氧化硅-硅胶混合膜中嵌入完整的气体囊泡。在足够高浓度的囊泡时,囊泡会相互接触,形成穿过膜的通道。将会有一个临界浓度[253],超过这个浓度就会有很多这样的通道。35% 的囊泡悬浮液[252]高于这个临界渗滤压概率。这样的薄膜将是纳米复合材料的一个例子[254-259]。制造具有精细控制形貌的多孔聚合物膜[260]可能会产生这种杂交膜。

不幸的是,没有关于气泡的水渗透性测量的报道。人们提出了两种相反的假设:如果含有一层非极性分子,极性水分子通过膜的通道就会受到限制;非极性的 CO_2 不会因此受到这种影响。气体液泡膜可能不必完全不渗透于水分子,以防止液态水在气体液泡内积累[261]。

无论如何水不凝结的原因是在如此小的体积内,水分子数量均匀成核的细胞寿命比异质成核细胞寿命长的多,蛋白质的内表面太疏水,粗糙,不适合异质成核[262]。但气体囊泡是否透水仍有待实验确定。在表 18.3 的最后一行极端假设为不透水。

对于一些合成膜(如 Nafion),扩散系数会随着薄膜厚度的变化而很大[263]。Nafion 是一种磺化四氟乙烯基含氟聚合物-共聚物[264],这些薄膜的氧和二氧化碳扩散系数尚未被研究(见表 18.3)。

奇怪的是,人类皮肤的顶层,角质层(见表 18.3),既是不透水的[265],也不透氧[266-267]。众所周知,二氧化碳会排出人体皮肤[268],吸引蚊子[269-270]。对其结构的纳米级模仿可能提供一种具有所需渗透性的"人工皮肤"。例如,70% 硅胶和 30% 肉豆蔻酸异丙酯(IPM)的膜与许多药物化合物的渗透性相匹配[271],但必须对常见分子 O_2、CO_2 和水进行测试,以确定其泡沫养殖的前景。

鉴于我们可以在表 18.3 中找到的不完整信息,我们初步得出结论,Nafion[264]是一个很好的候选者,而一些受生物启发的膜可能做得更好。然而,Nafion 的纳米结构会随着暴露于液态水中而发生变化[272]。Nakamura 等[273]在需要从水中去除 O_2 和 CO_2 的工业应用中解决了这个普遍问题:本研究首次研究干氧和 CO_2 的渗透性,以及那些通过疏水性高自由体积含硅无孔聚合物膜溶解在水中的膜,包括聚(三甲基硅甲基丙烯酸甲酯)(PTMSMMA)、PTMSP 和含氟无孔聚合物膜,如 4,4-(六氟异丙基)二苯二甲酐-4,4-(9-氟苯基)二苯胺(6FDA-FDA)和 4,4-[六氟二丙基二酐-2、3,5,6-四甲基-1,4-苯二胺(6FDA-TeDMPD)][273]。

Nafion 也对太阳敏感[274],因此需要一个保护层。其他前景是"混合基质膜"[275],也称为"复合膜",如含有疏水纳米颗粒的膜[276]。其影响可能会很大。例如,与纯 Pebax 相比,混合基质膜的 CO_2 的扩散系数增加了 7 倍[277]。

因此，我们可以寻找 CO_2 和 O_2 渗透膜的 H_2O 不渗透的参数空间相当大，可能有许多可接受的材料用于泡沫农业。标准，高质量的气泡包装含有尼龙，明确降低气体渗透性，相比聚乙烯气泡包装，其气泡变得松弛。由于水/气比为9.2（见表18.3），现成的聚乙烯气泡翘曲可能是一个很好的实验开始。那些我们有足够的数据来绘制表18.3中定义的比例的材料被绘制在图18.13中。

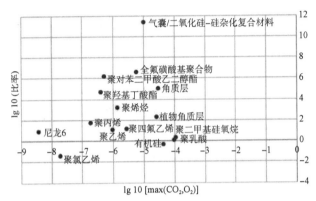

图 18.13 表 18.3 标题中定义的水渗透率与气体渗透率之比的对数曲线图[137]

作为渗透率特性的最后一个考虑，对于图18.12的最终方案，气泡包装塑料的顶层应该保留油。由于最终目标是原位生产高辛烷值的石油，这意味着烷烃分子相对较小，易挥发，可作为某些塑料的溶剂。例如，正己烷在聚四氟乙烯薄膜[278]和聚偏二氟乙烯（PVDF）中的扩散系数为 3×10^{-8} cm^2/s，因此是禁用物质，因为它可溶于有机溶剂，而聚丙烯（PP）、聚乙烯（PE）和聚四氟乙烯（PTFE）则不同[279]。

18.5 毒性限制

测量了原油浮油对硅藻的一般负面影响[280-283]，但没有新鲜硅藻油对硅藻的影响。至少有一种硅藻已被证明可以在细胞存活的同时渗出油[146]。一些天然的浮油被认为是由硅藻造成的[284]，尽管目前还没有确定产生这些浮油的硅藻是否仍然存活。需要一种筛选方法来检测超过20万种硅藻的自渗现象[285-286]。

在之前的一篇讨论将活硅藻纳入太阳能电池板的手稿中[89]，作者表明，在它们产生的许多化合物中有烷烃和烯烃（见图18.14），并提出选择性育种或基因工程可以提高它们的产量。在17-烷烃的蓝藻细菌中，在一定程度上达到了这一目标[287-288]。据报道，在天然水华阶段，硅藻可以产生20倍的4-烷烃，即丁烷[289]。

有时硅藻在石油泄漏后会膨胀，中间生物的表现比白藻表现得更好[291]。这可能是在气泡微世界中选择表现良好的硅藻的一个考虑因素，特别是如果硅藻被选择来生产烷烃。然而，至少对一种硅藻，原油中的石蜡（烷烃）部分几乎没有影响或根本没

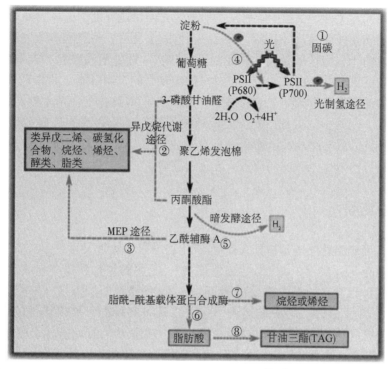

图 18.14 藻类中烷烃的代谢途径[290]

有影响[292]。因此,产生烷烃的硅藻不太可能毒害自己。这值得进一步测试,特别是看看高烷烃浓度是否抑制硅藻产生烷烃,尽管受到它们在水中分配系数的限制[293-294]。对于酵母种酿酒酵母,已经获得了一种接近辛烷值的化合物,1-辛醇,但它对该物种有毒[295]。

藻类在塑料变质中的作用需要进一步调查[296-299]。已知一些塑料对藻类有毒[300-301],因此在选择泡沫养殖的塑料时必须考虑毒性。一些塑料已经被认为是无害的,例如 PDMS,它被用于制造用于毒性测试的设备[302],以及作为一种无毒的防污表面[303]。

18.5.1 藻类油液滴特性

对于自榨取,外部化的、渗出的油滴应该尽可能大,因为较大的液滴比较小的液滴更容易从水介质中进行重力分离。对于第一个假设,这是斯托克定律的一个简单结果,它解释了每天观察到的大气泡比小气泡上升得快得多。很明显,细胞的大小限制了它可以包含的油滴的大小。硅藻的线性大小从 1.5 μm~5 mm 不等[304]。每个细胞的体积范围至少从 3 μm^3 到 4.71×10^9 μm^3[305]。对于渗出油的藻类,人们可以假设细胞内液滴的大小与外部液滴的初始大小高度相关。因此,重点可能是放在具有较大细胞体积的硅藻上[306-313]。需要调查硅藻液泡的大小[314]，和它所包含的油滴的大小之间的相关性。藻类中一种新的脂质荧光团可能有助于选择物种[315]。

一旦油滴在硅藻外，它们实际上就是一种油/水乳液。虽然很多研究都是关于热力学不稳定的[316]，但这里我们感兴趣的是"打破"这种乳剂或脱乳[317]。较大的飞滴去乳化的速度更高[318]，这表明较大的硅藻渗出较大的飞滴可能是最好的。

渗出的油滴迅速上升到气泡的顶部，在那里它们应该作为一个单独的相结合，与下面的水有一个最小的界面，以尽量减少油被硅藻和/或存在的任何细菌代谢的机会。采用倒置的漏斗结构可以增强分离效果(见图 18.12)。对于顶部板，刚性或紧度更大的塑料可以减少风对搅拌油重新乳化的影响。

18.6　生物膜

生物膜可能比悬浮培养物具有优势，其生物量产量从每天每平方米 2～20 g[319]：一些生物膜在 54 g m^{-2} · d^{-1} 时达到峰值[320]。更厚的生物膜具有捕获更多的光能的优势，而且，与普通的光生物反应器不同，生物膜通常被认为是一种障碍，需要清洁[321-322]，藻类通过生长、竞争和运动[323]，可以调整其种群以获得最佳的阳光捕获。生物膜的建模[324]可能需要扩展以适应这些现象。通过简单的吸光度，光穿透生物膜的程度大于预期[325]，这可能与我们关于硅藻和硅藻菌落作为光管的假设有关[201,326]。不需要使用生物膜消耗能量来提供水[327]，生物膜光生物反应器的水量可能提供，与悬浮培养相比，水含量低 100～1 000 倍[325]。这将大大减少泡沫养殖的用水需求。

光生物反应器的一种变体是使用微孔基质将散装水从藻类中分离出来[328]。这对于低含水量的生物质生产具有优势，但不能提供卵出油与藻类的相分离。它还需要每个气泡的三层结构，而不是更简单的两层结构的气泡包裹。然而，一种带有孔的可捕获单个细胞的塑料薄膜[329]可能有助于将藻类从油中分离出来。

硅藻和细菌的混合物种，可能会形成复杂的生物膜，从而减少从藻类中取水的成本[99,209,330]。事实上，在收获时，气泡膜中的每个气泡都被刺破并泄漏，而生物膜藻类的损失很小。运输减少的质量，只是气泡包裹及其潮湿的生物膜内容，将比充满水的气泡更具成本效益。生物膜的去除和进一步的处理可以在中央生物精炼厂进行。

18.7　细菌共生体

如果细菌存在于气泡的微观世界中，如果它们消化了产生的硅藻油，将会适得其反。然而，能够消化石油的细菌不会被石油泄漏所影响[331]，所以即使存在这些情况，硅藻油也可能存活到收获。但从积极的方面来看，需氧细菌不仅会利用光合作用藻类产生的部分 O_2，而且细菌/藻类直接共生的生物燃料生产也有了新的前景[118,332]。硅藻与细菌[97]和蓝藻细菌[98]的共生关系自然发生。

而将光合作用从典型的10％光子效率提高到约20％的一般方法是利用基因工程[333-334]，另一种方法是将细菌添加到含有硫化镉纳米粒子的微观世界中[335-336]达到超过80％的光子效率，这些小分子可以被硅藻代谢成油。相比之下，虽然还没有趋于稳定，但迄今为止，太阳能电池发电的最高效率为46％[337]。

18.7.1 土壤作为 CO_2 的来源

在全球范围内，通过土壤呼吸排放的年 CO_2 大约是目前化石燃料排放的10倍[338-339]。因此，被气泡膜覆盖的农田可以对藻类的碳捕获做出重大贡献[340]，同时直接利用捕获的 CO_2 作为生物燃料和/或藻类生物量。这将使夏季休耕拥有真正的双重作物，作为藻类的饲料，和藻类生物燃料的生产，同时改善土壤和减少休耕农田的贡献[341]，给泡沫养殖产生碳负足迹[342]。因此，藻类生物燃料可能成为作物轮作中的多一种作物[343]。然而，一些人表示担心，用塑料片覆盖土地会降低土壤质量，不良影响可能来自塑料添加剂，增强农药径流和塑料残留可能碎片成微塑料但保持化学完整和积累在土壤中可以连续吸附农药[117]。

如果藻类释放的 O_2 可以被定向进入土壤中，那么通过增加土壤细菌的呼吸作用，藻类就可以获得更多的 CO_2。如果底部塑料片同时具有 CO_2 和 O_2 渗透，而顶部塑料片只有 CO_2 渗透，就可以做到这一点。一种极端的做法是使顶部的塑料薄片气体，也许是不透水的，仅仅依赖土壤 CO_2 排放和 O_2 吸收与土壤微生物与气泡限制的藻类共生。CO_2 从土壤中释放的速率取决于其含水量[344]和有氧与厌氧条件[345]，土壤中 CO_2 的浓度有时几乎是空气中的两倍[346]。因此，现场实验寻找土壤空气和 CO_2 来源之间的最佳平衡可能是必要的。在上面的空气和下面的土壤之间有一些直接的气体交换，可以通过在气泡之间的空间中制造空隙。这些洞也会允许雨水进入土壤。只有雨水打开的孔，有助于保持土壤中的水分（见图18.11）。

氮氧化物通过过度施肥的土壤释放到大气中，它们被视为空气污染物[347]。如果选择底部的塑料可以渗透这些气体，一氧化氮（NO）[348]和二氧化氮（NO_2）[349]，它们可能会变成亚硝酸盐和硝酸盐[350]，并被藻类吸收。NO影响硅藻的粘附性，但不影响硅藻的生长[351]。因此，藻类的泡沫养殖可以改善这种空气污染源。它还可以消除多余的肥料径流，因为所有的藻类营养物质都保存在气泡中。如果水中的肥料适量，在收获时可能都在藻类中，因此与标准作物和免耕作物不同[352]，肥料不会被浪费。

泡沫膜实际上为下面的土壤创造了一个寒冷的框架[353-354]，以及里面生长的任何东西。因此，它可以延长任何能够在充满水的气泡膜的重量下生长的低洼作物的生长季节。例如用于固氮的三叶草[355]和一些根茎作物[356]。在泡沫膜下，作物可以选择在减少的阳光下生长[357]。这种作物可以减少土壤侵蚀，如这个类似的例子："反射器上的小孔传输足够的阳光，控制侵蚀的植物可以生长在阿雷西博射电望远镜的下面"[358]。如果使用这种方法，可能不需要气泡膜顶层的红外反射率。此外，也可以避免用普通塑料片覆盖土地所造成的损害[117]。

18.8 需求

全球范围内消费者和行业对能源的需求似乎正在向目前美国和加拿大的人均消费增长(见图 18.15)。考虑到世界平均水平,我们可以预计最终全球总需求是目前的 4 倍,随着人口增长需求将会更多。虽然许多国家正在考虑转向电动汽车[19],即"积极电气化"[359]。因此,液体燃料将存在相当长一段时间,如果以泡沫养殖汽油的形式廉价和可持续,甚至可能削弱现在大量补贴的电动汽车选择[360-361]。

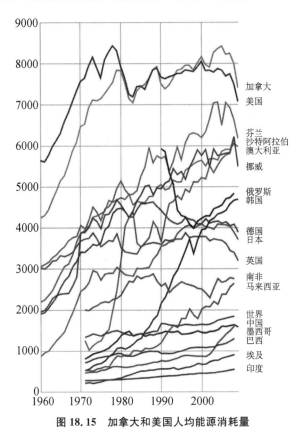

图 18.15　加拿大和美国人均能源消耗量

18.8.1　硅藻和其他藻类的选择

硅藻是一种具有硅基细胞壁和高生产力的单细胞微藻。硅藻属于一个叫做异养生物的大群体,包括自养生物和异养生物,它可以通过光合作用或在小有机分子上,或两者兼具[362]。如果不考虑硅藻壳,一些硅藻比其他藻类产生的石油生物量比例更高[89]。由埃克森美孚资助的微拟球藻最近成功地将石油产量增加了 80 亿美元,随后又转向硅藻[333]。关于"硅藻中特殊碳代谢"的特刊[363]包括一篇来自文特尔研究所的文章[364]。鉴于硅藻在进化上的成功,目前其产量约占地球上固定碳总量的25%[305],以及它们生产相当一部分化石石油的责任[365],这种兴趣的转变并不令人惊讶。我们知道如何大规模培养它们[366]并繁殖它们[143],它们可能适合稳定的基因工程[367],这导致"三酰基甘油积累增加45倍"[368]。有一些关于硅藻作为生物燃料来源的综述[89,369-374],以及基本的碳和石油[90,375-376]物理和化学正在研究中。硅藻是原生生物,通常含有被生物燃料包围的次生质体,如葡萄球菌膜,其他藻类(特别是绿藻)有很大区别,后者含有被两层膜包围的简单质体[377-378]。硅藻是地球上最具生产力、环境灵活性的真核微藻之一,已知的一些硅藻会积累大量的中性脂质[373]。最近关于硅藻转化(基因工程)的研究主要集中在增加石油的产量和蛋白质的分泌上,如人类抗体(见表18.1)。由于硅藻通常不分泌蛋白质,因此该产品的纯度较高[379-380]。因为硅藻油小体通常包括中性脂质[375],我们可以设想将两者结合起来来设计自发分泌(自我榨取)的油。使用细胞膜[381]来促进硅藻油的表达[145]尚未研究[382]。

一些硅藻需要外源性维生素[383],这占实验室培养基成本的50%。因此,一个新的研究领域将是在生物量、生物燃料或高价值产品和硅藻物种的选择以减少营养成本之间的权衡。此外,硅藻种类在脂质产生方面当然有所不同(见图18.16),还有许

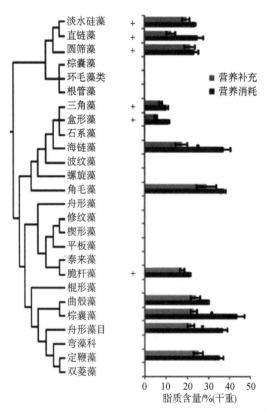

图18.16 硅藻在营养素缺乏和营养素耗尽条件下的脂质含量[385-386]

多类群有待探索[384-385]。

当然，其他用于生物燃料试点生产的藻类，如 *Botryococcus*[87,387]，*Nannochloropsis*[388] 和许多其他的藻类[382]值得认真考虑。考虑到封闭气泡膜微群落内部过热的可能性[207]，嗜热硅藻[389-395]或其他嗜热海藻[396-397]值得考虑。对嗜热藻类产油的研究已经开始[398]。然而，在耐热应激条件下，耐热硅藻增加了脂质的数量，并降低了它们的饱和度[399-400]。可能需要菌团来支持高达60℃的硅藻生长[401]。

以硅藻为基础的藻类燃料开始显示出前景[89,143,371,402-405]，考虑到他们对天然石油储量的贡献，这是合理的[365]。因为，硅藻有产生大量油的能力。例如，据报道，浮游硅藻产生脂质(中性和磷脂)，数量可达 40%[406]。培养的海洋硅藻的脂肪干重为 30%~45%[407]。据报道，单个硅藻中的油量可达藻类生物量的 25%[371]，可能达到非硅藻干质量的 60%[89]。对于某些硅藻物种，预计每公顷的油产量将达到大豆的 100~200 倍。硅藻也有快速繁殖的能力，并且可以创造非常大的生物量。在实验室条件下，硅藻可以在环境中繁殖 5 小时，在两天内繁殖并翻倍[90]。

18.9 指数增长与平稳阶段

典型藻类生长阶段的特征如下[408-410]（见图 18.17）：①初始诱导阶段；②快速增长的速度阶段（即指数）；③增长率下降的阶段；④生产（固定式）；⑤最后的死亡或"崩溃"阶段。

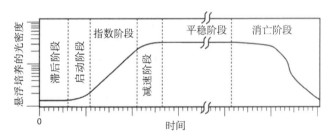

图 18.17 微生物培养物的生长阶段[410]

可以观察到，在固定阶段，由于氮气耗尽[412]，油滴几乎填满整个细胞，而指数增长的恢复导致脂肪的减少[413]。这个实验需要一个碳源[65]。在正常和限制条件下生长的淡水和海洋产油硅藻，其平均脂质含量的统计分析为 22.7% 干细胞重量（DCW），在胁迫条件下增加到 44.6%DCW[414]。老化的藻类细胞或在各种应激条件下（如氮消耗）维持的细胞中总脂质的增加主要由中性脂质组成，主要是三酰基甘油（TAG）[415]。一种藻类在固定阶段每干重产生的脂质比在指数增长阶段多 60%[416]。这些结果对于生物燃料来说看起来一直很有前景[378]：在固定相培养中，高脂质含量通常伴随着储存脂质的形成[三酰基甘油酯，通常是棕榈酸（C16：0）和棕榈酸

(C16∶1)],而不是膜极性脂质,如糖脂和磷脂[417-418]。大多数用于水产养殖的微藻是批量或半连续对数和早期固定阶段的培养,高极性脂质(例如,长链营养 ω-3 脂肪酸 C20∶5 和 C22∶6)[419]。

因此,人们想知道,为什么在完全固定阶段的收获不是常态?原因基本上很简单,就像生活中的一切一样,你不能既吃蛋糕同时又拥有蛋糕(也许除了榨取)。应激和未应激的藻类细胞之间的脂肪酸组成存在显著差异,所以如果你想为生物燃料最大限度地提高中性脂质,那么在营养耗尽(应激)条件下在完全固定阶段收获是一个合适的策略。然而,如果目标是营养性的(即食品/饲料/营养食品 ω-3 油),那么在营养充足的条件下,在后期—对数/早期固定阶段的收获是适当的策略[420]。

搅拌的、固定的硅藻培养物将以多快的速度取代它们细胞中的油尚未得到测试[90]。一个研究项目来了解固定阶段,当最佳时间发生,以最大化生物燃料脂质,以及榨取的最佳时间和频率,需要探索。同样地,如果有更多的自我榨取的硅藻可以被发现或创造(见图 18.12),它会发生在什么生长阶段,能否安排其时间以最大限度地产生非极性脂质?

生长动态的变化发生在(除其外):(1)细胞是否能储存营养物质[421];(2)与蓝藻细菌的竞争[422];(3)与细菌的相互作用[423-424];(4)光生物反应器中的水循环和"藻类细胞经历的时间-辐照度剖面"[425];(5)昼间温度和光照、食草动物和在跑道上沉降[426]。

已经制定了具有 9 个[347]到 27 个[427]参数的模型。因此,决定这五种增长阶段的类别是否适用于给定的增长情景(如泡沫种植),将需要一些研究。

18.10 碳回收

商业气泡膜的质量约为每平方米 65 克。如果用油制成的塑料只能使用一年就被丢弃,我们将需要 65 克/平方米/年的生物燃料产量来实现收支平衡。通过紫外线保护延长塑料的使用寿命,允许塑料的再利用或回收将大大降低这一要求。在任何情况下,年生物燃料产量报告为每平方米 5.87~13.69 L[428]或密度为 0.875 g/cm^3[429],每平方米 5 140~11 980 g,远远超过气泡膜的质量。

18.11 包装

可在现场充气的气泡膜的发明[173,430]特别是因为水的重量使其运输量[431]减少了约 40 倍。我们想要运输未膨胀的产品。然而,气泡中可能含有干燥形式的藻类、细菌和营养物。这应该是可能的,因为许多硅藻在干燥状态下存活[432]。此外,硅藻的休息阶段可以持续几十年[433],因此保质期应该很长。像其他微藻一样,干燥会导致硅藻中油滴的积累[434]。硅藻在干燥中生存的能力随着干燥而缓慢增加,并且在不同物种之间有很大差异[435]。

硅藻的一种基本营养物质是二氧化硅,用来建造它们的外壳。这可以是硅藻土(硅藻土)[201]或植物岩[436]的形式。植物岩的一个优点是,它们存在于所有草作物的秸秆中,如小麦和水稻,因此可以作为干草在当地供农民使用。此外,在叶植硅石中[437-438],添加硅藻土或植硅石的光散射特性可以提高硅藻利用阳光的效率,并可能保护它们免受紫外线的影响[439]。在一些硅藻中,外壳含有自身的紫外线屏蔽[440]。

18.11.1　农民的作物选择

独立的农民选择他们种植的作物,以及他们希望承担的合同。用于石油或其他产品的泡沫养殖藻类将允许不同的选择(见表18.1)。如果泡沫农业在不损害它的情况下取得成功,它可能会避免与其他作物的竞争,除非它更有利可图。一个自由的农业市场,以及向农民提供接种气泡包装的内容,或接种泡沫本身的能力,将开启这种新的农业方式的创新。因此,藻类生物燃料只能在非耕地上种植的预先观念[441-444]可以被消除。此外,大多数农民都居住在可耕地上。

18.11.2　泡沫养殖与光生物反应器和跑道之间

葡萄牙的藻类农场被宣传为世界上最大的光生物反应器设施[29,322,445-446]。它占地1公顷。在表18.4中,我们将其与大型跑道设施、质量健康和泡沫养殖进行了比较。

表18.4　葡萄牙帕塔亚斯基于光生物反应器的AlgaFarm[322,451](见图18.3)和美国新墨西哥州哥伦布基于赛道的Qualitas Health[452](见图18.2)与使用1平方米深10厘米的气泡养殖的对比(见表18.2)

特点	微藻农场反应器	质量健康/绿色流农场跑道	气泡养殖
面积	1.2公顷	39.2公顷	259公顷
人员	18工程师,生物学家,化学家,技术人员	14现场操作人员	1农民
成本	15 000 000 欧元 = US $18 000 000 US $15 000 000/公顷	US $85 000 000 US $2 000 000/公顷	US $5 000/公顷 + 灌装、铺设和收获设备成本
体积	1 300 m^3	105 000 m^3	1 000 m^3/公顷
生物燃料	否	否	可以
现场加工	可以	否	可能
透气性	否	可以	可以
二氧化碳源	食品级(待售)混凝土工厂	管道	空气和土壤

(续表)

特点	微藻农场反应器	质量健康/绿色流农场跑道	气泡养殖
日常维护	可以	可以	否
模块	11	68	≥1万/公顷每个气泡大小
易损性	聚甲基丙烯酸甲酯管	无开放水	柔性塑料薄膜
液体不停流动	可以	可以	否
污染	持续监测	每周监测	避免收获污染的气泡
蒸发	否	可以	轻微
土壤再利用	否,混凝土覆盖	否,水覆盖	可以,固氮作物下可行,轮作可行
地点	城市	沙漠	农田
清洁	可以	可以:每年清除沙坝	否
接种	阶段	每2~3年取下一次	在生产过程中
围栏	否	可以	可以,把牛挡在外面
温度控制	可以	否	否
可以冷冻	否,加热	可以	可以
可以过热	否,冷却	使用全年生存和生长的拟微球藻,观察到的最高温度为3℃	否,但可能是嗜热菌向红外反向反射
硅藻	可以(2017年1种,计划增加)	可以	可以
年产量		专有	10~30吨/公顷
水培	否	否	可以

18.12 结论

从概念上讲,藻类的气泡养殖有潜力解决开放式水沟池的污染问题,而成本只是封闭光生物反应器的一小部分。泡沫农业为农民带来了一种新的经济作物,它有可能让非耕土地同时带来经济作物。在气泡膜下种植像三叶草这样的固氮作物,藻类的气泡养殖既可以利用土壤中的 CO_2,又不能耗尽土壤,同时保护土壤免受风的侵蚀和干燥。主要的物理要求是一种成本效益高、可重复使用/可回收的塑料,可渗透 CO_2 和 O_2,同时足够不渗透水,使每个气泡中的水持续收获或整个生长季节。泡沫养殖对农民有额外的好处是,在泡沫充满后,就没有必要依靠降雨,也就是说,它是抗旱的。当然,它需要在填充气泡时提供足够的水,但所需的水量可以减少到1000倍。

最初,泡沫养殖可以用于种植藻类生物量,作为生物燃料和高价值的产品。后者可能包括基因工程藻类中的特定蛋白质,如人类抗体,以及用于纳米技术应用的硅藻壳[447],如太阳能电池板和电池[448-450](见表18.1)。作为后来的目标,一些藻类似乎可以被选择性地繁殖和/或改造来直接生产汽油,因为藻类已经有了合成烷烃的途径。

在用泡沫包装的农场里:
(1) 农民们会种植藻类。
(2) 不需要共同生产高价值的产品来抵消生物燃料的成本。
(3) 生物质生产并不是针对生物燃料的目标,但石油生产是。
(4) 有利用生物膜的形成,而不是被认为是一个问题。

从藻类,特别是硅藻中获得具有价格竞争力的汽油/柴油燃料的研究机会包括以下步骤和目标:

(1) 藻类可以在固定阶段生长,而不是呈指数增长,原则上不需要过量添加硝酸盐和磷肥,因此,避免了由于径流而导致的附近水体的富营养化。
(2) 通过对藻类微生物的控制和泡沫养殖的模块化,隔离被污染的气泡,可以将其影响降至最低,接种量足够大时,可以使用天然的、未经消毒的水。
(3) 气泡内生物膜的形成需要在生物量、石油产量和所需水的减少等方面进行测试。
(4) 应该寻求自榨取的藻类物种,以消除昂贵的下游处理步骤,即脱水和从石油中分离生物质(这可以通过重力来完成)。
(5) 一个具有光合细菌的联合体可以与藻类(可能是硅藻)一起使用,以提高太阳光子使用的光合效率,使其能够更快地生长。
(6) 选择性育种或基因工程可用于选择生产低分子量烷烃而不是甘油三酯。
(7) 通过让藻类自己进行生物燃料的加工,可能可以避免生物燃料的所有加工步骤,因为烷烃产品预计对藻类无毒。

泡沫养殖是一种大规模种植作物的新方式。它有可能从微藻扩展到其他作物植物。每个气泡都可以被认为是一个完全封闭的水培容器。它可能只部分充满水,如果水被保留一个完整的生长季节,雨水太多或太少都不会影响它。同样,肥料可以在每个气泡中,以缓慢释放的形式[344],避免肥料径流导致水富营养化等问题[453-455]和过量系统,在这种系统中,植物向大气开放,它们的根浸没在循环水中,经过各种水处理,但同时也向大气开放[456],即没有真正闭合的。与普通耕作技术相比,普通水培系统的产量高达24倍(见表18.5)。部分封闭的系统已经被考虑用于在空间中种植植物,当然,太空飞船本身是实际的完全封闭的系统[457]。因此,泡沫水培农业可能远远超出微藻种植,并被证明是一场新的绿色革命的开始[458-459]。因为气泡中的植物和藻类受环境的保护,接种物中没有其他东西生长人们可以从无病植物开始[460],气泡养殖实际上是一种有机农业的形式。

表 18.5　普通水培作物产量与每公顷农业平均产量的比率[456]

作物类型	水培/农业	作物类型	水培/农业
番茄	24	燕麦	3.5
水稻	14.5	大豆	2.5
马铃薯	8.7	花椰菜	2.4
小麦	8.3	生菜	2.3
豌豆	7.0	甜菜根	2.2
玉米	5.3	卷心菜	1.4
黄瓜	4.0		

我们设想有四个步骤,有一个序列或产品,每一个都应该是有利可图的:①大量的藻类,转化为高价值产品,从生化到纳米技术;②大量的藻类,将被提炼为生物燃料;③大规模水培法;④在气泡中自行提炼生物燃料的藻类。

①排除了生物燃料的共同生产,以保持工艺优化的 HVPs。②排除了 HVPs 的共同生产,以允许优化生物燃料的生产。在这两种情况下,藻类生物量都在生物炼厂加工。③可以允许某些非藻类作物以高产的方式种植。④从符合或超过最低质量标准的藻类中直接收获汽油/汽油或柴油是我们的梦想[461-465]。

一般而言,泡沫养殖不包括:昆虫、真菌、干旱、冰雹、风对农作物的破坏、霜冻、洪水、蜗牛、线虫、交叉授粉、肥料/农药径流、土壤 NO_x 污染、土壤风侵蚀。

未来不需要炼油厂的泡沫养殖生物燃料(见图 18.12)将需要大量研究才能实现其潜力:需要选择自身榨取分泌油的硅藻或其他微藻;石油上升到顶部,能让我们实现重力分离;烷烃是由硅藻自然产生的,可能不会自我抑制它们的产生,因为纯烷烃对硅藻(和其他藻类)无毒;可能需要选择性育种或基因操作;需要找到适应当地温度范围的物种和菌株;形成生物膜的物种可能有助于从油中分离出藻类;由于硅藻处于固定阶段,不需要多余的肥料,但接种剂的设计变得至关重要;我们能否在场外不加工,只是收获准备使用的生物燃料。

泡沫养殖解决了需要的关键改进[466]:降低藻类农场建设成本;降低藻类养殖成本;最大限度地生产生物质;最大限度地减少养分利用;减少用水减少;生物燃料的成本低于汽油。

有人提出,如果价格能下降 10 倍,微藻就可以养活世界,并以北美的消费率提供生物燃料(见图 18.15)[467]。随着石油地缘政治的重大调整,泡沫养殖可能使这成为可能[80]。因此,我们为地球提供了一个潜在的新视角(见图 18.18)。

图 18.18 对泡沫农业潜在影响

参考文献

[1] Wikipedia. (2018). Barrel of oil equivalent. https://en.wikipedia.org/wiki/Barrel_of_oil_equivalent.
[2] Acien, F. G., Fernandez, J. M., Molina-Grima, E. (2014). Economics of microalgae bio-mass production. In: Biofuels from Algae.
[3] Amer, L., Adhikari, B. and Pellegrino, J. (2011). Technoeconomic analysis of five mic-roalgae-tobiofuels processes of varying complexity. Bioresour. Technol. 102(20), 9350-9359.
[4] Hoffman, J., Pate, R. C., Drennen, T. and Quinn, J. C. (2017). Techno-economic assessment of open microalgae production systems. Algal Res. 23, 51-57.
[5] Rhodes, C. J. (2010). Biofuel from algae: Salvation from peak oil? In: Seaweeds and their Role in Globally Changing Environments.
[6] Xin, C., Addy, M. M., Zhao, J., Cheng, Y., Cheng, S., Mu, D., Liu, Y., Ding, R., Chen, P. and Ruan, R. (2016). Comprehensive techno-economic analysis of wastewater-based algal biofuel production: A case study. Bioresour. Technol. 211, 584-593.
[7] Anonymous (2009a). Biofuels match performance of Jet A in Boeing's flight tests. Pro-f. Eng. 22(11), 46.
[8] Anonymous (2009b). Algal biofuel powers jet airliner. Prof. Eng. 22(1), 44.
[9] Bwapwa, J. K., Anandraj, A. and Trois, C. (2017). Possibilities for conversion of micro-algae oil into aviation fuel: A review. Renew. Sust. Energ. Rev. 80, 1345-1354.
[10] Karthikeyan, S., Kalaimurugan, K., Prathima, A. and Somasundaram, D. (2018). Novel microemulsion fuel additive Ce-Ru-O catalysts with algae biofuel on diesel engine test-ing. Energy Sources Part A-Recovery Util. Environ. Eff. 40(6), 630-637.
[11] Pujan, R., Hauschild, S. and Gröngröft, A. (2017). Process simulation of a fluidized-bed catalytic cracking process for the conversion of algae oil to biokerosene.
[12] Kerschner, C. and Hubacek, K. (2009). Assessing the suitability of input-output analysis for enhancing our understanding of potential economic effects of Peak Oil. Energy 34(3), 284-290.
[13] Murphy, D. J. and Hall, C. A. S. (2011). Energy return on investment, peak oil, and the end of economic growth. In: Ecological Economics Reviews.
[14] Richardson, J. W., Johnson, M. D. and Outlaw, J. L. (2012). Economic comparison of open pond raceways to photo bio-reactors for profitable production of algae for transportation fuels in the Southwest. Algal Res. 1(1), 93-100.
[15] Currie, J. and Siewert, J. (2015). Oil: Lower for even longer [podcast]. http://video.goldmansachs.com/gsvideo/delivery/da/3a/da3a4d95-8fc8-42aa-b0f1-a4b7fb4dc7da/X15-21894_v01_2193-15-020_Oil_Lower_for_Even_Longer-YouTube_1500k.mp4.
[16] Apajalahti, E. L., Temmes, A. and Lempiälä, T. (2018). Incumbent organisations shaping emerging technological fields: cases of solar photovoltaic and electric vehicle charging. Technol. Anal. Strateg. Manage. 30(1), 44-57.
[17] Cohen, N. and Naor, M. (2017). Entrepreneurial failure in the transition to electric vehicles: A case study

of support for sustainability policy in Israel. Policy Soc. 36(4), 595-610.
[18] Stokes, L. C. and Breetz, H. L. (2018). Politics in the US energy transition: Case studies of solar, wind, biofuels and electric vehicles policy. Energy Policy 113, 76-86.
[19] Vinayak, V., Joshi, K. B. and Sharma, P. M. (2019). Diafuel (diatom biofuel) vs electric vehicles, a basic comparison: A high potential renewable energy source to make India energy independent [Chapter 21]. In: Diatoms: Fundamentals & Applications [DIFA, Volume 1 in the series: Diatoms: Biology & Applications, series editors: Richard Gordon & Joseph Seckbach]. J. Seckbach and R. Gordon, (eds.) Wiley-Scrivener, Beverly, MA, USA: 537-582.
[20] YCHARTS. (2017). Average Crude Oil Spot Price per World Bank. https://ycharts.com/indicators/average_crude_oil_spot_price.
[21] Market Reports Center. (2017). Global Biodiesel Market 2017-2021. https://marketreportscenter.com/reports/554756/global-biodiesel-market-2017-2021[cobweb site].
[22] Matsumoto, H., Shioji, N., Hamasaki, A. and Ikuta, Y. (1996). Basic study on optimization of raceway-type algal cultivator. J. Chem. Eng. Jpn. 29(3), 541-543.
[23] Mendoza Martín, J. L. (2016). Raceway system requirements for low-cost energy-efficient algal biomass cultivation [Ph. D. thesis], Department of Civil. Maritime and Environmental Engineering and Science, University of Southampton.
[24] Olivieri, G., Salatino, P. and Marzocchella, A. (2014). Advances in photobioreactors for intensive microalgal production: configurations, operating strategies and applications.
[25] Wolf, J., Stephens, E., Steinbusch, S., Yarnold, J., Ross, I. L., Steinweg, C., Doebbe, A., Krolovitsch, C., Müller, S., Jakob, G., Kruse, O., Posten, C. and Hankamer, B. (2016). Multifactorial comparison of photobioreactor geometries in parallel microalgae cultivations. Algal Res. 15, 187-201.
[26] Zittelli, G. C., Rodolfi, L., Bassi, N., Biondi, N. and Tredici, M. R. (2013). Photobioreactors for microalgal biofuel production. In: Algae for Biofuels and Energy. M. A. Borowitzka and N. R. Moheimani, (eds.) Springer, Dordrecht: 115-131.
[27] Singh, V., Tiwari, A. and Das, M. (2016). Phyco-remediation of industrial waste-water and flue gases with algal-diesel engenderment from micro-algae: A review. Fuel 173, 90-97.
[28] Nordrum, A. and White, R. (2018). Growing Algae in New Mexico. https://www.youtube.com/watch?time_continue=170&v=gA9LMjHoh_I.
[29] Secil. (2017). Algafarm: SECIL—Allmicroalgae's production unit. https://www.youtube.com/watch?v=QFIDEsbRMoM.
[30] Aqualia. (2017). Project All-Gas. http://www.all-gas.eu/en/home.
[31] Craggs, R. J., McAuley, P. J. and Smith, V. J. (1997). Wastewater nutrient removal by marine microalgae grown on a corrugated raceway. Water Res. 31(7), 1701-1707.
[32] Gani, P., Sunar, N. M., Matias-Peralta, H., Latiff, A. A. A. and Fuzi, S. F. Z. M. (2017). Growth of microalgae Botryococcus sp in domestic wastewater and application of statistical analysis for the optimization of flocculation using alum and chitosan. Prep. Biochem. Biotechnol. 47(4), 333-341.
[33] Hena, S., Abida, N. and Tabassum, S. (2015). Screening of facultative strains of high lipid producing microalgae for treating surfactant mediated municipal wastewater. RSC Adv. 5(120), 98805-98813.
[34] Juneja, A. and Murthy, G. S. (2017). Evaluating the potential of renewable diesel production from algae cultured on wastewater: techno-economic analysis and life cycle assessment. AIMS Energy 5(2), 239-257.
[35] Kang, Z., Kim, B.-H., Ramanan, R., Choi, J.-E., Yang, J.-W., Oh, H.-M. and Kim, H.-S. (2015). A cost analysis of microalgal biomass and biodiesel production in open raceways treating municipal wastewater and under optimum light wavelength. J. Microbiol. Biotechnol. 25(1), 109-118.
[36] Kesaano, M., Gardner, R. D., Moll, K., Lauchnor, E., Gerlach, R., Peyton, B. M. and Sims, R. C. (2015). Dissolved inorganic carbon enhanced growth, nutrient uptake, and lipid accumulation in wastewater grown microalgal biofilms. Bioresour. Technol. 180, 7-15.
[37] Lowrey, J., Brooks, M. S. and McGinn, P. J. (2015). Heterotrophic and mixotrophic cultivation of microalgae for biodiesel production in agricultural wastewaters and associated challenges-a critical review. Journal of Applied Phycology 27(4), 1485-1498.
[38] Mahapatra, D. M., Chanakya, H. N. and Ramachandra, T. V. (2014). Bioremediation and lipid synthesis through mixotrophic algal consortia in municipal wastewater. Bioresour. Technol. 168, 142-150.
[39] Ramachandra, T. V., Madhab, M. D., Shilpi, S. and Joshi, N. V. (2013). Algal biofuel from urban

wastewater in India: Scope and challenges. Renew. Sust. Energ. Rev. 21, 767-777.

[40] Shen, Y. (2014). Carbon dioxide bio fixation and wastewater treatment via algae photochemical synthesis for biofuels production. RSC Adv. 4(91), 49672-49722.

[41] Silva, N. F. P., Goncalves, A. L., Moreira, F. C., Silva, T. F. C. V., Martins, F. G., Alvim-Ferraz, M. C. M., Boaventura, R. A. R., Vilar, V. J. P. and Pires, J. C. M. (2015). Towards sustainable microalgal biomass production by phycoremediation of a synthetic wastewater: A kinetic study. Algal Res. 11, 350-358.

[42] Thokchom, A. K., Majumder, S. K. and Singh, A. (2016). Internal fluid motion and particle transport in externally heated sessile droplets. AIChE J. 62(4), 1308-1321.

[43] Bhola, V., Swalaha, F., Kumar, R. R., Singh, M. and Bux, F. (2014). Overview of the potential of microalgae for CO_2 sequestration. Int. J. Environ. Sci. Technol. 11(7), 2103-2118.

[44] Bilanovic, D., Holland, M. and Armon, R. (2012). Microalgal CO_2 sequestering—Modeling microalgae production costs. Energy Conv. Manag. 58, 104-109.

[45] Chung, I. K., Beardall, J., Mehta, S., Sahoo, D. and Stojkovic, S. (2011). Using marine macroalgae for carbon sequestration: A critical appraisal. Journal of Applied Phycology 23(5), 877-886.

[46] Kumar, K., Banerjee, D. and Das, D. (2014). Carbon dioxide sequestration from industrial flue gas by Chlorella sorokiniana. Bioresour. Technol. 152, 225-233.

[47] Ono, E. and Cuello, J. L. (2006). Feasibility assessment of microalgal carbon dioxide sequestration technology with photobioreactor and solar collector. Biosyst. Eng. 95(4), 597-606.

[48] Yue, D., Gong, J. and You, F. (2015). Synergies between geological sequestration and microalgae biofixation for greenhouse gas abatement: Life cycle design of carbon capture, utilization, and storage supply chains.

[49] Borowitzka, M. A. (1992). Algal biotechnology products and processes—matching science and economics. Journal of Applied Phycology 4(3), 267-279.

[50] Gateau, H., Solymosi, K., Marchand, J. and Schoefs, B. (2017). Carotenoids of microalgae used in food industry and medicine. Mini-Rev. Med. Chem. 17(13), 1140-1172.

[51] Harvey, P. J., Abubakar, A., Xu, Y., Bailey, D., Milledge, J. J., Swamy, R. A. R., Vieira, V. V., Harris, G., Hoekstra, H., Goacher, P., Crespo, J., Reinhardt, G., Martinelli, L., Pipe, R., Schroeder, D. C., Igl-Schmid, N., Kokossis, A., Ben-Amotz, A. and Persson, K. (2014). The CO_2 microalgae biorefinery: High value products from low value wastes using halophylic microalgae in the D-factory. Part1: Tackling cell harvesting. In: Papers of the 22nd European Biomass Conference: Setting the Course for a Biobased Economy. C. Hoffmann, D. Baxter, K. Maniatis, A. Grassi and P. Helm, (eds.) ETA-Florence Renewable Energies, Florence, Italy: 360-363.

[52] Heydarizadeh, P., Poirier, I., Loizeau, D., Ulmann, L., Mimouni, V., Schoefs, B. and Bertrand, M. (2013). Plastids of marine phytoplankton produce bioactive pigments and lipids. Mar. Drugs 11(9), 3425-3471.

[53] Mimouni, V., Ulmann, L., Pasquet, V., Mathieu, M., Picot, L., Bougaran, G., Cadoret, J.-P., Morant-Manceau, A. and Schoefs, B. (2012). The potential of microalgae for the production of bioactive molecules of pharmaceutical interest. Curr. Pharm. Biotechnol. 13(15), 2733-2750.

[54] Schoefs, B. (2011). The synthesis of secondary carotenoids in microalgae. In: BIT's 1st Annual Low Carbon Earth Summit 2011, October 19-22, Dalian, China. 057.

[55] Skjånes, K., Lindblad, P. and Muller, J. (2007). BiO CO_2—A multidisciplinary, biological approach using solar energy to capture CO_2 while producing H_2 and high value products. Biomol. Eng. 24(4), 405-413.

[56] Xiang, X. W., Ozkan, A., Chiriboga, O., Chotyakul, N. and Kelly, C. (2017). Techno-economic analysis of glucosamine and lipid production from marine diatom Cyclotella sp. Bioresour. Technol. 244, 1480-1488.

[57] Forrest, M. (2017). Federal government seeking partner to look at way to create domestic jet biofuel industry. http://nationalpost.com/news/politics/federal-governmentseeking-partner-to-look-at-creating-domestic-jet-biofuel-industry.

[58] Marowits, R. (2017). Canadian airlines aiming to become a biofuel superpower, reduce carbon footprint. http://nationalpost.com/pmn/news-pmn/canada-news-pmn/canadianairlines-aiming-to-become-a-biofuel-superpower-reduce-carbon-footprint.

[59] Alves da Cruz, R. V. and Oller do Nascimento, C. A. (2012). Process modeling and economic analysis of microalgal systems for CO_2 capture and production of chemicals. In: 11th International Symposium on Process Systems Engineering, Parts A and B. I. A. Karimi and R. Srinivasan, (eds.). 31: 490-494.

[60] Amanor-Boadu, V., Pfromm, P. H. and Nelson, R. (2014). Economic feasibility of algal biodiesel under alternative public policies. Renew. Energy 67, 136-142.

[61] van Beilen, J. B. (2010). Perspective: Why microalgal biofuels won't save the internal combustion machine. Biofuels Bioprod. Biorefining 4(1), 41-52.

[62] Bhatt, N. C., Panwar, A., Bisht, T. S. and Tamta, S. (2014). Coupling of algal biofuel production with wastewater. Sci. World J., #210504.

[63] Bohutskyi, P., Liu, K., Nasr, L. K., Byers, N., Rosenberg, J. N., Oyler, G. A., Betenbaugh, M. J. and Bouwer, E. J. (2015). Bioprospecting of microalgae for integrated biomass production and phytoremediation of unsterilized wastewater and anaerobic digestion centrate. Appl Microbiol Biotechnol, doi: 10.1007/s00253-00015-06603-00254.

[64] Drexler, I. L. C., Joustra, C., Prieto, A., Bair, R. and Yeh, D. H. (2014). AlgaeSim: A model for integrated algal biofuel production and wastewater treatment. Water Environ. Res. 86(2), 163-176.

[65] Heydarizadeh, P., Boureba, W., Zahedi, M., Huang, B., Moreau, B., Lukomska, E., CouzinetMossion, A., Wielgosz-Collin, G., Martin-Jezequel, V., Bougaran, G., Marchand, J. and Schoefs, B. (2017). Response of CO_2-starved diatom Phaeodactylum tricornutum to light intensity transition. Philosophical Transactions of the Royal Society B: Biological Sciences 372(1728), #20160396.

[66] Kern, J. D., Hise, A. M., Characklis, G. W., Gerlach, R., Viamajala, S. and Gardner, R. D. (2017). Using life cycle assessment and techno-economic analysis in a real options framework to inform the design of algal biofuel production facilities. Bioresour. Technol. 225, 418-428.

[67] Pokoo-Aikins, G., Nadim, A., El-Halwagi, M. M. and Mahalec, V. (2010). Design and analysis of biodiesel production from algae grown through carbon sequestration. Clean Technol. Environ. Policy 12(3), 239-254.

[68] García Prieto, C. V., Ramos, F. D., Estrada, V., Villar, M. A. and Diaz, M. S. (2017). Optimization of an integrated algae-based biorefinery for the production of biodiesel, astaxanthin and PHB. Energy 139, 1159-1172.

[69] Williamson, S. (2015). $500/BOE Profit from Algae? Yes We Can!, http://www.biofuelsdigest.com/bdigest/2015/02/09/500boe-profit-from-algae-yes-we-can/.

[70] Kerschner, C., Prell, C., Feng, K. and Hubacek, K. (2013). Economic vulnerability to Peak Oil. Global Environmental Change, http://dx.doi.org/10.1016/j.gloenvcha.2013.08.015.

[71] Kovarik, B. (2013). Oil "scarcity": We should have known better. http://www.environmentalhistory.org/billkovarik/2013/04/10/same-old-oil/.

[72] Verbruggen, A. and Al Marchohi, M. (2010). Views on peak oil and its relation to climate change policy. Energy Policy 38(10), 5572-5581.

[73] Wikipedia. (2019m). Peak oil. https://en.wikipedia.org/wiki/Peak_oil.

[74] Jaganathan, J. (2017). Goldman Sachs warns global oil demand could peak as early as 2024. http://www.businessinsider.com/goldman-sachs-oil-demand-expectation-2024-2017-7?op=1.

[75] Cooper, A., Ghaddar, A. and Zhdannikov, D. (2017). U. S. oil output may be set for last spike in 2018: Vitol. https://www.reuters.com/article/us-commodities-summit-vitol/u-s-oil-outputmay-be-set-for-last-spike-in-2018-vitol-idUSKBN1CF1MZ.

[76] Lane, J. (2014). The DOE's shifting worldview for biofuels deployment, now through 2030. http://www.biofuelsdigest.com/bdigest/2014/12/28/the-does-shifting-worldview-forbiofuels-deployment-now-through-2030/.

[77] Oil Sands Magazine. (2018). Oil prices explained: Putting a dollar value on a barrel of crude. http://www.oilsandsmagazine.com/market-insights/oil-prices-explained-how-to-value-a-barrel-of-crude.

[78] Khan, M. I., Shin, J. H. and Kim, J. D. (2018). The promising future of microalgae: current status, challenges, and optimization of a sustainable and renewable industry for biofuels, feed, and other products. Microb. Cell. Fact. 17, #36.

[79] Bullis, K. (2008). Oil Price Threatens Biofuel Firms. Falling oil prices could cause some alternative-fuel startups to fail. http://www.technologyreview.com/business/21775/.

[80] Gordon, R. and Poulin, B. J. (2012). Quitting cold turkey: Rapid oil independence for the USA. In: The Science of Algal Fuels: Phycology, Geology, Biophotonics, Genomics and Nanotechnology.

[81] Dutta, K., Daverey, A. and Lin, J. G. (2014). Evolution retrospective for alternative fuels: First to fourth generation. Renew. Energy 69, 114-122.
[82] Lü, J., Sheahan, C. and Fu, P. (2011). Metabolic engineering of algae for fourth generation biofuels production. Energy Environ. Sci. 4(7), 2451-2466.
[83] Wang, X. L., Ma, Y. J. and Su, Y. L. (1997). Determining surface areas of marine alga cells by acidbase titration method. Chemosphere 35(5), 1131-1141.
[84] De Angelis, R., Melino, S., Prosposito, P., Casalboni, M., Lamastra, F. R., Nanni, F., Bruno, L. and Congestri, R. (2016). The diatom Staurosirella pinnata for photoactive material production. PLoS One 11 (11), #e0165571.
[85] Ouwehand, J., Van Eynde, E., De Canck, E., Lenaerts, S., Verberckmoes, A. and Van Der Voort, P. (2018). Titania-functionalized diatom frustules as photocatalyst for indoor air purification. Applied Catalysis B: Environmental 226, 303-310.
[86] Chaudry, S., Bahri, P. A. and Moheimani, N. R. (2018). Techno-economic analysis of milking of Botryococcus braunii for renewable hydrocarbon production. Algal Res. 31, 194-203.
[87] Jackson, B. A., Bahri, P. A. and Moheimani, N. R. (2017). Repetitive non-destructive milking of hydrocarbons from Botryococcus braunii. Renewable and Sustainable Energy Reviews 79, 1229-1240.
[88] Jerney, J. and Spilling, K. (2018). Large scale cultivation of microalgae: Open and closed systems. Methods in Molecular Biology, doi:10.1007/7651_2018_1130.
[89] Ramachandra, T. V., Mahapatra, D. M., Karthick B. and Gordon, R. (2009). Milking diatoms for sustainable energy: Biochemical engineering versus gasoline-secreting diatom solar panels. Ind. Eng. Chem. Res. 48 (19, Complex Materials II special issue, October), 8769-8788.
[90] Vinayak, V., Manoylov, K. M., Gateau, H., Blanckaert, V., Pencréac'h, G., Hérault, J., Marchand, J., Gordon, R. and Schoefs, B. (2015). Diatom milking: a review and new approaches [with Correction]. Mar. Drugs 13(5,12), 2629-2665, 7301.
[91] Lee, S. Y., Cho, J. M., Chang, Y. K. and Oh, Y. K. (2017). Cell disruption and lipid extraction for microalgal biorefineries: A review. Bioresour. Technol. 244, 1317-1328.
[92] Econtrader. (2014). How much gasoline does a barrel of crude oil produce. http://www.econtrader.com/economics/explain/how-much-gasoline-one-barrel-crude-oil.htm.
[93] Groupe Anderson. (2017). HYBRID-X Xtractor. https://grpanderson.com/en/inline/wrappers/hybrid-x/.
[94] GroupAnderson. (2015). Anderson new IFX660 wrapper. https://www.youtube.com/watch?time_continue=13&v=sJe8CmPb4mk.
[95] Brandon, H. (2017). O. A. Cleveland: Cotton price near-term will depend on weather. http://www.deltafarmpress.com/cotton/oa-cleveland-cotton-price-near-term-will-depend-weather.
[96] Johnson, K. R. and Admassu, W. (2013). Mixed algae cultures for low cost environmental compensation in cultures grown for lipid production and wastewater remediation. J. Chem. Technol. Biotechnol. 88(6), 992-998.
[97] Amin, S. A., Hmelo, L. R., van Tol, H. M., Durham, B. P., Carlson, L. T., Heal, K. R., Morales, R. L., Berthiaume, C. T., Parker, M. S., Djunaedi, B., Ingalls, A. E., Parsek, M. R., Moran, M. A. and Armbrust, E. V. (2015). Interaction and signalling between a cosmopolitan phytoplankton and associated bacteria. Nature 522(7554), 98-101.
[98] Madhu, N. V., Paul, M., Ullas, N., Ashwini, R. and Rehitha, T. V. (2013). Occurrence of cyanobacteria (Richelia intracellularis)-diatom (Rhizosolenia hebetata) consortium in the Palk Bay, southeast coast of India. Indian J. Geo-Mar. Sci. 42(4), 453-457.
[99] Miranda, A. F., Ramkumar, N., Andriotis, C., Hoeltkemeier, T., Yasmin, A., Rochfort, S., Wlodkowic, D., Morrison, P., Roddick, F., Spangenberg, G., Lal, B., Subudhi, S. and Mouradov, A. (2017). Applications of microalgal biofilms for wastewater treatment and bioenergy production. Biotechnol. Biofuels 10, #120.
[100] Gutiérrez, M. H., Jara, A. M. and Pantoja, S. (2016). Fungal parasites infect marine diatoms in the upwelling ecosystem of the Humboldt current system off central Chile. Environ. Microbiol. 18(5), 1646-1653.
[101] Carotenuto, Y., Wichard, T., Pohnert, G. and Lampert, W. (2005). Life-history responses of Daphnia pulicaria to diets containing freshwater diatoms: Effects of nutritional quality versus polyunsaturated aldehydes. Limnol. Oceanogr. 50(2), 449-454.

[102] Sato, R., Maeda, Y., Yoshino, T., Tanaka, T. and Matsumoto, M. (2014). Seasonal variation of biomass and oil production of the oleaginous diatom Fistulifera sp in outdoor vertical bubble column and raceway-type bioreactors. J. Biosci. Bioeng. 117(6), 720-724.

[103] Bwambok, D. K., Christodouleas, D. C., Morin, S. A., Lange, H., Phillips, S. T. and Whitesides, G. M. (2014). Adaptive use of bubble wrap for storing liquid samples and performing analytical assays. Anal. Chem. 86(15), 7478-7485.

[104] McDonald, C. and McGloin, D. (2015). Bubble wrap for optical trapping and cell culturing. Biomed. Opt. Express 6(10), 3757-3764.

[105] Sealed Air. (2017b). Bubble Wrap IB. https://sealedair.com/product-care/product-care-products/bubble-wrap-ib.

[106] ULINE. (2013). Sealed Air NewAir I. B. Express Instant Bubble Machine [video]. https://www.youtube.com/watch?v=u1nPmwsNp24.

[107] Heredia-Guerrero, J. A. and Heredia, A. (2017). The plant cuticle: old challenges, new perspectives. J. Exp. Bot. 68(19), 5251-5255.

[108] Erickson, P. E. (2011). Hay Bale Wrapper [video]. https://www.youtube.com/watch?v=JUFyLrPiif0.

[109] Martinson, K., Coblentz, W. and Sheaffer, C. (2011). The effect of harvest moisture and bale wrapping on forage quality, temperature, and mold in orchardgrass hay.

[110] Rudstrom, M. (2006). The dollars and cents of bale wrapping haylage. https://www.extension.umn.edu/agriculture/dairy/forages/the-dollars-and-cents-of-bale-wrapping-haylage/[cobweb site].

[111] Adhikari, R., Bristow, K. L., Casey, P. S., Freischmidt, G., Hornbuckle, J. W., Adhikari, B. (2016). Preformed and sprayable polymeric mulch film to improve agricultural water use efficiency, Agricultural Water Management 169, 1-13.

[112] Briassoulis, D. and Dejean, C. (2010). Critical review of norms and standards for biodegradable agricultural plastics Part I. Biodegradation in soil. J. Polym. Environ. 18(3), 384-400.

[113] Briassoulis, D., Dejean, C. and Picuno, P. (2010). Critical review of norms and standards for biodegradable agricultural plastics Part II: Composting. J. Polym. Environ. 18(3), 364-383.

[114] Goldberger, J. R., Jones, R. E., Miles, C. A., Wallace, R. W. and Inglis, D. A. (2015). Barriers and bridges to the adoption of biodegradable plastic mulches for US specialty crop production. Renewable Agriculture and Food Systems 30(2), 143-153.

[115] Hayes, D. G., Dharmalingam, S., Wadsworth, L. C., Leonas, K. K., Miles, C. and Inglis, D. A. (2012). Biodegradable agricultural mulches derived from biopolymers. In: Degradable Polymers and Materials: Principles and Practice. K. Khemani and C. Scholz, (eds.). 1114: 201-223.

[116] Santagata, G., Malinconico, M., Immirzi, B., Schettini, E., Mugnozza, G. S. and Vox, G. (2014). An overview of biodegradable films and spray coatings as sustainable alternative to oil-based mulching films. In: International Symposium on New Technologies for Environment Control, Energy-Saving and Crop Production in Greenhouse and Plant Factory—Greensys 2013. J. E. Son, I. B. Lee and M. M. Oh, (eds.). 1037: 921-928.

[117] Steinmetz, Z., Wollmann, C., Schaefer, M., Buchmann, C., David, J., Troeger, J., Munoz, K., Fror, O. and Schaumann, G. E. (2016). Plastic mulching in agriculture. Trading short-term agronomic benefits for long-term soil degradation? Sci. Total Environ. 550, 690-705.

[118] Das, A. A. K., Esfahani, M. M. N., Velev, O. D., Pamme, N. and Paunov, V. N. (2015). Artificial leaf device for hydrogen generation from immobilised C. reinhardtii microalgae. J. Mater. Chem. A 3(41), 20698-20707.

[119] Janna Olmos, J. D. and Kargul, J. (2015). A quest for the artificial leaf. Int. J. Biochem. Cell Biol. 66, 37-44.

[120] Qiao, H. J., Fan, X., Xu, D., Ye, N. H., Wang, J. Y. and Cao, S. N. (2015). Artificial leaf aids analysis of chlorophyll fluorescence and P700 absorbance in studies involving microalgae. Phycol. Res. 63(1), 72-76.

[121] Peribere, J. and O'Keefe, B. (2015). Sealed Air CEO (bubble wrap) on material gains in a resourcechallenged world | Fortune. https://www.youtube.com/watch?v=Ozh0hKamczY.

[122] Davis, D. (2014). 4 Good Reasons for Injecting Bubble Wrap. http://beachpackagingdesign.com/boxvox/4-reasons-injecting-bubble-wrap.

[123] Huber, M. (2010). Petri Bubbles. http://www.sealedairprotects.com/yic/10/huber.htm.
[124] Van Denburg, H. (2010). Matthew Huber is the the Bubble Wrap inventor king of the world. http://www.citypages.com/news/matthew-huber-is-the-the-bubble-wrap-inventor-king-ofthe-world-6544993.
[125] Aeinehvand, M. M., Ibrahim, F. and Madou, M. J. (2016). A new approach for reagent storagereleasing on centrifugal microfluidic platforms using bubblewrap and latex membrane. In: International Conference for Innovation in Biomedical Engineering and Life Sciences, ICIBEL2015. F. Ibrahim, J. Usman, M. S. Mohktar and M. Y. Ahmad, (eds.). 56: 269-271.
[126] Harada, T. and Discher, D. E. (2011). Materials science: Bubble wrap of cell-like aggregates. Nature 471(7337), 172-173.
[127] Martinkova, P. and Pohanka, M. (2016a). Colorimetric sensor based on bubble wrap and camera phone for glucose determination. J. Appl. Biomed. 14(4), 315-319.
[128] Martinkova, P. and Pohanka, M. (2016b). Phone camera detection of glucose blood level based on magnetic particles entrapped inside bubble wrap. Neuroendocrinol. Lett. 37(Suppl. 1), 101-107.
[129] NPR. (2014). George M. Whitesides. https://gmwgroup.harvard.edu/people/george-m-whitesides.
[130] Pohanka, M. (2015). Photography by cameras integrated in smartphones as a tool for analytical chemistry represented by an butyrylcholinesterase activity assay. Sensors 15(6), 13752-13762.
[131] Pohanka, M. (2017). Small camera as a handheld colorimetric tool in the analytical chemistry. Chem. Pap. 71(9), 1553-1561.
[132] Marshall, J. (2014). Solar energy: Springtime for the artificial leaf. Nature 510(7503), 22-24.
[133] Dogaris, I., Welch, M., Meiser, A., Walmsley, L. and Philippidis, G. (2015). A novel horizontal photobioreactor for high-density cultivation of microalgae. Bioresour. Technol. 198, 316-324.
[134] Gressel, J. and Granot, M. (2014). Novel photobioreactor for enclosed horizontal cultivation of microalgae [International Patent Application WO 2014/064602 A2]. World Intellectual Property Organization, United Nations, Geneva, Switzerland.
[135] Gressel, J. and Granot, M. (2015). Novel photobioreactor for enclosed horizontal cultivation of microalgae [US Patent Application US 2015/0275161 A1]. US Patent Office, Washington, DC, USA.
[136] Settles, A. M. and Rathinasabapathi, B. (2018). Domesticating Algae for Sustainable Production of Feedstocks in Space. https://www.nasa.gov/mission_pages/station/research/experiments/explorer/Investigation.html?♯id=7446.
[137] Thompson, A. (2018). Plants, Algae and Other Weird Green Stuff Just Arrived at the Space Station. https://www.space.com/41071-space-plants-algae-microgreens-space-station.html.
[138] Hotta, Y., Zhang, Y. H. and Miki, N. (2012). A flexible capacitive sensor with encapsulated liquids as dielectrics. Micromachines 3(1), 137-149.
[139] Ishizuka, H. and Miki, N. (2015). Fabrication of hemispherical liquid encapsulated structures based on droplet molding. Journal of Micromechanics and Microengineering 25(12), ♯125010.
[140] Lee, J. K., Kim, H. R. and Kong, S. H. (2011). Hermetic sealing of liquid using Laplace pressure disparity induced by heterogeneous surface energy. Sens. Actuator A-Phys. 169(2), 333-340.
[141] Ninomiya, T., Okayama, Y., Matsumoto, Y., Arouette, X., Osawa, K. and Miki, N. (2011). MEMSbased hydraulic displacement amplification mechanism with completely encapsulated liquid. Sens. Actuator A-Phys. 166(2), 277-282.
[142] Okayama, Y., Nakahara, K., Arouette, X., Ninomiya, T., Matsumoto, Y., Orimo, Y., Hotta, A., Omiya, M. and Miki, N. (2010). Characterization of a bonding-in-liquid technique for liquid encapsulation into MEMS devices. Journal of Micromechanics and Microengineering 20(9), ♯095018.
[143] Chepurnov, V. A., Chaerle, P. and Mann, D. G. (2012). How to breed diatoms: examination of two species with contrasting reproductive biology. In: The Science of Algal Fuels: Phycology, Geology, Biophotonics, Genomics and Nanotechnology. R. Gordon and J. Seckbach, (eds.) Springer, Dordrecht: 323-340.
[144] Ceccarelli, S. (2009). Evolution, plant breeding and biodiversity. Journal of Agriculture and Environment for International Development 103(1/2), 131-145.
[145] Doshi, R., Tuan, N. and Chang, G. (2013). Transporter-mediated biofuel secretion. Proc. Natl. Acad. Sci. U. S. A. 110(19), 7642-7647.
[146] Vinayak, V., Gordon, R., Gautam, S. and Rai, A. (2014). Discovery of a diatom that oozes oil. Adv. Sci. Lett. 20(7-9, Special issue: National Conference on Nanotechnology and Renewable Energy, Jamia

第 18 章 泡沫农业：硅藻生物燃料和下一次绿色革命的可扩展缩影

Millia Islamia, New Delhi, April 28-29, 2014), 1256-1267.
[147] Chang, J. S., Mohan, S. V. and Lee, D. J. (2017). Preface: Special Issue on Algal Biorefinery. Bioresour. Technol. 244, 1197-1197.
[148] Canter, N. (2015). Higher-octane gasoline might lead to improved fuel economy. Tribol. Lubr. Technol. 71(1), 10-11.
[149] Speth, R. L., Chow, E. W., Malina, R., Barrett, S. R. H., Heywood, J. B. and Green, W. H. (2014). Economic and environmental benefits of higher-octane gasoline. Environ. Sci. Technol. 48 (12), 6561-6568.
[150] Heywood, J. B. (2018). Internal Combustion Engine Fundamentals. McGraw-Hill Education: 904, 2nd.
[151] Anderson, C. (2014). Agricultural drones. Technol. Rev. 117(3), 58-60.
[152] Balan, K. C. S. (2016). Robotic-based agriculture for rural renaissance: Drones and biosensors. In: Biosensors for Sustainable Food—New Opportunities and Technical Challenges. V. Scognamiglio, G. Rea, F. Arduini and G. Palleschi, (eds.). 74: 363-375.
[153] Patel, P. (2016). Agriculture drones are finally cleared for takeoff. IEEE Spectr. 53(11), 13-14.
[154] Reaganon, J. (2017). Report: Agriculture Drone Market May Exceed $4 billion. https://dronelife.com/2017/10/05/report-agriculture-drone-market-may-exceed-4-billion/.
[155] Shang, S. L., Lee, Z. P., Lin, G., Hu, C. M., Shi, L. H., Zhang, Y. N. A., Li, X. D., Wu, J. Y. and Yan, J. (2017). Sensing an intense phytoplankton bloom in the western Taiwan Strait from radiometric measurements on a UAV. Remote Sens. Environ. 198, 85-94.
[156] Liu, J., Song, Y. and Qiu, W. (2017). Oleaginous microalgae Nannochloropsis as a new model for biofuel production: Review & analysis. Renew. Sust. Energ. Rev. 72, 154-162.
[157] De La Bastide, D. (2017). This Project Used Drones To Farm An Entire Crop, And It's Ready For Harvesting. https://interestingengineering.com/this-project-used-drones-tofarm-an-entire-crop-and-its-ready-for-harvesting.
[158] ULINE. (2017). Instant Bubble Roll Machine. https://www.uline.ca/BL_174/Instant-Bubble-Roll-Machine.
[159] Sealed Air. (2018). Cryovac VPP 2045D Vertical Form-Fill Seal Machine. http://www.cryovac.com/EU/EN/pdf/model_2045.pdf.
[160] Bahadar, A. and Khan, M. B. (2013). Progress in energy from microalgae: A review. Renew. Sust. Energ. Rev. 27, 128-148.
[161] Fasaei, F., Bitter, J. H., Slegers, P. M. and van Boxtel, A. J. B. (2018). Techno-economic evaluation of microalgae harvesting and dewatering systems. Algal Res. 31, 347-362.
[162] Wileman, A., Ozkan, A. and Berberoglu, H. (2012). Rheological properties of algae slurries for minimizing harvesting energy requirements in biofuel production. Bioresour. Technol. 104, 432-439.
[163] Farmer, J., Zhang, B., Jin, X. X., Zhang, P. and Wang, J. K. (2017). Long-term effect of plastic film mulching and fertilization on bacterial communities in a brown soil revealed by high through-put sequencing. Archives of Agronomy and Soil Science 63(2), 230-241.
[164] Isaia, M., Bona, F. and Badino, G. (2006). Comparison of polyethylene bubble wrap and corrugated cardboard traps for sampling tree-inhabiting spiders. Environ. Entomol. 35(6), 1654-1660.
[165] Pinzón, J. and Spence, J. R. (2010). Bark-dwelling spider assemblages (Araneae) in the boreal forest: dominance, diversity, composition and life-histories. J. Insect Conserv. 14(5), 439-458.
[166] Roberts, D. J. and Roberts, M. J. (1988). Don't forget those trees! Newsletter, British Arachnological Society 52, 8.
[167] Umar, A., Caldwell, G. S. and Lee, J. G. M. (2018). Foam flotation can remove and eradicate ciliates contaminating algae culture systems. Algal Res. 29, 337-342.
[168] Wang, J. F. and Liu, T. Z. (2013). The contamination and control of biological pollutants in mass cultivation of microalgae. Bioresour. Technol. 128, 745-750.
[169] Wang, L., Yuan, D. N., Li, Y. H., Ma, M. Y., Hu, Q. and Gong, Y. C. (2016). Contaminating microzooplankton in outdoor microalgal mass culture systems: An ecological viewpoint. Algal Res. 20, 258-266.
[170] Wang, C., Trousdell, J., Conley, S., Faloona, I. and Houlton, B. Z. (2018). Agriculture is a major source of NOx pollution in California. Science Advances 4(1), #eaao3477.
[171] Johnston, L. (2017). Diatoms. http://www.gopetsamerica.com/bio/phytoplankton/diatoms.aspx.

[172] Neeson, M. J., Tabor, R. F., Grieser, F., Dagastine, R. R. and Chan, D. Y. C. (2012). Compound sessile drops. Soft Matter 8(43), 11042-11050.
[173] Augenstein, N. (2015). Shhhhh—new bubble wrap no longer pops. https://wtop.com/trending-now/2015/07/shhhhh-new-bubble-wrap-no-longer-pops/.
[174] Haan Jr., F. L., Sarkar, P. P. and Spencer-Berger, N. J. (2006). Development of an active gust generation mechanism on a wind tunnel for wind engineering and industrial aerodynamics applications. Wind and Structures 9(5), 369-386.
[175] Lancelot, P. M. G. J., Sodja, J., Werter, N. P. M. and De Breuker, R. (2017). Design and testing of a low subsonic wind tunnel gust generator. Advances in Aircraft and Spacecraft Science 4(2), 125-144.
[176] Lentink, D. and Quinn, D. B. (2017). From quiet laminar flow to turbulent gusts: A new wind tunnel for studying animal flight performance and control. Integr. Comp. Biol. 57, #e325.
[177] Tang, D. M., Cizmas, P. G. A. and Dowell, E. H. (1996). Experiments and analysis for a gust generator in a wind tunnel. Journal of Aircraft 33(1), 139-148.
[178] Volk, A., Rossi, M., Kahler, C. J., Hilgenfeldt, S. and Marin, A. (2015). Growth control of sessile microbubbles in PDMS devices. Lab Chip 15(24), 4607-4613.
[179] Wikipedia. (2019j). Electromagnetic absorption by water. https://en.wikipedia.org/wiki/Electromagnetic_absorption_by_water.
[180] Wheatley, J. A. and Schrenk, W. J. (1993). Visibly transparent infrared reflecting film with color masking. https://www.google.com/patents/US5233465.
[181] Bement, S. T., Nassar, A. R. and Mehta, K. (2013). The feasibility of rice bags as a low-cost and locally available alternative to greenhouse glazing. In: Proceedings of the Third 2013 IEEE Global Humanitarian Technology Conference. IEEE: 254-259.
[182] Gensin. (2017). Two-wall, 3-wall, 4-wall hollow plastic sheet. https://www.toppolycarbonatesheet.com/pools-used-solar-panel-perforated-plastic-sheet-for-sale.html[cobweb site].
[183] Huatao Group. (2017). Plastic corrugated board. http://www.huataogroup.com/productimage/43855608.html[cobweb site].
[184] Wikipedia. (2019k). Band-pass filter. https://en.wikipedia.org/wiki/Band-pass_filter.
[185] Wikipedia. (2019l). Circle packing. https://en.wikipedia.org/wiki/Circle_packing.
[186] Domps, A. and Roques-Carmes, T. (2011). Playing with water drops: from wetting to optics through electrostatics. Eur. J. Phys. 32(2), 559-570.
[187] Haußmann, A. (2017). Light scattering from sessile water drops and raindrop-shaped glass beads as a validation tool for rainbow simulations. Appl. Optics 56(19), G136-G144.
[188] Hejmadi, V., Shin, M., Kress, B. and Giliberto, A. (2011). Novel solar cogeneration trough system based on stretched microstructured Mylar film. Proc. SPIE 8065, #80650G.
[189] Sharaf, O. Z. and Orhan, M. E. (2015a). Concentrated photovoltaic thermal (CPVT) solar collector systems: Part I—Fundamentals, design considerations and current technologies. Renew. Sust. Energ. Rev. 50, 1500-1565.
[190] Sharaf, O. Z. and Orhan, M. F. (2015b). Concentrated photovoltaic thermal (CPVT) solar collector systems: Part II—Implemented systems, performance assessment, and future directions. Renew. Sust. Energ. Rev. 50, 1566-1633.
[191] Gebeshuber, I. C. and Lee, D. W. (2012). Nanostructures for coloration (organisms other than animals). In: Springer Encyclopedia of Nanotechnology. B. Bhushan, (ed.) Springer: 1790-1803.
[192] Ezequiel, J., Laviale, M., Frankenbach, S., Cartaxana, P. and Serôdio, J. (2015). Photoacclimation state determines the photobehaviour of motile microalgae: The case of a benthic diatom. Journal of Experimental Marine Biology and Ecology 468, 11-20.
[193] Laviale, M., Frankenbach, S. and Serôdio, J. (2016). The importance of being fast: comparative kinetics of vertical migration and non-photochemical quenching of benthic diatoms under light stress. Marine Biology 163(1), #10.
[194] McLachlan, D. H., Brownlee, C., Taylor, A. R., Geider, R. J. and Underwood, G. J. C. (2009). Light induced motile responses of the estuarine benthic diatoms Navicula perminuta and Cylindrotheca closterium (Bacillariophyceae). J. Phycol. 45(3), 592-599.
[195] Gordon, R. and Drum, R. W. (1970). A capillarity mechanism for diatom gliding locomotion. Proceedings of the National Academy of Sciences USA 67(1), 338-344.

[196] Cohn, S. A., Dunbar, S., Ragland, R., Schulze, J., Suchar, A., Weiss, J. and Wolske, A. (2016). Analysis of light quality and assemblage composition on diatom motility and accumulation rate. Diatom Res. 31(3), 173-184.
[197] Ljung, A. L. and Lundström, T. S. (2017). Heat and mass transfer boundary conditions at the surface of a heated sessile droplet. Heat Mass Transf. 53(12), 3581-3591.
[198] Thokchom, A. K., Gupta, A., Jaijus, P. J. and Singh, A. (2014). Analysis of fluid flow and particle transport in evaporating droplets exposed to infrared heating. Int. J. Heat Mass Transf. 68, 67-77.
[199] Mapes, M., Hseuh, H. C. and Jiang, W. S. (1994). Permeation of argon, carbon dioxide, helium, nitrogen, and oxygen through Mylar windows. J. Vac. Sci. Technol. A-Vac. Surf. Films 12(4), 1699-1704.
[200] Bianco, A., Perissinotto, S., Garbugli, M., Lanzani, G. and Bertarelli, C. (2011). Control of optical properties through photochromism: a promising approach to photonics. Laser Photon. Rev. 5(6), 711-736.
[201] Ghobara, M. M. and Mousa, A. M. (2019). Diatomite in use: Nature, modification, commercial applications, and prospective trends [Chapter 19]. In: Diatoms: Fundamentals & Applications [DIFA, Volume 1 in the series: Diatoms: Biology & Applications, series editors: Richard Gordon & Joseph Seckbach]. J. Seckbach and R. Gordon, (eds.) Wiley-Scrivener, Beverly, MA, USA: 471-510.
[202] Losic, D. (2008). Microstructured surfaces engineered using biological templates: A facile approach for the fabrication of superhydrophobic surfaces. J. Serb. Chem. Soc. 73(11), 1123-1135.
[203] Ardisson, G. B., Tosin, M., Barbale, M. and Degli-Innocenti, F. (2014). Biodegradation of plastics in soil and effects on nitrification activity. A laboratory approach. Front. Microbiol. 5, #710.
[204] Alles, M. and Rosenhahn, A. (2015). Microfluidic detachment assay to probe the adhesion strength of diatoms. Biofouling 31(5), 469-480.
[205] Holland, R., Dugdale, T. M., Wetherbee, R., Brennan, A. B., Finlay, J. A., Callow, J. A. and Callow, M. E. (2004). Adhesion and motility of fouling diatoms on a silicone elastomer. Biofouling 20(6), 323-329.
[206] Schilp, S., Kueller, A., Rosenhahn, A., Grunze, M., Pettitt, M. E., Callow, M. E. and Callow, J. A. (2007). Settlement and adhesion of algal cells to hexa(ethylene glycol)-containing self-assembled monolayers with systematically changed wetting properties. Biointerphases 2(4), 143-150.
[207] Wang, J. F., Liu, W. and Liu, T. Z. (2017). Biofilm based attached cultivation technology for microalgal biorefineries—A review. Bioresour. Technol. 244, 1245-1253.
[208] Blersch, D. M., Kardel, K., Carrano, A. L. and Kaur, M. (2017). Customized 3D-printed surface topography governs species attachment preferences in a fresh water periphyton community. Algal Res. 21, 52-57.
[209] Gross, M., Zhao, X. F., Mascarenhas, V. and Wen, Z. Y. (2016). Effects of the surface physicochemical properties and the surface textures on the initial colonization and the attached growth in algal biofilm. Biotechnol. Biofuels 9, #38.
[210] Mora-Gómez, J., Freixa, A., Perujo, N. and Barral-Fraga, L. (2016). Limits of the biofilm concept and types of aquatic biofilms. In: Aquatic Biofilms: Ecology, Water Quality and Wastewater Treatment. A. M. Romani, H. Guasch and D. M. Balaguer, (eds.) Caister Academic Press, Norfolk, UK: 3-27.
[211] Efimenko, K., Crowe, J. A., Manias, E., Schwark, D. W., Fischer, D. A. and Genzer, J. (2005). Rapid formation of soft hydrophilic silicone elastomer surfaces. Polymer 46(22), 9329-9341.
[212] Barillas, M. K., Enick, R. M., O'Brien, M., Perry, R., Luebke, D. R. and Morreale, B. D. (2011). The CO_2 permeability and mixed gas CO_2/H_2 selectivity of membranes composed of CO_2-philic polymers. J. Membr. Sci. 372(1-2), 29-39.
[213] Tremblay, P., Savard, M. M., Vermette, J. and Paquin, R. (2006). Gas permeability, diffusivity and solubility of nitrogen, helium, methane, carbon dioxide and formaldehyde in dense polymeric membranes using a new on-line permeation apparatus. J. Membr. Sci. 282(1-2), 245-256.
[214] Huang, Q. S., Jiang, F. H., Wang, L. Z. and Yang, C. (2017). Design of photobioreactors for mass cultivation of photosynthetic organisms. Engineering 3(3), 318-329.
[215] Ugwu, C. U., Aoyagi, H. and Uchiyama, H. (2008). Photobioreactors for mass cultivation of algae. Bioresour. Technol. 99(10), 4021-4028.
[216] Ugwu, C. U., Aoyagi, H. and Uchiyama, H. (2007). Influence of irradiance, dissolved oxygen

concentration, and temperature on the growth of Chlorella sorokiniana. Photosynthetica 45(2), 309-311.

[217] Kim, S. and Lee, Y. M. (2013). High performance polymer membranes for CO_2 separation. Curr. Opin. Chem. Eng. 2(2), 238-244.

[218] Börjesson, A., Erdtman, E., Ahlström, P., Berlin, M., Andersson, T. and Bolton, K. (2013). Molecular modelling of oxygen and water permeation in polyethylene. Polymer 54(12), 2988-2998.

[219] Mehta, G., Mehta, K., Sud, D., Song, J. W., Bersano-Begey, T., Futai, N., Heo, Y. S., Mycek, M. A., Linderman, J. J. and Takayama, S. (2007). Quantitative measurement and control of oxygen levels in microfluidic poly(dimethylsiloxane) bioreactors during cell culture. Biomed. Microdevices 9(2), 123-134.

[220] Zhang, Y., Ishida, M., Kazoe, Y., Sato, Y. and Miki, N. (2009). Water-vapor permeability control of PDMS by the dispersion of collagen powder. IEEJ Transactions on Electrical and Electronic Engineering 4(3), 442-449.

[221] Borde, A., Larsson, M., Odelberg, Y., Hagman, J., Löwenhielm, P. and Larsson, A. (2012). Increased water transport in PDMS silicone films by addition of excipients. Acta Biomaterialia 8(2), 579-588.

[222] Kettlewell, P. S. (2014). Waterproofing wheat—a re-evaluation of film antitranspirants in the context of reproductive drought physiology. Outlook Agric. 43(1), 25-29.

[223] Rottink, B. A. (1977). Effects of a silicone antitranspirant on tree growth. Forest Science 23(3), 361-362.

[224] Riederer, M. and Schreiber, L. (2001). Protecting against water loss: analysis of the barrier properties of plant cuticles. J. Exp. Bot. 52(363), 2023-2032.

[225] Schreiber, L. (2010). Transport barriers made of cutin, suberin and associated waxes. Trends Plant Sci. 15(10), 546-553.

[226] Zeisler-Diehl, V. V., Migdal, B. and Schreiber, L. (2017). Quantitative characterization of cuticular barrier properties: methods, requirements, and problems. J. Exp. Bot. 68(19), 5281-5291.

[227] Siracusa, V. (2012). Food packaging permeability behaviour: A report. International Journal of Polymer Science 2012, #302029.

[228] Van Krevelen, D. W. and Te Nijenhuis K., (2009). Polymer Properties. Elsevier, Amsterdam, The Netherlands, 4th.

[229] Eibl, D. and Eibl, R. (eds.) (2014). Disposable Bioreactors II.

[230] ThermoFisher Scientific. (2018). Gibco PE Film Data. https://www.thermofisher.com/ca/en/home/life-science/cell-culture/mammalian-cell-culture/classical-media/gibco-cell-culturebags/pe-film-data.html.

[231] Menzel, S., Finocchiaro, N., Donay, C., Thiebes, A. L., Hesselmann, F., Arens, J., Djeljadini, S., Wessling, M., Schmitz-Rode, T., Jockenhoevel, S. and Cornelissen, C. G. (2017). Towards a biohybrid lung: Endothelial cells promote oxygen transfer through gas permeable membranes. Biomed Research International 2017, #5258196.

[232] Sarstedt. (2018). Lumox technology. https://www.sarstedt.com/en/products/laboratory/celltissue-culture/lumox-technology/.

[233] Meyer, H. P. and Schmidhalter, D. (2014). Industrial Scale Suspension Culture of Living Cells. Wiley, Weinheim, Germany.

[234] Bajgain, P., Mucharla, R., Wilson, J., Welch, D., Anurathapan, U., Liang, B. T., Lu, X. H., Ripple, K., Centanni, J. M., Hall, C., Hsu, D., Couture, L. A., Gupta, S., Gee, A. P., Heslop, H. E., Leen, A. M., Rooney, C. M. and Vera, J. F. (2014). Optimizing the production of suspension cells using the G-Rex "M" series. Mol. Ther.-Methods Clin. Dev. 1, #14015.

[235] OriGen. (2018). PermaLife Cell Culture Bags. https://www.origen.com/products/cell-culture-products/permalife-cell-culture-bags.

[236] Becker, M., Kerstiens, G. and Schonherr, J. (1986). Water permeability of plant cuticles: permeance, diffusion and partition coefficients. Trees-Struct. Funct. 1(1), 54-60.

[237] Fernández, V., Guzmán-Delgado, P., Graça, J., Santos, S. and Gil, L. (2016). Cuticle structure in relation to chemical composition: Re-assessing the prevailing model. Front. Plant Sci. 7, #427.

[238] Boyer, J. S., Wong, S. C. and Farquhar, G. D. (1997). CO_2 and water vapor exchange across leaf cuticle (epidermis) at various water potentials. Plant Physiol. 114(1), 185-191.

[239] Dugger, W. M. (1952). The permeability of non-stomate leaf epidermis to carbon dioxide. Plant Physiol.

27(3), 489-499.
[240] De Giorgi, J., Piskurewicz, U., Loubery, S., Utz-Pugin, A., Bailly, C., Mène-Saffrané, L. and Lopez-Molina, L. (2015). An endosperm-associated cuticle is required for Arabidopsis seed viability, dormancy and early control of germination. PLoS Genet. 11(12), #e1005708.
[241] L'Haridon, F., Besson-Bard, A., Binda, M., Serrano, M., Abou-Mansour, E., Balet, F., Schoonbeek, H.-J., Hess, S., Mir, R., Léon, J., Lamotte, O. and Métraux, J.-P. (2011). A permeable cuticle is associated with the release of reactive oxygen species and induction of innate immunity. PLoS Pathogens 7 (7), #e1002148.
[242] Heredia-Guerrero, J. A., Heredia, A., Domínguez, E., Cingolani, R., Bayer, I. S., Athanassiou, A. and Benítez, J. J. (2017). Cutin from agro-waste as a raw material for the production of bioplastics. J. Exp. Bot. 68(19), 5401-5410.
[243] Wikipedia. (2019b). Clark electrode. https://en.wikipedia.org/wiki/Clark_electrode.
[244] Scholes, C. A., Kanehashi, S., Stevens, G. W. and Kentish, S. E. (2015). Water permeability and competitive permeation with CO_2 and CH_4 in perfluorinated polymeric membranes. Sep. Purif. Technol. 147, 203-209.
[245] Lee, L. P., Berger, S. A., Pruitt, L. and Liepmann, D. (2000). Key elements of a transparent Teflon microfluidic system. In: 3rd International Symposium on Micro-Total Analysis Systems (muTAS'98), Banff, Canada, October 13-16, 1998.
[246] Perschke, A., Podgorsek, R. and Franke, H. (1997). Optical detection of acetone and MEK in water using thin films of transparent Teflon. In: 55th Annual Technical Conference of the Society-of-Plastics-Engineers—Plastics Saving Planet Earth (ANTEC 97), Toronto, Canada, April 27-May 02, 1997.
[247] Lee, E., Zhang, H. B., Jackson, J. K., Lim, C. J. and Chiao, M. (2016). Janus films with stretchable and waterproof properties for wound care and drug delivery applications. RSC Adv. 6(83), 79900-79909.
[248] Sami, I. (2017). Flexible electronic devices with roll-to-roll overmolding technology. https://www.vttresearch.com/media/news/flexible-electronic-devices-with-roll-to-roll-overmolding-technology.
[249] Song, J., Liu, H., Wan, M. X., Zhu, Y. and Jiang, L. (2013). Bio-inspired isotropic and anisotropic wettability on a Janus free-standing polypyrrole film fabricated by interfacial electro-polymerization. J. Mater. Chem. A 1(5), 1740-1744.
[250] Takegami, S., Yamada, H. and Tsujii, S. (1992). Pervaporation of ethanol/water mixtures using novel hydrophobic membranes containing polydimethylsiloxane. J. Membr. Sci. 75(1-2), 93-105.
[251] Poschenrieder, S. T., Klermund, L., Langer, B. and Castiglione, K. (2017). Determination of permeability coefficients of polymersomal membranes for hydrophilic molecules. Langmuir 33(24), 6011-6020.
[252] Walsby, A. E., Revsbech, N. P. and Griffel, D. H. (1992). The gas permeability coefficient of the cyanobacterial gas vesicle wall. Journal of General Microbiology 138, 837-845.
[253] Hammersley, J. M. and Handscomb, D. C. (1964). Monte Carlo Methods. Methuen, London.
[254] Feldman, D. (2013). Polymer nanocomposite barriers. J. Macromol. Sci. Part A-Pure Appl. Chem. 50 (4), 441-448.
[255] Gholizadeh, B., Arefazar, A. and Barzin, J. (2012). Morphology and gas permeability of polymeric membrane by PC/PA/nanoclay ternary nanocomposite. Polym. Polym. Compos. 20(3), 271-278.
[256] Khani, M. M., Woo, D., Mumpower, E. L. and Benicewicz, B. C. (2017). Poly(alkyl methacrylate)-grafted silica nanoparticles in polyethylene nanocomposites. Polymer 109, 339-348.
[257] Omidi, S., Ahmadi, S., Ghorbani, P. and Moazezi, A. (2015). In situ polymerization of—caprolactam in the presence of polyester polyol and nanosilica toward amorphous polyamide6/SiO_2 nanocomposite. Iranian Polymer Journal 24(11), 945-952.
[258] Paul, M. T. Y. and Gates, B. D. (2016). Hierarchical surface coatings of polystyrene nanofibers and silica microparticles with rose petal wetting properties. Colloid Surf. A-Physicochem. Eng. Asp. 498, 42-49.
[259] Żenkiewicz, M. and Richert, J. (2008). Permeability of polylactide nanocomposite films for water vapour, oxygen and carbon dioxide. Polymer Testing 27(7), 835-840.
[260] Chen, S., Gao, S., Jing, J. and Lu, Q. (2018). Designing 3D biological surfaces via the breath-figure method. Advanced Healthcare Materials, #1701043.
[261] Walsby, A. E. (1969). The permeability of blue-green algal gas-vacuole membranes to gas. Proceedings of the Royal Society of London Series B-Biological Sciences 173(1031), 235-255.

[262] Herzfeld, J. (2014). Gas vesicles. In: The Cell Biology of Cyanobacteria. E. Flores and A. Herrero, (eds.) Caister Academic Press.

[263] Davis, E. M., Stafford, C. M. and Page, K. A. (2014). Elucidating water transport mechanisms in Nafion thin films. ACS Macro Lett. 3(10), 1029-1035.

[264] Wikipedia. (2019f). Nafion. https://en.wikipedia.org/wiki/Nafion.

[265] Li, T., Piltz, B., Podola, B., Dron, A., de Beer, D. and Melkonian, M. (2016). Microscale profiling of photosynthesis-related variables in a highly productive biofilm photobioreactor. Biotechnol. Bioeng. 113(5), 1046-1055.

[266] Androjna, C., Gatica, J. E., Belovich, J. M. and Derwin, K. A. (2008). Oxygen diffusion through natural extracellular matrices: Implications for estimating "Critical thickness" values in tendon tissue engineering. Tissue Eng. Part A 14(4), 559-569.

[267] Roe, D. F., Gibbins, B. L. and Ladizinsky, D. A. (2010). Topical dissolved oxygen penetrates skin: Model and method. J. Surg. Res. 159(1), E29-E36.

[268] Evans, N. J. and Rutter, N. (1986). Percutaneous respiration in the newborn infant. Journal of Pediatrics 108(2), 282-286.

[269] Carlson, D. A., Schreck, C. E. and Brenner, R. J. (1992). Carbon dioxide released from human skin: Effect of temperature and insect repellents. J. Med. Entomol. 29(2), 165-170.

[270] van Loon, J. J. A., Smallegange, R. C., Bukovinszkiné-Kiss, G., Jacobs, F., De Rijk, M., Mukabana, W. R., Verhulst, N. O., Menger, D. J. and Takken, W. (2015). Mosquito attraction: Crucial role of carbon dioxide in formulation of a five-component blend of human-derived volatiles. J. Chem. Ecol. 41(6), 567-573.

[271] Ottaviani, G., Martel, S. and Carrupt, P. A. (2006). Parallel artificial membrane permeability assay: A new membrane for the fast prediction of passive human skin permeability. J. Med. Chem. 49(13), 3948-3954.

[272] Fumagalli, M., Lyonnard, S., Prajapati, G., Berrod, Q., Porcar, L., Guillermo, A. and Gebel, G. (2015). Fast water diffusion and long-term polymer reorganization during Nafion membrane hydration evidenced by time-resolved small-angle neutron scattering. Journal of Physical Chemistry B, 119(23), 7068-7076.

[273] Nakamura, K., Kitagawa, T., Nara, S., Wakamatsu, T., Ishiba, Y., Kanehashi, S., Sato, S. and Nagai, K. (2013). Permeability of dry gases and those dissolved in water through hydrophobic high free-volume silicon-or fluorine-containing nonporous glassy polymer membranes. Ind. Eng. Chem. Res. 52(3), 1133-1140.

[274] Chemours (2016). NafionTM N115, N117, N1110, Ion Exchange Materials, Extrusion Cast Membranes [Product Bulletin P-12].

[275] Zhu, X., Tian, C. C., Do-Thanh, C. L. and Dai, S. (2017). Two-dimensional materials as prospective scaffolds for mixed-matrix membrane-based CO_2 separation. ChemSusChem 10(17), 3304-3316.

[276] Fernández-Barquin, A., Rea, R., Venturi, D., Giacinti-Baschetti, M., De Angelis, M. G., CasadoCoterillo, C. and Irabien, A. (2018). Effect of relative humidity on the gas transport properties of zeolite A/PTMSP mixed matrix membranes. RSC Adv. 8(7), 3536-3546.

[277] Zhao, D., Ren, J. Z., Wang, Y., Qiu, Y. T., Li, H., Hua, K. S., Li, X. X., Ji, J. M. and Deng, M. C. (2017). High CO_2 separation performance of Pebax/CNTs/GTA mixed matrix membranes. J. Membr. Sci. 521, 104-113.

[278] Sun, Y. M., Wu, C. H. and Lin, A. (2006). Permeation and sorption properties of benzene, cyclohexane, and n-hexane vapors in poly bis(2,2,2-trifluoroethoxy)phosphazene (PTFEP) membranes. Polymer 47(2), 602-610.

[279] Karatay, E. and Lammertink, R. G. (2012). Oxygenation by a superhydrophobic slip G/L contactor. Lab Chip 12(16), 2922-2929.

[280] Dahl, E., Laake, M., Tjessem, K., Eberlein, K. and Bøhle, B. (1983). Effects of Ekofisk crude oil on an enclosed planktonic ecosystem. Marine Ecology—Progress Series 14, 81-91.

[281] Nomura, H., Toyoda, K., Yamada, M., Okamoto, K., Wada, M., Nishimura, M., Yoshida, A., Shibata, A., Takada, H. and Ohwada, K. (2007). Mesocosm studies of phytoplankton community succession after inputs of the water-soluble fraction of Bunker A oil. La mer 45, 105-116.

[282] O'Brien, P. Y. and Dixon, P. S. (1976). The effects of oils and oil components on algae: a review.

British Phycological Journal 11(2), 115-142.
[283] Østgaard, K., Eide, I. and Jensen, A. (1984). Exposure of phytoplankton to ekofisk crude oil. Mar. Environ. Res. 11(3), 183-200.
[284] Dietz, R. S. and Lafond, E. C. (1950). Natural slicks on the ocean. J. Mar. Res. 9(2), 69-76.
[285] Julius, M. L. and Theriot, E. C. (2010). The diatoms: a primer. In: The Diatoms: Applications for the Environmental and Earth Sciences. J. P. Smol and E. F. Stoermer, (eds.) Cambridge University Press, Cambridge: 8-22. 2nd.
[286] Mann, D. G. and Droop, S. J. M. (1996). Biodiversity, biogeography and conservation of diatoms. Hydrobiologia 336(1-3), 19-32.
[287] Lu, X. and Wang, W. (2017). Engineering of alkane production in cyanobacteria. In: Cyanobacteria: Omics and Manipulation. D. A. Los, (ed.) Caister Academic Press, Wymondham, England: 219-234.
[288] Peramuna, A., Morton, R. and Summers, M. L. (2015). Enhancing alkane production in cyanobacterial lipid droplets: A model platform for industrially relevant compound production. Life-Basel 5(2), 1111-1126.
[289] Kameyama, S., Tsunogai, U., Nakagawa, F., Sasakawa, M., Komatsu, D. D., Ijiri, A., Yamaguchi, J., Horiguchi, T., Kawamura, H., Yamaguchi, A. and Tsuda, A. (2009). Enrichment of alkanes within a phytoplankton bloom during an in situ iron enrichment experiment in the western subarctic Pacific. Mar. Chem. 115(1-2), 92-101.
[290] Subramanian, V., Dubini, A. and Seibert, M. (2012). Metabolic pathways in green algae with potential value for biofuel production: Algal fuel metabolism. In: The Science of Algal Fuels: Phycology, Geology, Biophotonics, Genomics and Nanotechnology. R. Gordon and J. Seckbach, (eds.) Springer, Dordrecht: 399-422.
[291] Hallare, A. V., Lasafin, K. J. A. and Magallanes, J. R. (2011). Shift in phytoplankton community structure in a tropical marine reserve before and after a major oil spill event. Int. J. Environ. Res. 5(3), 651-660.
[292] Karydis, M. and Fogg, G. E. (1980). Physiological effects of hydrocarbons on the marine diatom Cyclotella cryptica. Microb. Ecol. 6(4), 281-290.
[293] Jönsson, J. A., Vejrosta, J. and Novák, J. (1982). Air/water partition coefficients for normal alkanes (n-pentane to n-nonane). Fluid Phase Equilib. 9(3), 279-286.
[294] Kenny, P. W., Montanari, C. A. and Prokopczyk, I. M. (2013). ClogPalk: a method for predicting alkane/water partition coefficient. J. Comput.-Aided Mol. Des. 27(5), 389-402.
[295] Henritzi, S., Fischer, M., Grininger, M., Oreb, M. and Boles, E. (2018). An engineered fatty acid synthase combined with a carboxylic acid reductase enables de novo production of 1-octanol in Saccharomyces cerevisiae. Biotechnol. Biofuels 11(1), #150.
[296] Eriksen, M., Lebreton, L. C. M., Carson, H. S., Thiel, M., Moore, C. J., Borerro, J. C., Galgani, F., Ryan, P. G. and Reisser, J. (2014). Plastic pollution in the world's oceans: More than 5 trillion plastic pieces weighing over 250,000 tons afloat at sea. PLoS One 9(12), #e111913.
[297] Gaylarde, C. C. and Morton, L. H. G. (1997). The importance of biofilms in microbial deterioration of constructional materials. Revista de Microbiologia 28(4), 221-229.
[298] Siebert, J. (1994). Microbial deterioration of materials: Case histories and countermeasures for plastics and natural materials: Coating systems [Mikrobielle Werkstoffzerstörung—Schadensfälle und Gegenmaßnahmen für Kunst-und Naturstoffe: Beschichtungssysteme] [German]. Werkstoffe und Korrosion-Materials and Corrosion 45(3), 172-177.
[299] Zettler, E. R., Mincer, T. J. and Amaral-Zettler, L. A. (2013). Life in the "plastisphere": Microbial communities on plastic marine debris. Environ. Sci. Technol. 47(13), 7137-7146.
[300] Nolte, T. M., Hartmann, N. B., Kleijn, J. M., Garnaes, J., van de Meent, D., Hendriks, A. J. and Baun, A. (2017). The toxicity of plastic nanoparticles to green algae as influenced by surface modification, medium hardness and cellular adsorption. Aquat. Toxicol. 183, 11-20.
[301] Zhang, C., Chen, X. H., Lu, J. T. and Tan, L. J. (2017a). Toxic effects of microplastic on marine microalgae Skeletonema costatum: Interactions between microplastic and algae. Environ. Pollut. 220, 1282-1288.
[302] Ziolkowska, K., Jedrych, E., Kwapiszewski, R., Lopacinska, J., Skolimowski, M. and Chudy, M. (2010). PDMS/glass microfluidic cell culture system for cytotoxicity tests and cells passage. Sens.

Actuator B-Chem. 145(1), 533–542.

[303] Wynne, K. J., Swain, G. W., Fox, R. B., Bullock, S. and Uilk, J. (2000). Two silicone nontoxic fouling release coatings: Hydrosilation cured PDMS and $CaCO_3$ filled, ethoxysiloxane cured RTV11. Biofouling 16(2-4), 277–288.

[304] Ghobara, M. M., Mazumder, N., Vinayak, V., Reissig, L., Gebeshuber, I. C., Tiffany, M. A. and Gordon, R. (2019). On light and diatoms: A photonics and photobiology review [Chapter 7]. In: Diatoms: Fundamentals & Applications [DIFA, Volume 1 in the series: Diatoms: Biology & Applications, series editors: Richard Gordon & Joseph Seckbach]. J. Seckbach and R. Gordon, (eds.) Wiley-Scrivener, Beverly, MA, USA: 129–190.

[305] Leblanc, K., Aristegui, J., Armand, L., Assmy, P., Beker, B., Bode, A., Breton, E., Cornet, V., Gibson, J., Gosselin, M. P., Kopczynska, E., Marshall, H., Peloquin, J., Piontkovski, S., Poulton, A. J., Queguiner, B., Schiebel, R., Shipe, R., Stefels, J., van Leeuwe, M. A., Varela, M., Widdicombe, C. and Yallop, M. (2012). A global diatom database-abundance, biovolume and biomass in the world ocean. Earth Syst. Sci. Data 4(1), 149–165.

[306] Connolly, J. A., Oliver, M. J., Beaulieu, J. M., Knight, C. A., Tomanek, L. and Moline, M. A. (2008). Correlated evolution of genome size and cell volume in diatoms (Bacillariophyceae). J. Phycol. 44 (1), 124–131.

[307] Cornet-Barthaux, V., Armand, L. and Quéguiner, B. (2007). Biovolume and biomass estimates of key diatoms in the Southern Ocean. Aquat. Microb. Ecol. 48(3), 295–308.

[308] Davidovich, N. A. (1991). Propagation rate of cells in relation to their volume in cultures of pennate diatoms. Soviet Plant Physiology 38(3), 419–423.

[309] Kumar, M. S. R., Ramaiah, N. and Tang, D. (2009). Morphometry and cell volumes of diatoms from a tropical estuary of India. Indian J. Mar. Sci. 38(2), 160–165.

[310] Lopez Fuerte, F. O., Siqueiros Beltrones, D. A. and Agüero de la Cruz, G. (2007). Biovolumen ponderado: índice para estimar la contribución de especies en asociaciones de diatomeas bentónicas [Weighed biovolume; an index for estimating the contribution of species in benthic diatom assemblages] [Spanish]. Hidrobiologica 17(1), 83–86.

[311] Lyakh, A. M. (2007). A new method for accurate estimation of diatom biovolume and surface area. In: Proceedings of the 1st Central European Diatom Meeting 2007. W. H. Kusber and R. Jahn, (eds.) Botanic Garden and Botanical Museum Berlin-Dahlem, Freie Universität Berlin Berlin: 113–116.

[312] Mizuno, M. (1991). Influence of cell-volume on the growth and size-reduction of marine and estuarine diatoms. J. Phycol. 27(4), 473–478.

[313] Nakov, T., Theriot, E. C. and Alverson, A. J. (2014). Using phylogeny to model cell size evolution in marine and freshwater diatoms. Limnol. Oceanogr. 59(1), 79–86.

[314] Hitchcock, G. L. (1983). An examination of diatom area: volume ratios and their influence on estimates of plasma volume. Journal of Plankton Research 5(3), 311–324.

[315] Harchouni, S., Field, B. and Menand, B. (2018). AC-202, a highly effective fluorophore for the visualization of lipid droplets in green algae and diatoms. Biotechnol. Biofuels 11, #120.

[316] McClements, D. J. and Jafari, S. M. (2018). Improving emulsion formation, stability and performance using mixed emulsifiers: A review. Adv. Colloid Interface Sci. 251, 55–79.

[317] Shehzad, F., Hussein, I. A., Kamal, M. S., Ahmad, W., Sultan, A. S. and Nasser, M. S. (2018). Polymeric surfactants and emerging alternatives used in the demulsification of produced water: A review. Polym. Rev. 58(1), 63–101.

[318] Zolfaghari, R., Fakhru'l-Razi, A., Abdullah, L. C., Elnashaie, S. and Pendashteh, A. (2016). Demulsification techniques of water-in-oil and oil-in-water emulsions in petroleum industry. Sep. Purif. Technol. 170, 377–407.

[319] Gross, M., Jarboe, D. and Wen, Z. (2015). Biofilm-based algal cultivation systems. Appl. Microbiol. Biotechnol. 99(14), 5781–5789.

[320] Shen, Y., Zhu, W., Chen, C., Nie, Y. and Lin, X. (2016). Biofilm formation in attached microalgal reactors. Bioprocess. Biosyst. Eng. 39(8), 1281–1288.

[321] Abodeely, J., Stevens, D., Ray, A., Schaller, K., Newby, D. (2013). Algal Supply System Design—Harmonized Version [Report INL/EXT-13-28890]. Idaho National Laboratory (INL), Idaho Falls, Idaho, USA. A. Pandey, D. J Lee, Y Chisti and C. R Soccol, pp. 313–325.

[322] da Fonseca, D. B., Guerra, L. T., Santos, E. T., Mendonça, S. H., Silva, J. G., Costa, L. A. and Navalho, J. C. (2016). AlgaFarm: A case study of industrial Chlorella production. In: Microalgal Production for Biomass and High-Value Products. S. P. Slocombe and J. R. Benemann, (eds.) CRC Press, Boca Raton, Florida, USA. 1: 295-310.

[323] Sathe, P. and Dobretsov, S. (2016). Interactions and communication within marine biofilms. In: Aquatic Biofilms: Ecology, Water Quality and Wastewater Treatment. A. M. Romani, H. Guasch and D. M. Balaguer, (eds.) Caister Academic Press, Norfolk, UK: 47-61.

[324] Polizzi, B., Bernard, O. and Ribot, M. (2017). A time-space model for the growth of microalgae biofilms for biofuel production. J. Theor. Biol. 432, 55-79.

[325] Li, X., Johnson, R. and Kasting, G. B. (2016). On the variation of water diffusion coefficient in stratum corneum with water content. J. Pharm. Sci. 105(3), 1141-1147.

[326] Gordon, R., Losic, D., Tiffany, M. A., Nagy, S. S. and Sterrenburg, F. A. S. (2009). The Glass Menagerie: diatoms for novel applications in nanotechnology. Trends Biotechnol. 27(2), 116-127.

[327] Hamano, H., Nakamura, S., Hayakawa, J., Miyashita, H. and Harayama, S. (2017). Biofilm-based photobioreactor absorbing water and nutrients by capillary action. Bioresour. Technol. 223, 307-311.

[328] Podola, B., Li, T. and Melkonian, M. (2017). Porous substrate bioreactors: A paradigm shift in microalgal biotechnology? Trends Biotechnol. 35(2), 121-132.

[329] Tahk, D., Paik, S. M., Lim, J., Bang, S., Oh, S., Ryu, H. and Jeon, N. L. (2017). Rapid large area fabrication of multiscale through-hole membranes. Lab Chip 17(10), 1817-1825.

[330] Ozkan, A., Kinney, K., Katz, L. and Berberoglu, H. (2012). Reduction of water and energy requirement of algae cultivation using an algae biofilm photobioreactor. Bioresour. Technol. 114, 542-548.

[331] Brito, E. M. S., Duran, R., Guyoneaud, R., Goñi-Urriza, M., de Oteyza, T. G., Crapez, M. A. C., Aleluia, I. and Wasserman, J. C. A. (2009). A case study of in situ oil contamination in a mangrove swamp (Rio De Janeiro, Brazil). Mar. Pollut. Bull. 58(3), 418-423.

[332] Das, A. A. K., Bovill, J., Ayesh, M., Stoyanov, S. D. and Paunov, V. N. (2016). Fabrication of living soft matter by symbiotic growth of unicellular microorganisms. J. Mat. Chem. B 4(21), 3685-3694.

[333] Ajjawi, I., Verruto, J., Aqui, M., Soriaga, L. B., Coppersmith, J., Kwok, K., Peach, L., Orchard, E., Kalb, R., Xu, W., Carlson, T. J., Francis, K., Konigsfeld, K., Bartalis, J., Schultz, A., Lambert, W., Schwartz, A. S., Brown, R. and Moellering, E. R. (2017). Lipid production in Nannochloropsis gaditana is doubled by decreasing expression of a single transcriptional regulator. Nat Biotech, doi: 10.1038/nbt.3865.

[334] Yang, B., Liu, J., Ma, X., Guo, B., Liu, B., Wu, T., Jiang, Y. and Chen, F. (2017). Genetic engineering of the Calvin cycle toward enhanced photosynthetic CO_2 fixation in microalgae. Biotechnol. Biofuels 10(1), #229.

[335] American Chemical Society. (2017). Cyborg bacteria outperform plants when turning sunlight into useful compounds (video). https://www.acs.org/content/acs/en/pressroom/newsreleases/2017/august/cyborg-bacteria-outperform-plants-when-turning-sunlight-into-usefulcompounds-video.html.

[336] Sakimoto, K. K. and Yang, P. (2017). Cyborg bacteria: Inorganic-biological hybrid organisms for solar-to-chemical production. In: 254th American Chemical Society National Meeting & Exposition, August 20-24, 2017, Washington, DC. Chemistry's Impact on the Global Economy. https://www.youtube.com/watch?v=opl5CnDA_2c&feature=youtu.be.

[337] Wikipedia. (2019c). Clover. https://en.wikipedia.org/wiki/Clover.

[338] Bond-Lamberty, B. and Thomson, A. (2010). A global database of soil respiration data. Biogeosciences 7(6), 1915-1926.

[339] Xu, M. and Shang, H. (2016). Contribution of soil respiration to the global carbon equation. J. Plant Physiol. 203, 16-28.

[340] Sayre, R. (2010). Microalgae: The potential for carbon capture. Bioscience 60(9), 722-727.

[341] Liu, L. T., Hu, C. S., Yang, P. P., Ju, Z. Q., Olesen, J. E. and Tang, J. W. (2015). Effects of experimental warming and nitrogen addition on soil respiration and CH_4 fluxes from crop rotations of winter wheat-soybean/fallow. Agric. For. Meteorol. 207, 38-47.

[342] Licht, S. (2017). Co-production of cement and carbon nanotubes with a carbon negative footprint. J. CO_2 Util. 18, 378-389.

[343] Wikipedia. (2019i). Crop rotation. https://en.wikipedia.org/wiki/Crop_rotation.
[344] Bao, X. Y., Ali, A., Qiao, D. L., Liu, H. S., Chen, L. and Yu, L. (2015). Application of polymer materials in developing slow/control release fertilizer. Acta Polym. Sin. (9), 1010-1019.
[345] Xu, M. and Shang, H. (2016). Contribution of soil respiration to the global carbon equation. J. Plant Physiol. 203, 16-28.
[346] Brummell, M. E. and Siciliano, S. D. (2011). Measurement of carbon dioxide, methane, nitrous oxide, and water potential in soil ecosystems. Method Enzymol. 496, 115-137.
[347] Almaraz, M., Bai, E., Amini, H., Wang, L. J., Hashemisohi, A., Shahbazi, A., Bikdash, M., Dukka, K. C. and Yuan, W. Q. (2018). An integrated growth kinetics and computational fluid dynamics model for the analysis of algal productivity in open raceway ponds. Comput. Electron. Agric. 145, 363-372.
[348] Mowery, K. A. and Meyerhoff, M. E. (1999). The transport of nitric oxide through various polymeric matrices. Polymer 40(22), 6203-6207.
[349] Pasternak, R. A., Christensen, M. V. and Heller, J. (1970). Diffusion and permeation of oxygen, nitrogen, carbon dioxide, and nitrogen dioxide through polytetrafluoroethylene. Macromolecules 3(3), 366-371.
[350] Misra, A. N., Misra, M. and Singh, R. (2010). Nitric oxide biochemistry, mode of action and signaling in plants. J. Med. Plants Res. 4(25), 2729-2739.
[351] Thompson, S. E. M., Taylor, A. R., Brownlee, C., Callow, M. E. and Callow, J. A. (2008). The role of nitric oxide in diatom adhesion in relation to substratum properties. J. Phycol. 44(4), 967-976.
[352] Trewavas, A. (2004). Fertilizer: no-till farming could reduce run-off. Nature 427(6970), 99.
[353] Dearborn, C. H. (1948). Preliminary notes on frost prevention under cold frame glass by sprinkling the glass with cold water. Proceedings of the American Society for Horticultural Science 51(JUN), 493-496.
[354] Wikipedia. (2017). Cold frame. https://en.wikipedia.org/wiki/Cold_frame.
[355] Wikipedia. (2019c). Clover. https://en.wikipedia.org/wiki/Clover.
[356] Wikipedia. (2019g). List of root vegetables. https://en.wikipedia.org/wiki/List_of_root_vegetables.
[357] Mauromicale, G., Occhipinti, A. and Mauro, R. P. (2010). Selection of shade-adapted subterranean clover species for cover cropping in orchards. Agron. Sustain. Dev. 30(2), 473-480.
[358] Condon, J. J. and Ransom, S. M. (2016). Essential Radio Astronomy: Radio Telescopes. http://www.cv.nrao.edu/course/astr534/RadioTelescopes.html.
[359] Zhao, S. J. and Heywood, J. B. (2017). Projected pathways and environmental impact of China's electrified passenger vehicles. Transport. Res. Part D-Transport. Environ. 53, 334-353.
[360] Abhyankar, N., Gopal, A., Sheppard, C., Park, W. Y., Phadke, A. (2017). Techno-Economic Assessment of Deep Electrification of Passenger Vehicles in India [Report LBNL-1007121]. Energy Analysis and Environmental Impacts Division, Lawrence Berkeley National Laboratory, Berkeley, California, USA.
[361] Zhang, X. P., Liang, Y. N., Yu, E. H., Rao, R. and Xie, J. (2017b). Review of electric vehicle policies in China: Content summary and effect analysis. Renew. Sust. Energ. Rev. 70, 698-714.
[362] Morales-Sánchez, D., Martinez-Rodriguez, O. A., Kyndt, J. and Martinez, A. (2015). Heterotrophic growth of microalgae: metabolic aspects. World J. Microbiol. Biotechnol. 31(1), 1-9.
[363] Schoefs, B., Hu, H. and Kroth, P. G. (2017). The peculiar carbon metabolism in diatoms. Philosophical Transactions of the Royal Society B: Biological Sciences 372(1728, The peculiar carbon metabolism in diatoms, Eds. Benoît Schoefs, Hanhua Hu & Peter G. Kroth), #20160405.
[364] Matsuda, Y., Hopkinson, B. M., Nakajima, K., Dupont, C. L. and Tsuji, Y. (2017). Mechanisms of carbon dioxide acquisition and CO_2 sensing in marine diatoms: a gateway to carbon metabolism. Philosophical Transactions of the Royal Society B: Biological Sciences 372(1728, The peculiar carbon metabolism in diatoms, Eds. Benoît Schoefs, Hanhua Hu & Peter G. Kroth), #20160403.
[365] Shukla, S. K. and Mohan, R. (2012). The contribution of diatoms to worldwide crude oil deposits. In: The Science of Algal Fuels: Phycology, Geology, Biophotonics, Genomics and Nanotechnology. R. Gordon and J. Seckbach, (eds.) Springer, Dordrecht: 355-382.
[366] Wang, J. K. and Seibert, M. (2017). Prospects for commercial production of diatoms. Biotechnol. Biofuels 10, #16.
[367] Huang, W. and Daboussi, F. (2017). Genetic and metabolic engineering in diatoms. Philosophical Transactions of the Royal Society B: Biological Sciences 372(1728, The peculiar carbon metabolism in

diatoms, Eds. Benoît Schoefs, Hanhua Hu & Peter G. Kroth), #20160411.
[368] Daboussi, F., Leduc, S., Maréchal, A., Dubois, G., Guyot, V., Perez-Michaut, C., Amato, A., Falciatore, A., Juillerat, A., Beurdeley, M., Voytas, D. F., Cavarec, L. and Duchateau, P. (2014). Genome engineering empowers the diatom Phaeodactylum tricornutum for biotechnology. Nat. Commun. 5, #3831.
[369] d'Ippolito, G., Sardo, A., Paris, D., Vella, F. M., Adelfi, M. G., Botte, P., Gallo, C. and Fontana, A. (2015). Potential of lipid metabolism in marine diatoms for biofuel production. Biotechnol. Biofuels 8, doi: 10.1186/s13068-13015-10212-13064.
[370] Graham, J. M., Graham, L. E., ZulkiXy, S. B., Pxeger, B. F., Hoover, S. W. and Yoshitani, J. (2012). Freshwater diatoms as a source of lipids for biofuels. J. Ind. Microbiol. Biotechnol. 39(3), 419-428.
[371] Hildebrand, M., Davis, A. K., Smith, S. R., Traller, J. C. and Abbriano, R. (2012). The place of diatoms in the biofuels industry. Biofuels 3(2), 221-240.
[372] Joseph, M. M., Renjith, K. R., John, G., Nair, S. M. and Chandramohanakumar, N. (2017). Biodiesel prospective of five diatom strains using growth parameters and fatty acid profiles. Biofuels-UK 8(1), 81-89.
[373] Merz, C. R. and Main, K. L. (2014). Microalgae (diatom) production—The aquaculture and biofuel nexus. In: Oceans-St. John's, 2014. IEEE: 1-10.
[374] Tokushima, H., Inoue-Kashino, N., Nakazato, Y., Masuda, A., Ifuku, K. and Kashino, Y. (2016). Advantageous characteristics of the diatom Chaetoceros gracilis as a sustainable biofuel producer. Biotechnol. Biofuels 9, #235.
[375] Maeda, Y., Nojima, D., Yoshino, T. and Tanaka, T. (2017). Structure and properties of oil bodies in diatoms. Philosophical Transactions of the Royal Society B: Biological Sciences 372(1728, The peculiar carbon metabolism in diatoms, Eds. Benoît Schoefs, Hanhua Hu & Peter G. Kroth), #20160408.
[376] Sayanova, O., Mimouni, V., Ulmann, L., Morant-Manceau, A., Pasquet, V., Schoefs, B. and Napier, J. A. (2017). Modulation of lipid biosynthesis by stress in diatoms. Philosophical Transactions of the Royal Society B: Biological Sciences 372: 20160407.
[377] Solymosi, K. (2012). Plastid structure, diversification and interconversions. I. Algae. Current Chemical Biology 6(3), 167-186.
[378] Yang, Z.-K., Niu, Y.-F., Ma, Y.-H., Xue, J., Zhang, M.-H., Yang, W.-D., Liu, J.-S., Lu, S.-H., Guan, Y. and Li, H.-Y. (2013). Molecular and cellular mechanisms of neutral lipid accumulation in diatom following nitrogen deprivation. Biotechnol. Biofuels 6, #67.
[379] Hempel, F. and Maier, U. G. (2012). An engineered diatom acting like a plasma cell secreting human IgG antibodies with high efficiency. Microb. Cell. Fact. 11, #126.
[380] Vanier, G., Hempel, F., Chan, P., Rodamer, M., Vaudry, D., Maier, U. G., Lerouge, P. and Bardor, M. (2015). Biochemical characterization of human anti-hepatitis B monoclonal antibody produced in the microalgae Phaeodactylum tricornutum. PLoS One 10(10), #e0139282.
[381] Wikipedia. (2019h). ATP-binding cassette transporter. https://en.wikipedia.org/wiki/ATP-binding_cassette_transporter.
[382] Hess, S. K., Lepetit, B., Kroth, P. G. and Mecking, S. (2018). Production of chemicals from microalgae lipids-status and perspectives. Eur. J. Lipid Sci. Technol. 120(1), #1700152.
[383] Croft, M. T., Warren, M. J. and Smith, A. G. (2006). Algae need their vitamins. Eukaryotic Cell 5(8), 1175-1183.
[384] Cointet, E., Meleder, V., Goncalvez, O. and Wielgosz-Collin, G. (2017). Exploration of the bioactive lipids diversity of the marine benthic diatoms of the Nantes Culture Collection (NCC) [abstract]. Phycologia 56(4, Supplement S), 34.
[385] Fields, F. and Kociolek, J. P. (2015). An evolutionary perspective on selecting highlipid-content diatoms (Bacillariophyta). Journal of Applied Phycology, doi: 10.1007/s10811-10014-10505-10811.
[386] Sims, P. A., Mann, D. G. and Medlin, L. K. (2006). Evolution of the diatoms: insights from fossil, biological and molecular data. Phycologia 45(4), 361-402.
[387] Prathima, A. and Karthikeyan, S. (2017). Characteristics of micro-algal biofuel from Botryococcus braunii. Energy Sources Part A-Recovery Util. Environ. Eff. 39(2), 206-212.
[388] Liu, B. X., Li, Y., Zhang, Q. and Han, L. (2017). Spectral characteristics of weathered oil films on

water surface and selection of potential sensitive bands in hyper-spectral images. J. Indian Soc. Remote Sens. 45(1), 171-177.

[389] Andrews, K. (1993). Thermophilic Diatoms from alkaline hot springs and geysers in Yellowstone National Park. https://irma.nps.gov/rprs/IAR/Profile/5993.

[390] Andrews, K. (1995). Thermophilic Diatoms from alkaline and acidic hot springs and geysers in Yellowstone National Park. https://irma.nps.gov/rprs/IAR/Profile/6393.

[391] Andrews, K. (1997). Thermophilic Diatoms from alkaline and acidic hot springs and geysers in Yellowstone National Park. https://irma.nps.gov/rprs/IAR/Profile/6819.

[392] Mpawenayo, B. and Mathooko, J. M. (2004). Diatom assemblages in the hotsprings associated with Lakes Elmenteita and Baringo in Kenya. Afr. J. Ecol. 42(4), 363-367.

[393] Nikulina, T. V. and Kociolek, J. P. (2011). Diatoms from hot springs from Kuril and Sakhalin Islands (Far East, Russia). In: The Diatom World. J. Seckbach and J. P. Kociolek, (eds.) Springer, Dordrecht, The Netherlands: 333-363.

[394] Owen, R. B., Renaut, R. W. and Jones, B. (2008). Geothermal diatoms: a comparative study of floras in hot spring systems of Iceland, New Zealand, and Kenya. Hydrobiologia 610, 175-192.

[395] Smith, T., Manoylov, K. and Packard, A. (2013). Algal extremophile community persistence from Hot Springs National Park (Arkansas, USA). International Journal on Algae 15(1), 65-76.

[396] McKee, S., Hinkley, R. L., Andrews, B. A. and Andrews, K. J. (1981). Thermophilic cyanobacteria (blue-green algae) from three Colorado thermal environments [abstract]. J. Phycol. 17(1 Suppl.), 12.

[397] Yu, J., Wang, C., Su, Z. Y., Xiong, P. and Liu, J. Q. (2014). Response of microalgae growth and cell characteristics to various temperatures. Asian J. Chem. 26(11), 3366-3370.

[398] Pruetiworanan, S., Duangjan, K., Pekkoh, J., Peerapornpisal, Y. and Pumas, C. (2018). Effect of pH on heat tolerance of hot spring diatom Achnanthidium exiguum AARL D025-2 in cultivation. Journal of Applied Phycology 30(1), 47-53.

[399] Dodson, V. J., Mouget, J.-L., Dahmen, J. L. and Leblond, J. D. (2014). The long and short of it: Temperature-dependent modifications of fatty acid chain length and unsaturation in the galactolipid profiles of the diatoms Haslea ostrearia and Phaeodactylum tricornutum. Hydrobiologia 727(1), 95-107.

[400] Rousch, J. M., Bingham, S. E. and Sommerfeld, M. R. (2003). Changes in fatty acid profiles of thermo-intolerant and thermo-tolerant marine diatoms during temperature stress. Journal of Experimental Marine Biology and Ecology 295(2), 145-156.

[401] Davis, A. and Andrews, K. J. (1987). A microscopic analysis of the thermophilic bacteria and diatoms from a Colorado USA hot springs [abstract]. Abstracts of the Annual Meeting of the American Society for Microbiology 87, 190.

[402] Dahiya, A. (2012). Integrated approach to algae production for biofuel utilizing robust algal species In: The Science of Algal Fuels: Phycology, Geology, Biophotonics, Genomics and Nanotechnology. R. Gordon and J. Seckbach, (eds.) Springer, Dordrecht: 83-100.

[403] Day, J. G. and Stanley, M. S. (2012). Biological constraints on the production of microalgalbased biofuels. In: The Science of Algal Fuels: Phycology, Geology, Biophotonics, Genomics and Nanotechnology. R. Gordon and J. Seckbach, (eds.) Springer, Dordrecht: 101-130.

[404] McNichol, J. and McGinn, P. (2012). Adapting mass algaculture for a northern climate. In: The Science of Algal Fuels: Phycology, Geology, Biophotonics, Genomics and Nanotechnology. R. Gordon and J. Seckbach, (eds.) Springer, Dordrecht: 131-146.

[405] Topf, M., Tavasi, M., Kinel-Tahan, Y., Iluz, D., Dubinsky, Z. and Yehoshua, Y. (2012). Algal oils: Biosynthesis and uses. In: The Science of Algal Fuels: Phycology, Geology, Biophotonics, Genomics and Nanotechnology. R. Gordon and J. Seckbach, (eds.) Springer, Dordrecht: 193-214.

[406] Shifrin, N. S. and Chisholm, S. W. (1981). Phytoplankton lipids: interspecific differences and effects of nitrate, silicate and light-dark cycles. J. Phycol. 17(4), 374-384.

[407] Chen, Y.-C. (2012). The biomass and total lipid content and composition of twelve species of marine diatoms cultured under various environments. Food Chem. 131(1), 211-219.

[408] Coutteau, P. (1996). 2. Micro-algae: 2.3.2. Growth dynamics. In: Manual on the Production and Use of Live Food for Aquaculture. P. Lavens and P. Sorgeloos, (eds.) Food and Agriculture Organization of the United Nations, Rome, Italy: 7-48.

[409] Merz, C. R. and Main, K. L. (2017). Microalgae bioproduction: Feeds, foods, nutraceuticals, and

polymers. In: Fuels, Chemicals and Materials from the Oceans and Aquatic Sources. F. M. Kerton and N. Yan, (eds.) John Wiley & Sons Ltd.: 83-112.

[410] Widdel, F. (2010). Theory and Measurement of Bacterial Growth [https://www.mpi-bremen.de/Binaries/Binary307/Wachstumsversuch.pdf].

[411] Finkel, S. E. (2006). Long-term survival during stationary phase: evolution and the GASP phenotype. Nat. Rev. Microbiol., 4(2), 113-120.

[412] Fogg, G. E. (1956). Photosynthesis and formation of fats in a diatom. Ann. Bot. 20(78), 265-285.

[413] Badour, S. S. and Gergis, M. S. (1965). Cell division and fat accumulation in Nitzschia sp. grown in continuously illuminated mass cultures. Archiv für Mikrobiologie 51(1), 94-102.

[414] Hu, Q., Zhang, C. and Sommerfeld, M. (2006). Biodiesel from algae: Lessons learned over the past 60 years and future perspectives [abstract]. J. Phycol. 42(Supplement 1), 12.

[415] Hu, Q., Sommerfeld, M., Jarvis, E., Ghirardi, M., Posewitz, M., Seibert, M. and Darzins, A. (2008). Microalgal triacylglycerols as feedstocks for biofuel production: perspectives and advances. Plant J. 54(4), 621-639.

[416] Sánchez-Saavedra, M. D. P. and Voltolina, D. (2006). The growth rate, biomass production and composition of Chaetoceros sp grown with different light sources. Aquacultural Engineering 35(2), 161-165.

[417] Brown, M. R., Dunstan, G. A., Norwood, S. J. and Miller, K. A. (1996). Effects of harvest stage and light on the biochemical composition of the diatom Thalassiosira pseudonana. J. Phycol. 32(1), 64-73.

[418] Dunstan, G. A., Volkman, J. K., Barrett, S. M. and Garland, C. D. (1993). Changes in the lipid composition and maximization of the polyunsaturated fatty acid content of three microalgae grown in mass-culture. Journal of Applied Phycology 5(1), 71-83.

[419] Mansour, M. P., Frampton, D. M. F., Nichols, P. D., Volkman, J. K. and Blackburn, S. I. (2005). Lipid and fatty acid yield of nine stationary-phase microalgae: Applications and unusual C24-C28 polyunsaturated fatty acids. Journal of Applied Phycology 17(4), 287-300.

[420] Roessler, P. G. (1988). Effects of silicon deficiency on lipid composition and metabolism in the diatom Cyclotella cryptica. J. Phycol. 24(3), 394-400.

[421] Collos, Y. (1986). Time-lag algal growth dynamics: Biological constraints on primary production in aquatic environments. Mar. Ecol.-Prog. Ser. 33(2), 193-206.

[422] Giani, A. and Delgado, P. C. S. (1998). Growth dynamics and competitive ability of a green (Oocystis lacustris) and a blue-green alga (Synechocystis sp.) under different N: P ratios. In: International Association of Theoretical and Applied Limnology-Proceedings. W. D. Williams and A. Sladeckova, (eds.). 26: 1693-1697.

[423] Grossart, H. P. and Simon, M. (2007). Interactions of planktonic algae and bacteria: effects on algal growth and organic matter dynamics. Aquat. Microb. Ecol. 47(2), 163-176.

[424] Shriwastav, A., Ashok, V., Thomas, J. and Bose, P. (2018). A comprehensive mechanistic model for simulating algal-bacterial growth dynamics in photobioreactors. Bioresour. Technol. 247, 640-651.

[425] Nauha, E. K. and Alopaeus, V. (2013). Modeling method for combining fluid dynamics and algal growth in a bubble column photobioreactor. Chem. Eng. J. 229, 559-568.

[426] Drewry, J. L., Choi, C. Y., An, L. and Gharagozloo, P. E. (2015). A computational fluid dynamics model of algal growth: Development and validation. Trans. ASABE 58(2), 203-213.

[427] Shriwastav, A., Thomas, J. and Bose, P. (2017). A comprehensive mechanistic model for simulating algal growth dynamics in photobioreactors. Bioresour. Technol. 233, 7-14.

[428] Mata, T. M., Martins, A. A. and Caetano, N. S. (2010). Microalgae for biodiesel production and other applications: A review. Renew. Sust. Energ. Rev. 14(1), 217-232.

[429] Vilens, A. (2018). AVCalc: Biodiesel density. https://www.aqua-calc.com/page/density-table/substance/biodiesel.

[430] Burke, M. (2006). Wrap star. Forbes 177(10), 108-109.

[431] Sealed Air. (2017a). Newair I. B. Express. https://sealedair.com/product-care/product-careproducts/newair-ib-express.

[432] Hostetter, H. and Hoshaw, R. (1970). Environmental factors affecting resistance to desiccation in the diatom Stauroneis anceps. Am. J. Bot., 512-518.

[433] McQuoid, M. R. and Hobson, L. A. (1996). Diatom resting stages. J. Phycol. 32(6), 889-902.

[434] Evans, J. H. (1958). The survival of fresh-water algae during dry periods. Part I. An investigation of the algae of 5 small ponds. J. Ecol. 46(1), 149-167.
[435] Evans, J. H. (1959). The survival of fresh-water algae during dry periods. Part II. drying experiments. Part III. Stratification of algae in pond margin litter and mud. J. Ecol. 47(1), 55-81.
[436] Neethirajan, S., Gordon, R. and Wang, L. (2009). Potential of silica bodies (phytoliths) for nanotechnology. Trends Biotechnol. 27(8), 461-467.
[437] Ball, P. (2012). Why leaves have stones. Nat. Mater. 11(4), 271.
[438] Gal, A., Brumfeld, V., Weiner, S., Addadi, L. and Oron, D. (2012). Certain biominerals in leaves function as light scatterers. Adv. Mater. 24(10), OP77-OP83.
[439] Schaller, J., Brackhage, C., Baucker, E. and Dudel, E. G. (2013). UV-screening of grasses by plant silica layer? J. Biosci. 38(2), 413-416.
[440] Ingalls, A. E., Whitehead, K. and Bridoux, M. C. (2010). Tinted windows: The presence of the UV absorbing compounds called mycosporine-like amino acids embedded in the frustules of marine diatoms. Geochim. Cosmochim. Acta 74(1), 104-115.
[441] Borowitzka, M. A. and Moheimani, N. R. (2013). Sustainable biofuels from algae. Mitigation and Adaptation Strategies for Global Change 18(1), 13-25.
[442] Hallenbeck, P. C., Grogger, M., Mraz, M. and Veverka, D. (2016). Solar biofuels production with microalgae. Appl. Energy 179, 136-145.
[443] Abdelaziz, A. E. M. and Hallenbeck, P. C. (2013). Algal biofuels: Challenges and opportunities. Bioresour. Technol. 145, 134-141.
[444] Stephens, E., Ross, I. L., Mussgnug, J. H., Wagner, L. D., Borowitzka, M. A., Posten, C., Kruse, O. and Hankamer, B. (2010). Future prospects of microalgal biofuel production systems. Trends Plant Sci. 15(10), 554-564.
[445] AlgaeIndustryMagazine.com. (2014). Secil and A4F form AlgaFarm JV. http://www.algaeindustrymagazine.com/secil-a4f-form-algafarm-jv/[cobweb site].
[446] AlgaeIndustryMagazine.com. (2017). AlgaFarm Revisited. http://www.algaeindustrymagazine.com/algafarm-revisited/[cobweb site].
[447] Prabhu, C. (2018). Agreement to support algae cultivation in Oman. http://www.omanobserver.om/agreement-support-algae-cultivation-oman/Fuel Process. Technol. 167, 582-607.
[448] Chandrasekaran, S. and Voelcker, N. H. (2018). The potential of modified diatom frustules for solar energy conversion. In: Diatom Nanotechnology: Progress and Emerging Applications. D. Losic, (ed.) Royal Society of Chemistry, Cambridge: 150-174.
[449] Gautam, S., Kashyap, M., Gupta, S., Kumar, V., Schoefs, B., Gordon, R., Joshi, K. B., Vinayak, V. and Jeffryes, C. (2016). Metabolic engineering of TiO_2 nanoparticles in Nitzschia palea to form diatom nanotubes: an ingredient for solar cells to produce electricity and biofuel. RSC Adv. 6(99), 97276-97284.
[450] Jeffryes, C., Campbell, J., Li, H., Jiao, J. and Rorrer, G. (2011). The potential of diatom nanobiotechnology for applications in solar cells, batteries, and electroluminescent devices. Energy Environ. Sci. 4(10), 3930-3941.
[451] Allmicroalgae. (2017). Algafarm. http://www.allmicroalgae.com/.
[452] Qualitas Health. (2017). Our Farms. https://www.qualitas-health.com A. Israel, R. Einav and J. Seckbach, (eds.) Springer, Dordrecht: 229-248.
[453] Eghball, B., Gilley, J. E., Baltensperger, D. D. and Blumenthal, J. M. (2002). Long-term manure and fertilizer application effects on phosphorus and nitrogen in runoff. Trans. ASAE 45(3), 687-694.
[454] Khan, F. A. and Ansari, A. A. (2005). Eutrophication: An ecological vision. Bot. Rev. 71(4), 449-482.
[455] Sharpley, A. (2016). Managing agricultural phosphorus to minimize water quality impacts. Sci. Agric. 73(1), 1-8.
[456] Hosseinzadeh, S., Verheust, Y., Bonarrigo, G. and Van Hulle, S. (2017). Closed hydroponic systems: operational parameters, root exudates occurrence and related water treatment. Rev. Environ. Sci. Bio-Technol. 16(1), 59-79.
[457] Poulet, L., Fontaine, J. P. and Dussap, C. G. (2016). Plant's response to space environment: a comprehensive review including mechanistic modelling for future space gardeners. Botany Lett. 163(3), 337-347.

[458] Evenson, R. E. and Gollin, D. (2003). Assessing the impact of the Green Revolution, 1960 to 2000. Science 300(5620), 758-762.
[459] Wikipedia. (2019d). Green Revolution. https://en.wikipedia.org/wiki/Green_Revolution.
[460] Aryal, M. R., Paudel, N. and Ranjit, M. (2018). Elimination of Chhirkey and Foorkey viruses from meristem culture of large Cardamom (Amomum subulatum Roxb.). European Online Journal of Natural and Social Sciences 7(2), 424-443.
[461] ASTM International. (2018). Oil & Gas Standards. https://www.astm.org/industry/oil-and-gasstandards.html.
[462] DOE. (2018). ASTM Biodiesel Specifications. https://www.afdc.energy.gov/fuels/biodiesel_specifications.html.
[463] EPA. (2018). Gasoline Standards: Federal Gasoline Regulations. https://www.epa.gov/gasoline-standards/federal-gasoline-regulations.
[464] Shin, Y. S., Choi, H. I., Choi, J. W., Lee, J. S., Sung, Y. J. and Sim, S. J. (2018). Multilateral approach on enhancing economic viability of lipid production from microalgae: A review. Bioresour. Technol. 258, 335-344.
[465] Talebi, A. F., Mohtashami, S. K., Tabatabaei, M., Tohidfar, M., Bagheri, A., Zeinalabedini, M., Mirzaei, H. H., Mirzajanzadeh, M., Shafaroudi, S. M. and Bakhtiari, S. (2013). Fatty acids profiling: A selective criterion for screening microalgae strains for biodiesel production. Algal Res. 2(3), 258-267.
[466] Olivares, J. A. (2014). National Alliance for Advanced Biofuels and Bioproducts (NAABB): Synopsis. https://www.energy.gov/sites/prod/files/2014/06/f16/naabb_synopsis_report.pdf.
[467] Draaisma, R. B., Wijffels, R. H., Slegers, P. M., Brentner, L. B., Roy, A. and Barbosa, M. J. (2013). Food commodities from microalgae. Curr. Opin. Biotechnol. 24(2), 169-177.

附 录

附表　藻类名称对照表

藻类	中文名
Trigonium arcticum	北冰洋三角藻
Aulacoseira Twaites	沟链藻
Cyclotella	小环藻
Toxarium Bailey	沟鞭藻
Synedra acus subsp. radians	尖针杆藻
Rhaphoneis amphiceros	具尾鳍藻
Achnanthidium sibiricum	沟鞭藻
Plagiogrammopsis vanheurckii	棕囊藻
Aulacoseira baicalensis	沟链藻
Thalassiosira eccentrica	海链藻
Coscinodiscus wailesii	威氏圆筛藻
Gephyria media Arnott	中间弓桥藻
Cyclotella baicalensis	小环藻黄芩属
Pinnularia sp.	羽纹藻属
Hantzschia amphioxys	双尖菱板藻
Surirella Robusta	粗壮双菱藻
Navicula saprophila	舟形藻
Rhizosolenia setigera Brightwell	刚毛根管藻
Chaetoceros decipiens Cleve	角毛藻
Cyclotella cryptica Reimann	隐秘小环藻
Coscinodiscus granii Gough	格氏圆筛藻
Surirella sp. Turpin/Surirella	双菱藻

(续表)

藻类	中文名
Craspedostauros australis Cox	芽孢杆菌
Nitzschia sp. Hassal	菱形藻
Pinnularia sp.	羽纹藻
Craticula spp.	格形藻
Gomphonema parvulum	小形异极藻
Thalassiosira eccentrica	离心列海链藻
Grammatophora	斑条藻属
Diploneis	双壁藻
Nitzschioid/Nitzschia	尼茨希亚属硅藻
Biremis	双贝壳硅藻
Diadesmidaceae	镜背藻科
Biremis lucens(Hustedt)/Biremis-lucens/B. lucens	双贝壳硅藻
Olifantiella mascarenica/O. mascarenica	马斯卡伦奥利凡藻
Diadesmis	二孔硅藻
Luticola	泥壳藻
Simonsenia sp./Simonsenia	西蒙氏藻属
Didymosphenia	二生槽藻
Caloneis	凸边藻
Pleurosigma	侧刺藻
Psammodictyon	沙生藻
Haslea	哈斯藻
Entomoneis sp.	翼内茧型藻
Coscinodiscus sp.	圆筛藻属
Thalassiosira pseudonana	海链藻
Skeletonema	骨条藻
Cf. Coscinodiscus wailesii	威氏圆筛藻
Coscinodiscus granii	格氏圆筛藻
Melosira varians	直链藻
Coscinodiscus centralis	中心圆筛藻

(续表)

藻类	中文名
Cyclotella meneghiniana	梅尼小环藻
Arachnoidiscus ehrenbergii	蜘蛛脉盘藻
Arachnoidiscus sp.	蛛网藻
Amphora coffeaeformis	咖啡豆形双眉藻
Gomphocymbella sp.	楔桥弯藻
Skeletonema costatum	中肋骨条藻
Chaetoceros spp.	角毛藻
Aulacoseira subarctica	北极镜藻
Aulacoseira baikalensis	贝加尔镜藻
Craticula cuspidata	格形藻
Stauroneis phoenicenteron	紫心辐节藻
Nitzschia linearis	形菱形藻
Pinnularia viridis	微绿羽纹藻
Nitzschia palea	谷皮菱形藻
Navicula grimmei	舟形藻
Pleurosira laevis	光滑侧链藻
Biddulphia pellucida	盒形藻
Ditylum brightwellii	布氏双尾藻
Lauderia borealis	北方娄氏藻
Ceratopteris richardii	美洲水蕨
Coscinodiscus gigas	巨圆筛藻
Pseudo-nitzschia multiseries	多列拟菱形藻
Achnanthes exigua	微小曲壳藻
apicomplexa	顶复门
P. tricornutum	三角褐指藻
T. pseudonanais	伪矮海链藻
T. weissflogii	威氏海链藻
Cyanidiophycean	温泉红藻
Cyanidium caldarium	类蓝藻

（续表）

藻类	中文名
Richelia	中华植生藻
rhopalodiaceae	棒杆藻科
Hemiaulaceae	半管藻科
Epithemia	窗纹藻属
Denticula	细齿藻
Ostreoccus	橄榄球菌
Halsea	海氏藻
Streptococcus alivarius	唾液链球菌
Streptococcus sanguinis	血链球菌(SN1)
Aeromonas hydrophila	嗜水气单胞菌
Alaria	边翅藻
infusorial earth	硅藻土
Fistulifera saprophila	舟形藻
Pinnularia saprophila	羽纹藻